ADVANCES IN X-RAY ANALYSIS

Volume 11

ADVANCES IN X-RAY ANALYSIS

Volume 11

Edited by
John B. Newkirk and Gavin R. Mallett
Denver Research Institute
The University of Denver
Denver, Colorado
and
Heinz G. Pfeiffer
Research and Development Center
General Electric Company
Schenectady, New York

Proceedings of the Sixteenth Annual Conference on Applications of X-Ray Analysis Held August 9-11, 1967

Sponsored by
University of Denver
Denver Research Institute
Metallurgy Division

SPRINGER SCIENCE+BUSINESS MEDIA, LLC
1968

ISBN 978-1-4684-8678-0 ISBN 978-1-4684-8676-6 (eBook)
DOI 10.1007/978-1-4684-8676-6

Library of Congress Catalog Card Number 58-35928

Originally published by Plenum Press in 1968
Softcover reprint of the hardcover 1st edition 1968

FOREWORD

X-ray emission spectrography, while based on Moseley's work, as a generally useful analytical method had its genesis in the work of Friedman, Birks, and Brooks 30 years ago. The central theme of this conference, quantitative methods in X-ray spectrometric analysis, and the large number of papers on that subject attest to the growth of the application and usefulness of X-ray emission. It is a privilege to have as an invited speaker Laverne Birks, one of the original group that put X-ray emission into analytical chemistry.

Determination of elements above titanium in the periodic table was considered the province of X-ray fluorescence, and most of the early development was aimed at the analysis of alloys. The papers in this volume on metals analysis accept most operational features as routine and have concentrated on the improved treatment of the observed data in order to convert them to more accurate results. As the treatment of matrix effects, geometry, and stability have been better understood, corrections have become routine. For most elements that are present in amounts greater than a few parts per million, determinations can now be done with accuracies rivaling wet methods.

Trace quantities are being determined to lower and lower amounts, largely owing to improvement of equipment and development of concentration techniques. For most trace elements, X-ray spectrography has become the preferred analytical method. The development of improved methods for separating signals from noise should lead to major reductions in minimum detection levels.

The analysis of light elements seems to have reached a limit at sodium by conventional X-ray methods. The first-row elements are being attacked by several techniques that show high promise. The Auger electron analysis and the formation of heavier-element compounds are both exciting new approaches.

The large amount of work on data handling in order to make the various corrections necessary for quantitative results should not obscure the need for fundamental research on the causes of variations. Even where we depend on empirical corrections, a knowledge of the underlying physical theories enables these corrections to be applied with more confidence.

The equipment is being steadily improved. Portable units, automatization, and remote readout have allowed field analysis of ores, have routinized continuous monitoring of industrial materials, and have made possible determinations in environments that do not support life.

The papers at this meeting provide ample evidence that new developments in X-ray fluorescence are continually forthcoming. The years ahead will surely bring even more progress.

H. G. Pfeiffer
Invited Co-Chairman

PREFACE

The resident co-chairmen wish to express their gratitude to Dr. H. G. Pfeiffer for his wholehearted and enthusiastic participation as co-chairman of this Conference. His extensive knowledge of the field of X-ray emission spectrography and of the people who are currently most active in it largely accounts for its success. We also thank all of the other technical conferees, authors of papers, discussors, and session chairmen whose contributions comprise the substance of this three-day meeting. Recognition to session chairmen is given with the following list of session titles:

Quantitative Methods in X-Ray Spectrometric Analysis
H. G. Pfeiffer, *General Electric Co., Schenectady, New York*

Basic Aspects of Fluorescence Analysis
H. J. Rose, Jr., *U.S. Geological Survey, Washington, D.C.*

Microanalysis and Long-Wavelength X-Rays
H. Steffen Peiser, *Bureau of Standards, Gaithersburgh, Maryland*

Fluorescence Techniques
K. F. J. Heinrich, *National Bureau of Standards, Washington, D.C.*

Diffraction Techniques
W. M. Mueller, *American Society for Metals, Metals Park, Ohio*

Fluorescence Analysis Applied to Nonmetallic Systems
Merlyn L. Salmon, *Fluo-X-Spec Laboratory, Denver, Colorado*

Microstructural Observations by X-Ray Diffraction
W. C. Hagel, *University of Denver, Denver, Colorado*

Special Applications and Instrumentation
W. L. Grube, *General Motors Corporation, Research Laboratory, Warren, Michigan*

Our sincere thanks go to those students and members of the staff of the Denver Research Institute for their respective parts in causing the mechanics of the Conference to run smoothly. The main secretarial burdens in coordinating the arrangements, in maintaining communication with the speakers and other participants, in maintaining records during the Conference, and in coordinating the final preparation of the manuscripts for the printer were ably handled by Mrs. Elaine Mason with the help of Mrs. Mildred Cain and other office staff members. Again, Mr. Frank T. Rivera lent his exceptional talents and experience to promote the general coordination of activities at conference time. His careful foresight resulted in the efficient functioning of the technical assistants and of the audiovisual equipment. Finally, we are indebted to Mr. John Silver and his staff for their help in designing and producing the advance program and the program abstract book.

G. R. Mallett and J. B. Newkirk
Co-Chairmen

CONTENTS

RECENT ADVANCES IN QUANTITATIVE X-RAY SPECTROMETRIC ANALYSIS BY SOLUTION TECHNIQUES

Eugene P. Bertin

Radio Corporation of America
Electronic Components and Devices
Harrison, New Jersey

ABSTRACT

Usually X-ray spectrometric analyses of samples in their original forms—solid, powder, small fabricated parts, liquid, *etc.*—are more rapid and convenient than analyses by any other method. Thus, the analyst is well advised to strive to analyze samples in their original form whenever practical. However, it is often necessary to reduce nonliquid samples to some other form, for example, when standards are unavailable in the original form, when the same substance is received for analysis in a variety of forms, when homogeneity or matrix effects are severe, or when an internal standard must be added. In such cases, samples and standards are often reduced to a powder, a fusion product, or a solution. If the decision is made to put the sample into solution, the many well-known advantages of solution techniques are realized, including: (1) homogeneity; (2) easy preparation of standards and blanks; (3) easy concentration, dilution, separation, and other treatment; (4) reduced matrix effects and wide choice of ways to deal with matrix effects; (5) wide choice of ways to present the specimen to the spectrometer; and (6) applicability of internal-standard, standard-addition or -dilution, indirect, absorption, and scatter methods. This paper reviews work reported since 1960 in which the specimen is presented to the spectrometer in liquid form. The review is not particularly critical and stresses experimental techniques and treatment of data rather than specific materials and results. The work is discussed in the following categories: (1) sensitivity; (2) liquid-specimen cells; (3) interaction of primary beam and liquid specimens; (4) matrix effects; (5) indirect (association) analysis; (6) X-ray scatter methods; and (7) X-ray absorption-edge spectrometry.

INTRODUCTION

Usually X-ray spectrometric analyses of samples in their original forms—solid, powder, small fabricated parts, liquid, *etc.*—are more rapid and convenient than analyses by any other method. Thus, the analyst is well advised to strive to analyze samples in their original form whenever practical. However, it is often necessary to reduce nonliquid samples to some other form, for example, when standards are unavailable in the original form, when the same substance is received for analysis in a variety of forms, when homogeneity or matrix effects are severe, or when an internal standard must be added. In such cases, standards and samples are often reduced to a powder, a fusion product, or a solution. If the decision is made to put the sample into solution, the many well-known advantages of solution techniques are realized, including: (1) homogeneity; (2) easy preparation of standards and blanks; (3) easy concentration, dilution, separation, and other treatment; (4) reduced matrix effects and wide choice of ways to deal with matrix effects; (5) wide choice of ways to present the specimen to the spectrometer; and (6) applicability of internal-standard, standard-addition or -dilution, indirect, absorption, and scatter methods.

In late 1964, the writer reviewed the literature of solution techniques[1] from a very broad point of view, including not only work in which the specimen is presented to the spectrometer in liquid form, but also that in which the specimen was in solution at any stage of its preparation. The present review updates the earlier one, but is more limited in scope, including only papers published during the 1960's reporting quantitative analyses in which the specimen is presented to the instrument in liquid form. The review is not critical and stresses experimental techniques and treatment of data rather than specific materials and results.

Other reviews of solution techniques have been published by Campbell and his colleagues,[2,3] Bertin and Longobucco,[4] and Gunn,[5] and applications of solution techniques to petrochemicals have been reviewed by Louis.[6,7] Cullen[8] has presented a comprehensive discussion of ways of dealing with matrix effects and interactions of the primary beam with liquid specimens. Because of its value and the lamentable fact that the paper was not published, its content will be reported here in considerable detail.

SENSITIVITY

The most notable improvements in sensitivity have been realized in analysis of petroleum products. Jones[9] determined sulfur in concentration down to 0.002% in gasoline by a standard-addition method. Louis[10,11] claimed sensitivities for phosphorus, sulfur, and heavier elements in oils of, respectively, 3, 1, and $< 1\ \mu$g/ml by addition of heavy metal naphthenates as sources of internal-standard lines. Kang, Keel, and Solomon[12] determined vanadium, iron, and nickel in concentration down to 1 ppm in residual petroleum stocks by use of mathematical matrix corrections.

Hale and King[13] determined nickel in oils in a concentration of 0–0.6 ppm with a precision to within ± 0.035 ppm (1 sigma) by simultaneous integration of Ni $K\alpha$ and background 0.01 Å away from Ni $K\alpha$ in a multichannel ARL curved-crystal spectrometer. The technique included use of very narrow secondary slit apertures, long integration times, and integration of the detector output signal rather than scaling.

Gunn[14,15] determined nickel in oils in a concentration of 0.2–10 ppm with a precision to within ± 0.1 ppm (1 sigma). He used the intensity ratio of Ni $K\alpha$ (48.7° 2θ) to scattered background at 49.5° 2θ *versus* nickel concentration with an enhancement filter in the primary beam. Zinc, gallium, germanium, selenium, and bromine as powdered element or compound deposited on 0.001-in. aluminum foil were evaluated as enhancement filters. Zinc reduced Ni $K\alpha$ intensity without increasing the Ni $K\alpha$ to background ratio. Germanium increased the ratio ~1.6 times but reduced the intensity to ~17% of its value without enhancement. Strontium was most effective, increasing the ratio ~1.5 times, while reducing the intensity to only ~40%.

Croke, Pfoser, and Solazzi[16] achieved sensitivity down to 10 ppm for magnesium, aluminum, silicon, and lead and down to 1 ppm for iron, nickel, copper, and silver in lubricating oils, using a Norelco Autrometer.

Siemes[17] suggests increasing the sensitivity for X-ray spectrometric determination of light elements in low concentration in water solution by the selection of optimum angle between primary and secondary beams and by increasing the areas of the specimen, collimator apertures, crystal, and detector.

Champion and Whittem[18] suggest a novel means of reducing the detection limit by reducing background in instruments in which the axis of rotation of the crystal is parallel to the specimen plane. Since background consists mostly of scattered radiation and since

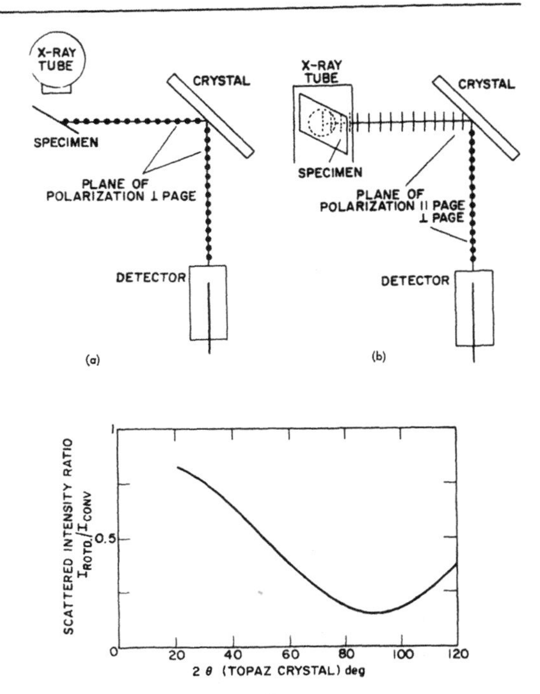

Figure 1. Background reduction by polarization analysis (Champion and Whittem).[18] (a) Conventional geometry: The specimen and crystal surfaces and their planes of polarization for scattered background radiation are substantially parallel at 90° 2θ. (b) Rotated geometry: The specimen and crystal surfaces and their planes of polarization are perpendicular. (c) Ratio of intensities scattered by water with rotated and conventional geometry.

the scattering angle intercepted by the collimator–crystal system is typically ~90°, the scattered radiation is highly plane-polarized. Conversely, the emitted characteristic radiation is completely random in direction and unpolarized. Consequently, substantial improvement in line-to-background ratio would be effected by a polarization analyzer, and, in fact, the analyzer crystal constitutes just such a device. Unfortunately, the geometry of instruments of the type noted above is such that, when the crystal is at optimum position for polarization (90° 2θ), the specimen and crystal planes are parallel, and the polarized scattered radiation is passed by the crystal very efficiently. If the X-ray tube and specimen plane were rotated ~90° about a line joining the centers of the specimen and crystal, the crystal would substantially discriminate against the polarized scattered radiation in the 2θ range ~90°. The principle is illustrated in Figure 1. It would be interesting to compare line-to-background ratios in commercial instruments having the geometry shown in Figure 1(a) and in those having the geometry shown in Figure 1(b).

LIQUID-SPECIMEN CELLS

Cells for Emission Spectrometry

Cell for Inclined Upright Specimen Plane (Figure 2a). Bertin and Longobucco[19,20] describe a completely clean, leak-free cell for instruments having inclined upright specimen planes. The construction is obvious in Figure 2a. The cell body *A* is $\frac{3}{8}$ in. deep and may be of Lucite, polyethylene, or Teflon. The depression *B* is $\frac{1}{4}$ in. deep and holds 2 to 3 ml. The Mylar cover *C* may be 0.00025 to 0.005 in. thick, depending on the wavelength to be measured, and is cemented to the cell with Pliobond rubber cement. When the cell is in place in the specimen drawer, the triangular space at the neck of the cell is

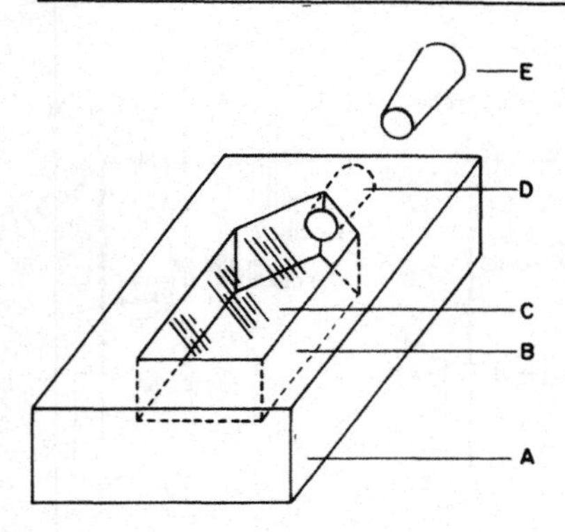

Figure 2a. Liquid-specimen cell. Leak-free cell for conventional or inverted geometry (Bertin).[19,20] *A*, cell body; *B*, depression; *C*, Mylar cover; *D*, tapered filling port; *E*, tapered plug.

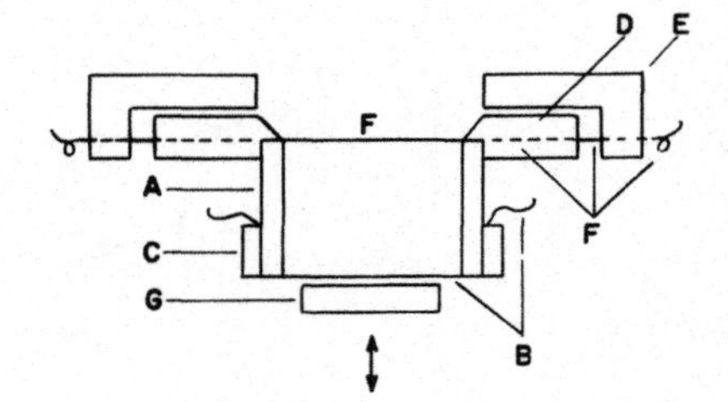

Figure 2b. Liquid-specimen cell. Uncovered liquid cell (Beard and Proctor).[21] *A*, cylindrical Plexiglas cell body ($1\frac{1}{8}$ in. ID, $\frac{7}{8}$ in. high, $\frac{1}{8}$-in. wall thickness); *B*, polyethylene film; *C*, cylindrical Plexiglas retaining ring; *D*, Micarta specimen mask; *E*, removable frame; *F*, nylon threads stretched on *E*, lying in grooves in *D*; *G*, movable piston to adjust liquid level by deforming *B*.

upward and covered by the specimen mask. This space serves to conceal the air volume at the filling port *D* and provides an expansion chamber for the liquid. The plug *E* is required only for volatile or corrosive liquids. Cells having 0.00025-in. Mylar covers have been used up to 150 times before the cover required replacement; thicker Mylar covers should last indefinitely.

Uncovered Liquid Cell (Figure 2b). Mylar cell covers trap bubbles, scatter the primary beam, and absorb long-wavelength spectral lines of elements of low atomic number. Elimination of the cover should improve precision, line-to-background ratio, and detection limit or permit reduced counting times for a given precision with consequent reduction of thermal or radiolytic effects. Beard and Proctor[21] describe an uncovered cell for use with the Picker spectrometer, in which an end-window X-ray tube and spectrogoniometer constitute an integral unit which may be mounted in any position. Ordinarily, the instrument is mounted with the X-ray tube axis vertical; the specimen plane is then inclined to the horizontal at 25°. Beard and Proctor tipped the instrument so that the specimen plane is horizontal.

The cell (Figure 2b) consists of a cylindrical Plexiglas body *A* having a polyethylene film *B* stretched across its bottom and secured with a retaining ring *C*, and mounted on a Micarta specimen mask *D*. The cell is filled nearly to the brim, and a frame *E* fitted with two stretched, parallel nylon threads *F* is placed over the mask. The mask has grooves to receive the threads, which then lie in the plane of the rim of the cell. The movable disk *G* presses against the polyethylene cell bottom *B* and is adjusted to bring the liquid surface into contact with the threads. The frame *E* is then removed, and the cell is ready for X-ray measurements.

Comparison of results for solutions in the uncovered cells and in the same cells with 0.00025-in. Mylar covers showed net intensity gains for Al $K\alpha$, Si $K\alpha$, and S $K\alpha$ by factors of 5.7, 4.7, and 3.4, respectively.

Continuous-Flow Cell (Figure 2c). Campbell[22] describes a cell for the study of liquids circulated continuously through the cell by a small pump. The construction and function of the cell are obvious from Figure 2c.

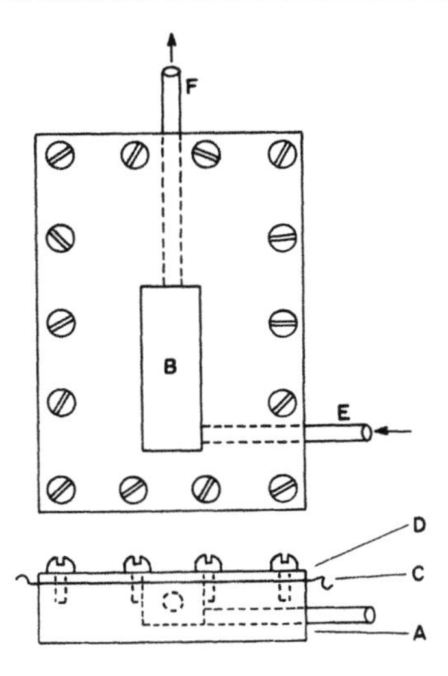

Figure 2c. Liquid-specimen cell. Cell for continuous flow of solutions (Campbell).[22] *A*, Plexiglas cell body, 8 × 5.5 × 1.3 cm; *B*, depression, 3.5 × 1.2 × 0.9 cm; *C*, 0.001-in. Mylar window; *D*, 1.5-mm aluminum cover plate; *E*, inlet; *F*, outlet.

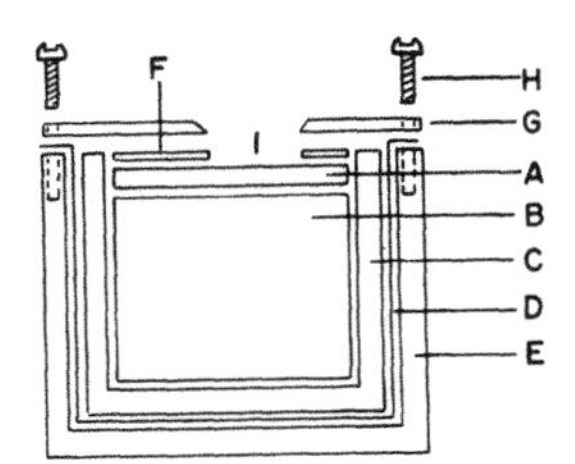

Figure 2d. Liquid-specimen cell. Frozen-specimen cell (Chan).[23] *A*, frozen specimen (see text); *B*, cooled aluminum cylinder; *C*, inner Plexiglas thermal-insulation cup; *D*, aluminum reflector cup; *E*, outer Plexiglas thermal-insulation cup; *F*, cork disk; *G*, metal specimen-mask disk; *H*, retaining screws; *I*, rectangular specimen window in *F* and *G*.

Frozen-Specimen Cell (Figure 2d). Freezing eliminates many of the disadvantages of liquid specimens, including bubble formation, expansion, radiolysis, inconvenience with inclined specimen planes and with vacuum spectrometers, and scatter from and absorption in Mylar cell covers. Chan[23] describes a cell for frozen specimens (Figure 2d). An aluminum cylinder *B*, previously cooled in liquid nitrogen or a freezing mixture, is placed in a thermal insulating cup consisting of a Plexiglas inner cup *C*, aluminum reflecting cup *D*, and Plexiglas outer cup *E*. The frozen specimen *A* (see below) is placed on *B*, followed by an insulating cork disk *F* and metal mask *G*. The assembly is secured with the screws *H*. Both *F* and *G* have rectangular windows *I*.

Three types of specimen are described: (1) A metal disk, 33 mm in diameter and 3 mm thick, has a rectangular depression of 20 × 12 × 2 mm. Liquids requiring no Mylar cover (low vapor pressure, noncorrosive) are placed in the depression. The disk is then placed between cold metal blocks to freeze the specimen. (2) A shallow metal planchet, 33 mm in diameter and 0.5 mm deep, holds a thick filter-paper disk impregnated with ~0.5 ml of solution. A 0.00025-in. Mylar film may be placed on the planchet and secured with a retaining ring. The specimen is frozen as before. (3) A shallow metal planchet, 33 mm in diameter and 3 mm deep, is filled with aqueous solution thickened with gelatin or with organic solution thickened with alkali metal stearate, covered with Mylar, and frozen.

All specimen preparation is done in a dry box fitted with rubber gloves and provided with dewar vessels containing the supply of aluminum cylinders and freezing blocks. A continuous stream of thoroughly dried nitrogen is passed through the box. The box communicates directly with the spectrometer specimen chamber through a flexible plastic tunnel. The helium and air inlets to the spectrometer are provided with drying trains.

Cell for Slurries. X-ray spectrometric analyses of powders and briquets are limited to thin surface layers and may be influenced by inhomogeneity and particle size. In such cases, especially if the powder is difficult to dissolve or fuse, it may be advantageous to analyze it as a slurry. Hudgens and Pish[24] describe a slurry cell consisting of a sealed body having 5- to 6-ml capacity, a 0.003-in. Mylar cover, an externally powdered stirrer mounted asymmetrically for maximum turbulence, and a bleeder valve to allow the escape of gases produced by radiolysis and heating. The sample powder and internal standard are placed in the disassembled cell, covered with liquid vehicle, and stirred with a toothpick to remove occluded air. The cell is then assembled and filled with vehicle through the bleeder hole with the use of a hypodermic syringe. The cell was evaluated for determination of the thorium and uranium in mixtures of their oxides by using $SrSO_4$ as internal standard.

High-Temperature Liquid-Specimen Cell (Figure 2e). Such a cell is described by Campbell and Leon.[25] A ceramic or metal crucible *A* rests in a depression in an Inconel block *B* surrounded by a Nichrome heating element *C* wound on a refractory ceramic cylinder *D* in a brass housing *F* provided with water cooling coils *G* and packed with high-temperature insulation *E*. A thermocouple well *H* and an inert-gas inlet (not shown) are provided. With this arrangement, specimen temperatures up to 900°C are attainable, but Campbell and Leon suggest that, with a platinum or alumina crucible, platinum block, platinum–rhodium heater coil, and alumina support, temperatures up to 1400°C should be attainable. The X-ray tube window was ½ in. away from the specimen surface and was protected by a thin sheet of mica. No adverse effect on the X-ray tube was observed after 50 runs of up to 5 hr at up to 800°C.

Miscellaneous Cells. Karttunen[26] describes an all-glass cell having a cemented-on borosilicate glass cover, 0.07 mm thick, and a filling port for use with a hypodermic syringe. Zimmerman and Heller[27] avoided the effects of thermal expansion and bubble

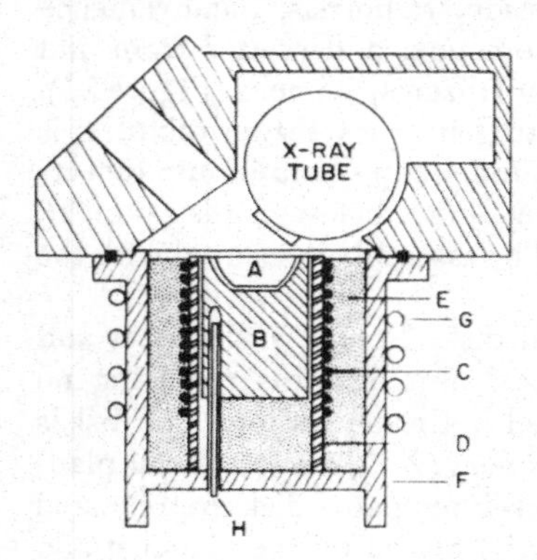

Figure 2e. Liquid-specimen cell. High-temperature liquid-specimen cell (Campbell and Leon).[25] *A*, ceramic or metal crucible; *B*, Inconel cylinder with depression for *A*; *C*, Nichrome heater coil; *D*, refractory cylinder, support for *C*; *E*, thermal insulation; *F*, brass shell; *G*, water cooling coils; *H*, thermocouple tube.

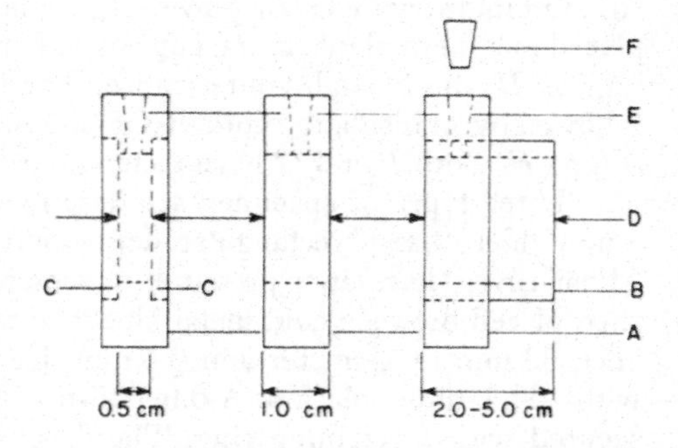

Figure 2f. Liquid-specimen cell. Absorption cells (Bertin, Longobucco, and Carver).[30] *A*, Lucite cell body, 1.0 cm thick; *B*, cylindrical Lucite insert, 2.0 to 5.0 cm long; *C*, cylindrical depressions; *D*, Mylar windows, 0.001 to 0.005 in. thick; *E*, tapered filling port; *F*, tapered Lucite plug.

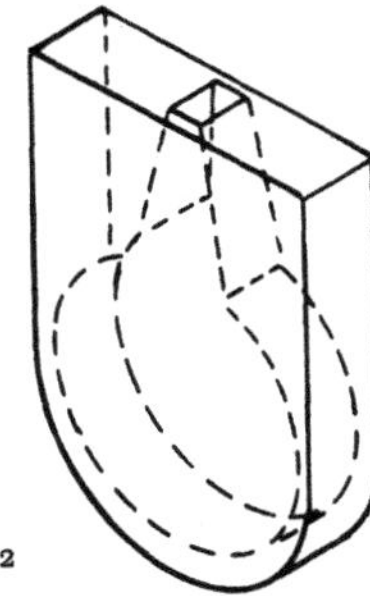

Figure 2g. Liquid specimen cell. Absorption cell (Hakkila and Waterbury).[31,32]

formation in distending thin Mylar cell windows in upright geometry by drilling small holes near the rims of the cells. The technique is applicable only to instruments having an inclined specimen plane. Care must be taken to keep the holes up when handling the filled cells and mounting them in the specimen drawer. Gunn[28] describes a liquid cell having a built-in internal intensity-reference standard in the form of a platinum disk. Provision is made to vary the distance of the disk from the Mylar cell window. Smith and Royer[29] give methods for storing, handling, and filling closed liquid cells with reactive organic liquids.

Cells for Absorption

Absorption cells described by Bertin, Longobucco, and Carver[30] are shown in Figure 2f. All the cells are made from Lucite blocks A, $1\frac{1}{2}$ in. square and 1 cm thick. The 0.5-cm cells are made by machining cylindrical holes, $\frac{3}{4}$ in. in diameter, through the blocks and concentric circular recesses C, $1\frac{1}{8}$ in. in diameter and 0.25 cm deep, on both sides of the blocks. The 1-cm cells are made the same way without the recesses. Cells having longer path length are made by machining through the blocks cylindrical holes, 1 in. in diameter, in which are cemented cylindrical Lucite inserts B having 1-in. OD, $\frac{3}{4}$-in. ID, and 2- to 5-cm length. Circular Mylar windows, 0.005 in. thick, are cemented on both sides of all cells with Pliobond rubber cement. All the cells have tapered filling ports E and plugs F. These workers also describe microcells for small specimen volumes and cells having path lengths of <0.5 cm for more X-ray-opaque solutions.

Hakkila and Waterbury[31,32] describe the all-polystyrene cells shown in Figure 2g. Path length is 0.12 to 3.0 cm. Dunn[33] describes an absorption cell for aqueous solutions consisting of a cylindrical polystyrene body with 0.010-in. polystyrene windows secured with screw caps. For organic solutions, the cell body is of gold-plated steel, and the windows are microscope-slide cover glasses. Griffin[34] describes an absorption cell having thin beryllium windows on a duralumin body.

INTERACTION OF THE PRIMARY BEAM WITH LIQUID SPECIMENS

In general, only liquid specimens interact with the primary beam in such a way as to affect the X-ray measurements. The primary beam causes heating and radiolysis with consequent expansion, bubble formation, precipitation, and displacement of the specimen plane. These interactions may be reduced by freezing the specimen[23] and by reduction of excitation power if the intensity is not reduced below values that give acceptable statistical precision. In absorption-edge spectrometry, another means of reducing incident flux is to derive the monochromatic bracketing wavelengths from secondary targets.[30] Other ways of dealing with interaction effects are given by Cullen,[8] and the following discussion is based largely on his paper.

Specimen Heating

Cullen reports that water in a liquid cell was warmed only 4°C in 10 min by a Machlett OEG-50 tungsten-target tube operating at 50 kV, 45 mA, in an air path. The corresponding expansion is only ~0.02%, and the effect on analytical results would be less than the statistical counting error. Specimen heating increases with concentration and absorption coefficient of solute but may be reduced by use of: (1) newer, cooler-operating X-ray tubes; (2) a water-cooled jacket on the X-ray tube head; (3) water cooling of the specimen drawer or compartment; (4) thermal insulation of the specimen with plastic cells and specimen masks to reduce heat conduction from the drawer or compartment; and (5) continuous helium flow.

Bubble Formation

Cullen suggests several ways to minimize bubble formation or its effects: (1) use of inverted optics; (2) with upright geometry, use of cells having a space concealed by the specimen mask into which bubbles may rise; (3) use of previously boiled water and acids for specimen dilution and processing; (4) avoidance of gas-forming reagents, such as peroxides and NH_4OH; and (5) use of gas inhibitors, for example, perchlorate, which reduces gas formation in $FeCl_3$ solutions. In addition, use of Desicote antiwetting agent on cell windows reduces bubble formation and adherence. The writer has successfully outgassed solutions by briefly exposing them to vacuum prior to placing them in cells. A brief preirradiation period before taking data, suggested by Cullen to allow radiolytic precipitation to equilibrate, might also apply to bubble formation.

Precipitation

Cullen has shown that precipitate formation may equilibrate after ~60 sec and recommends a preirradiation period prior to taking data. Avoidance of high concentrations of radiolytically labile constituents and use of complexing agents also reduce precipitation. In case there is any possibility of even minimal precipitation, cells must be used in which the window is replaced for each specimen, or the cell must be rinsed with a reagent that will remove any precipitate from the cell window. Zimmerman and Heller[27] report that cobalt, nickel, indium, and gold plating solutions were more stable on treatment with potassium citrate than with KCN.

Specimen Plane

Variation in specimen plane is a problem only in closed cells where thermal expansion and bubbles raise the pressure in the cell and distend the cell window. With upright geometry, cells having open filling ports[19,20] (Figure 2a) or breather holes,[27] or uncovered cells[21] (Figure 2b) may be used.

MATRIX EFFECTS

Often, determinations can be made without precautions for dealing with matrix effects. Pierron and Munch[35] determined titanium, vanadium, cobalt, nickel, copper, and molybdenum in the same solution, using only pulse-height selection. Visapaa[36,37] determined 10–300 μg Sr/ml and 10–500 μg Ba/ml in water and milk by measuring background on a blank. Hirokawa[38] determined manganese in ferromanganese and chromium in ferrochromium by a simple two-standard method, using the equation

$$\frac{(I_A/I_B)[(W_B/W_A) - 1]}{W_B - W_A} = \frac{(I_A/I_X)[(W_X/W_A) - 1]}{W_X - W_A} \qquad (1)$$

where I is net intensity of the analyte line and W is the weight fraction; A and B refer to the two standards, and X refers to the sample.

Kirchmayr and Mach[39] determined manganese, yttrium, samarium, and gadolinium in rare earth–manganese alloys in proportions of 1 : 100 to 100 : 1, using solutions containing 20 mg of sample in 10 ml and calibration curves of Mn $K\alpha$ or R.E. $L\alpha_1$ intensity *versus* concentration. Kullbom, Pollard, and Smith[40] determined alkylbenzene sulfonate and sodium sulfate in water slurries by a combination of X-ray spectrometric and infrared spectrophotometric methods. Total sulfur was determined by X-ray, sulfonate by infrared. Sulfur equivalent to the sulfonate was calculated and subtracted from the total sulfur to give Na_2SO_4 concentration. Caley[41] determined 13 elements in petroleum products in a Norelco Autrometer, using liquid and solid-plastic standards and simple analyte line *versus* concentration curves. Concentrations ranged from a few parts per million to 25%, and the precision and accuracy were very good.

Even complex matrix effects may often be compensated for by judicious preparation of standards. An example is provided by the work of Bertin and Longobucco,[20] who reduced the analysis time for samples from each of 50 plating baths from 3 man-days by chemical methods to 1 man-day by the X-ray spectrometric solution technique. The 50 baths were classified in 11 groups on the basis of plate metal, concentration, and matrix as follows: (1) nickel: 20–100 g/liter in chloride, high chloride–low sulfate, and high sulfate–low chloride media; (2) copper: 10–60 g/liter in NaCN–NaOH and KCN–KOH media; (3) silver: 0–3, 10–25, and 40–100 g/liter in NaCN–NaOH media, and 10–25 g/liter KCN–K_2CO_3 media; and (4) gold: 0–4 and 10–25 g/liter in KCN media. Eight stock plate-metal solutions in various media and seven matrix solutions were prepared from which 11 sets of four calibration standards were prepared. Simple $K\alpha$ peak intensities were measured for nickel and copper, and background ratio measurements were made for silver and gold.

However, when matrix effects are substantial, liquid specimens are subject to more means for dealing with them than specimens in any other form. These techniques have been reviewed by Cullen,[8] Campbell, Leon, and Thatcher,[3] and Gunn,[5,15] and recent applications of them are discussed below.

Separation of Analyte or Matrix

Karttunen[26] separated zirconium, molybdenum, ruthenium, rhodium, and palladium from uranium-base alloys by precipitation followed by liquid extraction and by elution from an ion-exchange column. The separated analytes were determined in their solutions by simple measurement of their $K\alpha$ lines. Tomkins, Borun, and Fahlbusch[42] separated zirconium, niobium, tantalum, and tungsten from high-temperature alloys and determined them by the background ratio method. Blank and Heller[43,44] determined zirconium and thorium in refined uranium by liquid extraction of the analytes. Thorium was determined by internal standardization with strontium, zirconium by background ratio.

Sets of Calibration Curves

Sets of curves may be established of analyte line intensity *versus* concentration, each curve for a different interfering matrix element concentration.[15] Boyd and Dryer[45] determined phosphorus, sulfur, chlorine, calcium, zinc, barium, and lead in oils, using an ARL Vacuum Production X-Ray Quantometer (VPXQ). Three techniques were evaluated: (1) The intensity ratio of the analyte line to an external standard was prone to error from mutual interference of mixed additives and from the carbon–hydrogen ratio of the base stock. However, satisfactory results were realized by using sets of curves. (2) The intensity ratio of the analyte line to the scattered background at 0.6 Å required fewer curves. (3) The intensity ratio of the analyte line to the scattered W $L\alpha_1$ line substantially eliminated the need for additional curves.

Dilution

The specimens may be diluted to the extent that matrix absorption is approximately the same in all specimens; this technique is applicable only if, on such dilution, usable intensities are still obtained. Cullen[8] recommends the following procedure for evaluation of this technique. Two solutions are prepared having the same analyte concentration, one in pure solvent, the other with maximum expected concentrations of matrix elements. The solutions are diluted stepwise until the analyte line intensity is the same for both. It is then determined whether analyte line intensity is still adequate for the required statistical precision in acceptable counting time. Table I shows data for copper in the presence of iron. Incidentally, Table I illustrates another important feature of solution techniques: On dilution of a solution, analyte line intensity decreases at a much lower rate than concentration, as is evident from comparison of the Cu $K\alpha$ intensities for 0 and 1 : 1 dilution and for 20 : 1 and 100 : 1 dilution at 50 kV, 50 mA.

Table I. Effect of Dilution on Cu $K\alpha$ Intensity in the Presence of Iron[8]

Dilution*	kV, mA	Cu $K\alpha$ intensity (counts/sec)		Cu $K\alpha$ intensity ratio
		Copper†	Copper + iron‡	I_{Cu}/I_{Cu+Fe}
None	30, 20	8707	7194	1.210
1 : 1	30, 20	6562	6072	1.081
10 : 1	50, 20	4709	4568	1.033
20 : 1	50, 35	4866	4858	1.002
20 : 1	50, 50	7059	7029	1.004
100 : 1	50, 50	3549	3541	1.002

* Volume of water per volume of solution (copper or copper and iron).
† 10 mg Cu/ml.
‡ 10 mg Cu + 10 mg Fe/ml.

Matrix-Masking Agent

A matrix-masking agent, in the form of a salt of a noninterfering, relatively light element, may be added to all specimens. Cullen[8] demonstrated this technique by adding $Ca(NO_3)_2$, in a concentration of 10 mg/ml, to solutions of copper only and copper with other elements. His data are given in Table II. Cullen also reports reduction of matrix effects in iron, nickel, copper, and indium plating solutions by adding $Ca(NO_3)_2$, and in gold plating solutions by adding KCN. Zimmerman and Heller[27] compensated matrix effects in cobalt, nickel, indium, and gold plating solutions by adding KCN or potassium citrate.

Matrix Equalization

A matrix element already present in the specimens in varying concentration may be added to all specimens to approximately equal concentration or to an excess concentration above which further increase has no effect on the analyte line intensity.

Table II. Effect of Matrix Masking Agent* on Cu $K\alpha$ Intensity in the Presence of Various Matrix Elements[8]

Element added and atomic number†	Cu $K\alpha$ net intensity (counts/sec)	Cu $K\alpha$ net intensity ratio I_{Cu}/I_{Cu+M}
None	6143 (I_{Cu})	1.000
19 K	6094	1.008
26 Fe	5999	1.024
28 Ni	6113	1.005
30 Zn	5964	1.030
47 Ag	6005	1.023
82 Pb	5895	1.042

* Matrix-masking agent $Ca(NO_3)_2$, 10 mg/ml.
† Copper concentration in all solutions, 2.0 mg/ml; concentration of all other elements, 2.0 mg/ml.

Internal Standardization

Blank and Heller[46] determined copper, tin, and uranium in bronze heat-treating material by internal standardization with the following net intensity ratios: Cu $K\alpha$/Zn $K\alpha$, Sn $K\alpha$/In $K\alpha$, and U $L\alpha_1$/In $K\alpha$. The net intensity of U $L\alpha_1$ was corrected for spectral interference by second-order In $K\beta$ with the following equation:

$$I_{\text{net}}(\text{U } L\alpha_1) = I_{\text{peak}}(\text{U } L\alpha_1) - I_B - \left\{\frac{I_{\text{net}}[\text{In } K\beta(2)]}{I_{\text{net}}(\text{In } K\alpha)}\right\}[I_{\text{net}}(\text{In } K\alpha)] \tag{2}$$

The net-intensity ratio of In $K\beta$ (2)/In $K\alpha$ was found to be constant at 0.0451. The same workers[43] determined thorium after liquid extraction from uranium by internal standardization with strontium.

Hakkila, Hurley, and Waterbury[47] determined zirconium and molybdenum in the presence of uranium, using niobium as the internal standard. For zirconium, the curve of Zr $K\alpha$/Nb $K\alpha$ intensity ratio R *versus* zirconium concentration is linear; as the uranium concentration changes, the slope changes, but the intercept on the R axis remains constant. The zirconium concentration is given by

$$\frac{\text{mg Zr}}{10 \text{ ml}} = \frac{R - b}{a - c(d - 100)} \tag{3}$$

where a is the slope of the curve for solutions containing 100 mg U/10 ml; b is the constant intercept on the R axis; c is the change in slope per 1 mg U/10 ml; d is the uranium concentration in the sample in milligrams per 10 ml. For molybdenum, the curve of Mo $K\beta$/Nb $K\beta$ intensity ratio R *versus* molybdenum concentration is linear; as the uranium concentration changes, the slope remains constant, but the intercept on the R axis changes. The molybdenum concentration is given by

$$\frac{\text{mg Mo}}{10 \text{ ml}} = \frac{R - b + c(100 - d)}{a} \tag{4}$$

where a is the constant slope of the curve; b is the intercept on the R axis for solutions containing 100 mg U/10 ml; *c* is the change in intercept per 1 mg U/10 ml; *d* is the uranium concentration in the sample in milligrams per 10 ml.

Strasheim and Wybenga[48] compensated matrix effects in determining iridium, platinum, and gold by two novel techniques involving background correction and a *variable internal-standard* technique. The Pt $L\alpha_1$ and background intensities were measured from solutions containing 400 ppm of platinum in pure 10% HCl and in various concentrations of various other matrix constituents. A curve was established of the Pt $L\alpha_1$ intensity ratio in 10% HCl and in other matrices (Pt_{HCl}/Pt_M) *versus* background intensity ratio in 10% HCl and in other matrices (B_{HCl}/B_M); this curve was linear. The analyses consisted of measurements of Pt $L\alpha_1$ and background intensities on the samples and on a standard in pure 10% HCl. The ratio B_{HCl}/B_M was calculated for each sample and applied to the curve to derive Pt_{HCl}/Pt_M, which was then used to correct the measured Pt $L\alpha_1$ intensity from the sample. Platinum was also determined by using zinc as an internal standard. Platinum was then used as a variable internal standard for the determination of iridium and gold. The Ir $L\alpha_1$/Pt $L\alpha_1$ and Au $L\alpha_1$/Pt $L\alpha_1$ intensity ratios were measured, and iridium and gold concentrations were determined relative to platinum.

In another paper,[49] the same authors determined ruthenium, rhodium, and palladium the same way. Palladium was determined the same way as platinum, then used as variable internal standard for the determination of ruthenium and rhodium.

Bartkiewicz and Hammatt,[50] Haycock,[51] and Louis[10,11] determined additives and trace impurities in organic liquids by the internal-standard technique. Internal standards were added as metal naphthenates.

Internal Control Standardization

An internal control standard is similar to an internal standard but does not necessarily have the same excitation, absorption, and enhancement properties as the analyte in the sample matrix. Mitchell and O'Hear[52] have developed a rapid, convenient, and accurate method for simultaneous determination of combinations of chromium, manganese, iron, cobalt, nickel, and copper in concentrations of 0.1–99.9% in manganese and chromium alloys, Alnico, refractory alloys, ores, and slags. The work was done on an ARL X-Ray Industrial Quantometer (XIQ). Sample concentration was 1 g/100 ml. Variations in volume, temperature, acid concentration, *etc.*, were compensated for by taking the intensity ratio of the analyte line and a line of an internal control standard, which may be added to the samples or already present therein. For each element, a single set of calibration standards suffices for all matrixes.

Background Ratio Method

Scott[53] determined antimony in pharmaceuticals, Blank and Heller[44] determined zirconium after liquid extraction from uranium, and Tomkins, Borun, and Fahlbusch[42] determined heavy elements separated from high-temperature alloys, all by background ratio techniques.

Scattered-Target-Line Ratio Method

The target line may be scattered coherently (Cullen[8,54]) or incoherently (Boyd and Dryer[45]). Cullen[54] gives the theory of the method and demonstrates its value for copper, nickel, and silver, using scattered W $L\beta_1$ radiation. Lopp and Claypool[55] determined vanadium and nickel in concentrations down to 5 ppm by measuring V $K\alpha$/W $L\alpha_1$ and Ni $K\alpha$/W $L\alpha_1$ intensity ratios from crude oil specimens thickened with aluminum stearate. Stever, Johnson, and Heady[56] determined tungsten, molybdenum, and impurity elements in tungsten–molybdenum alloys and electrolytes by solution techniques, using a platinum-target X-ray tube and analyte line-to-Pt $L\alpha_1$ intensity ratios.

Internal Intensity-Reference Standardization

In determining manganese in gasoline, Jones[57] compensated the varying carbon–hydrogen ratio by mounting an iron rod in the liquid-specimen cell to serve as an internal intensity-reference standard and using the Mn $K\alpha$/Fe $K\alpha$ intensity ratio *versus* manganese concentration. Gunn[28] applied the same technique in determining lead in gasoline, using a platinum disk in the cell and the Pb $L\alpha_1$/Pt $L\alpha_1$ intensity ratio. Ideal compensation is realized with the platinum disk at such a depth that the Pb $L\alpha_1$/Pt $L\alpha_1$ intensity ratio is the same for the same lead concentration in liquids having different carbon–hydrogen ratios. Gunn gives a method for establishing this optimum depth.[28,58]

Standard Addition and Dilution

These methods are particularly advantageous for very infrequent analyses because no standards are required. Jones[9] determined sulfur in gasoline in a concentration down to 0.002% by a combined standard-addition and density method. Two 50-ml portions A and B were taken of samples having $\leq 0.15\%$ sulfur. Portion A was treated with 5 ml of sulfur-free isooctane, B with 5 ml of a standard solution containing 0.13 g of sulfur. Net S $K\alpha$ intensities were measured from A and B, and the density D of the untreated sample was measured with a hydrometer. Sulfur concentration C in weight percent is given by

$$C = \frac{(2)[0.13 I_A/(I_B - I_A)]}{D} \tag{5}$$

Samples containing $> 0.15\%$ of sulfur were diluted and treated in the same way.

Mathematical Matrix Corrections and Treatment of Data

Most mathematical techniques are too elaborate to be presented here in any detail, but some of the outstanding features of recently reported techniques may be considered. Gunn[15] has given a summary of the subject.

Dwiggins[59] developed elaborate mathematical matrix corrections for the determination of sulfur, vanadium, and nickel in organic liquids, using the Norelco Autrometer. A feature of the method is the derivation of the background intensity ratio at and near the analyte line from the intensity ratio of a coherently and incoherently scattered target line. The scattered target lines have high intensities for liquid specimens and are measured quickly and with high precision.

Carter[60] developed simultaneous equations for determination of the common additive elements in oils. Only one set of standards is required for each element for all matrixes, including water. The accuracy of the results is limited by: (1) specimen depth less than infinite depth; (2) presence of an element having an absorption edge at a wavelength just less than that of an analyte line; (3) nonlinearity of the detector; (4) high background for low analyte line intensity; and (5) inaccuracy of values of absorption coefficients.

Kang, Keel, and Solomon,[12] using mathematical matrix corrections, were able to determine vanadium, iron, and nickel in concentrations down to ~1 ppm in residual petroleum stocks.

Hakkila and Waterbury[61] determined chromium, iron, and nickel in their ternary alloys by a method involving mathematical treatment of the measured data from samples and standards:

1. Six solutions were prepared containing 200 mg of, respectively, chromium, iron, nickel, Cr–Fe (1 : 1), Cr–Ni (1 : 1), and Fe–Ni (1 : 1).

2. Cr $K\alpha$, Fe $K\alpha$, and Ni $K\alpha$ net intensities were measured from all six solutions. The Cr $K\alpha$ background was obtained by averaging the intensities at 2θ for Cr $K\alpha$ from the pure iron and nickel solutions; Fe $K\alpha$ and Ni $K\alpha$ backgrounds were derived in a similar way.

3. Six net-intensity ratios were calculated: R_{CrFe}, R_{CrNi}, R_{FeCr}, R_{FeNi}, R_{NiCr}, and R_{NiFe}. These ratios had the form

$$R_{CrFe} = \frac{\text{Cr } K\alpha \text{ intensity from the (1 : 1) Cr–Fe solution}}{\text{Cr } K\alpha \text{ intensity from the pure Cr solution}} \tag{6}$$

4. Six interelement correction factors were then calculated: a_{CrFe}, a_{CrNi}, a_{FeCr}, a_{FeNi}, a_{NiCr}, and a_{NiFe}; for example, a_{CrFe} is the correction factor for the effect of iron on Cr $K\alpha$ intensity and is calculated from

$$\frac{1 - R_{CrFe}}{R_{CrFe}} = a_{CrFe}\left(\frac{1 - C_{Cr}}{C_{Cr}}\right) \tag{7}$$

where C is concentration, and, since C_{Cr} in the (1 : 1) Cr–Fe solution is 0.5, the parenthetic term becomes 1, which permits the calculation of a.

5. The Cr $K\alpha$, Fe $K\alpha$, and Ni $K\alpha$ net intensities were then measured from all sample solutions and from the pure chromium, iron, and nickel solutions.

6. Three intensity ratios were calculated for each sample: R_{Cr}, R_{Fe}, and R_{Ni}. These ratios had the form

$$R_{Cr} = \frac{\text{Cr } K\alpha \text{ intensity from the sample}}{\text{Cr } K\alpha \text{ intensity from the pure Cr solution}} \tag{8}$$

7. Chromium concentration was then calculated from the equation

$$\frac{1 - R_{Cr}}{R_{Cr}} = \left(\frac{a_{CrFe}R_{Fe} + a_{CrNi}R_{Ni}}{R_{Fe} + R_{Ni}}\right)\left(\frac{1 - C_{Cr}}{C_{Cr}}\right) \tag{9}$$

Nickel concentration was calculated from a similar equation.

8. By using these values for C_{Cr} and C_{Ni}, iron concentration was calculated from the equation

$$\frac{1 - R_{Fe}}{R_{Fe}} = \left(\frac{a_{FeCr}C_{Cr} + a_{FeNi}C_{Ni}}{C_{Cr} + C_{Ni}}\right)\left(\frac{1 - C_{Fe}}{C_{Fe}}\right) \tag{10}$$

Special Experimental Techniques

Schindler, Month, and Antalec[62] analyzed Ni–Fe films of various compositions and thicknesses by a solution technique involving a novel graphic treatment of the data: Four series of solutions were prepared containing, respectively, 5, 10, 15, and 20 mg of total nickel plus iron; each series consisted of six solutions having a Ni : Fe ratio of 10 : 0, 9 : 1, 8 : 2, 7 : 3, 6 : 4, and 5: 5. The Ni $K\alpha$ and Fe $K\alpha$ net intensities were measured from all 24 solutions, and the data were plotted on a graph of Ni $K\alpha$ *versus* Fe $K\alpha$ net intensity. Two series of lines were drawn on the plot, one joining points representing the same percent of nickel for different total weights of Ni–Fe, the other joining points representing the same Ni–Fe weight for different nickel concentrations. Combinations not represented by any of the lines were dealt with by interpolation.

Discussion

Some matrix correction methods may also compensate fluctuations in excitation power, variations in specimen plane and density, *etc.* Such methods include those involving an intensity ratio of the analyte line to radiation emitted from an internal standard, internal control standard, or internal intensity-reference standard or to scattered continuum or target-line radiation.

Table III. Comparison of Ratio Methods[8]

	Ratio $I_{\text{Cu } K\alpha}/I_{\text{Std}}$		
Iron concentration (mg/ml)	Internal standard 2.0 mg Zn/ml	Background ratio 0.6 Å	Coherently scattered target line W $L\beta_1$
0*	1.045	3.261	3.541
5	1.042	3.258	3.530
10	1.040	3.247	3.522
15	1.035	3.234	3.510
20	1.032	3.221	3.503

* Copper concentration in all solutions, 2.0 mg/ml.

Table III compares the internal-standard, background-ratio, and coherently scattered target-line ratio techniques for copper in a concentration of 2 mg/ml in the presence of up to 20 mg Fe/ml. In this case, all three methods are effective. The last two of these techniques and the external-standard method are compared by Boyd and Dryer;[45] their results are summarized in the section on sets of calibration curves above.

INDIRECT (ASSOCIATION) ANALYSIS

An extremely significant development is the indirect determination of elements not readily determined X-ray spectrometrically (sodium through chlorine), of elements not determinable at all in standard commercial spectrometers ($Z < 11$, sodium), and even of organic compounds by stoichiometric association with a readily determined element. Rudolph and Nadelin[63] determined trace chlorine in titanium by precipitating AgCl and measuring the silver. Smith[64] suggested determining silicon by forming silicomolybdate-8-quinolinol complex (Si : Mo = 1 : 12) and measuring the molybdenum. Mathies, Lund, and Eide[65] determined 0.2 to 5 μg of nitrogen in biological materials by converting it to ammonia, absorbing it in a filter-paper disk impregnated with Nessler's reagent, washing out the excess reagent, and determining the mercury in the mercury–iodine–amine complex. Papariello, Letterman, and Mader[66] evaluated three techniques to associate stoichiometrically a readily determinable element with an organic compound: bromination with bromine water, chelation with 5-chloro-7-iodo-8-quinolinol, and salt formation with lead.

These are just a few selected illustrative papers reporting indirect analyses. While, in most of this work, the specimen is not presented to the spectrometer in liquid form, it is processed in liquid form, and the indirect method is certainly applicable to true liquid-specimen techniques. Another example of the determination of an element, carbon, not ordinarily considered determinable by secondary-emission spectrometry is the work of Dwiggins described below.

The value of these methods cannot be overestimated. They extend the scope of conventional X-ray secondary-emission spectrometric equipment and techniques down to lithium and to the field of organic chemistry.

ANALYTICAL METHODS BASED ON X-RAY SCATTER

Coherent Scattering

McCue, Bird, Ziegler, and O'Connor[67] have extended their earlier work[68] on determination of heavy elements in light matrixes by coherent (Rayleigh) scatter. The technique may be regarded as X-ray nephelometry. A high-energy (80 to 150 kV) polychromatic X-ray beam irradiates the specimen solution, but, rather than an analyte spectral line, coherently scattered radiation is measured at an appropriate angle. Maximum differentiation of coherent and incoherent (Compton) scatter is realized by filtering the primary X-rays, measuring scatter at a large angle, and using a discriminator to pass only detector output pulses at or near the maximum primary beam energy. Since coherent and incoherent scatter are proportional, respectively, to Z^3 and Z, the scattered intensity is highly dependent on the concentration of high-Z elements.

McCue and his co-workers used the method to determine 1–10 g U/liter in solutions having high concentrations of aluminum or stainless steel. Their equation is

$$C_X = C_1 + (C_2 - C_1)\left(\frac{I_X - I_1}{I_2 - I_1}\right) \tag{11}$$

where C and I are concentration and intensity; 1 and 2 refer to standards having known uranium concentration; and X refers to the sample. The work apparently was done on X-ray radiography equipment, but it would be interesting to try the method on a 75- or 100-kV X-ray spectrometer.

Ratio of Coherently and Incoherently Scattered Target Line

Dwiggins[59,69] has made an extremely significant contribution with his determination of carbon in hydrocarbon liquids using an unmodified conventional X-ray spectrometer. The intensity ratio R of the coherently and incoherently scattered W $L\alpha_1$ line from the X-ray tube is measured from the sample(s) and from a series of pure hydrocarbon liquid compounds having various known carbon–hydrogen ratios. A calibration curve of R *versus* the percent of carbon is established from the known liquids; analytical concentrations are derived from this curve.

The C—H bond type has no effect. If a few percent of nitrogen or oxygen is present, the method gives %(C + N + O), and independent nitrogen and oxygen analyses are required to obtain the carbon concentration. In the presence of sulfur in concentrations up to ~10%, if the carbon–hydrogen ratio is constant, R is proportional to the sulfur concentration; if the sulfur concentration is constant, R is proportional to the carbon concentration. Dwiggins gives ways of correcting for sulfur.

Dwiggins used a water-cooled, uncovered specimen drawer in his original work,[69] but later used a Mylar-covered cell and inverted geometry in a Norelco Autrometer.[59] Toussaint and Vos[70] have applied the method to solid hydrocarbons.

X-RAY ABSORPTION-EDGE SPECTROMETRY

Dodd[71] has reviewed the principles of X-ray absorption-edge spectrometry, and routine applications have been reported by Hakkila and Waterbury[31,32] and Stewart.[72]

Dunn[33] has developed a routine procedure for absorption-edge spectrometric analysis on a diffractometer, deriving the bracketing wavelengths from the primary beam. Analyte concentration C is given by

$$C = \frac{2303}{\Delta\mu D}[m_S \log_{10} I_S - m_L \log_{10} I_L - a] \tag{12}$$

where $\Delta\mu$ is the difference in the analyte mass-absorption coefficients on the short and long sides of the analyte absorption edge; D is the cell length; I is the transmitted intensity; S and L refer to the bracketing wavelengths on the short and long wavelength sides of the analyte absorption edge; and m_S, m_L, and a are experimentally derived constants. Dunn gives the experimental procedure for evaluating m_S, m_L, a, and $\Delta\mu$. The method gives good results, but the preliminary work required to evaluate the constants can be time-consuming.

Bertin, Longobucco, and Carver[30] have developed a simplified, generally applicable, routine procedure for absorption-edge spectrometric analysis for use on standard commercial X-ray spectrometers with only minor modification. The bracketing wavelengths are derived from secondary targets. The only X-ray measurements required are the intensities transmitted by the sample(s) I, the empty cell I_0, a correction standard I_C, and the open secondary-beam tunnel I_T at each of two wavelengths λ_S and λ_L bracketing the analyte absorption edge. Analyte concentration is given by

$$C = \frac{2303}{(\mu_S - k\mu_L)D}\left[\log_{10}\left(\frac{I_0}{I}\right)_S - k \log_{10}\left(\frac{I_0}{I}\right)_L\right] \tag{13}$$

where μ is analyte mass-absorption coefficient and D is cell length. The value of k is derived experimentally from the intensities transmitted at λ_S and λ_L by an analyte-free correction standard (usually an aluminum plate) I_C and the empty secondary-beam tunnel I_T:

$$k = \frac{\log_{10}(I_T/I_C)_S}{\log_{10}(I_T/I_C)_L} \tag{14}$$

The absorption coefficients are obtained from the literature either directly or by graphic interpolation. The simplified procedure was evaluated for 10 elements having a wide range of atomic number (chromium to lead) in a wide range of concentration. The results for chromium had an average relative error of $\sim 10\%$, but, for all other elements, the relative error ranged from 0 to 8%, averaging 2.7%. These results compare surprisingly well with those from more elaborate methods.

The paper presents tables giving lines bracketing the K edges of elements 23 to 48 (vanadium to cerium) and the LIII edges of elements 56 to 92 (barium to uranium). Values of μ_S and μ_L are given for lines bracketing all K edges and some LIII edges. Since this paper appeared, more accurate values of absorption coefficients have become available.[73,74] Probably even better accuracy would result if μ_S and μ_L were derived from these sources.

Dodd[75] has applied pulse-height selection in a rapid, convenient absorption-edge method having improved sensitivity. The work is done on a diffractometer, and the two bracketing wavelengths are derived from the primary continuous spectrum. At each of the bracketing wavelengths, a differential pulse-height distribution curve is recorded with the use of an extremely narrow (0.1 V) window to give highly reproducible PHD peak height. This height was measured for the two wavelengths and used to give I_L/I_S in the equation

$$C = \frac{2.303 \log_{10}(I_L/I_S)}{\mu_S - \mu_L} \tag{15}$$

where C is analyte concentration; I is transmitted intensity; μ is analyte mass-absorption coefficient; and L and S refer to bracketing wavelengths. A detailed study was made of the effect of crystal (LiF or topaz), detector (proportional or scintillation), and amplifier gain.

Hakkila, Hurley, and Waterbury[76] have developed a three-wavelength method for determination of plutonium. Transmitted intensities are measured at three wavelengths, Nb $K\beta$ (λ_1) and Mo $K\alpha$ (λ_2) (which bracket the Pu LIII absorption edge) and Nb $K\alpha$ (λ_3), after passing through the absorption cell filled successively with water and a standard solution of known plutonium concentration. Then the intensities of the two bracketing lines only (λ_1 and λ_2) are measured after passing through the same cell filled with the sample(s). The measurements on water and standard are used to derive k_1/k_2:

$$k_1/k_2 = \left[\frac{\log_e (I_0/I)_3}{\log_e (I_0/I)_2}\right]^{\frac{\log_e (\lambda_2/\lambda_1)}{\log_e (\lambda_3/\lambda_2)}} \tag{16}$$

Then,

$$C = k_1 \log_e(I_0/I)_1 - k_2 \log_e(I_0/I)_2 \tag{17}$$

where I_0 is intensity transmitted through the water-filled cell and I is intensity transmitted through the cell filled with standard or sample. Plutonium was determined in a concentration of 10–25 mg Pu/ml with a relative standard deviation of $\pm 0{\cdot}6\%$.

Knapp, Lindahl, and Mabis[77] developed an absorption-edge spectrometric method for standard commercial X-ray spectrometers with only minor modification. The bracketing wavelengths are derived from secondary targets, usually the $K\alpha - K\beta$ lines of a single element. The transmitted intensity is measured at these two wavelengths on a series of standards in a certain solvent, and a calibration curve is established having the equation

$$\log_{10}(I_L/I_S) = K_A W_A + K_M \tag{18}$$

where I is transmitted intensity, K_A the slope of the curve, K_M the intercept, and W_A the weight fraction of analyte A in matrix M. The terms L and S refer to the bracketing lines. Calibration curves for other solvents are calculated simply from the absorption coefficients of the pure solvent at the bracketing wavelengths.

Cullen[78] describes three novel *internal differential absorption-edge* spectrometric methods in which the wavelengths bracketing the analyte absorption edge are either generated within or scattered by the sample itself: (1) An element having bracketing spectral lines may already be present in the sample; *e.g.*, determination of nickel in copper–nickel alloys, by using Cu $K\alpha$ and Cu $K\beta$. (2) The element having the bracketing lines may be added to the sample; *e.g.*, the determination of selenium by using Br $K\alpha$ and Br $K\beta$ from bromine added to the sample. (3) The two bracketing wavelengths may be derived from the continuous spectrum originating in the X-ray tube and scattered by the sample; *e.g.*, the determination of silver (λK_{abs} 0.49 Å) by using 0.46-Å and 0.63-Å scattered continuous radiation. Scattered characteristic target lines may be used if they have suitable wavelengths. In all three methods, Cullen prepared standards and established analytical curves of intensity ratio of the bracketing lines *versus* analyte concentration. These methods and that of Knapp, Lindahl, and Mabis above, lack one of the advantages of most absorption-edge methods in that they do require standards.

CONCLUSIONS

It would be nice to be able to conclude by citing several major breakthroughs in X-ray spectrometric analysis by solution techniques, but, in all honesty, with a single outstanding exception, this cannot be done. However, this does not mean that genuine and significant progress has not been made in recent years in all phases of this field. Sensitivity has been increased, perhaps by an order or two, but only for certain systems and only by exercise of great care and use of special techniques. Great ingenuity has been shown in the design of several sophisticated liquid cells for special applications. Most of the problems of interaction of the primary beam with the specimen—problems peculiar to liquid specimens—have been surmounted, and many ways of dealing with matrix effects have been developed and applied. Despite its many advantages, the application of X-ray absorption-edge spectrometry continues to lag behind that of emission spectrometry, but several novel techniques and simple routine procedures of wide general applicability have been reported.

In the writer's opinion, the most significant recent development in X-ray spectrometric analysis is the extension of its scope by the advent of indirect (association) techniques and the scatter-ratio technique for carbon. Previously, the practical low-Z limit of standard commercial X-ray spectrometers has been reduced at the rate of an element every few years, first to silicon and aluminum, then to magnesium, then to sodium, now possibly to fluorine. Only in elaborate laboratory-built instruments has it been possible to work with the K lines of elements down to beryllium. But, with these new indirect and scatter techniques, the scope of X-ray spectrometric analysis has been extended to lithium—conceivably even to helium, since the Group 0 gases can now be made to form well-defined compounds—and to organic chemistry. This is accomplished with standard commercial X-ray spectrometers, standard accessories, and conventional specimen preparation and presentation techniques.

REFERENCES

1. E. P. Bertin, "Solution Techniques in X-Ray Spectrometric Analysis," *Norelco Reptr.* **12**:15–26, 1965.
2. W. J. Campbell, "Fluorescent X-Ray Spectrographic Analysis of Trace Elements, Including Thin Films," *Am. Soc. Testing and Mater., Spec. Tech. Publ.* **349**: 48–69, 1964.
3. W. J. Campbell, M. Leon, and J. W. Thatcher, "Solution Techniques in Fluorescent X-Ray Spectrography," *U.S. Bur. Mines Rept. Invest.* **5497** (24 pp.), 1959.
4. E. P. Bertin and R. J. Longobucco, "Sample-Preparation Methods for X-Ray Fluorescence Emission Spectrometry," *Norelco Reptr.* **9**: 31–43, 1962.
5. E. L. Gunn, "X-Ray Spectrographic Analysis of Liquids and Solutions," *Am. Soc. Testing and Mater., Spec. Tech. Publ.* **349**: 70–85, 1964.
6. R. Louis, "X-Ray Analysis of Mineral Oil Products," *Erdoel Kohle* **17**: 360–367, 1964.
7. R. Louis, "X-Ray Analysis in the Petroleum Industry," *Erdoel Kohle* **18**: 187–190, 1965.
8. T. J. Cullen, "X-Ray Spectrometric Analysis of Solutions, a Comparison of Techniques," *Pittsburgh Conf. Anal. Chem. Appl. Spectry.*, 1963.
9. R. A. Jones, "Determination of Sulfur in Gasoline by X-Ray Emission Spectrography," *Anal. Chem.* **33**: 71–74, 1961.
10. R. Louis, "X-Ray Emission Analysis of Lubricating Oils," *Z. Anal. Chem.* **201**: 336–348, 1964.
11. R. Louis, "Detection Limits in X-Ray Emission Spectral Analysis of Mineral Oils," *Z. Anal. Chem.* **208**: 34–43, 1965.
12. C. C. Kang, E. W. Keel, and E. Solomon, "Determination of Traces of Vanadium, Iron, and Nickel in Petroleum Oils by X-Ray Emission Spectrography," *Anal. Chem.* **32**: 221–225, 1960.
13. C. C. Hale and W. H. King, Jr., "Direct Nickel Determinations in Petroleum Oils by X-Ray at the 0.1-P.P.M. Level," *Anal. Chem.* **33**: 74–77, 1961.
14. E. L. Gunn, "Problems of Direct Determination of Trace Nickel in Oil by X-Ray Emission Spectrography," *Anal. Chem.* **36**: 2086–2090, 1964.
15. E. L. Gunn, "Practical Methods of Solving Absorption and Enhancement Problems in X-Ray Emission Spectrography," *Develop. Appl. Spectry.* **3**: 69–96, 1964.
16. J. F. Croke, W. J. Pfoser, and M. J. Solazzi, "New Automatic X-Ray Spectrometer in Industrial Control," *Norelco Reptr.* **11**: 129–131, 1964.

17. H. Siemes, "X-Ray Fluorescence Spectrometric Determination of Light Elements in Low Concentrations in Water Solutions," *Z. Anal. Chem.* **199**: 321–334, 1964.
18. K. P. Champion and R. N. Whittem, "Utilization of Increased Sensitivity of X-Ray Fluorescence Spectrometry Due to Polarization of the Background Radiation," *Nature* **199**: 1082, 1963.
19. E. P. Bertin and R. J. Longobucco, "Some Special Sample Mounting Devices for the X-Ray Fluorescence Spectrometer," in: W. M. Mueller (ed.), *Advances in X-Ray Analysis, Vol. 5*, Plenum Press, New York, 1962, pp. 447–456.
20. E. P. Bertin and R. J. Longobucco, "X-Ray Spectrometric Analysis—Nickel, Copper, Silver, and Gold in Plating Baths," *Metal Finishing* **60**(3): 54–58, 1962.
21. D. W. Beard and E. M. Proctor, "A Method of Liquid Analyses Providing Increased Sensitivity for Light Elements," in: J. B. Newkirk and G. R. Mallett (eds.), *Advances in X-Ray Analysis, Vol. 10*, Plenum Press, New York, 1967, pp. 506–519.
22. W. J. Campbell, "Apparatus for Continuous Fluorescent X-Ray Spectrographic Analysis of Solutions," *Appl. Spectry.* **14**: 26–27, 1960.
23. F. L. Chan, "Apparatus for Analysis of Liquid Samples by the X-Ray Fluorescence Method with a Vacuum Spectrograph," *Develop. Appl. Spectry.* **5**: 59–75, 1966.
24. C. R. Hudgens and G. Pish, "X-Ray Fluorescence Emission Analysis of Slurries," *Develop. Appl. Spectry.* **5**: 25–29, 1966.
25. W. J. Campbell and M. Leon, "Fluorescent X-Ray Spectrograph for Dynamic Selective Oxidation-Rate Studies—Design and Applications," *U.S. Bur. Mines Rept. Invest.* **5739** (21 pp.), 1961.
26. J. O. Karttunen, "Separation and Fluorescent X-Ray Spectrometric Determination of Zirconium, Molybdenum, Ruthenium, Rhodium, and Palladium in Solution in Uranium-Base Fissium Alloys," *Anal. Chem.* **35**: 1044–1049, 1963.
27. R. H. Zimmerman and H. A. Heller, "Quantitative Determination of Major and Minor Constituents in Gold Plating Baths by X-Ray Fluorescence Analysis," Third National Meeting, Soc. Appl. Spectry., Cleveland, 1964.
28. E. L. Gunn, "Determination of Lead in Gasoline by X-Ray Fluorescence Using an Internal Intensity Reference," *Appl. Spectry.* **19**: 99–102, 1965.
29. H. F. Smith and R. A. Royer, "Determination of Aluminum in Organo-Aluminum Compounds by X-Ray Fluorescence," *Anal. Chem.* **35**: 1098–1099, 1963.
30. E. P. Bertin, R. J. Longobucco, and R. J. Carver, "Simplified Routine Method for X-Ray Absorption-Edge Spectrometric Analysis," *Anal. Chem.* **36**: 641–655, 1964.
31. E. A. Hakkila and G. R. Waterbury, "Applications of X-Ray Absorption-Edge Analysis," *Develop. Appl. Spectry.* **2**: 297–307, 1963.
32. E. A. Hakkila and G. R. Waterbury, "X-Ray Absorption-Edge Determination of Cobalt in Complex Mixtures," in: W. M. Mueller (ed.), *Advances in X-Ray Analysis, Vol. 5*, Plenum Press, New York, 1962, pp. 379–388.
33. H. W. Dunn, "X-Ray Absorption-Edge Analysis," *Anal. Chem.* **34**: 116–121, 1962.
34. L. H. Griffin, "Iron-55 X-Ray Absorption Analysis of Organically Bound Chlorine Using Conventional Proportional Counting Facilities," *Anal. Chem.* **34**: 606–609, 1962.
35. C. D. Pierron and R. H. Munch, "Rapid X-Ray Fluorescence Method for Determination of V, Cu, Mo, Ti, Co, and Ni," *Develop. Appl. Spectry.* **2**: 360–365, 1962.
36. A. Visapaa, "Determination by X-Ray Fluorescence of Microgram Amounts of Strontium in Water and Milk," *Teknillisen Kemian Aikakausilehti* **20**: 146–151, 1963; *C.A.* **59**: 8121c, 1963.
37. A. Visapaa, "Determination by X-Ray Fluorescence of Microgram Amounts of Barium in Water," *Teknillisen Kemian Aikakausilehti* **20**: 563–568, 1963; *C.A.* **60**: 7793b, 1964.
38. K. Hirokawa, T. Shimanuki, and H. Goto, "Determination of Manganese in Ferromanganese and Chromium in Ferrochromium by X-Ray Fluorescent Spectroscopy," *Sci. Rept. Res. Inst. Tohuku Univ., Ser. A* **15**: 127–132, 1963.
39. H. R. Kirchmayr and D. Mach, "X-Ray Fluorescence Analysis of Rare-Earth Metal–Manganese Alloys," *Z. Metallk.* **55**: 247–250, 1964.
40. S. D. Kullbom, W. K. Pollard, and H. F. Smith, "Combined Infrared and X-Ray Spectrometric Method for Determining Sulfonate and Sulfate Concentration of Detergent Range Alkylbenzene Sulfonate Solutions," *Anal. Chem.* **37**: 1031–1034, 1965.
41. J. L. Caley, "Use of X-Ray Emission Spectrography for Petroleum Product Quality and Process Control," in: W. M. Mueller and M. Fay (eds.), *Advances in X-Ray Analysis, Vol. 6*, Plenum Press, New York, 1963, pp. 396–402.
42. M. L. Tomkins, G. A. Borun, and W. A. Fahlbusch, "Quantitative Determination of Tantalum, Tungsten, Niobium, and Zirconium in High-Temperature Alloys by X-Ray Fluorescent Solution Method," *Anal. Chem.* **34**: 1260–1263, 1962.

43. G. R. Blank and H. A. Heller, "X-Ray Spectrometric Determination of Thorium in Refined Uranium Materials," *Norelco Reptr.* **8**: 112–115, 1961; *Develop. Appl. Spectry.* **1**: 3–15, 1962.
44. G. R. Blank and H. A. Heller, "X-Ray Spectrometric Determination of Zirconium in Refined Uranium Materials," *Norelco Reptr.* **9**: 23–27, 1962.
45. B. R. Boyd and H. T. Dryer, "Analysis of Nonmetallics by X-Ray Fluorescence Techniques," *Develop. Appl. Spectry.* **2**: 335–349, 1963.
46. G. R. Blank and H. A. Heller, "X-Ray Spectrometric Determination of Copper, Tin, and Uranium in Bronze Heat-Treating Material," in: W. M. Mueller (ed.), *Advances in X-Ray Analysis, Vol. 4*, Plenum Press, New York, 1961, pp. 457–473.
47. E. A. Hakkila, R. G. Hurley, and G. R. Waterbury, "X-Ray Fluorescence Spectrometric Determination of Zirconium and Molybdenum in the Presence of Uranium," *Anal. Chem.* **36**: 2094–2097, 1964.
48. A. Strasheim and F. T. Wybenga, "Determination of Certain Noble Metals in Solution by Means of X-Ray Fluorescence Spectroscopy. I. Determination of Platinum, Gold, and Iridium," *Appl. Spectry.* **18**: 16–20, 1964.
49. F. T. Wybenga and A. Strasheim, "Determination of Certain Noble Metals in Solution by Means of X-Ray Fluorescence Spectroscopy. II. Determination of Palladium, Rhodium, and Ruthenium," *Appl. Spectry.* **20**: 247–250, 1966.
50. S. A. Bartkiewicz and E. A. Hammatt, "X-Ray Fluorescence Determination of Cobalt, Zinc, and Iron in Organic Matrices," *Anal. Chem.* **36**: 833–836, 1964.
51. R. F. Haycock, "Determination of Barium, Calcium, and Zinc in Additives and Lubricating Oils by X-Ray Fluorescence Spectroscopy," *J. Inst. Petrol.* **50**: 123–128, 1964.
52. B. J. Mitchell and H. J. O'Hear, "General X-Ray Spectrographic Solution Method for Analysis of Iron-, Chromium-, and/or Manganese-Bearing Materials," *Anal. Chem.* **34**: 1620–1625, 1962.
53. R. B. Scott, "X-Ray Spectrographic Analysis of Antimonials," *Appl. Spectry.* **1**: 17–23, 1962.
54. T. J. Cullen, "Coherent Scattered Radiation Internal Standardization in X-Ray Spectrometric Analysis of Solutions," *Anal. Chem.* **34**: 812–814, 1962.
55. V. R. Lopp and C. G. Claypool, "Direct Determination of Vanadium and Nickel in Crude Oils by X-Ray Fluorescence," in: W. M. Mueller (ed.), *Advances in X-Ray Analysis, Vol. 3*, Plenum Press, New York, 1960, pp. 131–137.
56. K. R. Stever, J. L. Johnson, and H. H. Heady, "X-Ray Fluorescence Analysis of Tungsten–Molybdenum Metals and Electrolytes," in: W. M. Mueller (ed.), *Advances in X-Ray Analysis, Vol. 4*, Plenum Press, New York, 1961, pp. 474–487.
57. R. A. Jones, "Determination of Manganese in Gasoline by X-Ray Emission Spectrography," *Anal. Chem.* **31**: 1341–1344, 1959.
58. E. L. Gunn, "Absorption Effects in X-Ray Fluorescence Measurement of Elements in Oil," in: W. M. Mueller and M. Fay (eds.), *Advances in X-Ray Analysis, Vol. 6*, Plenum Press, New York, 1963, pp. 403–416.
59. C. W. Dwiggins, Jr., "Automated Determination of Elements in Organic Samples Using X-Ray Emission Spectrometry," *Anal. Chem.* **36**: 1577–1581, 1964.
60. H. V. Carter, "Simplified Mathematical Matrices for Solution Analysis," *Norelco Reptr.* **13**: 45–47, 1966.
61. E. A. Hakkila and G. R. Waterbury, "X-Ray Fluorescence Spectrometric Determination of Chromium, Iron, and Nickel in Ternary Alloys," *Anal. Chem.* **37**: 1773–1775, 1965.
62. M. J. Schindler, A. Month, and J. Antalec, "An X-Ray Fluorescence Technique for Analysis of Iron–Nickel Films over an Extremely Wide Range of Thickness and Composition," *Pittsburgh Conf. Anal. Chem. Appl. Spectry.*, 1961.
63. J. S. Rudolph and R. J. Nadelin, "Determination of Microgram Quantities of Chloride in High-Purity Titanium by X-Ray Spectrochemical Analysis," *Anal. Chem.* **36**:1815–1817, 1964.
64. G. S. Smith, "A Useful Technique in X-Ray Fluorescence Spectrography," *Chem. Ind. (London)* **1963**(22): 907–909.
65. J. C. Mathies, P. K. Lund, and W. Eide, "X-Ray Spectroscopy in Biology and Medicine. IV. A Simple, Indirect, Sensitive Procedure for the Determination of Nitrogen (Ammonia) at the Microgram and Submicrogram Level," *Anal. Biochem.* **3**:408–414, 1962; *Norelco Reptr.* **9**:93–95, 1962.
66. G. J. Papariello, H. Letterman, and W. J. Mader, "X-Ray Fluorescent Determination of Organic Substances Through Inorganic Association," *Anal. Chem.* **34**: 1251–1253, 1962.
67. J. C. McCue, L. L. Bird, C. A. Ziegler, and J. J. O'Connor, "X-Ray Rayleigh Scattering Method for Determination of Uranium in Solution," *Anal. Chem.* **33**: 41–43, 1961.

68. C. A. Ziegler, L. L. Bird, and D. J. Chleck, "X-Ray Rayleigh Scattering Method for Analysis of Heavy Atoms in Low-Z Media," *Anal. Chem.* **31**: 1794–1798, 1959.
69. C. W. Dwiggins, Jr., "Quantitative Determination of Low Atomic Number Elements Using Intensity Ratio of Coherent to Incoherent Scattering of X-Rays; Determination of Hydrogen and Carbon," *Anal. Chem.* **33**: 67–70, 1961.
70. C. J. Toussaint and G. Vos, "Quantitative Determination of Carbon in Solid Hydrocarbons Using the Intensity Ratio of Incoherent to Coherent Scattering of X-Rays," *Appl. Spectry.* **18**: 171–174, 1964.
71. C. G. Dodd, "X-Ray Absorption-Edge Spectrometry," in: *Proceedings of Symposium on Physics and Nondestructive Testing*, Southwest Research Institute, San Antonio, Tex., 1962, pp. 199–221.
72. J. H. Stewart, Jr., "Determination of Thorium by Monochromatic X-Ray Absorption," *Anal. Chem.* **32**: 1090–1092, 1960.
73. K. F. J. Heinrich, "X-Ray Absorption Uncertainty," in: E. D. McKinley, K. F. J. Heinrich, and D. B. Wittry (eds.), *The Electron Microprobe*, John Wiley and Sons, New York, 1966, pp. 296–377.
74. H. M. Stainer, "X-Ray Mass Absorption Coefficients, a Literature Survey," *U.S. Bur. Mines Inform. Circ.* **8166** (124 pp.), 1963.
75. C. G. Dodd, "Applications of Optical and Electronic Dispersion in X-Ray Absorption-Edge Spectrometry," in: W. M. Mueller (ed.), *Advances in X-Ray Analysis, Vol. 3*, Plenum Press, New York, 1960, pp. 11–39.
76. E. A. Hakkila, R. G. Hurley, and G. R. Waterbury, "A Three-Wavelength X-Ray Absorption-Edge Method for Determination of Plutonium in Nitrate Media," *Anal. Chem.* **38**: 425–428, 1966.
77. K. T. Knapp, R. H. Lindahl, and A. J. Mabis, "An X-Ray Absorption Method for Elemental Analysis," in: W. M. Mueller, G. R. Mallett, and M. J. Fay (eds.), *Advances in X-RayAnalysis, Vol. 7*, Plenum Press, New York, 1964, pp. 318–324.
78. T. J. Cullen, "Internal Differential X-Ray Absorption-Edge Spectrometry," *Anal. Chem.* **37**: 711–713, 1965.

DISCUSSION

B. S. Sanderson (National Lead Co.): Can you discuss the use of stretched polypropylene film as opposed to Mylar? Have you any experience with it for a liquid cell cover?

E. P. Bertin: No, I haven't had any experience with it. I think the reason they used polyethylene on this open-top cell was that it is more flexible. It can be stretched, and they wanted to be able to push in the bottom of that cell. Perhaps this is a little bit better than Mylar for that purpose.

B. S. Sanderson: With this special kind, you can stretch it very thin.

E. P. Bertin: You can stretch polyethylene pretty thin, and you can buy Mylar of 0.13 mil now. If you want to go below that, perhaps you should make formvar film by dropping a solution of formvar onto a liquid surface and fishing it out on a wire ring.

B. S. Sanderson: I think you can stretch polypropylene as thin as formvar. It's a question of comparisons of strength.

E. P. Bertin: I have no experience with it.

X-RAY FLUORESCENCE SPECTROSCOPY IN THE ANALYSIS OF ORES, MINERALS, AND WATERS*

Harry J. Rose, Jr., and Frank Cuttitta

U.S. Geological Survey
Washington, D.C.

ABSTRACT

X-ray fluorescence spectroscopy has been used in solving a wide variety of geologic problems involving mineral, ore, and water analysis. The technique has been a powerful analytical tool in the survey of mineral deposits, as a control to monitor ore flotation processes, for the semimicroanalysis of mineral separates and of rare new mineral species, and for the determination of trace elements in lake and saline waters. Many preparation techniques have been developed for the analysis of complex mineral systems, some combining X-Ray fluorescence with other analytical techniques to provide a complete analysis. These, coupled with improvements in instrumentation, have given the X-ray analyst a means of extending analytical ranges to the microgram level and to include elements that were previously not detectable. Significant advances in sample preparation and methods development have been made in the analysis of milligram quantities of complex geologic materials. The fusion and the solution of specimens appear to be the preferred methods of sample preparation. For samples that vary markedly in composition, the slope-ratio technique offers a new approach to solving matrix problems.

INTRODUCTION

The search for techniques that increase analytical capabilities, particularly in quantitative micro- and semimicroanalysis, is a continuing effort in many geochemical laboratories because of the rare occurrence of many geologic and extraterrestrial materials.

Methods development in X-ray fluorescence analysis have followed three techniques of preparation: powder, fusion, and solution methods. Once used almost exclusively, powder methods have been largely abandoned because of many inherent problems[1-3] associated with particle size, crystal structure, and matrix effects. The chief advantage of this method of preparation has been speed, but this often depends on the makeup of the sample and the length of grinding time required to reduce the sample to the optimum particle size. This time may often exceed that required for fusion or solution preparations, particularly when fibrous or micaceous components are present in the sample. However, despite inherent problems, powder techniques can be used advantageously provided that there is strict adherence to limited analytical ranges and to the use of standards of similar composition, particle size, and structure. This is mandatory for high precision and accuracy.

Fusion methods range from simple to complex procedures. Fusion techniques do not require the rigid observance of procedural techniques which characterizes powder methods and, also, they permit extension of usable analytical ranges. Dilution in a flux reduces enhancement and absorption effects, and some investigators[1,3] have sought to

* Publication authorized by the Director, U.S. Geological Survey.

reduce these effects even further by adding an element highly absorbent for those to be determined. Use of a heavy absorber minimizes interelement effects by effectively swamping out the influence of the sample composition on the determination, but it lowers counting rates. However, except in the determination of the lightest elements or of trace elements, the lower counting rates do not affect the accuracy of a determination.

Recently, X-ray fluorescence analytical research has been focused on the application of solution techniques to the determination of both trace constituents in macro samples and of major constituents in small (milligram size) geologic specimens. This method is the most intriguing of all preparation methods because of its great versatility. Solution techniques allow the separation of elements into groups or the concentration of trace constituents by precipitation with a carrier, extraction, ion-exchange, or microelectrolytic methods. Furthermore, the solution method can be completely independent of the need for chemically analyzed reference materials. Standards are easily prepared from solutions of pure metals or oxides, which may be combined in varying proportions to cover any anticipated analytical range.

A large variety of powder and fusion applications in X-ray fluorescence analysis are already described in the published literature. Only solution methods and some of the other analytical techniques that have been used in combination with them will be discussed here.

INSTRUMENTATION

Two single-channel X-ray spectrometers were used, one, an air or helium-path unit equipped with a platinum-target X-ray tube and, the other, a vacuum spectrometer having a dual-target (chromium and tungsten) X-ray tube. A pulse-height analyzer was used for all measurements, set with a base level of 3.5 V, a 10-V window, and a line pulse adjusted to 7.5 V. Each spectrometer was equipped with a piggyback detector, a dual-window flow counter and scintillation counter. Under optimum conditions, including choice of crystals, target, and detector, counting times were selected by determining the interval required to attain 10,000 counts on the lowest standard for each element measured.

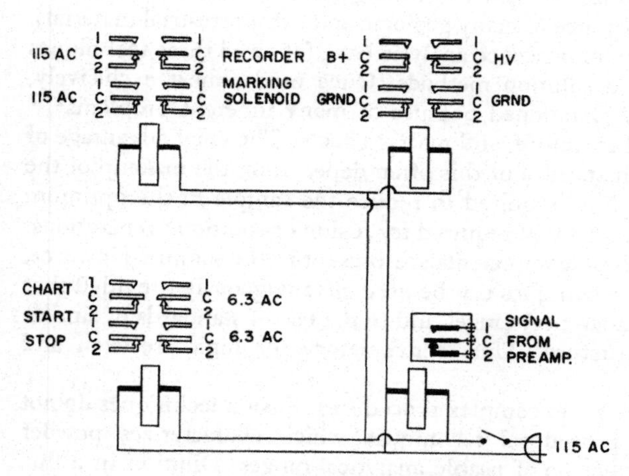

Figure 1. Diagram of relay switching circuit.

A single electronic console providing accessory power and electronic readout circuits was used for both spectrometers. Because the two spectrometers had identical electrical requirements, a relay-switching circuit was devised to provide an automatic and instantaneous switchover.

As shown in Figure 1, the switch consisted of three four-pole relays for the power and voltage circuits and a coaxial relay to handle the signal circuits. All the circuits were actuated from a single toggle switch mounted on the front panel of the console. Connectors were mounted on the back panel of the console to accommodate all the cables from the spectrometers.

SOLUTION-DILUTION METHOD

Dilution of the sample is a common technique used in quantitative X-ray fluorescence analysis to reduce matrix effects and present a more uniform sample to the X-ray beam. The dilution technique is applicable to powder, fusion, and solution preparations. For maximum versatility, solution-dilution methods were selected as the means for determining major and minor elements. The procedures described below were developed specifically as semimicro methods for the analysis of minerals and extraterrestrial materials[4] primarily because of the materials' scarce natural occurrence. However, these techniques are applicable to ores, alloys, and waters, and specific examples are given of each.

Determination of Major Elements in Silicates

The analysis of the major constituents of silicate specimens amounting to less than 10 mg has been of particular interest to earth scientists pursuing studies of formative and differentiation processes. Procedural techniques and experimental factors associated with the solution-dilution method are presented below.

Procedure. A 4- to 10-mg portion of the sample is weighed in a tared 15-ml platinum crucible to ± 0.01 mg and 1 ml of 5:2:3 $HF:HNO_3:H_2O$ is added. The crucible is covered and heated on a steam bath for several hours until the sample is decomposed. The resulting solution is evaporated to dryness on the steam bath three times, 1 ml of 1 : 1 HNO_3 being added to the residue after each evaporation. The silica-free nitrates are dissolved in 1.0 ml of 5 : 95 HNO_3. Nitric acid is used as the solvent medium in preference to acids containing heavier atoms (*e.g.*, H_2SO_4, HCl, $HClO_4$, and H_3PO_4) not only to minimize absorption of X-rays but to avoid adverse effects of acid upon the absorbent as well (*i.e.*, H_2SO_4 or $HClO_4$ on cellulose). This solution is absorbed onto 500 mg of powdered chromatographic paper, and the resulting moist pulp is mixed thoroughly with a small Teflon stirring rod and then dried overnight at 80°C. The dried paper is then mixed in a boron carbide mortar, and the ground powder is pressed into a pellet, 1 in. in diameter. For additional strength, the pellet is prepared as a double layer with powdered chromatographic paper as a backing. The pellet is then exposed to the X-ray beam.

Optimum Ratio of Solution to Absorbent. Figure 2 presents data for titania, showing the increase in counting rate as the ratio of absorbent to solution is decreased (curve *A*). The sample weight is kept constant by absorbing 1 ml of a solution containing 1.54 mg/ml of a 1 : 1 mixture of granite G-1 and diabase W-1 (0.67% TiO_2) on 100, 200, 300, 400, and 500 mg of powdered cellulose. Although the intensities are higher with smaller quantities of cellulose, the optimum ratio of solution to absorbent is 1 ml/500 mg, a ratio that just provides complete absorption of the sample solution.

Effect of Variable Sample Weight. The effect of a variable sample weight on counting rates with the use of a constant amount (500 mg) of powdered cellulose absorbent

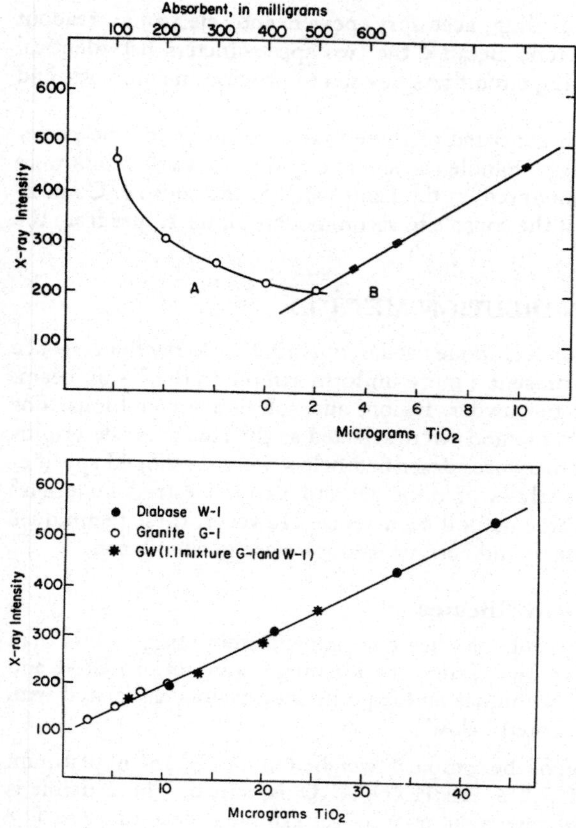

Figure 2. Effect of various amounts of absorbent and TiO_2 on X-ray intensities. Curve *A*: effect of varying amounts of absorbent on measured X-ray intensities for a constant sample weight of TiO_2. Curve *B*: same data on the X-ray intensities for TiO_2 as used for curve *A*, but calculated on the basis of 100 mg of absorbent.

Figure 3. X-ray intensity as a function of variable sample weight at constant weight of absorbent (500 mg).

is shown in Figure 3. Aliquots of 1 ml containing 1, 2, 3, and 4 mg of each of three reference standards (granite G-1, diabase W-1, and a 1 : 1 mixture of G-1 and W-1) are processed by the solution-pellet method, and the resultant X-ray counting rates are related to the titania content of the sample. The data show that even a fourfold change in sample weight has little or no effect upon the linearity of the concentration–intensity relation. This direct correlation is also implied in curve *B* of Figure 2, where the data shown in curve *A* have been replotted to represent the counting rates associated with the titania content (in micrograms) per 100 mg of the sample-impregnated absorbent.

Effect of Variable Composition. Effects arising from the variability of composition of silicate samples might have been expected to interfere significantly with the accurate determination of oxide concentrations in an unknown sample. A suite of analyzed silicates (granite, diabase, tektites, and synthetic glasses) was processed by using a constant amount (500 mg) of absorbent and sample weights ranging from 4 to 10 mg. These reference materials represent a variable silica content ranging from 45 to 87%. The data for titania are presented in Figure 4. The linearity of the curve indicates that variations in composition of silicate samples have little or no effect upon the direct correlation of the oxide concentration with the X-ray fluorescence intensity.

The lack of compositional effects may be attributed to the fact that the sample submitted to the X-ray beam is virtually powdered cellulose, and, as such, the role of the

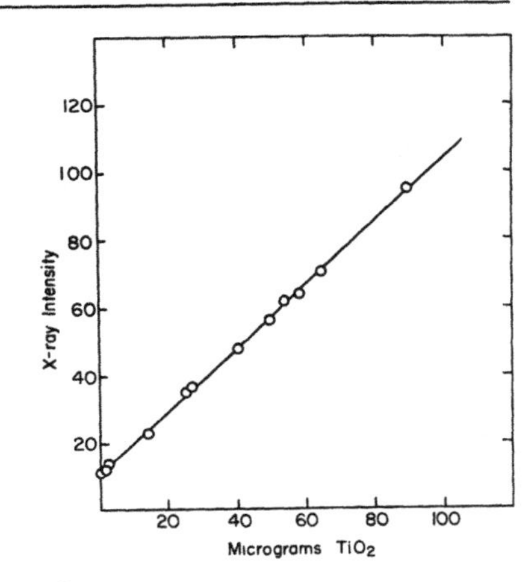

Figure 4. X-ray intensity of TiO_2 as a function of variable composition.

sample matrix on X-ray absorption or enhancement is reduced to a minimum. The experimental data and correlations shown for titania in Figures 2 and 3 hold for all the other oxides (Al_2O_3, total iron as Fe_2O_3, CaO, K_2O, and MnO) studied, and their presentation here would only be a duplication of the data already shown above.

Attainment of Infinite Thickness. Powdered cellulose impregnated with the silica-free residue of opal glass NBS-91 was prepared so that 100 mg of the mixture contained 3 mg of the residue. A series of pellets was prepared with varying amounts of the mixture as the sample layer. The data presented in Figure 5 are Zn $K\alpha$ intensities plotted as a function of the sample weight, which is used as an indirect measure of sample thickness. An approximately 650-mg weight of the mixture will provide a sample of at least critical depth. One can get by with less than the critical depth by taking an exactly weighed amount of mixture containing approximately the same proportion of sample weight to diluent to ensure no significant change in absorption of the final preparation. Relatively large changes in this ratio can alter the critical thickness of the sample, these being reflected in the variable intensities measured. Under stringently reproduced conditions, the intensity of the characteristic line will be proportional to the concentration of the element sought.

It might be well here to emphasize that the experimental factors given above hold as long as a high dilution of the sample is used. When the dilution is not maintained, the

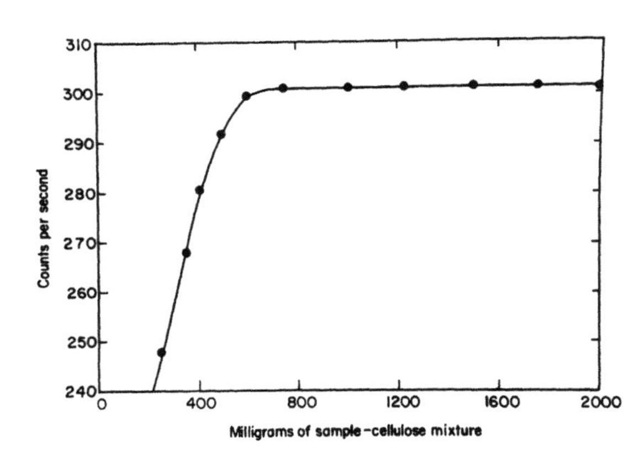

Figure 5. Variation of characteristic line intensity for zinc with thickness.

overall effect of the prepared pellet is no longer that of practically pure cellulose but becomes a function of the sample with which it is impregnated. When this occurs, several alternative corrective procedures can be used and will be discussed later.

When dilution is maintained at a high level, absorption effects are virtually eliminated, which makes possible the use of pure solution standards, *i.e.*, standards prepared from the pure metal or compounds of the element that is being determined. The main advantage in using single-element standards is an independence from chemically analyzed reference materials. As we shall see later, a simple, empirical correction procedure can be devised when interference occurs from other elements whose secondary emission lines are close to the wavelengths being measured. The correction procedure is indispensable to the analysis of a complex group such as the rare earths and is discussed under special techniques.

The solution-dilution method does not presently include the determination of silica, but several techniques are under investigation that will eventually permit determination of this element.

Analysis of Sulfide Minerals and Ores

The analysis of milligram amounts of sulfide minerals and ores can be a very difficult and time-consuming procedure by chemical methods. The solution-dilution method has proved useful in the analysis of a number of sulfide samples, one of which (a silver–copper sulfide) is given as an example below. The sample was dissolved in a sealed 25-ml volumetric flask containing 3 ml of fuming HNO_3. Three 0.25-ml portions of 30% H_2O_2 were slowly added by drops. The sample solution was stirred continuously and brought to 50°C to complete the dissolution of the sample. After cooling, the solution was diluted to 25 ml. The solution contained 1.04 mg of sample/ml.

The standards were prepared from copper metal, $AgNO_3$, and anhydrous Na_2SO_4. A series of five standards and a reagent blank were prepared as shown in Table I.

Three 3-ml aliquots of the sample solution were brought to dryness on a steambath and the residue dissolved in 1 ml of 5% HNO_3. Preparation of the pellets for both standards and samples is similar to that described above for silicates. The results of the analysis are given in Table II.

The results of the three runs are indicative of the precision of the determinations. Other X-ray fluorescence analyses of this sample made on two small, separate chips of 2 and 4 mg deviated in silver and sulfur content (silver, 75.9 and 82.2%; sulfur, 16.9 and 10.8%). Copper remained constant at 7.26 and 7.24% for the analyses of the two different samples. These discordant analyses were found to be due to the sensitivity of the silver sample to light—a photochemical reaction producing a change in sample composition.

Table I. Composition of Standards, Weight per Milliliter

	Silver (mg)	Copper (mg)	Sulfur (mg)
Blank	0	0	0
1	1	2.0	1.0
2	2	1.6	0.8
3	3	1.2	0.6
4	4	0.8	0.4
5	5	0.4	0.2

Table II. Determination of Copper, Silver, and Sulfur by Using 3.12 mg of Sample

	Run 1	Run 2	Run 3	Average
% Silver	79.0	78.6	78.8	78.8
% Copper	7.24	7.28	7.26	7.26
% Sulfur	13.5	13.5	13.6	13.5
Total	99.7	99.4	99.7	99.6

This analytical pitfall was circumvented by weighing and immediately dissolving a freshly isolated specimen.

Analysis of Alloys

The same solution-dilution technique used for the analysis of geologic materials can be applied to the analysis of metals and alloys. An example of this type of analysis is given below for a nickel–titanium alloy that was believed to be a fragment from an orbiting satellite that fell to earth.

Standards to cover the anticipated analytical range were prepared from pure TiO_2 and NiO and a titanium–nickel alloy (44.6% titanium and 55.4% nickel). The standards and sample (5.82 mg) were dissolved in platinum crucibles with 1.25 ml of 5 : 1 : 4 HF : HCl : H_2O on a steam bath. The resultant solutions were absorbed in 500 mg of powdered cellulose, and the mixtures were dried overnight at 80°C, ground, and pressed into pellets. Figure 6 illustrates the working curves and the results of analysis. The data indicate that the unknown is very similar to the alloy standard used.

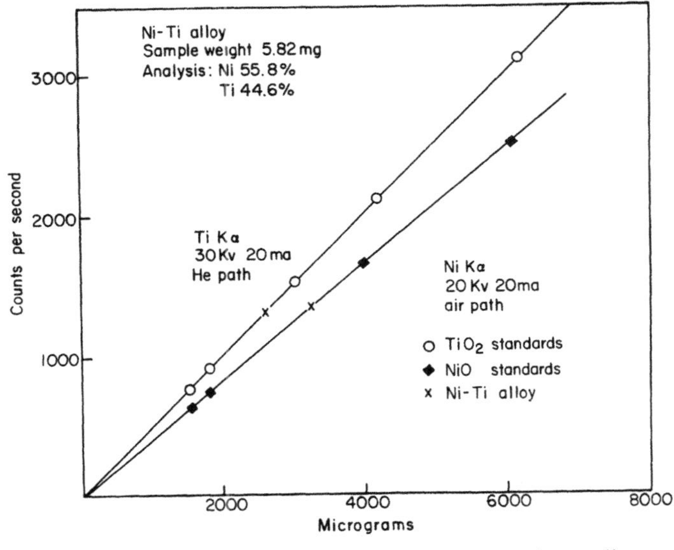

Figure 6. Working curves for nickel and titanium in an alloy.

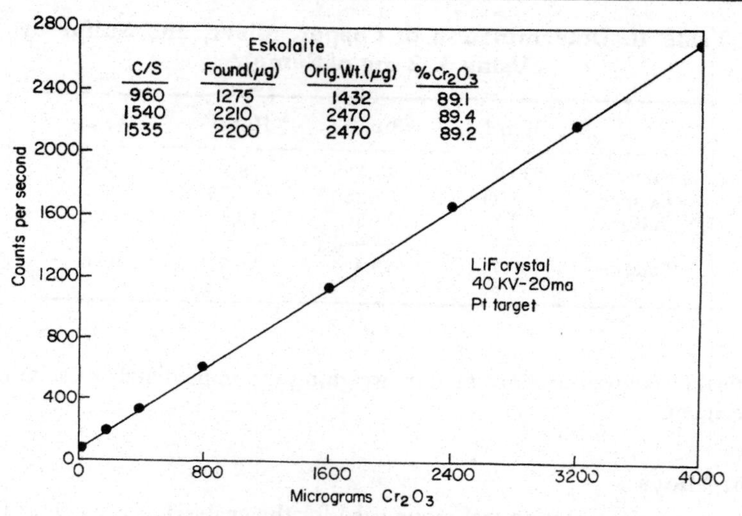

Figure 7. Determination of chromium in eskolaite.

Analysis of Minerals

The solution-dilution method provides an independent technique for the semi-microanalysis of minerals. Frequently, it is one of the few means by which such an analysis can be made (*e.g.*, the analysis of complex rare-earth minerals). The X-ray analysis of eskolaite, an unusual double oxide of chromium, serves as an example.

The sample (weighing 12.35 mg) was dissolved in 2 ml of fuming perchloric acid, and the resulting solution was diluted with 25 ml of H_2O. Cr_2O_3 was precipitated from the perchlorate solution with NaOH by using 20 mg of aluminum (as the nitrate) as a carrier in the presence of 5 ml of 5 wt.% Na_2SO_3 as the reducing agent and phenolphthalein as the indicator. The precipitate was separated by filtration and then dissolved with 10 ml of hot 5 : 95 HNO_3. The solution and hot acid washings were run directly into a 25-ml volumetric flask and diluted to volume with water. The sample solution was equivalent to 494 μg of eskolaite/ml. The analytical standards were solutions prepared from pure $K_2Cr_2O_7$ processed simultaneously with aliquots of the eskolaite solution from the aluminum-carrier precipitation step through the preparation of the pellets. Figure 7 summarizes the results obtained on three determinations. The titrimetric value of 88.8% Cr_2O_3 agrees well with the X-ray results.

Determinations such as these have been made on a large number of mineral samples. The preparation of the mineral can be made by any means at the disposal of the chemist. The one rule that should be observed, however, is that standards and samples be processed simultaneously through the various steps of the procedure.

Minor Element Analysis

The analysis of the minor or trace elements can most often follow the same procedures used for analysis of major elements, but the analytical approach will depend to a great extent on the concentration to be determined and, also, on the sensitivity required for the particular element. From the procedures given above, it can be concluded that sensitivity levels of 10 μg or less can be expected for most elements with the use of a 1-in. pellet composed of sample solution absorbed into 500 mg of powdered cellulose. The practical limit depends on the amount of sample available and the ingenuity of the chemist in devising methods to concentrate the element or group of elements required for analysis. The basic concept of using and maintaining high dilutions must be observed if matrix effects are to be minimal. Described below are several examples in which varying degrees of absorption affect the analytical determination because of a change in the dilution factor. Also described are two methods of coping with this effect, and the selection of one or the other depends on the degree to which the dilution factor has been disturbed.

SOLUTION-ADDITION METHOD (*e.g.*, DETERMINATION OF ZINC IN SILICATES)

The analysis of zinc in NBS-91 opal glass is used to illustrate the determination of a minor constituent (NBS certificate value = 0.1% ZnO). Prepare an SiO_2-free sample solution as described in the procedure for silicates. Also prepare a standard zinc solution by dissolving a weighed amount of pure zinc metal in dilute HNO_3 and diluting the solution to a specific volume. To a series of five 10-ml beakers, add aliquots of the standard solution containing increasing amounts of zinc (20, 40, 60, 80, and 100 μg of zinc). Evaporate the solutions to dryness on a steam bath. Add 1-ml aliquots of the unknown sample solution to each of the five beakers and also to a sixth beaker containing no zinc. Carry along a seventh preparation containing 1 ml of dilute HNO_3 as a reagent blank. Add 500 mg of powdered cellulose, and then process the sample–cellulose mixtures through the same mixing, drying, grinding, and pelletizing steps previously described. The application of the addition, or "spiking," technique for two different weight preparations of the opal glass (50 and 100 mg) is illustrated in Figure 8. Net counts (total counts less background) for Zn $K\alpha$ radiation are plotted against concentration. Extrapolation of the linear spiked curves to their intercepts with the abscissa locates the terminal points that provide the zinc values. These values for zinc are 665 and 670 ppm, which are in close agreement with 672 ppm for zinc obtained by atomic absorption. Also, observe in Figure 8 that there are two different slopes for the curves illustrated. The conclusion drawn is that there is a change in the overall absorption of the prepared pellet. These preparations represent a change in the

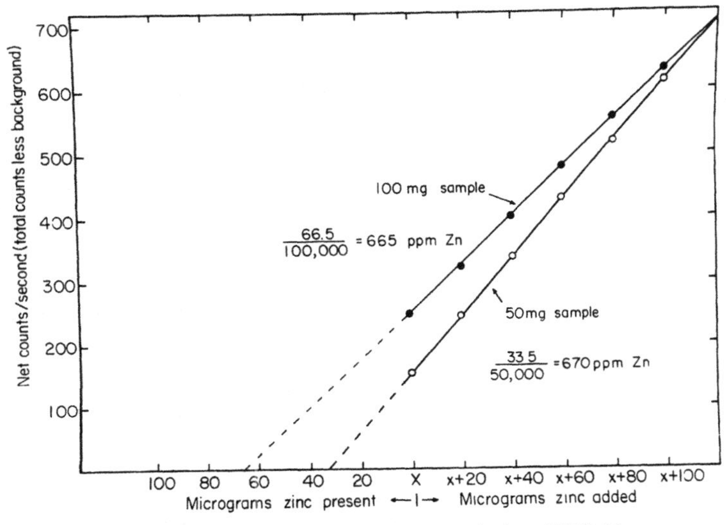

Figure 8. Determination of zinc in opal glass NBS-91.

ratio of sample to absorbent compared with those used for the determination of major elements by solution-dilution techniques, where all preparations were less than 10 mg in a total of 500 mg of absorbent. The silica-free matrix containing the desired zinc content is 10 to 20% of the total sample–absorbent mixture, and this brings about a marked change in the composition of the pellet submitted to X-ray excitation.

Calcium is a major constituent of the opal glass and has a large absorption effect on the Zn $K\alpha$ radiation. A tenfold change in calcium content reduces the emission intensity of zinc by 21%. The example represents a drastic change by an element highly absorbing for Zn $K\alpha$ radiation but illustrates the use of the addition technique, in which the sample acts as a self-corrective system as well as a standard.

SLOPE-RATIO METHOD (*e.g.*, DETERMINATION OF BROMINE IN SALINE WATERS)

In all the preparations described above, the determinations, because of dilution, have been virtually void of extraneous effects or have been easily corrected by the addition method. There has been no difficulty in establishing background counts through the simple use of carrying a reagent blank along during the processing of standards and samples. Except for the technique just described for zinc, the aim has been to dilute the role of the sample matrix. The determination of low-level bromine concentrations in saline waters represents an extreme problem because of a variability in both the composition and density of the brines. The evaporation of 1 ml of the brine onto 500 mg of cellulose powder completely voids the dilution effect, the brine itself assumes the dominant role, and its composition dictates the overall absorption of the prepared sample. This has a very large effect on the intensities of the element being measured as well as the background of the individual brine. While a true background can be obtained by exactly reproducing the composition of each sample exclusive of the element sought, the task is extremely complex. The easiest course is the use of the slope-ratio method[5] described below.

A standard solution of bromine is prepared from reagent-grade sodium bromide. Stock solutions containing 1 mg, 100 μg, and 10 μg/ml (as bromine) are used to spike the brines. The preparation of the spiked samples is similar to that described for the zinc analyses (solution-addition method). In addition, a series of pure bromine solutions are prepared as standards.

A corrected background (B_u) can be calculated for each bromine determination by comparing the slope of the spiked-sample curve with that of the pure solutions through use of the slope-ratio expression

$$B_u = B_p \frac{S_u}{S_p}$$

where B_p is the background emission of the reagent blank; S_u, the slope of the unknown spiked curve; and S_p, the slope of the pure solution curve.

Substantiation of the slope-ratio expression is given in Figure 9. Line *DE* is the curve prepared from pure bromine solutions. Line *GF* is the curve for a brine spiked with known concentrations of bromine. Extensions of *DE* and *GF* intersect at point *A*. Intersect *A* represents zero radiation characterized by complete freedom from any effect, no matter what the source. Although data for only one saline water are compared with that of pure bromine, all the saline waters tested show point *A* (zero radiation) as a common intercept for extensions of all their curves. The ordinate intercept of the standard solution as well as the saline water represents the background radiation for these solutions and varies from sample to sample.

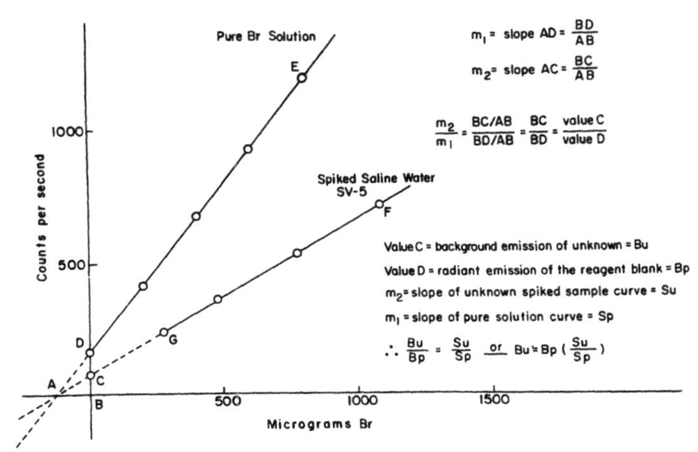

Figure 9. Experimental substantiation of the slope-ratio expression, $B_u = B_p(S_u/S_p)$.

Figure 10 illustrates the use of the slope-ratio technique in the determination of bromine in three brines. On multiplying the background radiation of the pure bromine solution (point *a*) by the ratio of the slope of each spiked-sample curve to the slope of the pure bromine solution, the background radiation for each sample is obtained (points *b*, *c*, and *d*). These calculated backgrounds should not differ significantly from those of the cellulose diluent impregnated with bromine-free matrices. The amount of bromine present in the sample aliquot taken is the distance along the abscissa between X and these points (*b*, *c*, and *d*). Simple extrapolation of the curves to *a* will yield results much lower than those obtained by conventional chemical analyses. The smaller slopes of the brine

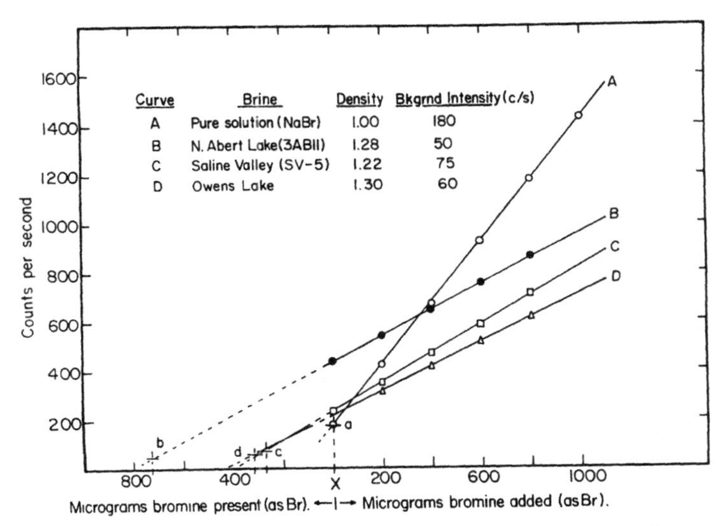

Figure 10. Application of slope-ratio–addition techniques to the determination of bromine in saline waters. One-milliliter portions of brines spiked with indicated amounts of bromine.

Table III. Comparison of Analytical Results for Bromide in Saline Waters

Brine	Density, saline water	Bromide			
		X-ray fluorescence			
		Uncorrected	Corrected		Chemical,*
		μg/ml	μg/ml	ppm	ppm
Deep Springs Lake, Calif. (DL2J)	1.291	390	695	538	526
N. Abert Lake, Ore. (3AB11)	1.284	480	725	565	548
Deep Springs Lake, Calif. (DL2P)	1.279	320	600	468	440
Owens Lake, Calif.	1.304	50	315	241	238
Saline Valley, Calif. (SV-5)	1.219	112	265	217	228
Synthetic brine*†	1.176	390	660	528	500†

* Analyst: Shirley L. Rettig, U.S. Geological Survey.

† Synthetic brine spiked with 500 ppm of bromide was prepared from CP reagents. Bromide content of reagents used to prepare synthetic brine was later determined as 35 ppm which provided a total bromide content of 535 ppm.

curves are due to absorption effects even though the spiked samples have higher bromine content than the pure solution curve.

Table III presents a comparison of analytical data on several brines. The data are presented to show both the uncorrected and the corrected values. Although not shown in this tabulation, several saline water samples have been analyzed which contained less than 10 ppm of bromine.

COMBINED CHEMICO-XRF METHODS

In this section, emphasis is given to special methods combining chemical techniques with X-ray spectroscopy. Space does not permit a general description of the various applications of enrichment and separation methods to XRF, such as amalgamation, ashing, chromatography, electrolysis, evaporation, fire assay, ion exchange, liquid–liquid extraction, magnetic separation, precipitation (of impurities or matrix), sublimation, distillation, and zone refining. However, there are advantages and disadvantages in combining chemical processes with X-ray spectroscopy. The major disadvantages are that the analysts must have a firm grasp of both chemical and X-ray techniques, that there are risks of losses or contamination of standards and samples, and that there is an increased time for analysis. Nonetheless, combined chemico-XRF techniques offer a number of advantages over direct XRF analysis. Among these are the easier preparation of standards to cover the anticipated analytical range, freedom from interelement effects and interference due to matrix or overlapping X-ray lines, increased sensitivities, use of group separations (rare earths, Nb–Ta, and Zr–Hf), and possible analyses at ultratrace levels. Several examples are offered below to illustrate the successful use of chemical techniques (precipitation, extraction, ion exchange, and microelectrolysis) in combination with the semimicro X-ray–solution method.

Group Separations (*e.g.*, Determination of Individual Rare Earths)

Using the solution–dilution method, the analysis of the lanthanides, yttrium, and scandium can be made on less than 2 mg of separated oxides. This group is chosen as an example because of the complex X-ray spectrum it produces, in which the $L\beta$ lines of other rare-earth elements interfere with the analytical emission $L\alpha$ lines. It has been

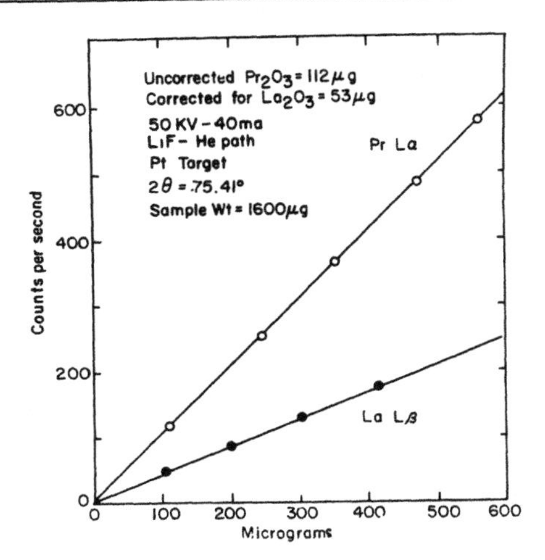

Figure 11. Correction curves for the effect of La $L\beta$ on the Pr $L\alpha$ radiation.

shown previously that, at high dilution of the sample, absorption effects are minimized and the problem of correcting for coinciding wavelengths is greatly facilitated. Figure 11 illustrates the correction procedure. The element of interest is praseodymium. The upper analytical curve was prepared from aliquots of the pure solution of this element. The $L\beta$ line of lanthanum (2.458 Å) interferes with the $L\alpha$ line of praseodymium (2.463 Å). The amount of interference by a given concentration of lanthanum is indicated by the lower curve which was prepared by using aliquots of a pure solution of lanthanum. These data

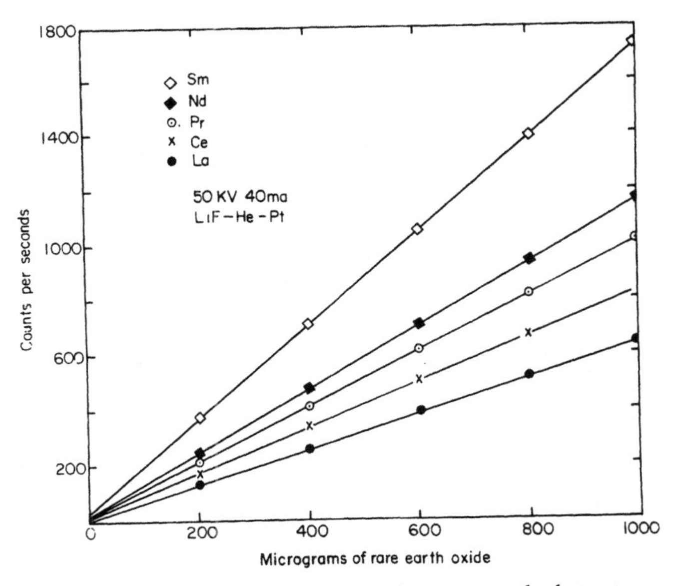

Figure 12. Calibration curves for cerium rare-earth elements.

were obtained at the wavelength of praseodymium (2.463 Å $L\alpha$ line). A prior determination of the lanthanum content in the unknown fixes exactly the contribution of this element to the total intensity observed at the praseodymium wavelength. The size of the correction can be quite large, as indicated by the example given. Because the interferences are due to $L\beta$ emission lines of elements of lower atomic number, the analytical scheme should begin with lanthanum and continue, in turn, through the series to lutetium. Yttrium and scandium, although associated with the lanthanides, are separated by atomic number and can be determined at the convenience of the analyst.

Figure 12 summarizes data on the sensitivity of these elements by this method. The family of curves shown were prepared from solutions of the pure oxides of each element. The slope of the curves increase with atomic number. This increase is not unexpected for, as Z increases, λ decreases, and the radiation becomes less susceptible to absorption by air and the detector window.

Solvent Extraction (*e.g.*, Determination of Thallium in Manganese Ores)

One of the advantages of combining techniques is that ultratrace analysis becomes possible. An example of combining separation, concentration, and X-ray fluorescence analysis is the determination of thallium in manganese ores. Thallium can be separated from iron, manganese, bismuth, lead, tin, and indium from a 0.75 N HBr solution by extraction with ethyl ether. Determinations of microgram quantities of thallium have been made on many samples with excellent agreement with chemical values.

Ion-Exchange Techniques

Ion-exchange resins in bead form, ion-exchange membranes, and ion-exchange resin-loaded paper disks have been found not only to be excellent support media for X-ray spectroscopy but also to provide the means to separate and isolate ions at the submicrogram to milligram levels.[6–9] Such combinations of ion exchange with X-ray spectroscopy have been used for separating and determining trace quantities of metallic impurities in high purity tungsten and WO_3[10] in molybdenum,[11] for determining major constituents in complex alloys, in microanalyzing rare museum specimens, and in determining trace metallic constituents in biologic materials.[9] In the last instance, a resin-loaded paper technique was used to determine calcium, cadmium, cobalt, copper, iron, manganese, nickel, and zinc at 1- to 90-μg levels in solutions prepared from 2-g samples of rat-kidney tissue. Ion-exchange–X-ray techniques also offer a powerful analytical tool in water analysis for stream-pollution studies, hydrogeochemical prospecting, or determination of elemental concentrations in sea waters in order to obtain concentration-depth profiles.

Hubbard and Green[10] determined trace element impurities (copper, nickel, and zinc from 0.3 to 20 ppm, lead from 1.0 to 20 ppm, and cobalt in the 0.1- to 1.0-ppm range) in high-purity tungsten by coupling dithizone-extraction–ion-exchange separation procedures to a determination by XRF. Results comparable to other determination techniques such as extraction–colorimetric were obtained. An X-ray fluorescence method[11] in which trace metal impurities (cobalt, copper, iron, manganese, nickel, lead, and zinc) in molybdenum were collected on ion-exchange resin-loaded paper disks provided the means to determine this large number of elements in samples weighing up to 10 g with a precision within $\pm 10\%$ of the amount present.

Luke[8] developed a combined chemico-XRF method for the ultratrace analysis of 28 or more cationic metals and of 15 or more anionic metals by using an X-ray milliprobe with fully focused, curved-crystal optics and tiny ($\frac{1}{8}$-in.) membrane disks of a strong-acid type of cation exchange resin or a strong-base type of anion exchange resin as support and separation media. The determination of as little as 0.01 μg of such elements as silver and copper were reported.

Electrochemical Separations

Procedures employing both potentiometric and amperostatic electrochemical deposition techniques (with pyrolytic graphite, platinum, nickel, or copper electrodes) have provided quantitative separations of microgram to milligram quantities of silver, gold, cobalt, and copper. Many other metals and nonmetals can be concentrated by electrolysis as strongly adherent deposits from solutions.

Mitchell and co-workers[12] have described a procedure for the electrolytic deposition of small amounts of gold onto copper electrodes and the subsequent measurement of the intensity of the $L\beta_1$ line for gold. Although the method was not applied to the determination of gold in a naturally occurring rock or mineral, the technique nevertheless shows promise of applicability.

Natelson and De[13] have worked out a practical procedure for the XRF estimation of microgram quantities of the heavy elements gold, lead, and bismuth in whole blood by using combined ashing and electroplating techniques. Although these investigators included mercury and thallium in their study, these elements were almost entirely lost on ashing. Natelson and De used a molybdenum X-ray target, a neon counter tube, and a copper or nickel sample holder.

Vassos and co-workers[14] have developed procedures for the electrochemical preparation of thin metal films on a substrate of pyrolytic graphite. Potential application of such quantitatively prepared high-purity carbon disks to XRF measurement is particularly appealing, especially in the analysis of mineral separates from iron meteorites, which, because of the small amounts available, require efficient systems for handling, characterizing, and determining the composition and minor-element distributions. Studies of such analytical techniques are at present in progress in the laboratories of the United States Geological Survey.

CONCLUSIONS

X-ray spectroscopy is a useful, specialized, and highly selective technique which often makes possible analyses considered impossible with other techniques. Like many other analytical techniques, X-ray fluorescence has limitations, but careful selection of conditions and standards will minimize or overcome virtually all these difficulties. The technique offers the chemist a tool of great utility and potential, particularly for the study of complex systems such as geologic materials.

REFERENCES

1. F. Claisse, "Accurate X-Ray Analysis Without Internal Standard," *Quebec Dept. Mines, Prog. Rept.* 327 (16 pp.), 1956.
2. G. Andermann, "Improvements in the X-Ray Analysis of Cement Raw Mix," *Anal. Chem.* **33**: 1689, 1961.
3. H. J. Rose, Jr., I. Adler, and F. J. Flanagan, "X-Ray Fluorescence Analysis of the Light Elements in Rocks and Minerals," *Appl. Spectry.* **17**: 81, 1963.
4. H. J. Rose, Jr., F. Cuttitta, and R. R. Larson, "Use of X-Ray Fluorescence in Determination of Selected Major Constituents in Silicates," *U.S. Geol. Surv. Profess. Papers* 525-B, p. B155, 1965.
5. F. Cuttitta and H. J. Rose, Jr., "The Slope-Ratio Technique for the Determination of Trace Elements by X-Ray Spectroscopy: A New Approach to Matrix Problems," *Appl. Spectry.*, in press.

6. E. F. Spano, T. E. Green, and W. J. Campbell, "Determination of Cobalt and Nickel in Tungsten by a Combined Ion-Exchange X-Ray Spectrographic Method," *U.S. Bur. Mines Rept. Invest.* 6308 (20 pp.), 1963.
7. E. F. Spano, T. E. Green, and W. J. Campbell, "Evaluation of a Combined Ion-Exchange X-Ray Spectrographic Method for Determining Trace Metals in Tungsten," *U.S. Bur. Mines Rept. Invest.* 6565 (16 pp.), 1964.
8. C. L. Luke, "Ultratrace Analysis of Metals with a Curved-Crystal X-Ray Milliprobe," *Anal. Chem.* **36**: 318, 1964.
9. W. J. Campbell, E. F. Spano, and T. E. Green, "Micro and Trace Analysis by a Combination of Ion-Exchange Resin-Loaded Papers and X-Ray Spectrography," *Anal. Chem.* **38**: 987, 1966.
10. G. L. Hubbard and T. E. Green, "Dithizone Extraction and X-Ray Spectrographic Determination of Trace Metals in High-Purity Tungsten or Tungsten Oxide," *Anal. Chem.* **38**: 428, 1966.
11. E. F. Spano and T. E. Green, "Determination of Metallic Impurities in Molybdenum by a Combined Ion-Exchange–X-Ray Spectrographic Method," *Anal. Chem.* **38**: 1341, 1966.
12. I. W. Mitchell, N. M. Saum, and C. L. Hiltrop, "Combined Electrolytic and X-Ray Spectrochemical Method for the Analysis of Gold," *Norelco Rept.* **11**: 39, 1964.
13. S. Natelson and P. K. De, "Application of X-Ray Emission Spectrometry to the Estimation of the Heavy Metals (At. No. 79–83)," *Microchem. J.* **7**: 448, 1963.
14. B. H. Vassos, F. J. Berlandi, T. E. Neal, and H. B. Mark, Jr., "Electrochemical Preparation of Thin Metal Films as Standards on Pyrolytic Graphite," *Anal. Chem.* **37**: 1653, 1965.

DISCUSSION

M. Boris (American Steel Foundries): Have you noticed that some of these cellulose powders have as much as 5% water and this water content varies from box to box?

H. J. Rose: Yes, there can be some variation.

M. Boris: How do you come to a standard matrix with that material?

H. J. Rose: You can let it stabilize by letting it sit in open air.

M. Boris: How do you control your humidity that way?

H. J. Rose: We have pretty good control in our lab.

M. Boris: You have an air-conditioned lab?

H. J. Rose: Yes. This is true; you must be careful of this.

M. Boris: How do you check the water? Do you use a vacuum oven to see that you do have a constant concentration of water?

H. J. Rose: Yes, you can check this. We do this periodically.

M. Boris: Cellulose powder is a good binder, but the water content affects the size of the briquette or the thickness and this is where you get big variations.

H. J. Rose: Obviously, then, you haven't any control in your lab, humidity or temperature. Is that right?

M. Boris: Yes. When you put this in a vacuum spectrometer, the water pumps out and you are being affected on the size of the sample.

H. J. Rose: Yes, this is true.

M. Boris: Well, then you are not using vacuum in most of your work, you are using helium. Is that correct?

H. J. Rose: I am using both. Our laboratory is pretty well controlled. Our humidity is quite low. We don't have any problem with the expansion of the pellet itself. Another technique, of course, would be to keep it in some dry atmosphere or set them actually in the vacuum chamber if it is large enough.

M. Boris: I've come close to getting polyethylene powder that has no water to work fairly well, but it has a low melting point and this makes it difficult to form good briquettes.

H. J. Rose: No, I haven't used any of those binders, but I don't seem to have any problem with cellulose powders.

W. Pojasek (Kennecott Copper Corporation): In the analysis of various minerals by X-ray fluorescence, do you feel that a separation of the minerals, for example, the ore silver–copper sulfide that you referred to, by using a previously performed heavy-liquid separation on the whole rock would eliminate some of the enhancement and absorption effects and thereby lead to greater analytical accuracy, and is the time spent worth doing that?

H. J. Rose: It certainly is. There is nothing that bothers me more than to analyze a sample that is a mixture. I try to get the geologists and mineralogists when they submit their samples to give me as clean a sample as possible. The cleaner they make it means less work for me to try to clear it up. When you stop to think about it, when you are analyzing a mixture, how much is the geologist or mineralogist getting out of the data? It seems he himself has to make some sort of an analytical determination to figure out how much contamination, how much silica, for example, is attributed to some other phase. It really is lost time when you consider that, if he made a good separation to begin with, he would be working with pure materials and the numbers would mean something.

COMMON SOURCES OF ERROR IN ELECTRON PROBE MICROANALYSIS

Kurt F. J. Heinrich

Analytical Chemistry Division
Institute for Materials Research
National Bureau of Standards
Washington, D.C.

ABSTRACT

In order to reduce the error of quantitative electron probe microanalysis, error sources in the preparation and measurement of specimens and standards must be minimized. These sources of error are described, and literature references for detailed study are given. A critical analysis is made of 150 analytical measurements of binary specimens previously discussed by Poole and Thomas.[4,5] It is shown that the cases of serious errors reported by these authors are mainly due to poorly characterized or measured specimens and, in some cases, to the omission of characteristic fluorescence corrections. If these sources of error are eliminated, a much more favorable error distribution can be obtained through relatively simple correction calculations. Further progress in quantitative microprobe analysis is dependent upon measurements under well-controlled conditions and standard materials of experimentally proven microhomogeneity and reliably determined composition.

INTRODUCTION

In his doctoral thesis,[1] Castaing gave the first full description of the electron probe microanalyzer which he constructed. He claimed that quantitative analysis could be performed with this instrument, and he corroborated this with the analysis of several alloys. A great number of applications have been published since,[2] and the importance of this method is now firmly established. The use of the microprobe has been extended to a much wider range of X-ray wavelengths and to specimens of diverse nature, ranging from alloys and minerals to biological materials. It was inevitable that in this process difficulties would be encountered. Many investigators have therefore reinvestigated the procedures to be followed in the analysis and in the interpretation of the measurements, and several modifications of the techniques originally established by Castaing and his collaborators[3] have been proposed. Most workers agree that Castaing's correction procedures are admirably sound and remain valid in principle. There is, however, considerable disagreement on several aspects of their application.

In the course of their investigation of the atomic number effect, Poole and Thomas[4,5] subjected the results of 150 measurements on binary specimens, obtained by several laboratories, to a variety of proposed correction procedures. The calculated concentrations were compared with the concentrations assumed to be true, on the basis of chemical analysis or assumed stoichiometry. The results show a disappointingly wide spread of error with all procedures, including that proposed by the authors (Figure 1). Since the

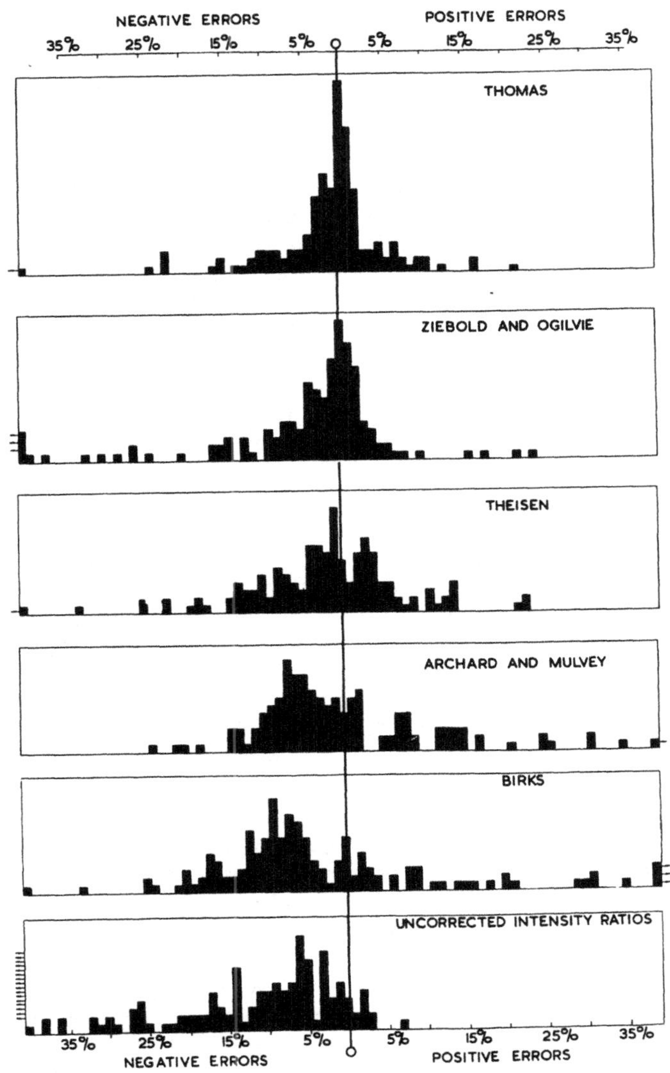

Figure 1. Error-distribution diagram of 150 analyses of binary specimens (relative error), reported by Poole and Thomas.[5]

specimens considered in this investigation represent simple binary systems, some doubt of the efficacy of quantitative microprobe analysis seems justified. While the available information is insufficient to pinpoint the sources of all errors, it can be shown that the conditions of analysis could be changed to advantage in several instances and that improvement in the calculation procedures is also possible. It must also be admitted that several aspects of X-ray generation and absorption are still in need of further investigation. The purpose of the present publication is to pinpoint the sources of common errors which can be avoided at the present state of knowledge and to indicate in which areas further experimental research is most urgently needed. It will also be shown that there is strong evidence, in several cases, that the assumed concentrations are themselves in error, so that Poole and Thomas' error distributions represent really a combination of the errors in microprobe analysis with those committed in the characterization of the test specimens.

MEASUREMENTS AND CALCULATIONS

The procedure followed in the analyses which were the object of Poole's study is as follows. The specimen is prepared to have a flat surface in the area to be examined. If it is a poor conductor of electricity or heat, it is coated with a thin layer of conductive material (usually carbon or aluminum). The intensity of a characteristic line of the element to be determined is measured. This measurement is corrected for the effects of dead time (coincidence losses) and for the background contribution. The corrected intensity is compared with the intensity emitted by a standard specimen of known composition excited under identical conditions. The surface preparation of the standard and the corrections applied to its measurement are the same as for the specimen.

It is difficult in most cases to provide homogeneous standards of known composition close to that of the specimen to be analyzed. For this reason, Castaing[1] used for his thesis research pure elements as standards. When the composition of the standard differs significantly from that of the specimen, the characteristics of X-ray emission and absorption are not the same. In such cases, corrective procedures must be applied which take into account the effects of the atomic number of the components upon the primary emission (atomic number effect), the emission of secondary X-rays (fluorescence corrections), and the absorption of the X-rays produced within the specimen (absorption correction). The magnitude of these corrections depends upon the composition of the specimen, which cannot be assumed to be known *a priori*. It is therefore necessary to calculate the specimen composition by a method of successive approximations.

The Specimen

The problem of obtaining a flat surface, the composition of which is representative of the original specimen, is by no means trivial. The main difficulties which arise in the specimen preparation are deviation from flatness (especially at edges and at interfaces of phases differing in hardness), smearing, imbedding of polishing compounds in soft surfaces, and transport or removal of elements in etching procedures. This subject is covered in recent publications by Yakowitz[6] and by Picklesimer and Hallerman.[7]

Attention must be paid to the possibility of changes of the specimen conditions during exposure to the electron beam. Deposition of carbonaceous or sulfur-containing materials may occur.[8,9] This may produce a progressive loss of signal intensity, particularly when X-rays of long wavelength are measured and when long exposures are necessary to observe trace amounts. Decomposition of specimens and volatilization of elements having a relatively high vapor pressure (*e.g.*, potassium)[10] have also been observed.

The Standard

Since the relative accuracy of X-ray intensity measurements is usually not better than 0.2 to 0.5%, obtaining elementary standards is not difficult provided that the absence of substantial amounts of impurities has been verified. Standards other than pure elements must be used, however, when the pure elements are not solids having low vapor pressure, when the analysis is performed with the aid of empirical calibration procedures,[11] or when such standards are used to test the accuracy of proposed correction procedures.[4,5,12,13] Many arguments have arisen concerning the quality of standards of this type. Most stoichiometric compounds have low electrical conductivity or other undesirable properties. On the other side, the stoichiometry and the narrow range of solubility of intermetallic compounds indicated in many equilibrium phase diagrams cannot be taken for granted. In alloys within the solubility range of a phase, homogeneity cannot be assumed unless proven by microprobe analysis of many points, and the composition of such alloys must be established by careful chemical analysis. When data obtained on such standards are published, the means of characterization of the standards should be explicitly stated. If this is not done or if the technique of characterization is subject to any doubt, no conclusions can be drawn from the subsequent microprobe measurements. This is the main reason the empirical method of Ziebold and Ogilvie[11] is not applied more generally.

An advance in this area was achieved by Goldstein, Majeske, and Yakowitz,[14] who prepared microscopically homogeneous alloys by modification of a splat cooling technique previously developed by Duwez.[15] The molten alloy is cooled quickly by contact with a massive cold plate. Goldstein *et al.* showed by microprobe analysis that specimens of acceptable homogeneity can be obtained far outside the normal solubility range (*e.g.*, in the gold–silicon system). They also observed, however, that some of the metastable alloys (*e.g.*, in the magnesium–aluminum system) segregate at room temperature within a few weeks. In spite of this difficulty, splat cooling, with careful chemical analysis of the product, is a promising way to obtain composite standards. Further applications of this technique are under way both at NASA (Goddard Space Flight Center) and at the National Bureau of Standards.

The Measurement

Although the technique of the intensity measurement of X-ray emission in the electron probe is seldom discussed in the literature, this subject should be given careful attention. The most obvious limitation of accuracy is set by the statistics of X-ray counting. Since the Poisson statistics[16] apply to the counting of photons produced randomly in time, the square root of the number of counts accumulated is an estimate of the standard deviation of a photon count *provided that all other sources of error are of negligible magnitude*. When major concentrations are measured, it is usually easy to accumulate sufficient counts to keep this source of uncertainty small. The consideration of Poisson statistics becomes important, however, when trace concentration levels are measured; in this case, the combination of errors in the measurement of both line and background must be taken into account.[17,18] The counting statistics also become a limiting factor in techniques involving the measurement at many points of the specimen[19] and in concentration mapping,[20] since the number of counts accumulated for each point may be quite limited.

When large numbers of photons are accumulated and particularly in protracted operations, errors due to drift become apparent. Besides specimen contamination, drift of X-ray emission intensity may be due to warping of the gun filament, to variations in the power supply outputs for operating voltage and condenser lens, and to variations in the detector efficiency (particularly in flow detectors not provided with pressure regulation).[21] Temperature changes in the spectrometers can also change the spacing of the crystals and hence produce defocusing.[22]

Repositioning of both the spectrometer and the specimen stage may cause variation in the observed intensities. Since curved-crystal spectrometers focus on a point on the Rowland circle,[23] variations of specimen elevation produce intensity variations of the signal. In order to minimize this effect, the spectrometer should be aligned in such a way that the emitted intensity reaches a maximum at the elevation corresponding to focusing conditions of the light microscope. In subsequent operations, the light image of the specimen through the microscope should remain sharply focused when the specimen is translated mechanically. In operations involving displacement of the electron beam, defocusing of the spectrometer must also be avoided.[24]

Defocusing of the X-ray spectrometers may also occur owing to chemical shifts of lines and bands in the soft X-ray region.[25] Neglect of this effect may produce very serious errors in the analysis of low-atomic-number elements (*e.g.*, silicon, aluminum, and magnesium) if the composition of the standard differs appreciably from that of the specimen.

If, in a series of measurements, the observed signal intensities vary considerably, the possibility of pulse-height shrinkage at high counting rates[26–28] must be considered. Problems due to this effect can be avoided by limiting the maximum counting rate and by raising sufficiently the mean pulse height of the amplified signal above the threshold of the discriminator or pulse-height analyzer, so that a slight pulse-height shrinkage does not result in loss of pulse detection.

If the mentioned causes of errors are taken into consideration, it will be possible to maintain the precision of the measurement within acceptable limits. It is a good practice however, to take these limits into account when results are reported. To publish or transmit results of microprobe analysis to four significant places is inappropriate and misleading.

Corrections for Dead Time and Background

During a short period after the detection of an X-ray photon, the detector system is inactive, so that pulses arriving within this interval are not registered. This period is called *dead time* or *paralysis time.* Since the probability of a second pulse arriving within the period of inactivity increases with increasing counting rates, the resulting coincidence losses also increase with increasing counting rates. The technique of determining the dead time of a counter system and applying the correction is described in detail in a previous publication.[29] Most procedures assume that the dead-time period is invariant with respect to counting rates. This must be proved in each case, as described in that earlier paper.[29]

In order to obtain the net intensity ratios of characteristic radiation, a background correction must be applied. Except for line interferences, the background registered by the X-ray spectrometer should be mainly due to continuous radiation, although background due to detector noise and scattered radiation may also be present. Since the intensity of the continuous radiation varies with both the atomic number of the target and the corresponding X-ray absorption coefficient, it is necessary to determine the background level for the materials emitting the line of interest. It was proposed to measure this level close to the analytic line[30] or on materials of atomic number similar to that of the specimen.[31] The possibility of effects of satellite lines[30] and absorption edges must be considered. In serial measurements, it is inconvenient to make background measurements for each point, and it would be desirable to develop methods for calculating the background levels of composite targets. Some research in this direction was performed by Moreau and Calais, who investigated the intensity of continuous radiation of binary targets,[32] but further work on the background problem is needed.

Absorption Correction

Several authors proposed algebraic models for correcting the effects of X-ray absorption within the target.[4,33,34] A comparison with experimental data concerning $f(\chi)$, the absorption function,[35] showed that Philibert's equation[33]

$$\frac{1}{f(\chi)} = \left(1 + \frac{\chi}{\sigma}\right)\left(1 + a\frac{\chi}{\sigma}\right) \tag{1}$$

yields satisfactory results. Notation is as follows: $\chi = (\mu/\rho)$ cosec θ; μ/ρ is the mass attenuation coefficient of the specimen for the measured radiation; θ is the X-ray emergence angle; $a = 6A/(5Z^2 + 6A)$; Z is the atomic number of the target; and A is its atomic mass. For complex targets, it can be assumed that

$$a = \sum C_i a_i \tag{2}$$

where C_i are the mass fractions and a_i are the values of a of the elements composing the target. The term σ is an *electron absorption coefficient*, dependent upon both the operating voltage and the minimum excitation potential of the atomic level producing the radiation in question. Duncumb and Shields[36] proposed the following expression for σ:

$$\sigma = \frac{2.39 \times 10^5}{V^{1.5} - E^{1.5}} \tag{3}$$

A study of experimental data on the function $f(\chi)$ indicated[35] that the equation

$$\sigma = \frac{4.25 \times 10^5}{V^{1.67} - E^{1.67}} \tag{4}$$

produces slightly better results for elements of low atomic number. In both expressions, V is the operating voltage and E is the minimum excitation potential. Errors in the absorption correction may arise from errors in the model or in the input parameters. Since published data on X-ray mass-attenuation coefficients are frequently in poor agreement, we produced a table of mass-attenuation coefficients covering the main range of interest for microprobe analysis.[37] The values contained in this table were obtained by interpolation from selected experimental data and are believed to be accurate to 10% or better; further experimental work in this area is presently under way. A study of the error propagation in the absorption correction by Yakowitz and Heinrich[38] shows that, even with the best available parameters, significant errors may result from this correction unless the value of the factor $f(\chi)$ can be maintained above 0.8. This can be achieved in most practical cases by using high X-ray emergence angles and relatively low operating voltages. The possibility of serious errors in the absorption correction cannot be eliminated when radiation of wavelength above 10 Å is used.

Poole and Thomas[4] use a modified version of Philibert's absorption correction. A comparison of the results obtained by these authors with the results of the original correction as proposed by Philibert, with the use of either Duncumb and Shield's equation for σ or that proposed by us, suggests that the original Philibert correction is capable of the same or better accuracy.

Fluorescence Correction

Several models for correcting the effects of secondary excitation by characteristic emission have been proposed.[39,40] The results are significantly affected by the choice of a model,[39,41] but their experimental evaluation cannot be considered conclusive. Error propagation[41] indicates that, besides the choice of a model, the uncertainty in the values of the fluorescence yields may introduce significant errors. A recent publication of Fink *et al.*[42] appears to be the most reliable source of fluorescence yield values.

The correction for fluorescence of L lines excited by K lines and *vice versa* can be performed by using the model proposed by Reed.[40] The values obtained by this model indicate that these effects can be quite appreciable, although they are frequently not taken into account. This correction has been neglected, for instance, in the systems of nickel–platinum and copper–gold in the previously mentioned work of Poole and Thomas.[5] The accuracy of the correction procedure cannot be expected to be very high, since fluorescence yields in the L shell are only poorly known and since the possibility of significant deficiencies in the model cannot be discounted. This is an area in which further experimental and theoretical studies are needed.

Secondary emission can also be excited by continuous radiation. This subject has been treated by Green [43] and by Hénoc *et al.*,[44] who stated that the effects of fluorescence excited by the continuum can in general be neglected. Since the calculation of this fluorescence contribution is rather involved, the corresponding correction is seldom applied in practice; however, since the accuracy requirements tend to become more exacting, a critical review of this field would be desirable.

Atomic Number Correction

It was assumed originally that the number of primary X-ray photons generated within the specimen was, within experimental error, proportional to the mass concentration of the emitting element.[1] A closer analysis of the processes of electron scattering and deceleration and of the mechanism of X-ray generation indicated that this "first approximation" could not be strictly valid. Castaing had proposed[3] that, in targets composed of elements differing considerably in atomic numbers, the X-ray emission would be related to concentration by

$$k = \frac{\alpha_A C_A}{\Sigma \alpha_i C_i} \tag{5}$$

where k is the intensity ratio of primary X-rays generated within the specimen (*i.e.*, the ratio obtained after applying all the corrections mentioned previously) and C_A is the concentration of the emitting element. The factor α_i may be assumed to be formed by the relation

$$\alpha_i = \frac{S_i}{R_i} \tag{6}$$

in which S_i represents the "electron stopping power" and R_i is the loss of potential X-ray production due to the escape of backscattered electrons. Poole and Thomas[4,5] derived values of R from experimental data on the energy distribution of backscattered electrons and used for S the stopping-power values of Nelms' table,[45] at an electron potential halfway between the operating voltage and the minimum excitation voltage.

This model of the atomic number effect is a drastic simplification of a very complex situation in which the deceleration and change of direction of the electrons, the variation of the ionization cross section as a function of electron energy, and the loss of backscattered electrons covering a wide energy range combine in a sequence of randomly occurring events. A more satisfactory description of these target events can be given through extensive calculations of the Monte Carlo type.[46] Since these cannot be applied to routine analyses at the present state of the art of computation, it is worthwhile to reinvestigate whether simpler models are capable of providing the corrections needed in practice.

For this purpose, we have reexamined the data used by Poole and Thomas. It will be seen in what follows that a considerable fraction of the large errors remaining in the set of results obtained by Poole and Thomas is due to errors not connected with the atomic number effect. After removing these errors, a more satisfactory error distribution can be obtained with only slight modification of the procedure advocated by Poole and Thomas.

Reexamination of the Data Used by Poole and Thomas

Since we have no information concerning the techniques of measurement applied to obtain the data used by Poole and Thomas, we can draw no general conclusions as to the possible effects of errors in the characterization and measurement of the standard used. Consistency tests can be applied, however, to groups of measurements with change in one variable only. When a series of binaries of the same end members is analyzed under identical conditions, it is advantageous to plot for this purpose the ratio k/C as a function of k. Here, k is the background and dead-time corrected signal-intensity ratio, and C is the concentration of the element measured. According to Ziebold and Ogilvie,[11] this plot should, in the absence of strong fluorescence effects, yield a straight line tending to unity for $C = 1$. At any event, a smooth curve should be obtained. When the same alloy is analyzed at varying voltages, k/C should vary smoothly as a function of operating voltage. A test of the systems used in Poole and Thomas's tabulation shows some lack of smoothness for varying concentrations (Figures 2 to 5), while the corresponding curves for specimens analyzed at varying voltage are quite smooth (Figure 6). This suggests the presence of errors in the chemical determination of composition. Systematic errors in either the chemical or the microprobe analysis are not detected by this procedure. In view of the importance of the absorption correction, we have recalculated this correction for all cases, using the equation proposed by Philibert, equation (1). The mass absorption coefficients were those previously published.[37] The value of σ was calculated in each case according to both equations (3) and (4).

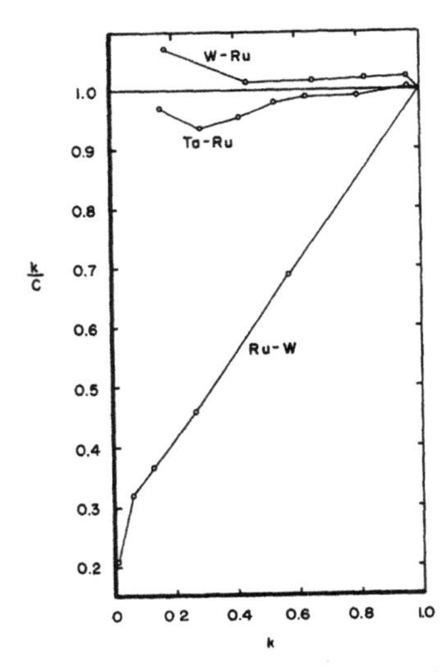

Figure 2. Consistency test of data obtained on the binary systems W–Ru, Ta–Ru, and Ru–W[5] for the element listed first.

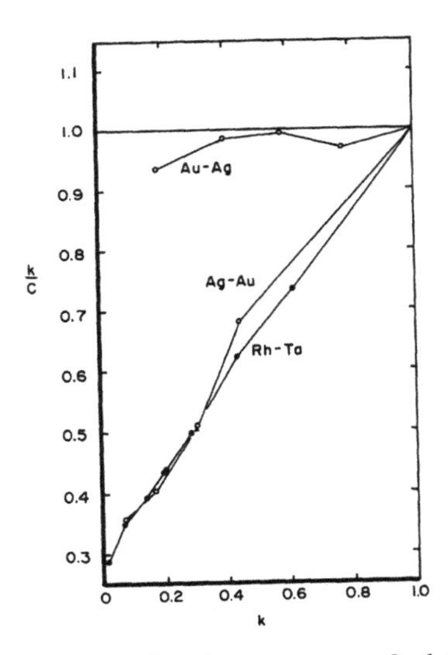

Figure 3. Consistency test of data obtained on the binary systems Au–Ag, Ag–Au, and Rh–Ta[5] for the element listed first.

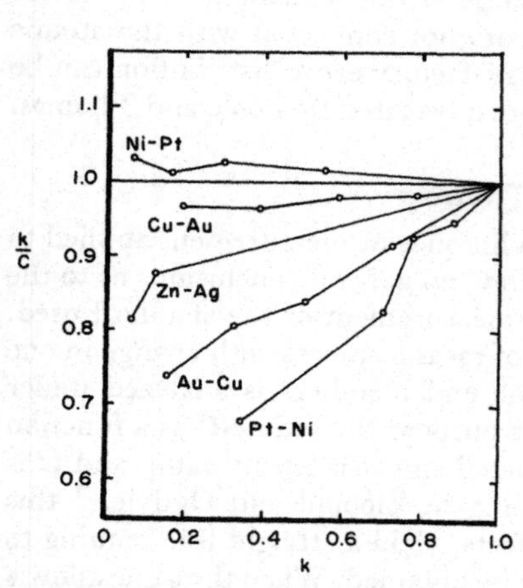

Figure 4. Consistency test of data obtained on the binary systems Ni–Pt, Pt–Ni, Cu–Au, Au–Cu, and Zn–Ag[5] for the element listed first.

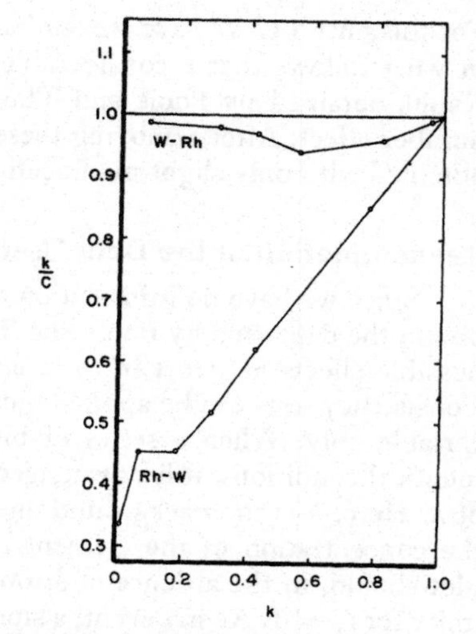

Figure 5. Consistency test of data obtained on the binary system W–Rh[5] for the element listed first.

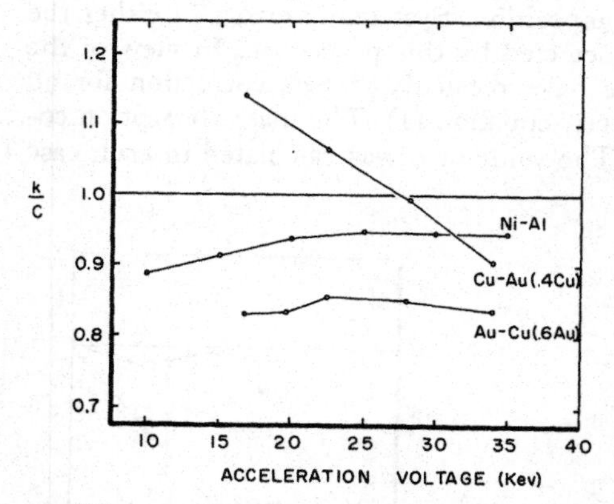

Figure 6. Consistency test of data obtained on the binary alloys $NiAl_3$ and 0.6 Au–0.4 Cu for measurements of the element listed first, at various accelerating voltages.[5]

The set of data employed contains several instances of fluorescence of *K* lines by *L* lines. We performed the fluorescence correction omitted by Poole and Thomas, using the equation proposed by Reed. The factor P_{ij} contained in Reed's equation[40] to adapt the general case to the *L–K* type of fluorescence is 4.2. We have changed this factor to 2.4, on the basis of considerations derived from Green's thesis.[47] Further experimental work on this subject will show if our modification was correct.

The atomic number correction was performed by the approach proposed by Poole and Thomas, but the values for *S* and *R* were obtained in a slightly different way. The values for the backscatter coefficients are those obtained by Duncumb and Reed,[48] who used the backscatter coefficients obtained by Bishop and the energy distributions of backscattered electrons published by Kulenkampff and Spyra. For the stopping power we used the same source as Poole and Thomas.[4,5] These authors observed some difficulties in determining the stopping power for elements not tabulated by Nelms,[45] since Nelms' data do not vary in a continuous fashion as a function of atomic number. Bethe's

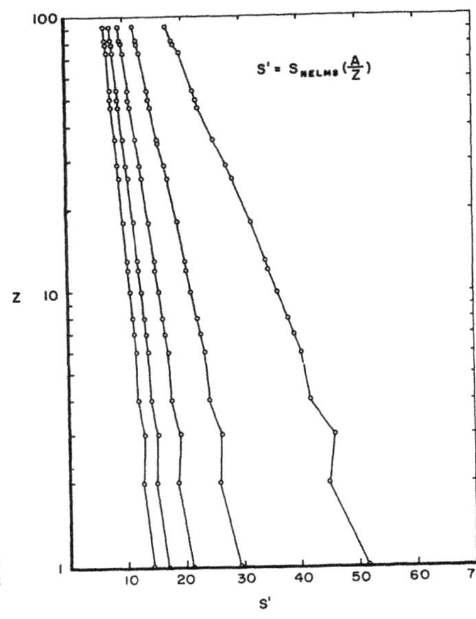

Figure 7. Plot of the stopping-power coefficient S', derived from Nelms' data,[45] as a function of atomic number.

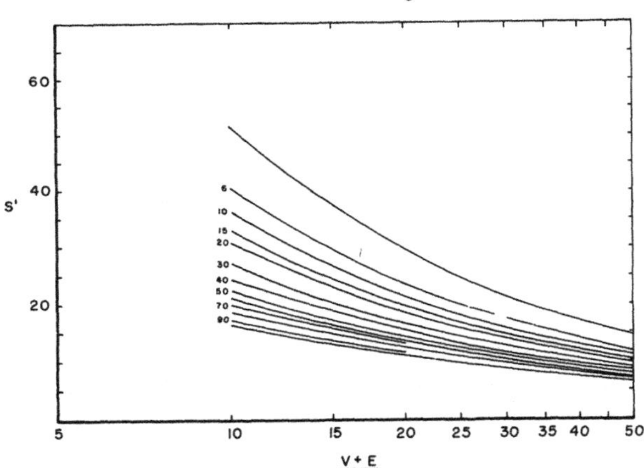

Figure 8. Plot of S' as a function of the mean electron energy in kilo-electron volts.

equation which was used by Nelms for computing the stopping power contains the factor (Z/A). By multiplying Nelms' value by (A/Z), a function which varies smoothly with atomic number is obtained (Figures 7 and 8). We used this function S' to interpolate for the elements not obtained in Nelms' values, and then multiplied all values by (Z/A), so that α can be expressed by

$$\alpha_i = \frac{S_i' Z_i}{R_i A_i} \tag{7}$$

A marked improvement is obtained in this fashion for the analysis of $TiAl_3$, for which, by Poole's method, one obtains an unexpectedly high α factor (1.003 for 15 keV). Our procedure yields an α factor equal to 0.889 at 15 keV, which is more in line with the factors obtained for similar binaries. The resulting analytical error is greatly diminished.

The error distribution of the new sets of calculated results (Figure 9) shows some

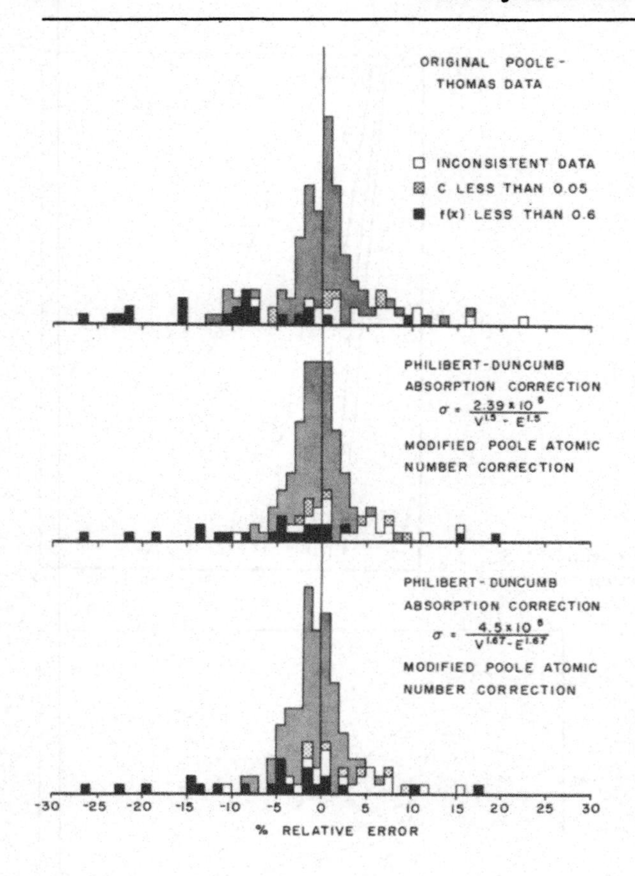

Figure 9. Error distribution of the data shown in Figure 1 with the results of our recalculation. The adverse effect of excessive absorption correction and of inconsistent data groups can be observed.

improvement over the results obtained by Poole and Thomas. However, an alarming percentage of gross errors is still present.

In view of the results of the previously described tests for consistency on alloy series, we decided to observe the effects of excluding inconsistent groups of data from the error distribution test. Similarly, on the basis of error propagation in the absorption correction, we separated all measurements in which the factor $f(\chi)$ was smaller than 0.6. Although this criterion is less strict than the one proposed previously [$f(\chi) \geq 0.8$], this should exclude the most serious errors due to X-ray absorption. Finally, measurements of concentration below 5% were excluded since, at this level, the same relative accuracy as for higher concentrations cannot be expected. It should be noted that none of the measurements we have excluded represents very large atomic number differences.

The error distribution after exclusion of the aforementioned categories (Figure 10) shows a very significant improvement over both the original results of Poole and Thomas and the modified procedure used by us. The differences arising from the use of equations (3) or (4) in calculating σ are not significant since measurements on low-atomic-number elements are not represented in this group. It appears that the Philibert equation with Duncumb and Shield's expression for σ [equation (3)] can be used with confidence within the limits for the value of $f(\chi)$ indicated by Yakowitz and Heinrich[38] [$f(\chi) \geq 0.8$]. In fact,

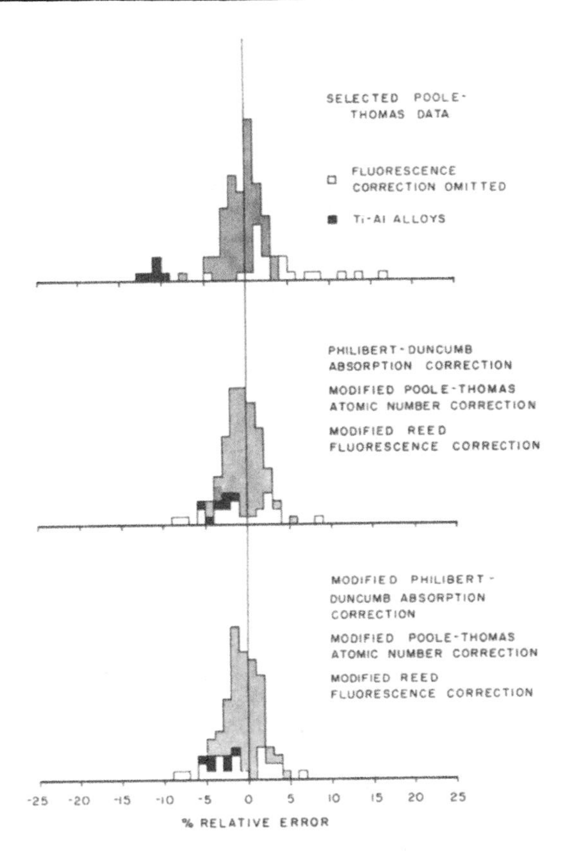

Figure 10. Error distribution of selected data from Poole and Thomas.[5] The effects of the fluorescence correction and of the improved atomic number correction can be observed.

the original results obtained by Poole and Thomas for the cases not rejected for excessive absorption correction or inconsistency are not significantly worse than those of the two other methods, with exception of the analysis of $TiAl_3$ and the cases where fluorescence corrections were omitted. It should also be noted (Table I) that there is no evidence of increasing errors at low operating voltages due to failure of the atomic number correction. It is thus clear that the most severe errors in the original sets of results are due, not to the atomic number correction, but to errors in the characterization or measurement of the standards, to the omission of the fluorescence correction, and to an excessive absorption correction. The last error might have been avoided if low voltages had been used in the analysis.

A plot of the errors of the residual group of measurements as a function of the atomic numbers of the constituents fails to indicate significant trends (Figure 11). We believe that the prospects for further progress with this set of data are quite remote. Since any proposed method of data treatment requires an experimental test, it is concluded that further progress in quantitative microprobe analysis is contingent upon measurements with well-controlled and documented conditions and standard materials of experimentally proved microhomogeneity and reliably determined composition.

Table I. Relative Error As a Function of Operating Voltage*

System	C_A	kV	ϵ_{pt}	ϵ_{pd}	ϵ_h
Au–Cu	0.600	16.8	0.000	−0.008	−0.010
		19.8	0.000	−0.010	−0.012
		22.5	0.017	0.008	0.008
		28.2	0.007	0.015	0.015
		33.9	0.007	0.015	0.015
Cu–Au	0.400	16.8	0.018	−0.018	−0.023
		22.5	0.013	−0.028	−0.033
		28.2	0.020	−0.040	−0.043
		33.9	−0.005	−0.081	−0.083
Ni–Al	0.421	10	−0.017	−0.031	−0.031
		15	−0.029	−0.040	−0.040
		20	−0.017	−0.029	−0.029
		25	−0.010	−0.021	−0.021
		30	−0.017	−0.036	−0.036
		35	−0.024	−0.026	−0.026
	0.592	10	−0.017	−0.027	−0.029
		15	−0.021	−0.029	−0.029
		20	−0.004	−0.013	−0.012
		25	0.002	−0.008	−0.008
		30	0.004	−0.017	−0.019
		35	−0.009	−0.010	−0.010
Fe–Al	0.408	10	−0.020	−0.005	−0.005
		15	−0.042	−0.020	−0.020
		20	−0.037	−0.017	−0.020
		25	−0.027	−0.010	−0.012
		30	−0.020	−0.002	−0.002
		35	−0.015	0.001	0.002
Ti–Al	0.372	10	−0.110	−0.024	−0.030
		15	−0.110	−0.019	−0.016
		20	−0.094	−0.022	−0.027
		25	−0.105	−0.038	−0.043
		30	−0.116	−0.043	−0.046
		35	−0.124	−0.056	−0.056
Fe_2O_3	0.697	12	−0.011	0.010	0.010
		15	−0.009	0.010	0.010
		20	−0.009	0.007	0.009
		25	−0.001	0.011	0.011
		29	−0.001	0.011	0.011

* The element listed first under the column "System" is the one which was determined. The $L\alpha_1$ line was used for gold, and the $K\alpha$ line, in all other cases. The concentrations are expressed as fractions, and the errors, by dividing the absolute error by the nominal composition. The listed errors correspond to the following methods: ϵ_{pt}, results reported by Thomas;[4] ϵ_{pd}, Philibert–Duncumb absorption correction, modified Reed fluorescence correction where applicable, and modified Poole–Thomas atomic number correction; ϵ_h, Philibert-modified-Duncumb absorption correction with the use of equation (4), modified Reed fluorescence correction where applicable, and modified Poole–Thomas atomic number correction.

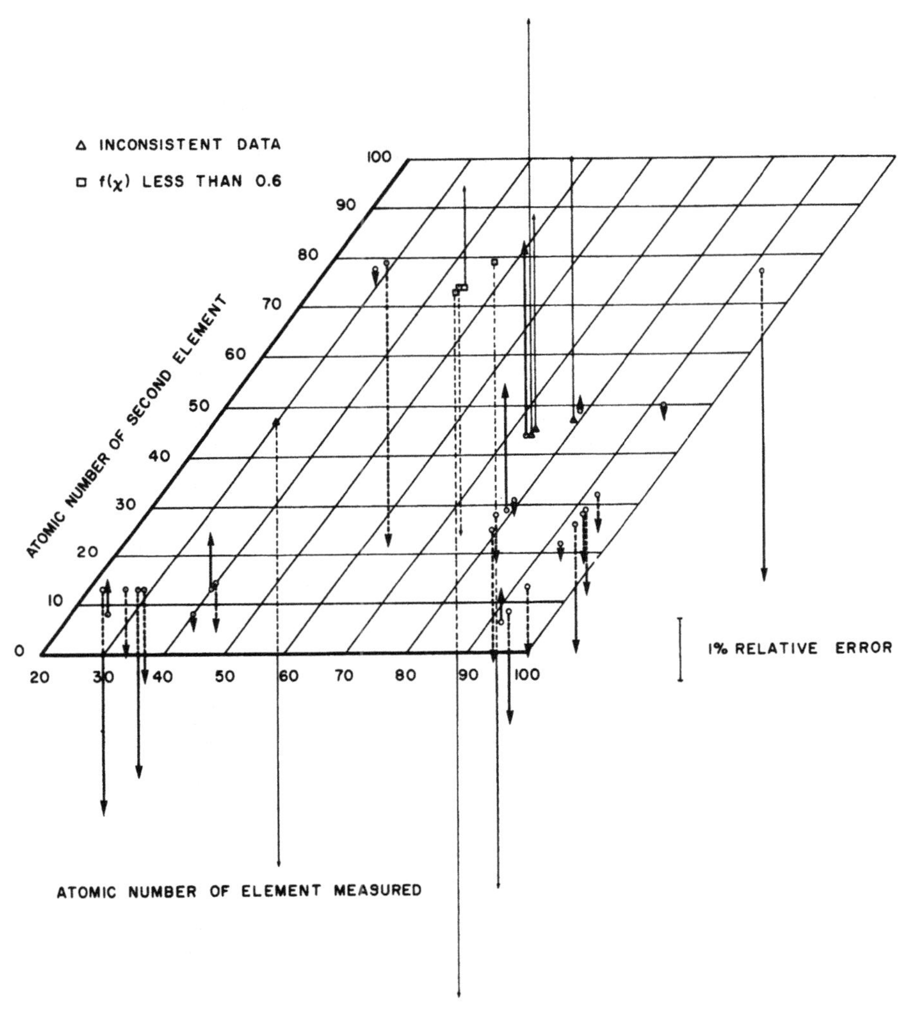

Figure 11. Average errors of binary systems as a function of atomic numbers of the components, modified Philibert–Duncumb absorption correction, modified Poole–Thomas atomic number correction, and modified Reed fluorescence correction.

REFERENCES

1. R. Castaing, "Application des sondes électroniques à une méthode d'analyse ponctuelle chimique et cristallographique" (Application of Electron Beams to a Method for Local Chemical and Crystallographical Analysis), Thesis, University of Paris, 1951.
2. K. Heinrich, *Bibliography on Electron Probe Microanalysis and Related Subjects*, third edition, E. I. DuPont de Nemours & Co., Inc., Wilmington, Delaware, 1963; Supplement, 1965. (Obtainable from the author.)
3. R. Castaing and J. Descamps, "Sur les bases physiques de l'analyse ponctuelle par spectrographie x" (On the Physical Bases of Point Analysis by X-Ray Spectrography), *J. Phys. Radium* **16**: 304, 1955.
4. P. M. Thomas, "A Method for Correcting for Atomic Number Effects in Electron Probe Microanalysis," *At. Energy Res. Estab. (Gt. Brit.) Rept.* 4593, 1964.
5. D. M. Poole and P. M. Thomas, "Correction of Atomic Number Effects in Microprobe Analysis," in: T. D. McKinley, K. F. J. Heinrich, and D. M. Wittry (eds.), *The Electron Microprobe*, John Wiley & Sons, Inc., New York, 1966, p. 269.

6. H. Yakowitz, "Evaluation of Specimen Preparation and the Use of Standards in Electron Probe Microanalysis," *ASTM Spec. Tech. Publ.* **430**: 383, 1968.
7. M. L. Picklesimer and G. Hallerman, "The Influence of the Preparation of Metal Specimens on the Precision of Electron Probe Microanalysis," *U.S. At. Energy Comm. Rept.* ORNL-TM-1591, 1966.
8. I. Adler, E. J. Dwornik, and H. J. Rose, Jr., "The Detection of Sulphur in Contamination Spots in Electron Probe X-Ray Microanalysis," *Brit. J. Appl. Phys.* **13**: 245, 1962.
9. G. V. T. Ranzetta and V. D. Scott, "Specimen Contamination in Electron Probe Microanalysis and its Prevention Using a Cold Trap," *J. Sci. Instr.* **43**: 816, 1966.
10. M. P. Borom and R. E. Hanneman, "Local Compositional Changes in Alkali Silicate Glasses During Electron Microprobe Analysis," *General Electric Rept.* 66-C-484, 1966; *J. Appl. Phys.* **38**: 2406, 1967.
11. T. O. Ziebold and R. E. Ogilvie, "An Empirical Method for Electron Microanalysis," *Anal. Chem.* **36**: 322, 1964.
12. J. W. Colby, "The Applicability of Theoretically Calculated Intensity Corrections in Microprobe Analysis," in: T. D. McKinley, K. F. J. Heinrich, and D. B. Wittry (eds.), *The Electron Microprobe*, John Wiley & Sons, Inc., New York, 1966, p. 95.
13. D. R. Beaman, "Evaluation of Correction Procedures Used in Electron Probe Microanalysis with Emphasis on Atomic Number Interval 13 to 33," *Anal. Chem.* **39**: 418, 1967.
14. J. I. Goldstein, F. J. Majeske, and H. Yakowitz, "Preparation of Electron Probe Microanalyzer Standards Using a Rapid Quench Method," in: J. B. Newkirk and G. R. Mallett (eds.), *Advances in X-Ray Analysis, Vol. 10*, Plenum Press, New York, 1967, p. 431.
15. J. Duwez, R. H. Willens, and W. Klement, Jr., "Continuous Series of Metastable Solid Solutions in Ag–Cu Alloys," *J. Appl. Phys.* **31**: 1136, 1960.
16. H. A. Liebhafsky, H. G. Pfeiffer, and P. D. Zemany, "Precision in X-Ray Emission Spectrography," *Anal. Chem.* **26**: 1257, 1955.
17. K. F. J. Heinrich, "Count Distribution and Precision in X-Ray Fluorescence Analysis," in: W. M. Mueller (ed.), *Advances in X-Ray Analysis, Vol. 3*, Plenum Press, New York, 1959, p. 95.
18. M. Mack and N. Spielberg, "Statistical Factors in X-Ray Intensity Measurements," *Spectrochim. Acta* **12**: 169, 1958.
19. L. S. Birks and A. P. Batt, "Use of a Multichannel Analyzer for Electron Probe Analysis," *Anal. Chem.* **35**: 778, 1963.
20. K. F. J. Heinrich, "Concentration Mapping Device for the Scanning Electron Probe Microanalyzer," *Rev. Sci. Instr.* **33**: 884, 1962.
21. R. D. Deslattes, B. G. Simson, and R. E. LaVilla, "Gas Density Stabilizer for Flow Proportional Counters," *Rev. Sci. Instr.* **37**: 596, 1966.
22. T. A. Davies, "The Effect of Variations in Ambient Temperature upon the Optical Alinement of an X-Ray Fluorescence Spectrometer," *J. Sci. Instr.* **35**: 407, 1958.
23. R. E. Ogilvie, "X-Ray Optics in Electron Microanalysis," *ASTM Spec. Tech. Publ.* **349**: 17, 1963.
24. H. Malissa, *Elektronenstrahl-Mikroanalyse*, Springer-Verlag, Vienna and New York, 1966, p. 98.
25. W. L. Baun and D. W. Fischer, "The Effect of Valence and Coordination on *K* Series Diagram and Nondiagram Lines of Magnesium, Aluminum, and Silicon," in: W. M. Mueller, G. R. Mallett, and M. J. Fay (eds.), *Advances in X-Ray Analysis, Vol. 8*, Plenum Press, New York, 1965, p. 371.
26. S. L. Bender and E. J. Rapperport, "Nonproportional Behavior of the Flow Proportional Detector," in: T. D. McKinley, K. F. J. Heinrich, and D. B. Wittry (eds.), *The Electron Microprobe*, John Wiley & Sons, Inc., New York, 1966, p. 405.
27. N. Spielberg, "Elimination of Intensity Dependent Shifts in Proportional Counter Pulse Height Distributions," *Rev. Sci. Instr.* **37**: 1268, 1966.
28. N. Spielberg, "Effect of Anode Material on Intensity Dependent Shifts in Proportional Counter Pulse Height Distributions," *Rev. Sci. Instr.* **38**: 291, 1967.
29. K. F. J. Heinrich, D. Vieth, and H. Yakowitz, "Correction for Non-Linearity of Proportional Counter Systems in Electron Probe X-Ray Microanalysis," in: G. R. Mallett, M. J. Fay, and W. M. Mueller (eds.), *Advances in X-Ray Analysis, Vol. 9*, Plenum Press, New York, 1966, p. 208.
30. J. Philibert, "L'analyse quantitative en microanalyse par sonde électronique, troisième partie (Quantitative Analysis in Microanalysis by the Electron Probe), *Metaux Corrosion-Ind.* **40**: 325, 1964.
31. T. O. Ziebold, "The Electron Microanalyzer and Its Applications," Lecture Notes, Massachusetts Institute of Technology, Summer Session, 1965, S-5.

32. G. Moreau and D. Calais, "Détermination du numéro atomique moyen d'un binaire homogène AB (solution solide ou composé défini)" [Determination of the Mean Atomic Number of a Homogeneous Binary AB (Solid Solution or Definite Composition)], *J. Phys. Radium* **25**: 83A, 1964.
33. J. Philibert, "A Method for Calculating the Absorption Correction in Electron-Probe Microanalysis," in: H. H. Pattee, V. E. Cosslett, and A. Engström (eds.), *X-Ray Optics and X-Ray Microanalysis*, Academic Press, New York, 1963, p. 379.
34. R. Theisen, *Quantitative Electron Microprobe Analysis*, Springer-Verlag, New York, 1965.
35. K. F. J. Heinrich (to be published).
36. P. Duncumb and P. K. Shields, "Effect of Critical Excitation Potential on the Absorption Correction," in: T. D. McKinley, K. F. J. Heinrich, and D. B. Wittry (eds.), *The Electron Microprobe*, John Wiley & Sons, Inc., New York, 1966, p. 284.
37. K. F. J. Heinrich, "X-Ray Absorption Uncertainty," in: T. D. McKinley, K. F. J. Heinrich, and D. B. Wittry (eds.), *The Electron Microprobe*, John Wiley & Sons, Inc., New York, 1966, p. 296.
38. H. Yakowitz and K. F. J. Heinrich, "Quantitative Electron Probe Microanalysis: Absorption Correction Uncertainty," *Mikrochim. Acta*, p. 182, 1968.
39. P. Duncumb and P. K. Shields, "Calculation of Fluorescence Excited by Characteristic Radiation in the X-Ray Microanalyzer," in: H. H. Pattee, V. E. Cosslett, and A. Engström (eds.), *X-Ray Optics and X-Ray Microanalysis*, Academic Press, New York, 1963, p. 329.
40. S. J. B. Reed, "Characteristic Fluorescence Corrections in Electron-Probe Microanalysis," *Brit. J. Appl. Phys.* **16**: 913, 1965.
41. K. F. J. Heinrich and H. Yakowitz, "Quantitative Electron Probe Microanalysis: Fluorescence Correction Uncertainty," *Mikrochim. Acta*, 1968 (in press).
42. R. F. Fink, R. C. Jopson, H. Mark, and C. D. Swift, "Atomic Fluorescence Yields," *Rev. Mod. Phys.* **38**: 513, 1966.
43. M. Green, "The Efficiency of Production of Characteristic X-Radiation," Thesis, University of Cambridge, Great Britain, 1962.
44. J. Hénoc, F. Maurice, and A. Kirianenko, "Microanalyseur à sonde électronique, étude de la correction de fluorescence due au spectre continu" (Electron Probe Microanalyzer: Study of the Correction for Fluorescence Due to the Continuous Spectrum), *Comm. Énergie At. (France), Rappt.* CEA-R 2421, 1964.
45. A. T. Nelms, "Energy Loss and Range of Electrons and Positrons," *NBS Circ.* 577, 1956; *Suppl. NBS Circ.* 577, 1958.
46. H. E. Bishop, "Calculations of Electron Penetration and X-Ray Production in a Solid Target," in: R. Castaing, P. Deschamps, and J. Philibert (eds.), *X-Ray Optics and Microanalysis*, Hermann, Paris, 1966, p. 112.
47. H. Yakowitz, private communication.
48. P. Duncumb and S. J. B. Reed, "Progress in the Calculation of Stopping Power and Backscatter Effects," *Quantitative Electron Probe Microanalysis, Natl. Bur. Std. Spec. Publ.* **289**, April 1968 (in press).

X-RAY SPECTROGRAPHIC ANALYSIS OF TRACES IN METALS BY PRECONCENTRATION TECHNIQUES

C. M. Davis, Keith E. Burke, and M. M. Yanak

The International Nickel Company, Inc.
Paul D. Merica Research Laboratory
Sterling Forest, Suffern, New York

ABSTRACT

Chemical separation techniques with their roots in classical analysis have been highly developed since the turn of the century. During the last two decades, X-ray spectrography has proven to be a very acceptable method of analysis because of the relative ease and rapidity of measurement of the intensity of characteristic wavelengths, the ready knowledge of the precision of the measurement, the facility of automating the analysis, and the nondestructive nature of the method. When chemical separation techniques are combined with X-ray spectrography, the problem of matrix effects is eliminated and the element being analyzed is substantially concentrated, which affords a means of performing trace element analyses. Published examples of preconcentration followed by X-ray measurement both outside and in the field of metallurgy are cited.

INTRODUCTION

In any discussion of trace element analysis, it should be pointed out that, because of the tremendous importance of trace elements in the metallurgical field, a wide variety of techniques have been developed and are being used successfully. The optical emission spectrograph with the DC arc and, more recently, controlled atmospheres around the excitation source has made an outstanding contribution to this challenging problem. Polarography has come into prominence in certain laboratories. The solids mass spectrometer and activation analysis as well are proving to be of great assistance in the determination of trace elements below the part per million level. Because of the relative cost of these techniques, they are not generally available at the present time. The reported accuracy of 200 to 300% in the case of the solids mass spectrometer also leaves something to be desired.[1] There are some elements that do not respond to polarographic analysis at trace concentrations, just as certain elements do not respond to spectrographic excitation.

X-ray spectroscopy has become a widely accepted method of elemental analysis in the metallurgical field for both major and minor elements. Many reviews attest to this acceptance.[2–4] There are several reasons for this popularity. The relative ease and rapidity of measurement of the intensity of the characteristic wavelength, the ready knowledge of the precision of the measurement, the facility of automating the analysis,[5] and the nondestructive nature of X-rays are among these reasons. The fact that the limits of detection may be lowered by using the optimum counters, crystals, and tube target materials also contributes to this wide acceptance of X-ray spectroscopy. Optimum conditions for the direct analysis of trace elements in NBS steels are described by Loomis.[6] By using the definition of the minimum detectable limit published by Birks[7]—concentration for which the intensity of the analytical line exceeds the background intensity by three times the standard deviation of the background—a limit of detection of 1 ppm was reported for one element. A few additional elements were reported detectable below the 10-ppm range without chemical concentration.

Like other analytical techniques, the problem of the effect of the matrix on the measurement of the element being analyzed often impairs the accuracy unless precautions are taken. Generally, when standards of the same matrix are available, these are used to prepare calibration curves, which compensates for the enhancement and absorption effects of the major elements in the alloy. When similar materials are not available as standards, one method of handling this problem is by mathematical corrections for these matrix effects. This normally involves a computer because of the complexity of the corrections. Another method is to remove the matrix chemically. Preferably, the element being analyzed may be removed and the intensity of the characteristic radiation measured in the absence of a matrix. Also, a fixed and consistent quantity of a selected matrix may be added to the separated element. Either of these last two techniques allows the preparation of a set of standards which may be used subsequently for a wide variety of alloys and, as well, an alloy system wherein the concentration of the alloying elements is changed from day to day. Speed is sacrificed in carrying out a chemical separation, but, on the other hand, the element in question is concentrated, which substantially lowers the limit of detection and allows parts per million of many elements to be analyzed.

The purpose of this paper is to cite examples of trace element analysis by using chemical separation (or pre-concentration) followed by measurement of the characteristic X-radiation.

SEPARATION TECHNIQUES

Classically or historically, to analyze a material, it was necessary to separate the desired element or known compound of the element in a pure state and weigh the material separated. It could be said that the separation was performed to eliminate the matrix or its effect on the final measurement. In instrument terminology, the separation is required to eliminate the matrix effect. This is certainly the case in X-ray measurements. Separation techniques have become highly developed, and, when coupled with the features and advantages of X-ray measurements mentioned above, the combination affords an effective method of dealing with the difficult problem of trace element analysis.

Among the separation techniques generally available are ion exchange, solvent extraction, and precipitation. Each has its advantages and disadvantages. A few examples of each may demonstrate their respective usefulness.

Ion-Exchange Separations

At the conference held at the University of Exeter, England, in 1964, Minns[8] reported a detection limit for chromium of 0.1 μg with the use of a filter paper impregnated with an ion-exchange resin rather than a column of resin. He reported that chromium (VI) was quantitatively retained and the analysis could be performed with adequate speed and precision. Success with other metals separated as their anionic complexes were also reported. Mercury and platinum as their chloroanions were found to be quantitatively separated and measured down to 0.1 μg.

Using ion-exchange paper, Luke[9] isolated and analyzed up to 28 trace elements, using a curved-crystal X-ray milliprobe. He showed that 3-hr equilibration of the ion-exchange paper with the solution was sufficient for recoveries of greater than 90% for most of the 28 elements. Limits of detection of 0.01 μg for the most sensitive of these elements were reported.

Tin, down to the level of 1 μg in samples of glass, was extracted onto anion-exchange resin impregnated paper discs and subsequently measured by X-ray spectrography (Chamberlain and Leech).[10] At the 5-μg level, the precision attained was $\pm 10\%$.

Ion exchange is used to collect trace elements after solvent extraction from high-purity tungsten. Hubbard and Green[11] first removed the trace elements by using dithizone in chloroform, which serves as a group extractant. The traces were then back extracted with the acid aqueous phase and collected on ion-exchange resin impregnated paper discs. X-ray readings were taken on both sides of the discs and averaged. Trace elements determined in this manner were copper, nickel, lead, cobalt, and zinc. Limits of detection of less than $\frac{1}{2}$ ppm were found for all but lead.

Solvent Extraction

The reason that solvent extraction has become widely accepted as a separation technique is because of the simplicity and speed of carrying out the separation. A symposium held in 1958 on "Solvent Extraction in the Analysis of Metals" records the versatility of this technique.[12] The text by Morrison and Freiser[13] presents the theory, apparatus, and applications of solvent extraction. The recently published review of Freiser[14] lists elements determined and reagents employed in this useful technique.

This technique was successfully used by Luke[15] for the determination of traces of nickel and iron in a magnesium-base alloy. Dimethyglyoxime in chloroform was used to extract nickel, and cupferron in chloroform was used to extract iron. After separation, the solvent was removed by evaporation. The trace quantities were fused in borax and cast into buttons $\frac{1}{16}$ in. thick. The buttons or discs were measured by using the X-ray spectrograph. The reported limit of detection for a 1-g sample was from 1 to 10 ppm, depending on the element.

It can be readily seen that, with the use of this technique, particularly in view of the variety of reagents available for solvent extraction, coupled with X-ray measurement, trace element determinations can be made with acceptable speed and precision.

Separation by Precipitation

Precipitation followed by filtering and washing has had the reputation of being both long and tedious. However, since the advent of the micropore filter, filtering can be done in minutes. This means that precipitation can compete with solvent extraction as an acceptable method of separation. With care, the precipitate can be quantitatively deposited on the filter disc and, with a brief drying period, can be available for measurement on the X-ray spectrograph. If coprecipitated with a known amount of a carrier element, quantitative handling of the coprecipitated elements is not essential since X-radiation is measured for both the trace element and the carrier element and the ratio of these two measurements (which is independent of the recovery) is plotted against the quantity of the desired element. For these reasons, most of the methods developed in our laboratory have involved precipitation.

Analysts outside the field of metallurgy have applied precipitation with subsequent X-ray measurements as a means of performing trace analysis. Stone[16] determined traces of strontium in tap water by coprecipitating calcium and strontium as their oxalates. The precipitate was collected with a filter stick. Rubidium added in a fixed quantity was used as an internal standard.

Lytel *et al.*,[17] after trying several separation techniques prior to X-ray measurement, concluded that the most successful scheme for analyzing plant material for a wide variety of trace metals, including iron, manganese, copper, zinc, cobalt, molybdenum, chromium, nickel, titanium, vanadium, and calcium, was to use an organic precipitant.

In the field of biology and medicine, Mathies *et al.*[18] reported an indirect method of analyzing organic materials for nitrogen below the microgram level. This was done by using a filter-paper disc impregnated with Nessler's reagent to trap the ammonia developed from the nitrogen. Excess reagent was washed from the filter paper, and the amount of nitrogen in the brown, insoluble deposit was assayed by measurement of the mercury in the compound by X-ray spectrography.

In the metallurgical field, the analysis of heat-resisting and corrosion-resisting alloys has been performed for refractory elements, tungsten, niobium, tantalum, and molybdenum, by using precipitation followed by X-ray measurement.[19] In this report, Luke fused the separated elements in borax and subsequently measured the characteristic X-radiation, using the borax button as the sample form. Although this work was not specifically aimed at trace elements, it demonstrates the possibility of extending the method to trace quantities.

Rudolph and Nadalin[20] were able to determine microgram quantities of chloride in high-purity titanium by precipitation of the chloride with silver nitrate. The convenience of using the micropore filter pad to collect the silver chloride was cited. The pad mounted between two films of Mylar* was presented to the X-ray spectrograph. Silver $K\alpha$ radiation was measured because this is less absorbed by the Mylar film than the chlorine $K\alpha$ radiation.

This same procedure was applied by Garska[21] to the determination of traces of chloride in refractory solids after extraction of the soluble chlorides with water. A micropore filter was used to collect the silver chloride precipitate, which was subsequently exposed to X-rays for measurement. The possibility of extending the method to less than 1 ppm was discussed.

Part per million quantities of zirconium in cobalt, iron, and nickel alloys have been determined after concentrating the zirconium by precipitation with *p*-bromomandelic acid.[22] The precipitate was collected on a micropore filter, and the intensity of the characteristic zirconium wavelength was measured with the X-ray spectrograph. A salient advantage is pointed out in this paper: "One set of standards is sufficient for the determination of an element in a number of matrices." Micropore filters were used in a manner similar to that in the determination of traces of chloride.

We were recently faced with the problem of determining parts per million of tellurium in a variety of matrices, including nickel, nickel–iron, and copper–nickel. Because of the need to have several sets of standards, coupled with the poor sensitivity of tellurium below the 100-ppm level in a solid metal sample, chemical concentration followed by X-ray measurement was attempted. Burke *et al.*[23] in our laboratory, confirming previous reports, found that tellurium could be reduced to the elemental state in an acid solution with tin (II) as a reductant. In aqueous media, only a few elements, namely, tellurium, selenium, mercury, silver, bismuth, arsenic, and gold, are reduced to their elemental states by this reagent. Selenium often occurs with tellurium and has to be separated prior to the precipitation of tellurium. This was done by using hydroxylamine. Figure 1 verifies the reliability of the separation scheme for tellurium. The points on the calibration curve are the actual micropore filters. Although the gradation of these from gray to black makes it possible to estimate the tellurium concentration visually, this could not be done if other elements were present.

It occurred to us that perhaps a convenient method for selenium in a wide variety of matrices could be developed by using a similar scheme. Albright *et al.*[24] recently reported that, in fact, complete precipitation of selenium is obtained when 4 g of hydroxylamine

* Registered Trademark of E. I. DuPont deNemours and Company.

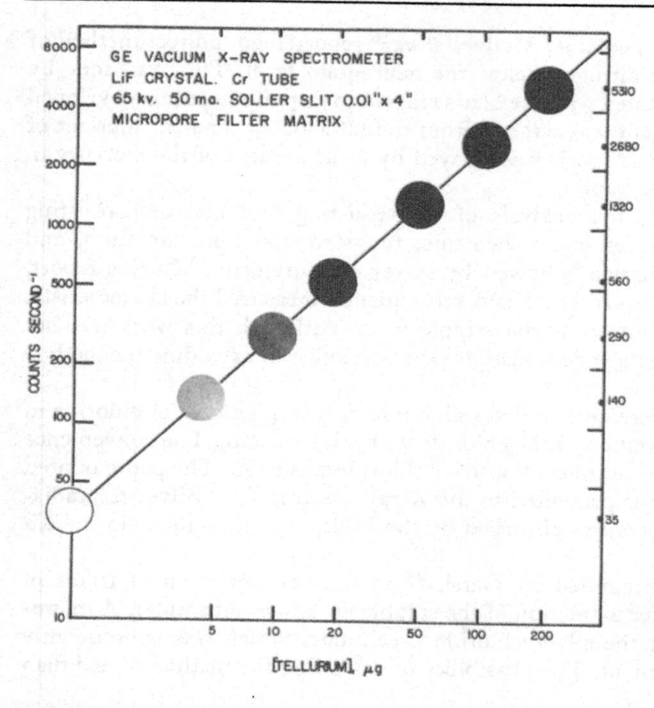

Figure 1. X-ray calibration curve for tellurium. (Reprinted from *Anal. Chem.* **39**: 14, 1967, by permission of the copyright owner.)

hydrochloride is heated for 2 to 5 min with 100 ml of a solution which is 2 to 8 M in hydrochloric acid. The precision for the selenium determination in a copper–nickel sample at the 60-ppm level was found to be ± 3.5 ppm.

A general method for traces of cerium was developed in our laboratory, which is applicable to a variety of nickel- and iron-base alloys. The mercury cathode was employed to remove the matrix elements. A measured quantity of iron was added to the electrolyte after electrolysis was complete. By using ammonium hydroxide as precipitant, the iron and traces of cerium were coprecipitated. After filtering and igniting, the oxides were mounted on a cellulose pad and submitted to the X-ray spectrograph. The intensities of the X-radiation of the cerium as well as the iron were measured. The ratio of the counts was plotted against the percent of cerium present. As stated earlier, quantitative handling of the precipitate in mounting for the X-ray measurement was not necessary since the ratio was independent of the quantity of the coprecipitate. This procedure has been in use for several years in our laboratory. Cerium contents below 10 ppm are often determined.

DISCUSSION OF RESULTS

In the technique of preconcentration followed by measurement of the characteristic X-radiation as a means of trace element analysis, it has already been pointed out that the matrix problem is dealt with by simply removing the matrix. In some cases, the element is measured in the elemental or combined state in the absence of a matrix, while, in other cases, as for cerium, a selected matrix is added and used as an internal standard. The limit of detection in the examples cited is lowered by simply concentrating the trace elements. In contrast to other techniques, the technique discussed here affords a means of analysis without reliance on wet chemical procedures to prepare standards. This is accomplished by using pure elements or compounds of the elements of known purity to prepare synthetic standards. As in any chemical procedure, the recovery has to be checked against the separation conditions, and the optimum conditions must be established. Of course, this has been done in all the examples cited.

The use of the X-ray spectrograph to complete the analysis is not absolutely essential since other measuring instruments are available. However, since the measuring system of the X-ray spectrograph is a counting system and the precision of the measurement is a function of the total count, the X-ray spectrograph was an obvious selection. It should be pointed out, however, that the ideal analytical method is one employing a minimum of handling and manipulation and, when precipitation is used to concentrate the trace element and the precipitate is collected on a micropore filter, very little handling is done in submitting the dried filter pad to the X-ray spectrograph.

SUMMARY

Trace element analysis has been a challenge to the analyst for many years. This challenge has been met in the past decade by coupling chemical separation techniques with X-ray spectrography. In the examples cited, the precision can be roughly stated as $\pm 10\%$ or better for concentrations between 1 and 100 ppm. The limits of detection and speed of analysis are also acceptable.

For elements not mentioned here, the analyst is still challenged to select a means of separation that is amenable to the element in question and convenient in his own laboratory.

REFERENCES

1. T. A. Davies, "Application of Solids Mass Spectroscopy to Metallurgical Analysis," in: P. W. West, A. M. G. MacDonald, and T. S. West (eds.), *Analytical Chemistry, 1962*, Elsevier Publishing Co., Amsterdam, 1963, p. 328.
2. C. M. Davis, "X-Ray Emission Spectrography in the Metal Industry," *Colloq. Spectroscopicum Intern.* **8**: 281, 1959.
3. K. I. Narbutt, "Present State of X-Ray Analysis," *Zavodsk. Lab.* **24**: 604, 1958.
4. W. J. Campbell, T. D. Brown, and J. W. Thatcher, "X-Ray Absorption and Emission," *Anal. Chem.* **38**: 416R, 1966.
5. C. M. Davis and M. M. Yanak, "Performance of an Unattended Automated X-Ray Spectrograph," in: W. M. Mueller, G. Mallett, and M. Fay (eds.), *Advances in X-Ray Analysis, Vol. 7*, Plenum Press, New York, 1964, p. 644.
6. T. C. Loomis, "X-Ray Spectroscopy as an Analytical Tool," *Ann. N.Y. Acad. Sci.* **137**: 284, 1966.
7. L. S. Birks, *X-Ray Spectrochemical Analysis*, Interscience Publishers, Inc., New York, 1966, p. 55.
8. R. E. Minns, "Reduction of Limits of Detection by Concentration—An Anion Exchange Technique," in: *Limitations of Detection in Spectrochemical Analysis*, Hilger & Watts, Ltd., London, 1964, p. 45.
9. C. L. Luke, "Ultratrace Analysis of Metals with a Curved Crystal X-Ray Milliprobe," *Anal. Chem.* **36**: 318, 1964.
10. B. R. Chamberlain and R. T. Leech, "Determination of Microgram Quantities of Tin(IV) by a Combined Ion-Exchange/X-Ray Fluorescence Technique," *Talanta* **14**: 597, 1967.
11. G. L. Hubbard and T. E. Green, "Dithizone Extraction and X-Ray Spectrographic Determination of Trace Metals in High-Purity Tungsten or Tungsten Trioxide," *Anal. Chem.* **38**: 428, 1966.
12. "Symposium on Solvent Extraction in the Analysis of Metals," *ASTM Spec. Tech. Publ. No. 238*, 1958.
13. G. H. Morrison and H. Freiser, *Solvent Extraction in Analytical Chemistry*, John Wiley & Sons, Inc., New York, 1957.
14. H. Freiser, "Extraction," *Anal. Chem.* **38**: 131R, 1966.
15. C. L. Luke, "Trace Analysis of Metals by Borax Disk X-Ray Spectrometry," *Anal. Chem.* **35**: 1551, 1963.

16. R. G. Stone, "The Determination of Strontium in Tap-Water by X-Ray Fluorescence Spectrometry," *Analyst* **88**: 56, 1963.
17. F. W. Lytel, W. B. Dye, and H. J. Seim, "Determination of Trace Elements in Plant Material by Fluorescent X-Ray Analysis," in: W. M. Mueller (ed.), *Advances in X-Ray Analysis, Vol. 5*, Plenum Press, New York, 1962, p. 433.
18. J. C. Mathies, P. K. Lund, and W. Eide, "A Simple, Indirect Sensitive Procedure for the Determination of Nitrogen (Ammonia) at the Microgram and Submicrogram Level," *Norelco Rept.* **9**: 93, 1962.
19. C. L. Luke, "Determination of Refractory Metals in Ferrous Alloys and High-Alloy Steel by the Borax Disk X-Ray Spectrochemical Method," *Anal. Chem.* **35**: 56, 1963.
20. J. S. Rudolph and R. J. Nadalin, "Determination of Microgram Quantities of Chloride in High Purity Titanium by X-Ray Spectrochemical Analysis," *Anal. Chem.* **36**: 1815, 1964.
21. K. J. Garska, "Determination of Microgram Quantities of Inorganic Chlorides in Refractory Solids," presented as paper No. 143 at the 18th Annual Mid-America Symposium on Spectroscopy, May 1967.
22. J. S. Rudolph, O. H. Kriege, and R. J. Nadalin, "Applications of Chemical Precipitation Methods for Improving Sensitivity in X-Ray Fluorescent Analysis," in: E. N. Davis (ed.), *Developments in Applied Spectroscopy, Vol. 4*, Plenum Press, New York, 1965, p. 57.
23. K. E. Burke, M. M. Yanak, and C. G. Albright, "Determination of Parts per Million Quantities of Tellurium in Various Alloys by X-Ray Spectrometry," *Anal. Chem.* **39**: 14, 1967.
24. C. H. Albright, K. E. Burke, and M. M. Yanak, "A Chemical Concentration X-Ray Determination of Selenium in Nickel- and Iron-Base Alloys," presented as paper No. 142 at the 18th Annual Mid-America Symposium on Spectroscopy, May 1967.

DISCUSSION

V. A. Lazar (U.S. Plant, Soil, and Nutrition Laboratory): I should like to make a plug for a common language in describing these results. I should like to see the parts-per-million term reserved for the original solid material and what you have in the solution as micrograms. If we have a standard solution, we like to use the term micrograms per milliliter.

C. M. Davis: I agree.

Chairman H. Pfeiffer: This trace element paper reminds me of the early efforts to determine trace elements. Once, we happened to be near Gary, Indiana, during a rain storm. We decided to see how much iron there was in the rain water. So, we caught ourselves a glass jugful of the water and suspended a little ion-exchange membrane—this was about 1954—in the jar of rain water, and we left it in there for about 24 hr. The X-ray spectrograph indicated that iron was present. Then we followed up by looking for silver after cloud seeding, with mixed results. With the improved techniques, it shouldn't be difficult to detect at all. Ion-exchange concentration techniques can be very powerful on some rather unusual samples.

THEORETICAL CORRECTION FOR COEXISTENT ELEMENTS IN FLUORESCENT X-RAY ANALYSIS OF ALLOY STEEL

Toshio Shiraiwa and Nobukatsu Fujino

Central Research Laboratories, Sumitomo Metal Industries
Amagasaki, Japan

ABSTRACT

With fluorescent X-ray analysis as routine work, the correction term for the coexistent element can be expressed as linear terms of the weight fraction of the element because the compositions of samples are limited to a small range. Usually those correction factors which require a great deal of work are obtained experimentally. The authors have obtained theoretical equations of fluorescent X-ray intensity which are in good agreement with experimental values. The linear correction factors are obtained from derivatives of those equations, and their values can be easily calculated with a computer. The experimental X-ray intensity *versus* the weight fractions is usually expressed as a line. However, the linear approximation is not correct over a wide range of the composition. The second derivative of the theoretical equation explains the deviation from the linear approximation and gives the range where the linear approximation is allowed. The calculations are applied to the analysis of stainless steels, several low-alloy steels, and iron ores, and experimental results are corrected by the calculated results. Correction factors for Ni $K\alpha$, Fe $K\alpha$, Cr $K\alpha$, Mn $K\alpha$, and Cu $K\alpha$ in stainless steels and Cr $K\alpha$ and Mn $K\alpha$ in low-alloy steels are calculated for coexistent elements such as carbon, silicon, titanium, chromium, manganese, copper, niobium, and molybdenum. For example, standard deviations of chromium and manganese analyzed results in low-alloy steels decrease from 0.169 and 0.044% to 0.030 and 0.023%, respectively, with theoretical corrections. In the analysis of iron ore, the fluorescent X-ray intensity of iron is affected by combined oxygen, which is different for the various compounds of iron oxides, and other impurities such as alumina, silica, and lime. The correction factors of these are obtained by calculation, and the standard deviation decreases from 1.70 to 0.44% for 55.1 to 68.5% iron. It is found by experiment that the theoretical values have about 1 or 2% of relative errors, and their derivatives also have relative errors of the same order of magnitude. But the ranges of coexistent elements are usually small, a few percent at most in routine work, and the theoretical values can be used in practical analyses.

INTRODUCTION

In fluorescent X-ray analysis, there are many interelement effects, and this problem must be solved for the practical analyst. Authors have reported the theoretical equations of the fluorescent X-ray intensities.[1] The fluorescent X-ray intensity is calculated theoretically with the use of well-known physical constants for X-rays. From these theoretical calculations, the interelement effects are estimated, and the fluorescent X-ray intensities can be given with an error of less than 1% for nickel–iron–chromium ternary alloys.[2] The errors are caused by the inaccuracy of the physical constants and approximations used and also by neglect of the coexistent minor elements contained in the alloys.

In the practical use of fluorescent X-ray analysis, the composition of the samples is usually limited to a narrow range, and a more simplified theoretical equation can be used. In steel analysis, this is useful because we must estimate the influence of many co-existent minor elements.

The simplified theoretical equation is obtained by Taylor's expansion of the full equation in the previous paper,[1] and the fluorescent X-ray intensity is expressed as a linear function of the minor component elements and the small deviations of the major elements. The expansion factors are obtained by calculation.[3]

A few works have been reported[4-7] to get the correction factors of the interelement effect in the linear formula, but the correction factors must be obtained from a series of experimental data by linear regression, which requires an enormous amount of work. However, the theoretical method gives the correction factors easily with a computer.

In the present paper, an outline of this calculation procedure and the application of the results are reported. The range where the linear expression approximation is valid and the errors due to the approximation are also discussed.

THEORY

The theoretical intensities of the fluorescent X-rays reported in the previous paper[1] are as follows. The definitions of the primary, secondary, and tertiary fluorescent X-rays are also detailed in the previous paper.[1]

Primary Fluorescent X-Rays

$$I_1(ip) = \frac{1}{\sin\Psi}\int_{\lambda_m}^{\lambda_e^i} \frac{Q_{ip}(\lambda) I_0(\lambda)}{\left[\dfrac{\mu(\lambda)}{\rho}\Big/\sin\Phi\right] + \left[\dfrac{\mu(ip)}{\rho}\Big/\sin\Psi\right]}\, d\lambda$$

$$Q_{ip}(\lambda) = \frac{\mu_i(\lambda)}{\rho_i} W_i \omega_i R_p{}^i K_i \tag{1}$$

$$\frac{\mu(\lambda)}{\rho} = \sum_j \frac{\mu_j(\lambda)}{\rho_j} W_j$$

Secondary Fluorescent X-Rays

$$I_2(ip) = \frac{1}{2\sin\Psi}\sum_{jq}\int_{\lambda_m}^{\lambda_e^j} \frac{Q_{jq}(\lambda) Q_{ip}(jq) I_0(\lambda)}{\left[\dfrac{\mu(\lambda)}{\rho}\Big/\sin\Phi\right] + \left[\dfrac{\mu(ip)}{\rho}\Big/\sin\Psi\right]}$$

$$\times\left(\frac{\sin\Phi}{[\mu(\lambda)]/\rho}\log\left\{1+\frac{\dfrac{\mu(\lambda)}{\rho}\Big/\sin\Phi}{[\mu(jq)]/\rho}\right\} + \frac{\sin\Psi}{[\mu(ip)]/\rho}\log\left\{1+\frac{\dfrac{\mu(ip)}{\rho}\Big/\sin\Psi}{[\mu(jq)]/\rho}\right\}\right) d\lambda \tag{2}$$

It is known that the tertiary fluorescent X-ray is small,[1] so it need not be considered here. Notation in the above equations is given at the end of this paper.

In routine analysis work, the range of compositions is limited and the theoretical equations can be expanded under the assumption that the variation of the composition is small.

Expansion of the Equation for Primary Fluorescent X-Ray Intensity

Let ΔW_j be an increment of W_j. Then $I_1(ip)$ is given by Taylor's expansion

$$I_1(ip) = I_1^0(ip) + \sum_j \left[\frac{\partial I_1(ip)}{\partial W_j}\right]\Delta W_j + \frac{1}{2!}\left(\sum_j \frac{\partial}{\partial W_j} \cdot \Delta W_j\right)^2 I_1(ip) + \cdots \tag{3}$$

so the first approximation of the increment is

$$\Delta I_1(ip) \doteqdot \sum_j \left[\frac{\partial I_1(ip)}{\partial W_j}\right]\Delta W_j \tag{4}$$

The differential coefficients are as follows: When $j \neq i$,

$$\frac{\partial I_1(ip)}{\partial W_j} = \frac{\omega_i R_p{}^i K_i}{\sin \Psi} \int_{\lambda_m}^{\lambda_e{}^i} \frac{[\mu_i(\lambda)/\rho_i] I_0(\lambda) W_i}{\left\{\left[\frac{\mu(\lambda)}{\rho}\Big/\sin \Phi\right] + \left[\frac{\mu(ip)}{\rho}\Big/\sin \Psi\right]\right\}^2}$$

$$\times \left\{\left[-\frac{\mu_j(\lambda)}{\rho_j}\Big/\sin \Phi\right] - \left[\frac{\mu_j(ip)}{\rho_j}\Big/\sin \Psi\right]\right\} d\lambda \equiv a_{i\cdot j} \tag{5}$$

When $j = i$,

$$\frac{\partial I_1(ip)}{\partial W_i} = \frac{I_1(ip)}{W_i} + \frac{\omega_i R_p{}^i K_i}{\sin \Psi} \int_{\lambda_m}^{\lambda_e{}^i} \frac{[\mu_i(\lambda)/\rho_i] I_0(\lambda) W_i}{\left\{\left[\frac{\mu(\lambda)}{\rho}\Big/\sin \Phi\right] + \left[\frac{\mu(ip)}{\rho}\Big/\sin \Psi\right]\right\}^2}$$

$$\times \left\{\left[-\frac{\mu_i(\lambda)}{\rho_i}\Big/\sin \Phi\right] - \left[\frac{\mu_i(ip)}{\rho_i}\Big/\sin \Psi\right]\right\} d\lambda \tag{6}$$

In equation (6), the first term is the increment proportional to the weight fraction, and the second term corresponds to the correction of the absorption effects.

From equations (5) and (6),

$$\Delta I_1(ip) = \frac{I_1(ip)}{W_i}\Delta W_i + \sum_j a_{i\cdot j}\Delta W_j \tag{7}$$

In equation (7), it is possible to omit an element k because

$$\sum_j \Delta W_j = 0$$

so,

$$\Delta I_1(ip) = \frac{I_1(ip)}{W_i}\Delta W_i + \sum_{j \neq k} (a_{i\cdot j} - a_{i\cdot k})\Delta W_j \tag{8}$$

and equation (8) gives the increment of the fluorescent X-ray intensity according to the exchange of the weight fraction ΔW_j between elements j and k.

In the analysis of steels, iron is not usually analyzed but shown by the balance of the other elements, and it is convenient to take iron as k.

When a weight fraction W_j is small, it can be considered that W_j is an increment ΔW_j from zero. The factors $a_{i\cdot j}$ can be obtained by theoretical calculations, but experiments give only the difference $a_{i\cdot j} - a_{i\cdot k}$. Dividing equation (8) by the intensity $I_1{}^0(ip) \simeq I_1(ip)$, we obtain

$$\frac{\Delta I_1(ip)}{I_1(ip)} = \frac{\Delta W_i}{W_i} + \sum_{j \neq k} \frac{a_{i\cdot j} - a_{i\cdot k}}{I_1(ip)} \Delta W_j \tag{9}$$

Defining the formula

$$A_{j-k}^{ip} \equiv \frac{a_{i\cdot j} - a_{i\cdot k}}{I_1(ip)} \tag{10}$$

we get the following equation

$$I_1(ip) = I_1{}^0(ip)\left[1 + \left(\frac{\Delta W_i}{W_i} + \sum_{j \neq k} A_{j-k}^{ip} \cdot \Delta W_j\right)\right] \tag{11}$$

When we obtain the calibration curve $I_1(ip)$ *versus* W_i, the term $\partial/\partial W_i$ in equations (8) and (11) is excluded and

$$I_1(ip) = I_1{}^0(ip)\left(1 + \sum_{j \neq i,k} A_{j-k}^{ip} \cdot \Delta W_j\right) \tag{12}$$

Then the factor A_{j-k}^{ip} expresses the correction factor. The notation *ip* is omitted hereafter.

Error of the Linear Approximation

An error of the linear approximation may be estimated from the magnitude of the second differential of equation (3). When $i \neq j$, the second differential is

$$\frac{1}{2!} \cdot \frac{\partial^2}{\partial W_j \partial W_k} I_1(ip) \Delta W_j \cdot \Delta W_k$$

For simplification, let us exchange the small part of the weight fraction between only two elements k and j, *i.e.*, $\Delta W_k = -\Delta W_j$,

$$\begin{aligned} &\frac{1}{2} \cdot \frac{\partial^2}{(\partial W_j)^2} I_1(ip)(\Delta W_j)^2 \\ &= \frac{\omega_i R_p{}^i K_i}{\sin \Psi} \int_{\lambda_m}^{\lambda_e{}^i} \frac{[\mu_i(\lambda)/\rho_i] I_0(\lambda) W_i}{\left\{\left[\frac{\mu(\lambda)}{\rho} \Big/ \sin \Phi\right] + \left[\frac{\mu(ip)}{\rho} \Big/ \sin \Psi\right]\right\}^3} \\ &\times \left\{\left[\frac{\mu_j(\lambda)}{\rho_j} \Big/ \sin \Phi\right] + \left[\frac{\mu_j(ip)}{\rho_j} \Big/ \sin \Psi\right]\right. \\ &\left. - \left[\frac{\mu_k(\lambda)}{\rho_k} \Big/ \sin \Phi\right] + \left[\frac{\mu_k(ip)}{\rho_k} \Big/ \sin \Psi\right]\right\}^2 (\Delta W_j)^2 \end{aligned} \tag{13}$$

In equation (5), if we assume monochromatic incident X-rays, the ratio of the first approximation term to $I_1(ip)$ is expressed approximately as follows:

$$\frac{\text{first correction term}}{I_1^0(ip)} = \frac{\left\{\left[-\frac{\mu_j(\lambda)}{\rho_j}\Big/\sin\Phi\right]-\left[\frac{\mu_j(ip)}{\rho_j}\Big/\sin\Psi\right]\right\}-\left\{\left[-\frac{\mu_k(\lambda)}{\rho_k}\Big/\sin\Phi\right]-\left[\frac{\mu_k(ip)}{\rho_k}\Big/\sin\Psi\right]\right\}}{\left[\frac{\mu(\lambda)}{\rho}\Big/\sin\Phi\right]+\left[\frac{\mu(ip)}{\rho}\Big/\sin\Psi\right]}\Delta W_j \tag{14}$$

And the ratio of the second differential is similarly expressed.

$$\frac{\text{second correction term}}{I_1^0(ip)} = \frac{\left(\left\{\left[\frac{\mu_j(\lambda)}{\rho_j}\Big/\sin\Phi\right]+\left[\frac{\mu_j(ip)}{\rho_j}\Big/\sin\Psi\right]\right\}-\left\{\left[\frac{\mu_k(\lambda)}{\rho_k}\Big/\sin\Phi\right]+\left[\frac{\mu_k(ip)}{\rho_k}\Big/\sin\Psi\right]\right\}\right)^2}{\left\{\left[\frac{\mu(\lambda)}{\rho}\Big/\sin\Phi\right]+\left[\frac{\mu(ip)}{\rho}\Big/\sin\Psi\right]\right\}^2}(\Delta W_j)^2 \tag{15}$$

Then,

$$\left[\frac{\text{first correction term}}{I_1(ip)}\right]^2 = \frac{\text{second correction term}}{I_1(ip)}$$

In the case of $i = j$,

$$\frac{1}{2}\cdot\frac{\partial^2 I_1(ip)}{\partial W_i^2}(\Delta W_i)^2 = \frac{\omega_i R_p{}^i K_i}{2\sin\Psi}\int_{\lambda_m}^{\lambda_e^i}\frac{[\mu_i(\lambda)/\rho_i]I_0(\lambda)}{\left\{\left[\frac{\mu(\lambda)}{\rho}\Big/\sin\Phi\right]+\left[\frac{\mu(ip)}{\rho}\Big/\sin\Psi\right]\right\}^2}$$

$$\times\left\{\left[-\frac{\mu_i(\lambda)}{\rho_i}\Big/\sin\Phi\right]-\left[\frac{\mu_i(ip)}{\rho_i}\Big/\sin\Psi\right]\right\}$$

$$-\left\{\left[-\frac{\mu_k(\lambda)}{\rho_k}\Big/\sin\Phi\right]-\left[\frac{\mu_k(ip)}{\rho_k}\Big/\sin\Psi\right]\right\}(\Delta W_i)^2\, d\lambda$$

$$+\frac{\omega_i R_p{}^i K_i}{\sin\Psi}\int_{\lambda_m}^{\lambda_e^i}\frac{[\mu_i(\lambda)/\rho_i]I_0(\lambda)W_i}{\left\{\left[\frac{\mu(\lambda)}{\rho}\Big/\sin\Phi\right]+\left[\frac{\mu(ip)}{\rho}\Big/\sin\Psi\right]\right\}^3}$$

$$\times\left(\left\{\frac{[\mu_i(\lambda)]/\rho_i}{\sin\Phi}+\frac{[\mu_i(ip)]/\rho_i}{\sin\Psi}\right\}-\left\{\frac{[\mu_k(\lambda)]/\rho_k}{\sin\Phi}+\frac{[\mu_k(ip)]/\rho_k}{\sin\Psi}\right\}\right)^2(\Delta W_i)^2\, d\lambda \tag{16}$$

On the assumption of monochromatic incident X-rays,

$$[\text{first term of equation (16)}] \simeq [\text{second term of equation (7)}] \times \frac{\Delta W_i}{W_i}$$

and

$$\left[\frac{\text{second term of equation (16)}}{I_1(ip)}\right] \simeq \left[\frac{\text{second term of equation (7)}}{I_1(ip)}\right]^2$$

Expansion of Secondary Fluorescent X-Ray Intensity Equation

The intensity of the secondary fluorescent X-rays is given by the next equation from the previous report[1]

$$I_2(ip) = W_i \cdot W_j \cdot f(W_k) \qquad k = 1, 2, \ldots, i, j, \ldots$$

where element j produces the primary fluorescent X-rays, which excite the secondary fluorescent X-rays of element i.

The expanded equations are expressed as follows, where the weight fraction of the element k varies:

Case 1—$k \neq i, j$

$$\Delta I_2(ip) = W_i \cdot W_j \frac{\partial f(W_1, W_2, \ldots, W_k, \ldots)}{\partial W_k} \Delta W_k \tag{17}$$

Case 2—$k = i$

$$\Delta I_2(ip) = I_2(ip)\left[\frac{1}{W_i} + \frac{1}{I_2(ip)} \frac{\partial f}{\partial W_i}\right] \Delta W_i$$

$$\frac{\partial f}{\partial W_i} \equiv b_{i \cdot i} \tag{18}$$

Case 3—$k = j$

$$\Delta I_2(ip) = I_2(ip)\left[\frac{1}{W_j} + \frac{1}{I_2(ip)} \frac{\partial f}{\partial W_j}\right] \Delta W_j$$

$$\frac{\partial f}{\partial W_j} \equiv b_{i \cdot j} \tag{19}$$

The first term in equations (18) and (19) is the term corresponding to the variation of the fluorescent X-ray production element itself, and the second term is the absorption effect.

The term $\partial f/\partial W_k$ is complex, and it is rather easy to compute the derivatives from the difference of the value of function f at two neighboring points. In that case, the correction value

$$B_{k-\mathrm{Fe}}^{ip} \equiv \frac{b_{i \cdot k} - b_{i \cdot \mathrm{Fe}}}{I_2(ip)}$$

is obtained.

COMPUTATION OF CORRECTION FACTORS

Computations of the correction factors are carried on Ni $K\alpha$, Fe $K\alpha$, and Cr $K\alpha$ from nickel–iron–chromium ternary alloys for the small deviation of weight fractions of nickel, iron, and chromium and for trace elements carbon, silicon, phosphorus, sulfur, titanium, manganese, copper, niobium, and molybdenum.

The calculated results are given in Figures 1, 2, and 3 for the primary fluorescent X-rays of nickel, iron and chromium and in Figures 4 and 5 for the secondary fluorescent X-rays, respectively. These magnitudes of $a_{i\cdot j}/I_1(ip)$ and the correction factors are mainly dependent on the mass absorption coefficient of the element which changes the weight fraction and are a larger value for the larger mass absorption coefficient. The values shown in Figures 1, 2, and 3 are 2 or 3 at most in the present calculation, and this means that the effect on variation of the fluorescent intensity is of the same order as that of the increment of the element.

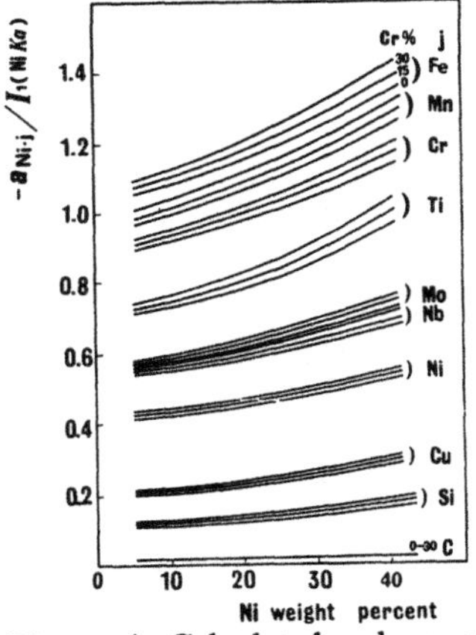

Figure 1. Calculated values of $-a_{Ni\cdot j}/I_1(\text{Ni } K\alpha)$.

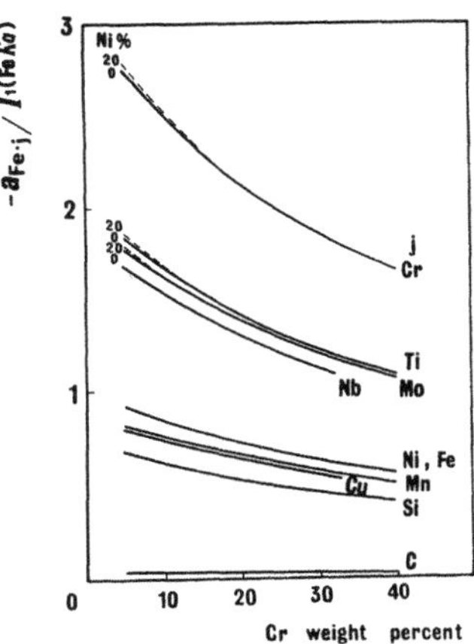

Figure 2. Calculated values of $-a_{Fe\cdot j}/I_1(\text{Fe } K\alpha)$.

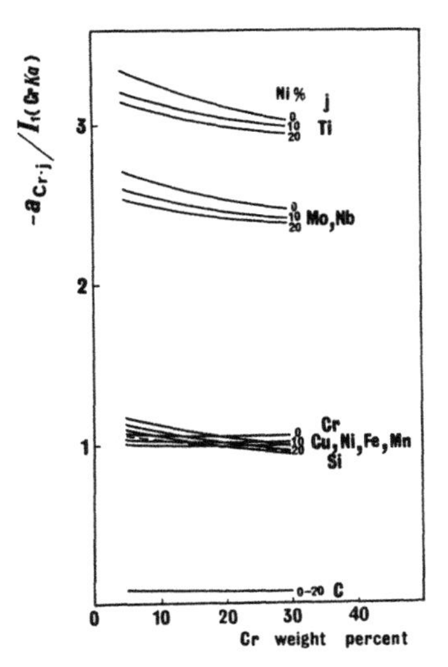

Figure 3. Calculated values of $-a_{Cr\cdot j}/I_1(\text{Cr } K\alpha)$.

Table I. Calculated Correction Values of

A. Element		Carbon	Silicon	Phosphorus
B. Composition range defined by JIS, in percent		<0.08	<1.00	<0.04
C. Maximum deviation of weight fraction	$-\Delta W_j$ (%)	0.08	1.00	0.04
D. Calculated correction factor	$\frac{a_{Ni \cdot j}}{I_1(\text{Ni } K\alpha)}$	−0.02	−0.22	−0.22
E. Calculated correction factor for exchange of element and iron	$\frac{a_{Ni \cdot j} - a_{Ni \cdot Fe}}{I_1}$	1.08	0.88	0.88
F. Calculated correction value of X-ray intensity	$-\frac{a_{Ni \cdot j} - a_{Ni \cdot Fe}}{I_1} \Delta W_j$ (%)	0.09	0.88	0.04
G. Correction value of nickel weight fraction	$-\Delta W_{Ni}$ (%)	0.009	0.079	0.004
H. Estimated correction factor of the second-order correction	$\left(\frac{a_{Ni \cdot j} - a_{Ni \cdot Fe}}{I_1} \Delta W_j\right)^2$ (%)	0.81×10^{-4}	62.4×10^{-4}	0.16×10^{-4}
I. Estimated value of second-order correction of W_{Ni}		0.72×10^{-5}	7.0×10^{-5}	0.14×10^{-5}

* $\frac{1}{W_{Ni}} = 10.53.$

† $-\left(\frac{a_{Ni \cdot j} - a_{Ni \cdot Fe}}{I_1}\right) \frac{\Delta W_{Ni}}{W_{Ni}} = 0.158$

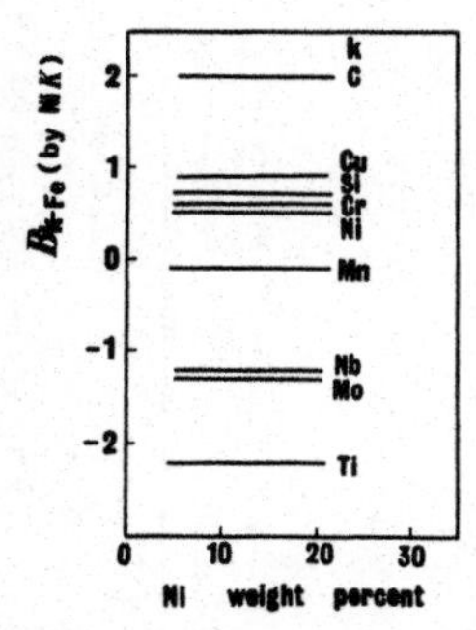

Figure 4. Calculated values of $B_{k-Fe}^{CrK\alpha}$ by Ni K.

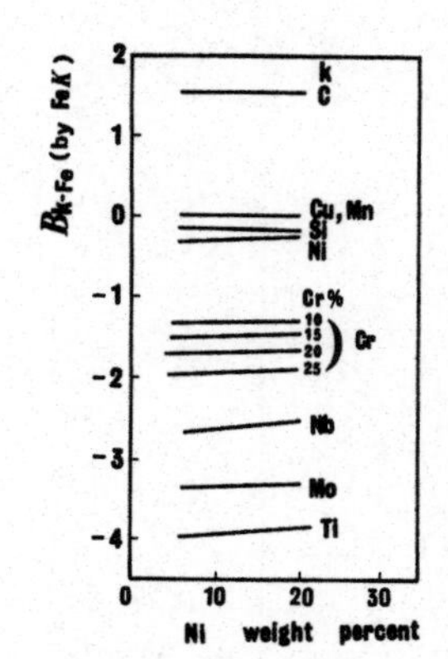

Figure 5. Calculated values of $B_{k-Fe}^{CrK\alpha}$ by Fe K.

Ni $K\alpha$ for SUS 27(Ni = 9.5%, Cr = 19.0%)

Sulfur	Manganese	Chromium	Iron	Nickel
<0.03	<2.00	18.00 to 20.00		8.00 to 11.00
0.03	2.00	1.00		1.50
−0.22	−1.01	−0.93	−1.10	−0.43
0.88	0.09	0.17		0.67 + 10.53* = 11.20
0.03	0.18	0.17		1.001
0.003	0.016	0.015		
0.09×10^{-4}	3.2×10^{-4}	0.9×10^{-4}		$1 \times 10^{-2} + 0.158† = 0.17$
0.08×10^{-5}	2.9×10^{-5}	0.8×10^{-5}		0.015

In routine work, any type of steel is limited to a narrow range of compositions by industrial standards, and whether the linear approximation is appropriate or not in such a narrow range will be determined.

For example, consider JIS SUS 27,* 18–8 stainless steel. The linear theoretical correction values and the estimated correction values of the second order given by equations (14)–(16) are listed in Tables I and II.

In the tables, the line B is the composition range defined by industrial standard, and C gives the maximum deviation of the composition from the mean value for the major element and the maximum allowed values of the trace elements. The line D is the theoretical correction value of the first order for each element and E shows those values for exchange with the iron component and F gives the correction for the fluorescent X-ray intensity for ΔW_j given in C. The line G gives the correction value for the weight fraction of element i. The lines H and I give the order of the secondary-derivative correction values.

The columns for nickel in Table I and chromium in Table II contain the term which is caused by the variation of the measured element itself.

* Japanese industrial standard.

Table II. Calculated Correction Values of

A. Element			Carbon	Silicon	Phosphorus
B. Composition range defined by JIS, in percent			<0.08	<1.00	<0.04
C. Maximum deviation of weight fraction	$-\Delta W_j$	(%)	0.08	1.00	0.04
D. Calculated correction factor	$\frac{a_{Cr \cdot j}}{I_1(\text{Cr } K\alpha)}$		−0.08	−1.00	−1.00
E. Calculated correction factor for exchange of element and iron	$\frac{a_{Cr \cdot j} - a_{Cr \cdot Fe}}{I_1}$		0.80	−0.12	−0.12
F. Calculated correction value of X-ray intensity	$-\frac{a_{Cr \cdot j} - a_{Cr \cdot Fe}}{I_1} \Delta W_j$	(%)	0.064	−0.12	−0.005
G. Correction value of weight fraction	$-\Delta W_{Cr}$	(%)	0.016	−0.030	−0.0013
H. Estimated correction factor of the second-order correction	$\left(\frac{a_{Cr \cdot j} - a_{Cr \cdot Fe}}{I_1} \Delta W_j\right)^2$	(%)	0.41×10^{-4}	1.44×10^{-4}	0.0025×10^{-4}
I. Estimated value of second-order correction of W_{Cr}			0.01×10^{-4}	0.36×10^{-4}	0.0006×10^{-4}

* $\frac{1}{W_{Cr}} = 5.13.$

† $-\left(\frac{a_{Cr \cdot j} - a_{Cr \cdot Fe}}{I}\right) \frac{\Delta W_{Cr}}{W_{Cr}} = 0.0108.$

From the values shown in Tables I and II, it is clear that the linear approximation is appropriate.

EXPERIMENTS

In order to verify the theory, the present correction method was applied to all the types of commercial stainless steels, many low-alloy steels, and iron ores, and good results were obtained.

Stainless Steel

All types of JIS stainless steels which were provided for the joint experiment of fluorescent X-ray analysis by the Iron and Steel Institute of Japan were examined for molybdenum, niobium, copper, nickel, manganese, chromium, and titanium. The chemical compositions of the samples are listed in Table III, in which their JIS numbers and the corresponding ASTM numbers are both shown.

In the present work, although composition ranges of the samples are wide, only one reference sample is used through a whole range, and the reference samples are marked in Table IV.

Cr $K\alpha$ for SUS 27 (Ni = 9.5%, Cr = 19.0%)

Sulfur	Manganese	Chromium	Iron	Nickel
<0.03	<2.00	18.00 to 20.00		8.00 to 11.00
0.03	2.00	1.00		1.50
−1.00	−0.92	−1.02	−0.88	−1.01
−0.12	−0.04	−0.14 + 5.13* = 3.99		−0.18
−0.004	−0.08	−0.14		−0.36
−0.001	−0.020	−0.035		−0.090
0.0016×10^{-4}	0.64×10^{-4}	$1.95 \times 10^{-4} + 0.0108\dagger = 0.0110$		1.3×10^{-4}
0.0004×10^{-4}	0.16×10^{-4}	0.0028		0.33×10^{-4}

Experimental results for molybdenum, niobium, copper, nickel, manganese, chromium, and titanium are shown in Figures 6 to 12, respectively. Fluorescent X-ray intensities expressed by the relative value to that of the reference sample are shown in Table III.

For molybdenum, niobium, and titanium, the correction is not necessary because their composition ranges are narrow and the results of fluorescent X-ray analysis are obtained by linear regression. However, for copper, nickel, manganese, and chromium, the corrections are necessary.

Correction for the overlapping line Cr $K\beta$ on Mn $K\alpha$ and Ni $K\beta$ on Cu $K\alpha$ is necessary before the theoretical correction for the coexistent elements. The correction method for the overlapping line is not given here because it is well known.

Theoretical corrections of coexistent elements are carried out by using equation (12), where $I_1{}^0(ip)$ is the observed X-ray intensity excluding background and overlapping X-rays.

The calculated correction factors $A_{j-\mathrm{Fe}}$ of each stainless steel type are given in Tables V, VII, and VIII for Ni $K\alpha$, Cu $K\alpha$, and Mn $K\alpha$, respectively. In Table VI, the correction factors for Cr $K\alpha$ contain $A_{j-\mathrm{Fe}}$ and $B_{j-\mathrm{Fe}}$.

For the major components nickel and chromium, their variation cannot be treated as a small change, and the X-ray intensities of Ni $K\alpha$ and Cr $K\alpha$ are reduced to those of fixed compositions by the complete theoretical equation. The fixed compositions are 0% nickel for Cr $K\alpha$ and 20% chromium for Ni $K\alpha$.

Table III. Compositions of the Examined Stainless Steels

Specimen No.	JIS No.	ASTM (AISI) No.	Carbon	Silicon	Manganese	Phosphorus	Sulfur	Copper	Nickel	Chromium	Molybdenum	Titanium	Niobium
1	SUS 33	TP 316L	0.029	0.51	1.47*	0.022	0.010	1.21*	14.44*	17.68*	2.62*		
2	23	420	0.32	0.52	0.69	0.027	0.005	0.19	0.19	12.95	0.03		
3	22	403	0.14	0.41	0.82	0.028	0.012	0.11	0.18	12.41	0.08		
4	24	430	0.09	0.35	0.76	0.028	0.016	0.10	0.42	16.62	0.07		
5	33	316L	0.02	0.73	1.62	0.022	0.013	0.17	15.10	18.34	3.48		
6	27	304	0.07	0.63	1.63	0.032	0.007	0.16	8.86	18.70	0.24		
7	27	304	0.08	0.64	1.77	0.032	0.244	0.16	9.36	17.65	0.30		
8	27	304	0.07	0.90	1.19	0.023	0.007	0.14	9.46	18.57	0.01		
9	28	304L	0.02	0.86	1.59	0.021	0.009	0.12	10.58	18·33	0.22		
10	29	321	0.07	0.65	1.59	0.021	0.006	0.10	9.75	17.27	0.22	0.39	
11	29	321	0.06	0.80	1.68	0.031	0.005	0.10	10.33	17.76	0.12	0.38	
12	32	316	0.07	0.45	1.39	0.024	0.013	0.21	10.87	17.32	2.16		
13	32	316	0.06	0.58	1.94	0.034	0.011	0.11	12.31	17.28	2.62		
14	32	316	0.07	0.58	1.85	0.028	0.017	0.20	10.27	16.25	2.55		
15	33	316L	0.02	0.81	1.62	0.021	0.007	0.30	14.10	16.92	2.48		
16	35	–	0.05	0.86	1.51	0.024	0.011	1.17	12.35	17.75	2.45		
17	35	–	0.05	0.52	1.67	0.035	0.009	2.07	21.84	17.20	2.37		
18	41	309	0.08	0.65	1.71	0.017	0.010	0.10	14.23	22.71	0.20		
19	41	309	0.07	0.54	1.81	0.031	0.006	0.07	13.28	22.61	0.08		
20	42	310	0.09	0.63	1.75	0.020	0.006	0.09	19.81	25.16	0.14		
21	43	347	0.07	0.44	1.77	0.024	0.007	0.27	11.04	18.00	0.32		0.72
22	39	301	0.05	0.52	0.61	0.030	0.004	4.05	4.05	16.03	0.11		0.36
NBS 845				0.52	0.77			0.065	0.28	13.31	0.92	0.03	0.11
846				1.19	0.53			0.19	9.11	18.35	0.43	0.34*	0.60*
847				0.37	0.23			0.19	13.26	23.72	0.059	0.02	0.03
848				1.25	2.13			0.16	0.52	9.09	0.33	0.23	0.49
849				0.68	1.63			0.21	6.62	5.48	0.15	0.11	0.31
850				0.12	–			0.36	24.8	2.99	–	0.05	0.05
1184				0.70	1.04			–	9.47	19.44	1.46	0.056	0.49
1185				0.40	1.22			0.067	13.18	17.09	2.01	<0.001	<0.001

* Reference sample.

Table IV. Experimental Conditions*

Spectrum	Operated condition of X-ray tube		Target	Analyzing crystal	X-ray path	Detector	Irradiated area	Intensity measurement†
	kVP	mA						
Mo $K\alpha$	45	40	Tungsten	LiF	Air	GM counter, Amperex	10-mm diameter	1×10^4
Nb $K\alpha$	45	40	Tungsten	LiF	Air	GM counter, Amperex	15 × 20 mm	1×10^4
Cu $K\alpha$	30	20	Tungsten	LiF	Air	GM counter, Amperex	10-mm diameter	1×10^4
Ni $K\alpha$	30	30	Tungsten	LiF	Air	GM counter, Amperex	7-mm diameter	4×10^4
Mn $K\alpha$	30	40	Tungsten	LiF	Air	GM counter, Amperex	15 × 20 mm	4×10^4
Cr $K\alpha$	30	30	Tungsten	LiF	Air	GM counter, Amperex	15 × 20 mm	4×10^4
Ti $K\alpha$	30	40	Tungsten	LiF	Vacuum	Gas-flow counter	15 × 20 mm	4×10^4

* Apparatus: Shimadzu FX-403 (flat crystal). Collimation: Sollar slit, 25′–2°25′. X-ray tube: Machlett OEG-50S. Incident and emitted X-ray angle: $\Phi = 60°$, $\Psi = 30°$.

† Fixed counts.

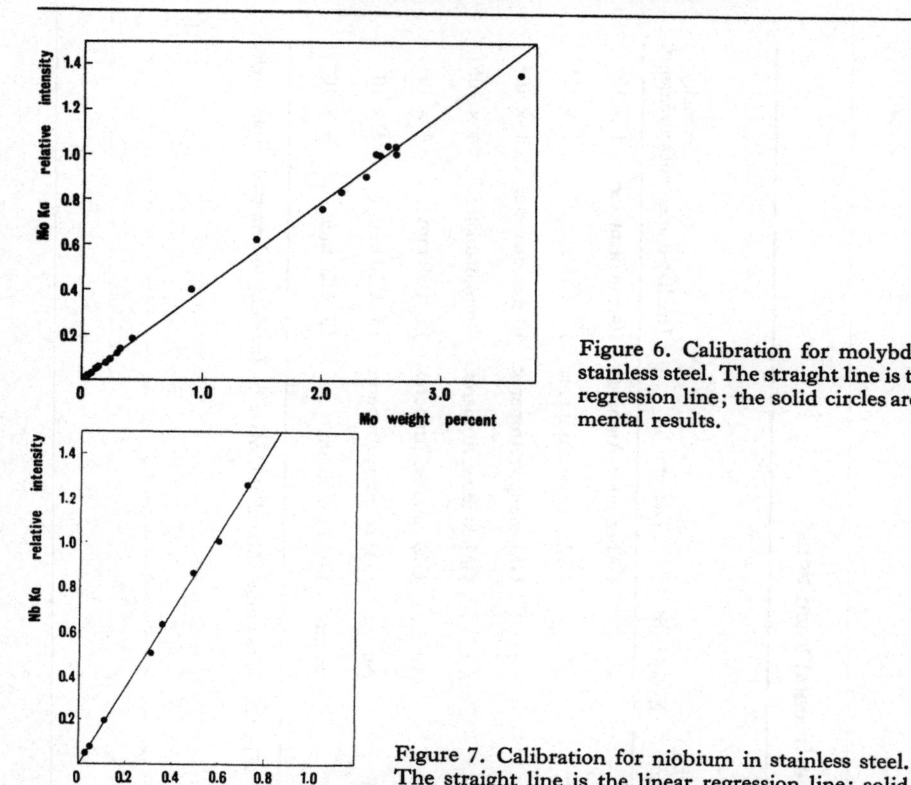

Figure 6. Calibration for molybdenum in stainless steel. The straight line is the linear regression line; the solid circles are experimental results.

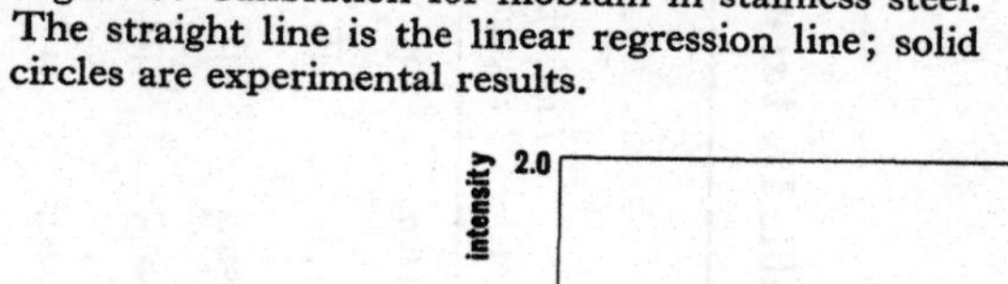

Figure 7. Calibration for niobium in stainless steel. The straight line is the linear regression line; solid circles are experimental results.

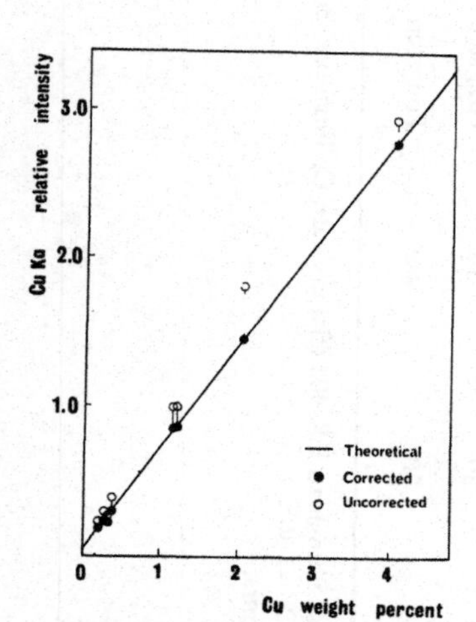

Figure 8. Calibration for copper in stainless steel. The curve is theoretical; open circles are uncorrected results; solid circles are corrected results.

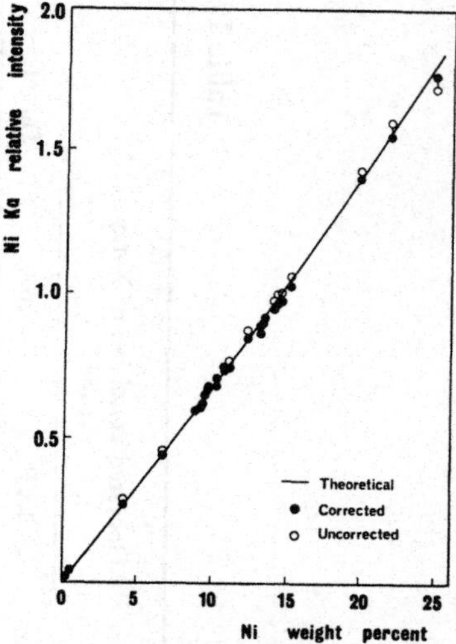

Figure 9. Calibration for nickel in stainless steel. The curve is theoretical; open circles are uncorrected results; solid circles are corrected results.

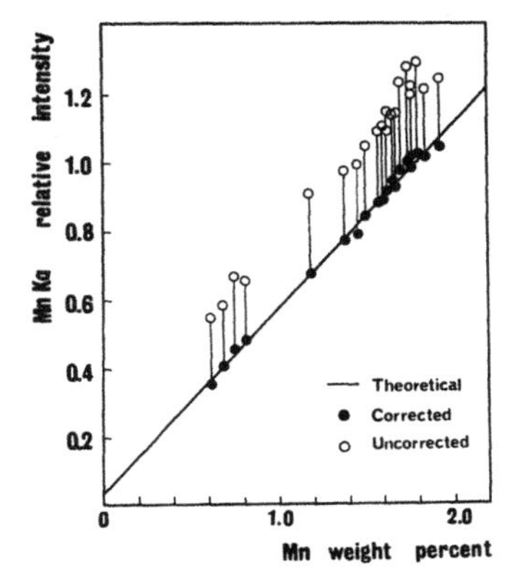

Figure 10. Calibration for manganese in stainless steel. The curve is theoretical; open circles are uncorrected results; solid circles are corrected results.

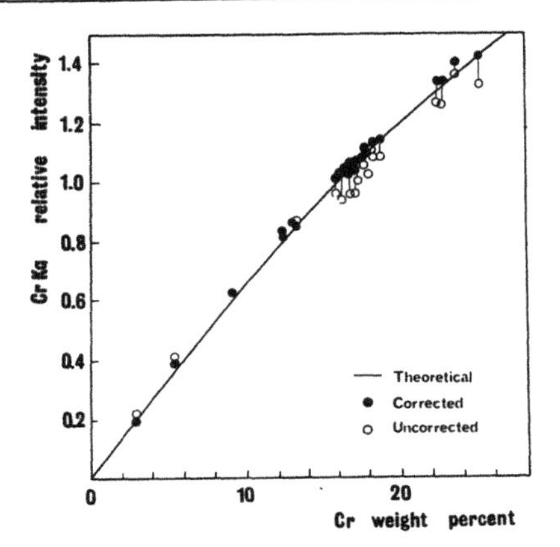

Figure 11. Calibration for chromium in stainless steel. The curve is theoretical; open circles are uncorrected results; solid circles are corrected results.

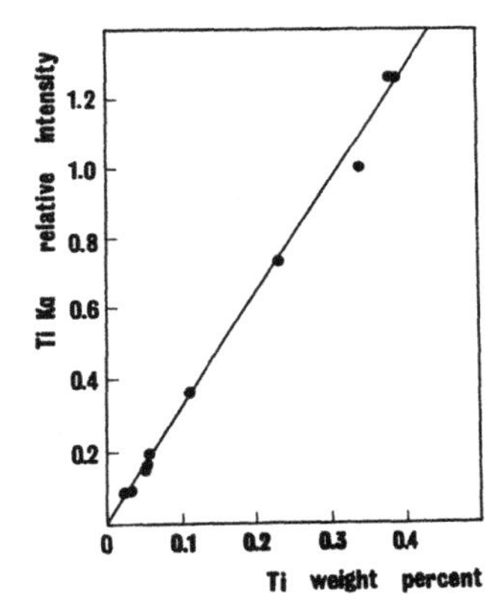

Figure 12. Calibration for titanium in stainless steel. The straight line is the linear regression line; solid circles are experimental results.

The observed intensities and results are shown in Tables IX to XII for nickel, chromium, copper, and manganese, respectively.

The standard deviations, listed as SD, of the differences between the chemical and the X-ray analysis are given in Table XIII, and the ratios of the standard deviation to the average composition, expressed as CV%, are also given.

Low-Alloy Steel

The experiments of the fluorescent X-ray analysis of low-alloy steels, pig iron, high-speed steels, and free-cutting steels are carried out on molybdenum, copper, nickel, manganese, chromium, vanadium, and titanium. The chemical compositions of the samples are listed in Table XIV. Experimental conditions are nearly the same as shown in Table IV, except for the analysis of vanadium and titanium with a chromium target tube. Experimental results for molybdenum, copper, nickel, manganese, chromium, vanadium, and titanium are shown in Figures 13 to 19, respectively. Fluorescent X-ray intensity is expressed by the value relative to that of the reference sample shown in Table XIV. The uncorrected results are obtained by linear regression.

Table V. Correction Factors A_{j-Fe} of Ni$K\alpha$ for Stainless Steels

SUS No.	Composition (wt.%)			Carbon	Silicon	Titanium	Manganese	Chromium	Copper	Niobium	Molybdenum
	Nickel	Chromium	Iron								
21	0	13	87	1.02	0.83	0.33	0.08	0.15	0.75	0.50	0.48
22	0	12.5	87.5	1.02	0.83	0.33	0.08	0.15	0.75	0.50	0.48
23	0	13	87	1.02	0.83	0.33	0.08	0.15	0.75	0.50	0.48
24	0	17	83	1.02	0.83	0.33	0.08	0.15	0.75	0.50	0.48
27	9.5	19	71.5	1.08	0.88	0.36	0.09	0.17	0.79	0.55	0.52
28	11	19	70	1.09	0.89	0.36	0.09	0.18	0.80	0.56	0.52
29	11	18.5	70.5	1.09	0.89	0.36	0.09	0.18	0.80	0.56	0.52
32	12	17	71	1.10	0.90	0.37	0.09	0.18	0.81	0.57	0.53
33	14	17	69	1.11	0.91	0.36	0.09	0.18	0.82	0.57	0.53
35	12	18	70	1.10	0.90	0.37	0.09	0.18	0.81	0.57	0.53
36	14	18	68	1.11	0.91	0.36	0.09	0.18	0.82	0.57	0.53
37	0	12.75	87.25	1.02	0.83	0.33	0.08	0.15	0.75	0.50	0.48
38	0	12.75	87.25	1.02	0.83	0.33	0.08	0.15	0.75	0.50	0.48
39	7	17	76	1.06	0.86	0.35	0.08	0.17	0.78	0.54	0.51
40	9.5	18	72.5	1.08	0.88	0.36	0.09	0.17	0.79	0.55	0.52
41	13.5	23	63.5	1.11	0.90	0.36	0.08	0.17	0.82	0.57	0.53
42	20.5	25	54.5	1.17	0.95	0.38	0.09	0.19	0.86	0.59	0.55
43	11	18.5	70.5	1.09	0.89	0.36	0.09	0.18	0.80	0.56	0.52
44	1.875	16	82.125	1.02	0.83	0.32	0.08	0.15	0.75	0.50	0.48

Table VI. Correction Factors $A_{j-\mathrm{Fe}} + B_{j-\mathrm{Fe}}$ of Cr $K\alpha$ for Stainless Steels

SUS No.	Composition (wt. %)			Carbon	Silicon	Titanium	Manganese	Nickel	Copper	Niobium	Molybdenum
	Nickel	Chromium	Iron								
21	0	13	87	0.72	−0.49	−3.15	−0.35	−0.33	−0.43	−2.26	−2.39
22	0	12.5	87.5	0.72	−0.49	−3.18	−0.35	−0.33	−0.43	−2.27	−2.40
23	0	13	87	0.72	−0.49	−3.15	−0.35	−0.32	−0.43	−2.26	−2.39
24	0	17	83	0.72	−0.47	−3.09	−0.34	−0.31	−0.42	−2.20	−2.32
27	9.5	19	71.5	0.67	−0.44	−2.87	−0.34	−0.28	−0.40	−2.03	−2.19
28	11	19	70	0.67	−0.44	−2.86	−0.34	−0.28	−0.40	−2.03	−2.18
29	11	18.5	70.5	0.67	−0.45	−2.89	−0.35	−0.29	−0.41	−2.05	−2.18
32	12	17	71	0.66	−0.46	−2.89	−0.35	−0.28	−0.41	−2.05	−2.19
33	14	17	69	0.66	−0.45	−2.86	−0.35	−0.29	−0.40	−2.03	−2.17
35	12	18	70	0.66	−0.44	−2.86	−0.34	−0.27	−0.40	−2.02	−2.17
36	14	18	68	0.67	−0.45	−2.86	−0.35	−0.29	−0.39	−2.02	−2.18
37	0	12.75	87.25	0.72	−0.49	−3.16	−0.35	−0.33	−0.43	−2.26	−2.38
38	0	12.75	87.25	0.72	−0.49	−3.16	−0.35	−0.33	−0.43	−2.26	−2.38
39	7	17	76	0.68	−0.47	−3.01	−0.35	−0.30	−0.42	−2.13	−2.26
40	9.5	18	72	0.65	−0.46	−2.90	−0.35	−0.29	−0.41	−2.03	−2.21
41	13.5	23	63.4	0.64	−0.42	−2.78	−0.35	−0.25	−0.38	−1.96	−2.12
42	20.5	25	54.5	0.63	−0.40	−2.65	−0.33	−0.22	−0.36	−1.87	−2.00
43	11	18.5	70.5	0.67	−0.45	−2.86	−0.35	−0.29	−0.41	−2.05	−2.18
44	1.875	16	82.125	0.69	−0.46	−3.07	−0.35	−0.34	−0.43	−2.19	−2.31

Table VII. Correction Factors A_{j-Fe} of Cu$K\alpha$ for Stainless Steels

Type of steel	Chromium	Nickel	Niobium	Molybdenum
SUS 21–44	0.16	0.63	0.55	0.52

Table VIII. Correction Factors A_{j-Fe} of Mn$K\alpha$ for Stainless Steels

Type of steel	Titanium	Niobium	Molybdenum
SUS 21–24 SUS 37–38		−0.95	−1.00
SUS 27–36 SUS 39–44	−1.60	−0.90	−0.95

Table IX. Experimental and Corrected Results for Nickel in Stainless Steel

				Uncorrected		Corrected
Specimen	No.	Chemical analysis (wt. %)	Ni $K\alpha$ relative intensity	Linear regression result (wt. %)	Parabolic regression result (wt. %)	Coexisting element (theoretical) (wt. %)
	1	14.44	1.000	14.44	14.38	14.31
	2	0.19	0.027	0.36	0.15	0.12
	3	0.18	0.026	0.34	0.13	0.11
	4	0.42	0.046	0.64	0.45	0.42
	5	15.10	1.055	15.24	15.12	14.97
	6	8.86	0.603	8.70	8.87	8.94
	7	9.36	0.632	9.11	9.29	9.36
	8	9.46	0.654	9.43	9.60	9.67
	9	10.58	0.715	10.31	10.46	10.55
	10	9.75	0.684	9.87	10.03	10.13
	11	10.33	0.694	10.02	10.17	10.26
	12	10.87	0.743	10.72	10.86	10.88
	13	12.31	0.840	12.12	12.21	12.19
	14	10.27	0.714	10.30	10.45	10.44
	15	14.10	0.973	14.05	14.02	14.03
	16	12.35	0.876	12.65	12.71	12.58
	17	21.84	1.595	23.05	21.88	21.65
	18	14.23	0.991	14.31	14.26	14.34
	19	13.28	0.917	13.24	13.26	13.33
	20	19.81	1.426	20.61	19.85	19.88
	21	11.04	0.759	10.95	11.08	11.14
	22	4.05	0.289	4.16	4.23	4.33
NBS	845	0.28	0.028	0.38	0.17	0.11
	846	9.11	0.612	8.82	9.00	9.00
	847	13.26	0.868	12.54	12.60	12.74
	848	0.52	0.049	0.67	0.50	0.32
	849	6.62	0.462	6.66	6.82	6.74
	850	24.80	1.664	24.05	23.28	24.16
	1184	9.47	0.664	9.58	9.75	9.74
	1185	13.18	0.899	12.99	13.02	13.20
SD				0.372	0.337	0.215
CV %				3.59	3.26	2.08

Table X. Experimental and Corrected Results for Chromium in Stainless Steel

Specimen No.		Chemical analysis (wt. %)	Cr $K\alpha$ relative intensity	Uncorrected: Linear regression result (wt. %)	Uncorrected: Parabolic regression result (wt. %)	Corrected: Coexisting element (theoretical) (wt. %)
	1	17.68	1.000	17.24	17.10	18.06
	2	12.95	0.865	14.52	14.32	13.53
	3	12.41	0.820	13.62	13.40	12.63
	4	16.62	1.031	17.87	17.74	16.84
	5	18.34	0.978	16.79	16.63	17.94
	6	18.70	1.093	19.12	19.04	18.60
	7	17.65	1.030	17.85	17.72	17.37
	8	18.57	1.097	19.19	19.12	18.55
	9	18.33	1.083	18.91	18.83	18.47
	10	17.27	1.002	17.29	17.14	16.91
	11	17.76	1.060	18.44	18.35	17.38
	12	17.32	0.992	17.08	16.93	17.36
	13	17.28	0.962	16.48	16.32	17.12
	14	16.25	0.937	15.97	15.79	16.35
	15	16.92	0.959	16.42	16.24	17.02
	16	17.75	0.988	17.01	16.86	17.59
	17	17.20	0.949	16.21	16.04	17.28
	18	22.71	1.264	22.55	22.64	22.67
	19	22.61	1.267	22.61	22.70	22.59
	20	25.16	1.329	23.85	24.02	24.58
	21	18.00	1.025	17.74	17.62	17.58
	22	16.03	0.962	16.47	16.31	15.87
NBS	845	13.31	0.871	14.64	14.45	13.79
	846	18.35	1.080	18.85	18.76	18.39
	847	23.72	1.362	24.52	24.72	24.36
	848	9.09	0.630	9.79	9.55	9.27
	849	5.48	0.401	5.20	5.03	5.28
	850	2.99	0.219	1.52	1.50	2.57
	1184	19.44				
	1185	17.09				
SD				0.847	0.845	0.311
CV %				5.08	5.07	1.87

In these experiments, the corrections are not necessary except for manganese and chromium. For manganese and chromium, it is obvious that there are influences of the coexistent elements upon the X-ray intensities and the relation between the fluorescent X-ray intensity and composition is not linear.

The theoretical corrections are applied to manganese and chromium in the same manner as for the stainless steel analysis.

The theoretical correction factors for Mn $K\alpha$ and Cr $K\alpha$ shown in Table XV are obtained for the composition of 5% manganese–95% iron and 5% chromium–95% iron, respectively.

Table XI. Experimental and Corrected Results for Copper in Stainless Steel

			Uncorrected		Corrected	
Specimen No.	Chemical analysis (wt. %)	Cu $K\alpha$ relative intensity	Linear regression result (wt. %)	Parabolic regression result (wt. %)	Overlapping (experi-mental) (wt. %)	Coexisting element (theoretical) (wt. %)
1	1.21	1.000	1.19	1.122	1.26	1.20
2	0.19	0.886	0.15	0.155	0.17	0.19
3	0.11	0.135	0.09	0.100	0.10	0.12
4	0.10	0.134	0.09	0.099	0.09	0.12
5	0.17	0.219	0.19	0.192	0.17	0.17
6	0.16	0.180	0.14	0.149	0.14	0.15
7	0.16	0.174	0.14	0.143	0.13	0.14
8	0.14	0.174	0.14	0.142	0.13	0.14
9	0.12	0.162	0.12	0.130	0.11	0.12
10	0.10	0.153	0.11	0.119	0.10	0.11
11	0.10	0.131	0.09	0.096	0.09	0.08
12	0.21	0.228	0.20	0.201	0.20	0.20
13	0.11	0.144	0.10	0.110	0.08	0.08
14	0.20	0.219	0.19	0.192	0.19	0.19
15	0.30	0.294	0.28	0.276	0.28	0.27
16	1.17	0.989	1.18	1.108	1.23	1.19
17	2.07	1.789	2.28	2.200	2.29	2.06
18	0.10	0.172	0.14	0.141	0.11	0.12
19	0.07	0.119	0.07	0.083	0.04	0.05
20	0.09	0.187	0.15	0.157	0.12	0.11
21	0.27	0.289	0.27	0.270	0.27	0.29
22	4.05	2.938	3.85	4.017	3.91	4.05
NBS 845	0.065	0.096	0.045	0.058	0.04	0.06
846	0.19	0.225	0.20	0.199	0.20	0.21
847	0.19	0.221	0.19	0.194	0.18	0.18
848	0.16	0.187	0.15	0.157	0.15	0.20
849	0.21	0.226	0.20	0.200	0.20	0.23
850	0.36	0.395	0.40	0.395	0.29	0.36
1184	–					
1185	0 067					
SD			0.043	0.039	0.055	0.017
CV %			9.77	8.86	12.5	3.87

In Tables XVI and XVII for manganese and chromium the uncorrected and the corrected X-ray intensities and the results of analysis are given. The final corrected intensities are shown Figures 16 and 17 with the observed intensities.

The standard deviations of the analysis are shown in Table XVIII.

Iron Ores

In the fluorescent X-ray analysis of iron in iron ore, it has been known that the working curves of iron are different according to the type of ore, and a theoretical calculation has revealed[3] the reason. In the calculation, oxygen is considered the coexistent

Table XII. Experimental and Corrected Results for Mn in Stainless Steel

		Uncorrected			Corrected	
Specimen No.	Chemical analysis (wt. %)	Mn $K\alpha$ relative intensity	Linear regression result (wt. %)	Parabolic regression result (wt. %)	Overlapping (experimental) (wt. %)	Coexisting element (theoretical) (wt. %)
1	1.47	1.000	1.40	1.44	1.41	1.43
2	0.69	0.589	0.67	0.66	0.69	0.69
3	0.82	0.657	0.79	0.81	0.83	0.84
4	0.76	0.669	0.81	0.83	0.77	0.78
5	1.62	1.090	1.57	1.58	1.59	1.68
6	1 63	1.145	1.66	1.67	1.65	1.63
7	1.77	1.220	1.80	1.77	1.82	1.80
8	1.19	0.910	1.24	1.28	1.20	1.20
9	1.59	1.106	1.59	1.61	1.58	1.56
10	1.59	1.095	1.57	1.59	1.59	1.59
11	1.68	1.143	1.66	1.66	1.66	1.65
12	1.39	0.977	1.36	1.40	1.37	1.36
13	1.94	1.241	1.83	1.80	1.88	1.90
14	1.85	1.212	1.78	1.76	1.84	1.86
15	1.62	1.092	1.57	1.58	1.60	1.62
16	1.51	1.047	1.49	1.51	1.51	1.53
17	1.67	1.134	1.64	1.65	1.69	1.70
18	1.71	1.234	1.82	1.79	1.75	1.73
19	1.81	1.288	1.92	1.87	1.85	1.83
20	1.75	1.276	1.90	1.85	1.80	1.78
21	1.77	1.197	1.76	1.74	1.77	1.77
22	0.61	0.551	0.60	0.58	0.58	0.59
NBS 845	0.77					
846	0.53					
847	0.23					
848	2.13					
849	1.63					
850	–					
1184	1.04					
1185	1.22					
SD			0.064	0.053	0.031	0.027
CV %			4.32	3.58	2.10	1.82

element, and it was shown that the difference in the coexistent elements (including oxygen) causes the difference in the iron fluorescent X-ray intensity of the same weight fraction of iron for various types of ore. In Figure 20, the solid curve shows the iron fluorescent X-ray intensity for the iron–oxygen binary system, obtained from theoretical calculation. If the ore contains impurities, such as SiO_2, Al_2O_3, and CaO, the iron X-ray intensity decreases for each of the iron oxides—magnetite Fe_3O_4, hematite Fe_2O_3, goethite $Fe_2O_3 \cdot H_2O$, and limonite $2Fe_2O_3 \cdot 3H_2O$—as shown along the broken curves.

In the actual ores, the quantity of impurity is up to 10 or 20%, and the influence from the difference in oxygen is not negligible. So different working curves are obtained for different types of ore.

Table XIII. Comparison of Uncorrected and Corrected Results of Stainless Steel

Element	Average composition (wt. %)	Uncorrected				Corrected			
		Linear regression		Parabolic regression		Overlapping (experimental)		Coexisting element (theoretical)	
		SD	CV %	SD	CV %	SD	CV %	SD	CV %
Molybdenum	0.97	0.049	5.05	–	–	–	–	–	–
Niobium	0.35	0.012	3.43	–	–	–	–	–	–
Copper	0.44	0.043	9.77	0.039	8.86	0.055	12.5	0.017	3.87
Nickel	10.34	0.372	3.59	0.337	3.26	–	–	0.215	2.08
Manganese	1.48	0.064	4.32	0.053	3.58	0.031	2.10	0.027	1.82
Chromium	16.68	0.847	5.08	0.845	5.07	–	–	0.311	1.87
Titanium	0.18	0.013	7.22	–	–	–	–	–	–

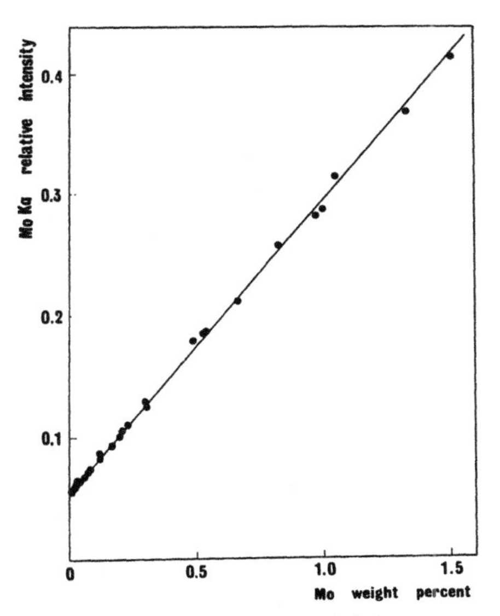

Figure 13. Calibration for molybdenum in low-alloy steel. The straight line is the linear regression line; solid circles are experimental results.

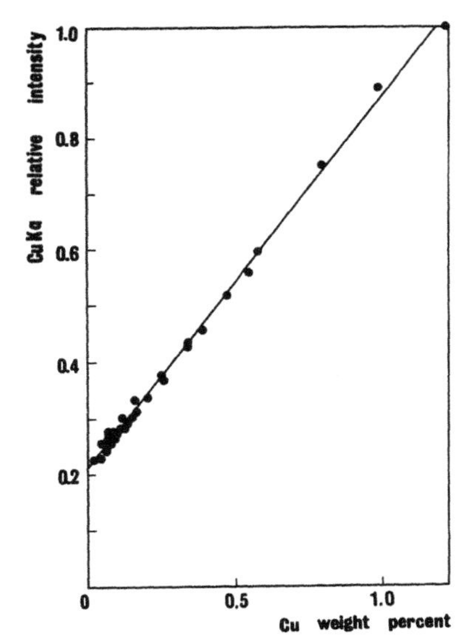

Figure 14. Calibration for copper in low-alloy steel. The straight line is the linear regression line; solid circles are experimental results.

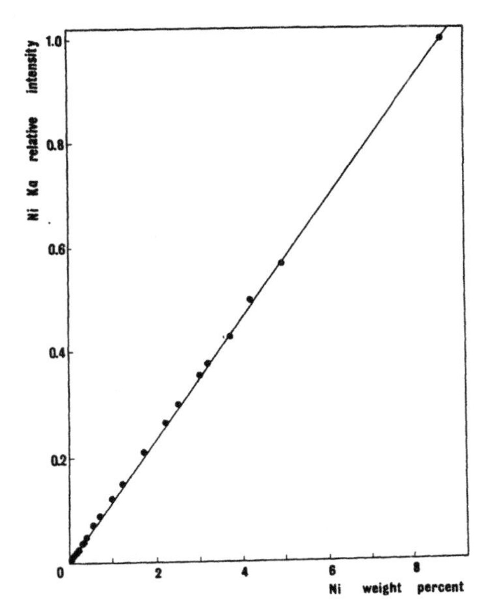

Figure 15. Calibration for nickel in low-alloy steel. The straight line is the linear regression line; solid circles are experimental results.

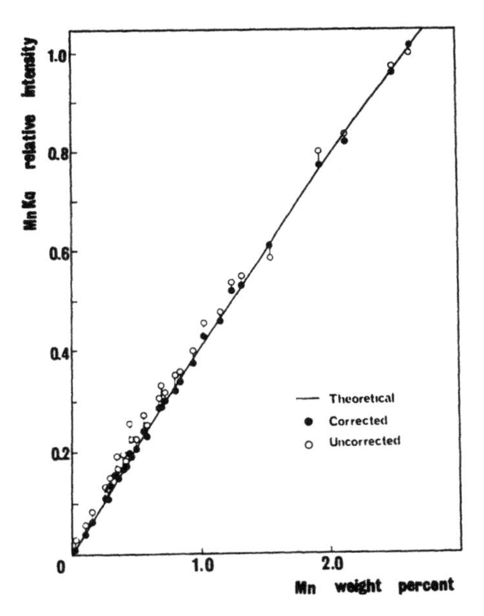

Figure 16. Calibration for manganese in low-alloy steel. The curve is theoretical; open circles are uncorrected results; solid circles are corrected results.

Table XIV. Compositions of the

Specimen No.	JIS No.	ASTM (AISI)	Carbon	Silicon	Manganese	Phosphorus	Sulfur	Nickel	Chromium
101			0.015	3.02	2.13	0.019	0.031	3.05	0.008
102			0.098	2.14	1.93	0.039	0.064	0.016	2.09
103			0.26	1.94	2.50	0.069	0.017	0.23	0.30
104			0.13	1.45	2.64*	0.071	0.040	0.041	
105	Pig iron		4.70	0.90	0.80	0.176	0.012	0.009	0.035
105′	Pig iron		4.16	0.32	0.49	0.162	0.052	0.020	0.017
106	Pig iron		3.49	3.01	0.58	0.125	0.036		
107	STBL46	A333-64	0.11	0.29	0.50	0.010	0.012	3.75	0.19
108	SNCM25	–	0.17	0.27	0.48	0.014	0.013	4.23	0.82
109	SNCM5	–	0.28	0.26	0.46	0.012	0.013	2.56	2.75
110	SKD61	H13	0.37	1.09	0.35	0.020	0.006	0.13	5.19
111	STBA26	A213-65	0.14	0.67	0.45	0.017	0.013	0.06	8.68*
112	SKD4	H20	0.32	0.23	0.30	0.010	0.010	0.05	2.61
113	SKS42	–	0.79	0.23	0.44	0.026	0.006	0.06	0.46
114	SKS2	O7	1.07	0.27	0.72	0.023	0.007	0.105	0.89
115	SKS11	F3	1.12	0.30	0.42	0.012	0.012	0.09	0.33
116	SKD12	A2	1.01	0.31	0.69	0.016	0.014	0.09	5.04
117	SUMIA	–	0.12	0.28	0.68	0.160	0.190	0.023	0.06
118	SFC3F	–	0.12	0.05	1.24	0.079	0.309	0.10	0.11
119	Low-temperature service steel		0.081	0.21	0.42	0.010	0.014	8.73*	0.065
120	SKH9	M2	0.85	0.28	0.30	0.020	0.014	0.08	4.02
121	Free-cutting steel		0.99	0.21	0.41	0.024	0.056	0.04	0.10
122			0.10	0.03	1.15	0.058	0.318	0.06	0.17
NBS 1161			0.15	0.047	0.36	0.053		1.73	0.13
1162			0.40	0.28	0.94	0.045		0.70	0.74
1163			0.19	0.41	1.15	0.031		0.39	0.26
1164			0.54	0.48	1.32	0.017		0.135	0.078
1165			0.04	0.029	0.032	0.008		0.026	0.044
1166			0.07	0.025	0.113	0.012		0.051	0.011
1167			0.11	0.26	0.275	0.033		0.088	0.036
1168			0.26	0.075	0.47	0.023		1.03	0.54
BAS SS 1/1				0.41	1.54			1.24	0.51
2/1				0.28	0.34			2.23	0.42
3/1				0.65	0.84			1.00	0.99
4/1				0.22	0.26			3.26	0.27
5/1				0.15	0.16			4.98	0.19
6/1				0.23	1.02			0.19	2.33
7/1				0.10	0.56			0.27	2.97
8/1				0.96	0.40			0.55	1.29

* Reference sample.

The theoretical correction is applied to iron ores. But the results of correction are not satisfactory because the gradient of the intensity curve is low and a small variation in the X-ray intensity corresponds to a large deviation in the iron weight fraction. The inaccuracy of X-ray intensity measurement is mainly due to a heterogeneity of powder size.

In order to make the gradient steep, a technique which is a sort of dilution method has been developed. Alumina is chosen as the diluent material because it has the qualities of reasonable cost and being commonly available, powdery, stable, dry, of high purity, and easy to briquette. The appropriate dilution rate, three times alumina to ore in weight, is chosen from the theoretical point. The inaccuracy of uncorrected values decreases because of the dilution of impurities, but the standard deviation of the error remains 1.70%. Then, the theoretical correction of the coexistent element is applied. In the

Examined Low-Alloy Steels

Molybdenum	Copper	Tungsten	Vanadium	Lead	Cobalt	Tin	Arsenic	Titanium	Niobium
0.009	0.58								
0.002	0.80								
0.49	0.99								
0.036	1.23*	0.89	0.51						
0.002	0.060								
0.002	0.044								
0.03	0.116								
0.21	0.07								
0.54	0.16								
1.33	0.11		0.89						
1.00	0.07								
0.08	0.07	5.06	0.34						
0.12	0.08	2.26	0.23						
0.06	0.08	1.45							
0.02	0.07	3.25	0.166						
0.97	0.12		0.415						
	0.04								
0.02	0.13			0.179					
0.001	0.061				0.019				
4.47*	0.056	6.70	1.90*						
0.02	0.09			0.20					
0.025	0.14			0.185					
0.030	0.34	0.012	0.024		0.26	0.022	0.028		0.011
0.080	0.20	0.053	0.058	0.006	0.11	0.066	0.046	0.037	0.096
0.12	0.47	0.105	0.10	0.012	0.013	0.013	0.10	0.010	0.195
0.029	0.094	0.022	0.295	0.020	0.028	0.043	0.018	0.004	0.037
0.005	0.019		0.002		0.008	0.001	0.010	0.20	
0.011	0.033		0.007		0.46	0.005	0.014	0.057	0.005
0.021	0.067	0.20	0.041	0.006	0.074	0.10	0.14	0.26*	0.29*
0.020	0.26	0.077	0.17		0.16	0.009	0.008	0.011	0.006
1.51	0.55		0.65						
1.05	0.20		0.23						
0.67	0.39		0.51						
0.23	0.34		0.15						
0.30	0.25		0.21						
0.53	0.16		0.18						
0.17	0.13		0.12						
0.83	0.15		0.35						

Table XV. Correction Factors $A_{j-\mathrm{Fe}}$ of Manganese and Chromium for Low-Alloy Steel (Correction Factors $B_{j-\mathrm{Fe}}$ are Contained for Chromium)

	Carbon	Silicon, phosphorus, and sulfur	Vanadium	Titanium	Molybdenum and niobium	Tungsten	Lead
Mn $K\alpha$*	0.8	−0.2	−2.4	−2.4	−1.7	−2.8	−2.8
Cr $K\alpha$†	0.8	−0.5	neg.	−3.2	−2.4	−3.1	−3.1

* Mn $K\alpha$: Copper, nickel, cobalt, and chromium are negligible.
† Cr $K\alpha$: Copper, nickel, cobalt, and manganese are negligible.

theoretical correction for iron ores, it is assumed that the exchange element is oxygen, and iron fluorescent X-ray intensity is determined on the diluted iron–oxygen binary system. The calculated correction factors are shown in Table XIX.

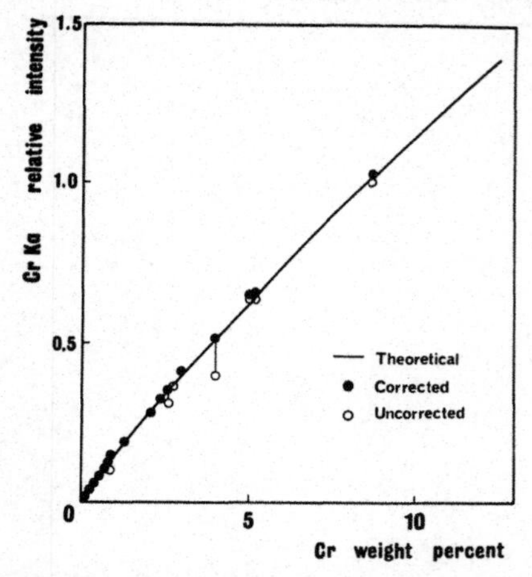

Figure 17. Calibration for chromium in low-alloy steel. The curve is theoretical; open circles are uncorrected results; solid circles are corrected results.

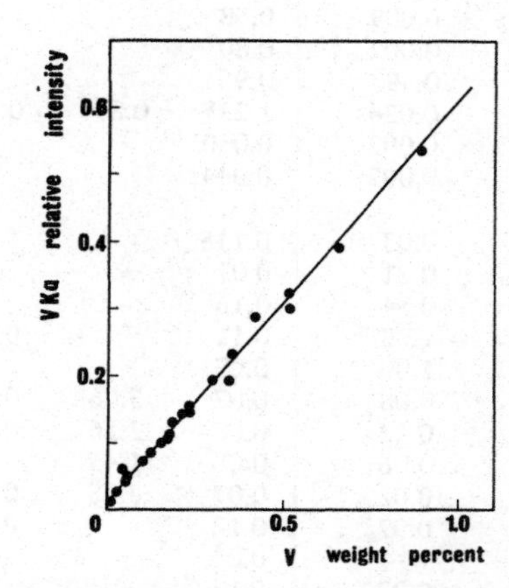

Figure 18. Calibration for vanadium in low-alloy steel. The straight line is the linear regression line; solid circles are experimental results.

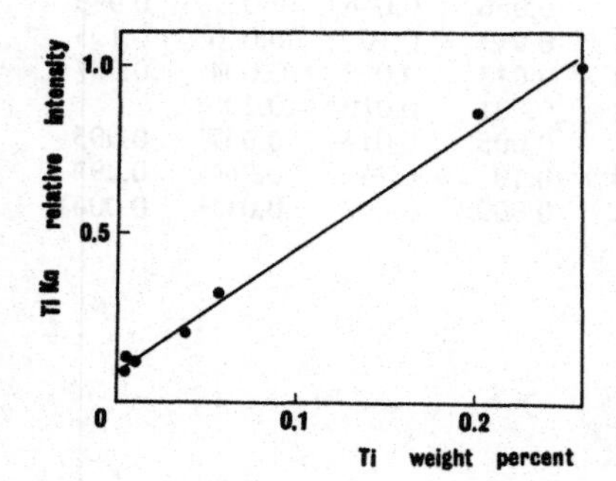

Figure 19. Calibration for titanium in low-alloy steel. The straight line is the linear regression line; solid circles are experimental results.

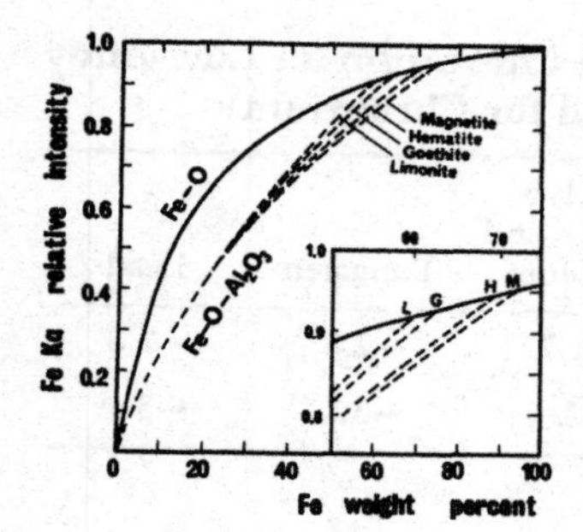

Figure 20. Theoretical intensity of Fe $K\alpha$ fluorescent X-rays from the iron–oxygen binary system shown by the solid curve; decreases of Fe $K\alpha$ intensity by impurity are shown as broken curves.

Table XVI. Experimental and Corrected Results for Manganese in Low-Alloy Steel

Specimen No.	Chemical analysis (wt. %)	Relative intensity	Uncorrected: Linear regression result (wt. %)	Uncorrected: Parabolic regression result (wt. %)	Corrected: Overlapping (experimental) (wt. %)	Corrected: Coexisting element (theoretical) (wt. %)
101	2.13	0.835	2.087	2.106	2.098	2.088
102	1.93	0.800	1.994	2.001	1.972	1.947
103	2.50	0.972	2.445	2.521	2.452	2.490
104	2.64	1.000	2.520	2.607	2.529	2.663
105	0.80	0.354	0.821	0.788	0.837	0.770
105′	0.49	0.234	0.507	0.494	0.524	0.493
106	0.58	0.257	0.566	0.549	0.584	0.552
107	0.50	0.230	0.495	0.483	0.509	0.496
108	0.48	0.228	0.491	0.480	0.495	0.484
109	0.46	0.230	0.497	0.485	0.472	0.466
110	0.35	0.196	0.407	0.404	0.349	0.336
111	0.45	0.260	0.574	0.555	0.471	0.469
112	0.30	0.151	0.289	0.298	0.271	0.326
113	0.44	0.199	0.414	0.410	0.424	0.446
114	0.72	0.320	0.731	0.703	0.733	0.727
115	0.42	0.185	0.378	0.377	0.391	0.422
116	0.69	0.333	0.768	0.737	0.707	0.691
117	0.68	0.309	0.703	0.676	0.719	0.690
118	1.24	0.536	1.301	1.262	1.313	1.267
119	0.42	0.195	0.403	0.400	0.420	0.415
120	0.30	0.133	0.241	0.255	0.212	0.286
121	0.41	0.193	0.400	0.397	0.416	0.411
122	1.15	0.522	1.263	1.223	1.274	1.230
NBS 1161	0.36	0.170	0.336	0.341	0.353	0.358
1162	0.94	0.401	0.945	0.907	0.948	0.911
1163	1.15	0.478	1.147	1.106	1.157	1.122
1164	1.32	0.549	1.336	1.297	1.348	1.305
1165	0.032	0.029	−0.032	0.179	−0.012	0.037
1166	0.113	0.059	0.046	0.087	0.065	0.103
1167	0.275	0.129	0.230	0.246	0.248	0.268
1168	0.47	0.212	0.449	0.442	0.458	0.453
BAS SS 1/1	1.54	0.609	1.491	1.459	1.496	1.510
2/1	0.34	0.160	0.312	0.319	0.323	0.336
3/1	0.84	0.361	0.840	0.806	0.839	0.822
4/1	0.26	0.134	0.243	0.257	0.256	0.272
5/1	0.16	0.085	0.114	0.146	0.130	0.161
6/1	1.02	0.456	1.091	1.051	1.068	1.035
7/1	0.56	0.274	0.612	0.591	0.602	0.584
8/1	0.40	0.198	0.411	0.407	0.407	0.414
SD			0.044	0.028	0.041	0.023
CV %			5.79	3.68	5.39	3.03

Table XVII. Experimental and Corrected Results for Chromium in Low-Alloy Steel

			Uncorrected		Corrected
Specimen No.	Chemical analysis (wt. %)	Relative intensity	Linear regression result (wt. %)	Parabolic regression result (wt. %)	Coexisting element (theoretical) (wt. %)
101	0.008	0.005	−0.061	−0.012	0.008
102	2.09	0.280	2.262	2.198	2.060
103	0.30	0.050	0.310	0.324	0.316
104					
105	0.035	0.009	−0.028	0.017	0.035
105′	0.017	0.007	−0.046	−0.017	0.006
106					
107	0.19	0.033	0.176	0.197	0.196
108	0.82	0.119	0.902	0.872	0.812
109	2.75	0.361	2.949	2.867	2.706
110	5.19	0.638	5.288	5.276	5.210
111	8.68	1.000	8.356	8.616	8.655
112	2.61	0.307	2.493	2.420	2.608
113	0.46	0.066	0.456	0.456	0.456
114	0.89	0.127	0.966	0.924	0.888
115	0.33	0.047	0.289	0.313	0.324
116	5.04	0.641	5.321	5.295	5.129
117	0.06	0.012	−0.008	0.022	0.039
118	0.11	0.021	0.072	0.104	0.113
119	0.065	0.013	−0.002	0.048	0.062
120	4.02	0.395	3.240	3.175	3.968
121	0.10	0.020	0.064	0.097	0.104
122	0.17	0.028	0.127	0.154	0.158
NBS 1161	0.13	0.022	0.083	0.119	0.127
1162	0.74	0.109	0.819	0.792	0.742
1163	0.26	0.042	0.251	0.269	0.264
1164	0.078	0.018	0.044	0.083	0.093
1165	0.004	0.005	−0.064	−0.024	−0.003
1166	0.011	0.005	−0.060	−0.008	0.012
1167	0.036	0.009	−0.032	0.017	0.034
1168	0.54	0.082	0.589	0.580	0.546
BAS SS 1/1	0.51	0.079	0.562	0.547	0.530
2/1	0.42	0.064	0.439	0.435	0.421
3/1	0.99	0.144	1.113	1.056	0.996
4/1	0.27	0.044	0.265	0.285	0.277
5/1	0.19	0.033	0.175	0.207	0.206
6/1	2.33	0.318	2.581	2.454	2.314
7/1	2.97	0.405	3.319	3.173	2.980
8/1	1.29	0.185	1.461	1.372	1.300
SD			0.169	0.158	0.030
CV %			13.95	13.05	2.48

Table XVIII. Comparison of Uncorrected and Corrected Results of Low-Alloy Steel

		Uncorrected				Corrected			
		Linear regression		Parabolic regression		Overlapping (experimental)		Coexisting element (theoretical)	
Element	Average composition (wt. %)	SD	CV %	SD	CV %	SD	CV %	SD	CV %
Molybdenum	0.320	0.013	4.06	–	–	–	–	–	–
Copper	0.232	0.018	7.80	–	–	–	–	–	–
Nickel	1.087	0.019	1.75	–	–	–	–	–	–
Manganese	0.760	0.044	5.79	0.028	3.68	0.041	5.39	0.023	3.03
Chromium	1.210	0.169	13.95	0.158	13.05	–	–	0.030	2.48
Vanadium	0.250	0.021	8.40	–	–	–	–	–	–
Titanium	0.083	0.0085	1.02	–	–	–	–	–	–

Table XIX. Correction Factors of A_{j-O} for the Alumina-Diluted Iron Ore

j \ W_{Fe}	Fe	Fe_3O_4	Fe_2O_3	$Fe_2O_3 \cdot H_2O$	$Fe_2O_3 \cdot \frac{3}{2}H_2O$	Fe_{55}–O_{45}	Fe_{50}–O_{50}
	1.000	0.724	0.700	0.629	0.599	0.550	0.500
Al_2O_3	−0.45	−0.50	−0.50	−0.52	−0.53	−0.54	−0.56
SiO_2	−0.52	−0.58	−0.59	−0.61	−0.62	−0.64	−0.65
CaF_2	−1.75	−1.96	−1.98	−2.04	−2.07	−2.12	−2.17
CaO	−2.35	−2.64	−2.67	−2.75	−2.79	−2.86	−2.93
TiO_2	−2.54	−2.85	−2.88	−2.97	−3.02	−3.09	−3.16
MnO	−0.88	−1.02	−1.03	−1.07	−1.08	−1.11	−1.14
Mn	−1.42	−1.61	−1.63	−1.69	−1.71	−1.76	−1.81
MgO	−0.37	−0.42	−0.43	−0.44	−0.45	−0.45	−0.46
S	−1.74	−1.95	−1.98	−2.04	−2.07	−2.13	−2.20
P	−1.41	−1.58	−1.60	−1.65	−1.67	−1.71	−1.75

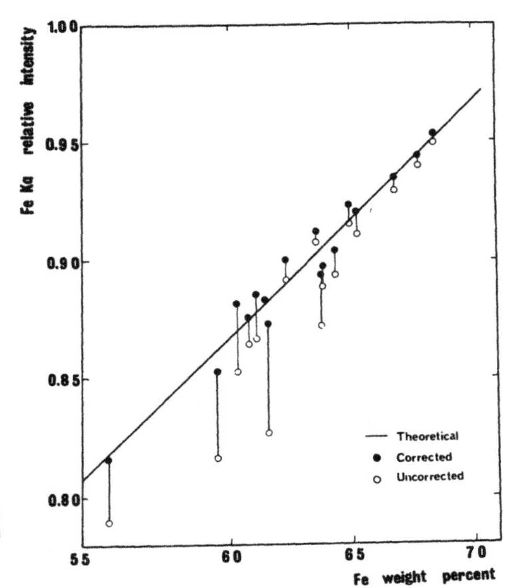

Figure 21. Calibration of iron in iron ores with the use of the alumina-dilution method. The curve is theoretical; open circles are uncorrected results; solid circles are corrected results.

Experimental results are shown in Figure 21. Standard deviation decreases from 1.70 to 0.44%.

CONCLUSION

The fluorescent X-ray intensity is affected by the variation of the coexistent element, and it has been shown that the effect is expressed in a linear correction term when the variation of such elements is small. The correction factors are theoretically given in derivatives of the theoretical formulas of the fluorescent X-ray intensity, and they are easily obtained by computation with an electronic computer and without experiments.

In order to know the magnitude of the correction term, those values for the nickel–iron–chromium ternary system and for iron ores are calculated when a tungsten target tube operated at 30 kVP was used. The validity of the linear approximation is also confirmed theoretically.

Applications of the present theoretical correction formula are carried out on stainless steels, low-alloy steels, and iron ores, and good results are obtained. The standard deviations between the chemical and the X-ray results decrease with the use of theoretical corrections. In stainless steel analysis, the standard deviations of copper, nickel, manganese, and chromium decrease from 0.055 to 0.017%, 0.372 to 0.215%, 0.064 to 0.027%, and 0.847 to 0.311%, respectively; in low-alloy steel analysis, those of manganese and chromium decrease from 0.044 to 0.023% and 0.169 to 0.030%, respectively; and in iron ore analysis, those of iron decrease from 1.70 to 0.44%.

It is found that the theoretical values have about 1 or 2% relative errors, and the derivatives of them also have relative errors of the same order of magnitude. But the composition ranges of coexistent elements are usually small, a few percent at most in routine work; thus, the theoretical correction values can be used in routine analysis.

NOTATION

Φ The angle made by the incident beam with the sample surface

Ψ The angle made by the emergent beam with the sample surface

$I_0(\lambda)$ The intensity of X-ray of wavelength λ

I_1 The intensity of the primary fluorescent X-rays (emitted by the incident X-rays)
I_2 The intensity of the secondary fluorescent X-rays (emitted by the primary fluorescent X-rays)
I_3 The intensity of the tertiary fluorescent X-rays (emitted by the secondary fluorescent X-rays)
$I(ip)$ The intensity of the X-ray of a p-line emitted from element i, where i refers to the element and p refers to the spectral line
$\mu(\lambda)$ The linear absorption coefficient of the specimen for the X-ray of wavelength λ
$\mu(ip)$ The linear absorption coefficient of the specimen for the characteristic X-ray p line of the element i
$\mu_j(\lambda)$ The linear absorption coefficient of the element j for the X-ray of wavelength λ
$\mu_j(ip)$ The linear absorption coefficient of element j for the characteristic X-ray p line of element i
ρ The density of the specimen
ρ_i The density of element i
W_i The weight fraction of element i in the specimen
ω_i Fluorescent yield of element i
R_p^i The intensity fraction of the ip line in the characteristic X-ray series to which ip line belongs
K_i The absorption jump of element i
λ_m The minimum wavelength of the exciting X-rays
λ_e^i The wavelength of the absorption edge of element i

ACKNOWLEDGMENT

The authors would like to acknowledge with sincere gratitude the encouragement given them by Dr. M. Sumitomo, Director of Central Research Laboratories, Sumitomo Metal Industries.

REFERENCES

1. T. Shiraiwa and N. Fujino, "Theoretical Calculation of Fluorescent X-Ray Intensities in Fluorescent X-Ray Spectrochemical Analysis," *Japan. J. Appl. Phys.* **5**: 886–899, 1966.
2. T. Shiraiwa and N. Fujino, "Theoretical Calculation of Fluorescent X-Ray Intensities of Nickel–Iron–Chromium Ternary Alloys," *Bull. Chem. Soc. Japan* **40**: 2289–2296, 1967.
3. T. Shiraiwa and N. Fujino, "Application of Theoretical Calculations to X-Ray Fluorescent Analysis," *The Pittsburgh Conference on Analytical Chemistry and Applied Spectroscopy, Paper No. 284*, 1967.
4. W. Marti, "Determination of the Interelement Effect in the X-Ray Fluorescent Analysis of Cr in Steels," *Spectrochim. Acta* **17**: 379–383, 1961.
5. L. S. Birks (Private communications and his lectures in Japan, 1963).
6. M. Sugimoto, "Correction Method for Matrix Effect in Fluorescent X-Ray Analysis I, II," *Bunseki Kagaku* (*Japan Analyst*) **11**: 1168–1176, 1962; **12**: 475–483, 1963.
7. T. Adachi and M. Ito, "On Variation of Correction Factors of Absorption and Enhancement Effects," *Denki Seiko* (*Electric Steel Making*) **36**: 68–74, 1965.

MICRO FLUORESCENT X-RAY ANALYZER

Toshio Shiraiwa and Nobukatsu Fujino

Central Research Laboratories, Sumitomo Metal Industries
Amagasaki, Japan

ABSTRACT

A micro fluorescent X-ray analyzer with a focusing type of spectrometer has been developed to analyze samples of small amounts such as extracted inclusions or precipitates from metals or small areas in samples from 0.1 to 2.0 mm in diameter. This instrument is expected especially to analyze powder samples of small quantity because average values from such samples can be obtained and because surface conditions of the samples scarcely affect the results compared with their effect in electron probe microanalysis. A commercial X-ray tube is combined with a device of slits limiting incident X-rays, a focusing spectrometer with a Rowland circle of 4-in. radius, and a microscope of low magnification for observing the analyzing point on the samples. The wavelength range of the spectrometer with LiF and ADP analyzing crystals is from 1.20 to 9.94 Å, and, therefore, higher elements than aluminum in atomic number can be analyzed. The authors exerted their efforts to obtain the higher X-ray intensities in order to analyze smaller areas. The X-ray intensities obtained are satisfactory, except for light elements. For example, the detected X-ray intensity of pure nickel is 1650 cps with the use of a 0.1-mm-diameter specimen, and that of pure sulfur is 52 cps with the use of a 0.1-mm-diameter specimen; however, with a 1-mm-diameter specimen, the intensity of pure nickel is over 5000 cps and that of pure sulfur is 1650 cps. These correspond to the intensities from 20-mm-diameter specimens of those elements when a flat-crystal spectrometer is used. The calibration curve for quantitative analysis generally varies with the sample area under analysis, but the same curves are obtained if the sample area is larger than 1 mm in diameter. Then, powder samples are analyzed quantitatively by using a plastic sample holder of 1-mm diameter and 0.3-mm depth. This instrument has good ability for microanalyzing trace elements by, for example, the ion-exchange membrane method. The sensitivity represented is nearly 5000 cps/μg for Ni $K\alpha$ from $NiSO_4$ that is soaked and dried in thin rice paper. Some applications of the micro fluorescent X-ray analyzer to precipitates in steels and corrosion products are reported.

INTRODUCTION

Fluorescent X-ray analysis is available for any type of sample, solid, powder, or liquid, and also is nondestructive. It enables a rapid analysis of elements above magnesium in atomic number, and it has been put to practical use in various fields. In fluorescent X-ray analysis, it is usually required that the irradiated area be more than 10 mm in diameter. Granting that the thickness of the specimen is less than 1 mm, the quantity of the sample is considerable and it is often difficult to obtain enough material, for example, for analyzing the precipitates or the nonmetallic inclusions extracted from steels. If fluorescent X-ray microanalysis is available, the sampling time of those extracted residues will be shortened and samples for X-ray or electron diffraction can be examined by fluorescent X-ray analysis. Its applications are further widened for the analysis of segregations, foreign material in steel, corrosion products, *etc.* Such applications of fluorescent X-ray microanalysis have been previously examined and the results reported.[1-5]

Whether an apparatus for fluorescent X-ray microanalysis can be used practically depends on whether the fluorescent X-ray intensity is sufficient. The present authors made a micro fluorescent X-ray analyzer with a focusing type of spectrometer that is available for practical use. The apparatus is manufactured by Shimadzu Seisakusho, Ltd. It is designed to analyze a minimum sample area of 0.1 mm in diameter.

At present, the electron probe microanalyzer is commonly used for microanalysis. The electron probe microanalyzer can analyze down to 1 μ, but the penetration depth of the electron beam is about 1 μ, and so the surface condition of the sample greatly affects the results and it is difficult to get information on the average composition of samples. In such cases, the micro fluorescent X-ray analyzer (MFX) is very useful because it is not affected by surface conditions. The probe size of the apparatus is above 0.1 mm, so it should be perhaps called "milli" probe analyzer rather than "micro," and it is thought to be useful that the MFX can fill up the gap between the electron probe microanalyzer and ordinary analyzing methods. In addition, when it is applied to the quantitative fluorescent X-ray analysis of trace elements after pretreatment,[6,7] a remarkable improvement in sensitivity can be expected.

SPECIFICATIONS OF THE MICRO FLUORESCENT X-RAY ANALYZER (MFX)

A general view of the MFX is shown in Figure 1, and details are illustrated by Figure 2. A schematic diagram of the MFX is shown in Figure 3.

Spectrometer

In order to obtain high X-ray intensities, the area of the crystal, radius of the Rowland circle, aberration of focusing, and resolving power required for analysis have been taken into consideration. The ARL type of spectrometer with a 4-in. Rowland radius has been adopted, in which the curved analyzing crystal moves on a straight line and the emergence angle of X-rays from the specimen surface is constant. The crystal can be moved automatically and manually and the scanning speed is changed in two stages. Operations and adjustments of the spectrometer can be performed while it is

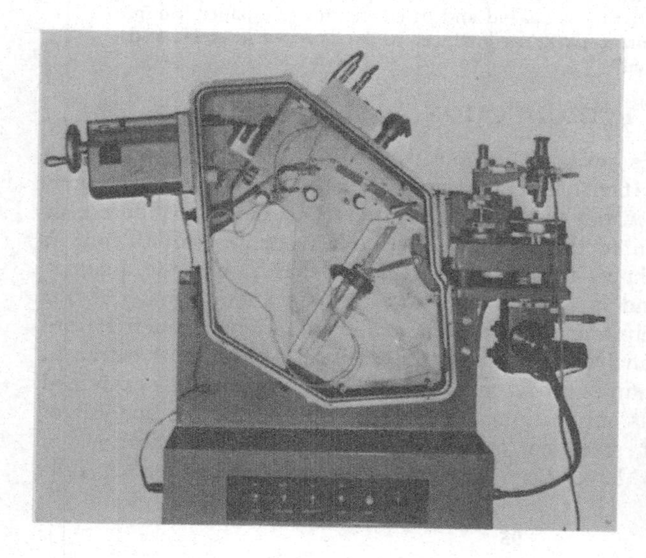

Figure 1. A general view of the micro fluorescent X-ray analyzer (MFX). The lid of the vacuum chamber of the spectrometer is removed.

Figure 2. A close-up view of the MFX. (1) Vacuum chamber of the spectrometer; (2) microscope; (3) vacuum lines; (4) X-ray tube; (5) sample chamber; and (6) scanning system of the sample.

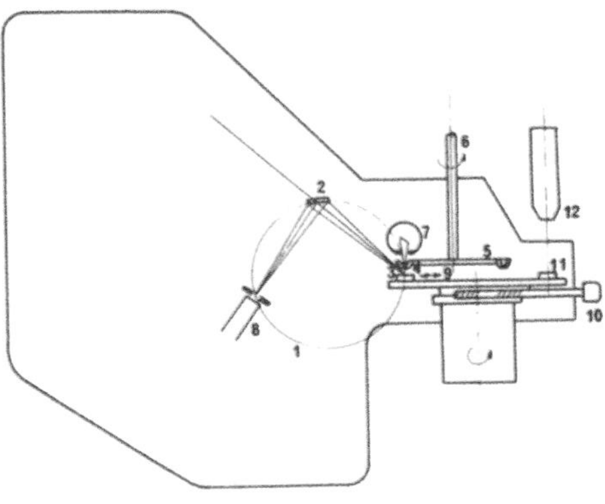

Figure 3. Schematic diagram of the MFX. (1) Rowland circle; (2) analyzing crystal; (3) specimen; (4) slit for X-ray probe; (5) slit holding the disk; (6) revolving shaft of slit holding the disk; (7) X-ray tube; (8) detector; (9) scanning stage of the sample; (10) scanning micrometer of the sample; (11) sample position of microscopic observation; (12) microscope.

in a vacuum. When the sample is to be exchanged, it can be replaced by breaking the vacuum of the sample chamber only.

Curved analyzing crystals LiF and ADP with 4-in. radius are provided for shorter and longer wavelengths. LiF covers 1.00 to 3.77 Å, and ADP covers 2.66 to 10.00 Å.

Detectors are sealed counters, krypton Multitron for shorter wavelengths, and neon Exatron (GM type) for longer wavelengths.

X-Ray Source

There are two methods of getting intense irradiating X-rays; one is the method of shortening the distance to the sample by using a microfocus X-ray tube, and the other is the method of using an ordinary sealed tube of high power. It is estimated that the former gives about twice as much intensity as the latter, but a commercial sealed tube has been adopted because the former is inferior to the latter in stability and is more complicated in construction. The tube is a Machlett (OEG-50S) with a tungsten target.

Slit System

The irradiated area of the sample is limited by a slit. Slits are held on a circular disk and they can be replaced by a revolving shaft outside the vacuum. The slits provided are pin holes of 0.1, 0.3, 0.5, 1.0, and 2.0 mm in diameter and rectangular slits of 0.1 by 2.5, 0.3 by 3.0, 0.5 by 4.0, 1.0 by 4.5, and 2.0 by 4.5 mm whose longer direction is vertical to the Rowland circle plane.

Table I. Intensities of Fluorescent X-Rays of Various Pure Elements in counts per second

		Machlett tube, tungsten target, 45 kV, 45 mA												
Detector:		Krypton Multitron, in air							Neon Exatron, in vacuum					
Analyzing crystal:		LiF							LiF		ADP			
Spectrum:		Pb $L\alpha$ (2)	Pt $L\alpha$ (2)	Mo $K\alpha$ (2)	Zr $K\alpha$ (2)	Cu $K\alpha$	Ni $K\alpha$	Fe $K\alpha$	Cr $K\alpha$	Ti $K\alpha$	Ca $K\alpha$ (2)	S $K\alpha$	Si $K\alpha$	Al $K\alpha$
Wavelength (Å):		1.175	1.313	0.710	0.788	1.542	1.659	1.937	2.291	2.750	3.360	5.373	7.125	8.339
Slip (mm):		× 2	× 2	× 2	× 2						× 2			
Nominal	Observed													
0.1	0.25 diameter	19	18	270	260	1800	1650	680	460	240	52	52	25	7
0.3	0.35 diameter	54	50	820	740	4400	4150	1850	1020	650	120	130	54	17
0.5	0.55 diameter	205	195	2800	2550	>5000	>5000	>5000	>3000	2320	580	610	245	76
1.0	1.0 diameter	700	550	>5000	4650	>5000	>5000	>5000	>3000	>3000	1540	1650	710	205
2.0	2.0 diameter	2200	1800	>5000	>5000	>5000	>5000	>5000	>3000	>3000	2750	>3000	1560	620
0.1L	0.25 × 2.5	100	84	1260	1140	>5000	>5000	2800	1920	1080	175	175	83	28
0.3L	0.5 × 3.0	360	305	3600	3350	>5000	>5000	>5000	>3000	>3000	840	860	350	110
0.5L	0.7 × 4.0	2050	1800	>5000	>5000	>5000	>5000	>5000	>3000	>3000	>3000	>3000	1800	670
1.0L	1.0 × 4.5	3200	2800	>5000	>5000	>5000	>5000	>5000	>3000	>3000	>3000	>3000	2000	1000
2.0L	2.0 × 4.5	4500	4100	>5000	>5000	>5000	>5000	>5000	>3000	>3000	>3000	>3000	>3000	1200

The actual areas of X-ray probe on the sample are different because of the sizes of the slits and because of their penumbra. The experimental results are shown in Table I for the different types of slit.

Sample

The MFX is designed so that the analyzed point shall be on the Rowland circle, and it is observed by the equipped microscope at the position where the sample holder is rotated by 90° from the position of analysis. The specimen can be moved up to 10 mm in the x and y directions by micrometers, and it can be automatically scanned in the x direction.

Samples are usually set in an aluminum sample holder, 4 mm thick, but a powder sample is packed into a hole of 1-mm diameter and 0.3-mm depth in a polymethylmethacrylate holder. It can be assumed that a thickness of 0.3 mm is infinite for ordinary materials, considering the absorption of X-rays.

PERFORMANCES OF MFX

The following results were obtained by experiments.

Intensity of Fluorescent X-Rays

Observed intensities of the fluorescent X-rays of some elements are shown in Table I for various slits.

The observed values of more than 5000 cps for krypton Multitron and 3000 cps for neon Exatron were not measured accurately because of their resolving time.

Relation Between Slit Size and Fluorescent X-Ray Intensity

The relation between slit diameters and observed fluorescent X-ray intensities is affected not only by the irradiating X-ray intensity distribution on the sample but also by the focusing conditions of the spectrometer. The results are shown in Figure 4.

Working Curve

It has been thought that the working curves in the limited sample area differ from the theoretical curve of infinite area, and so the actual measurements were made. The

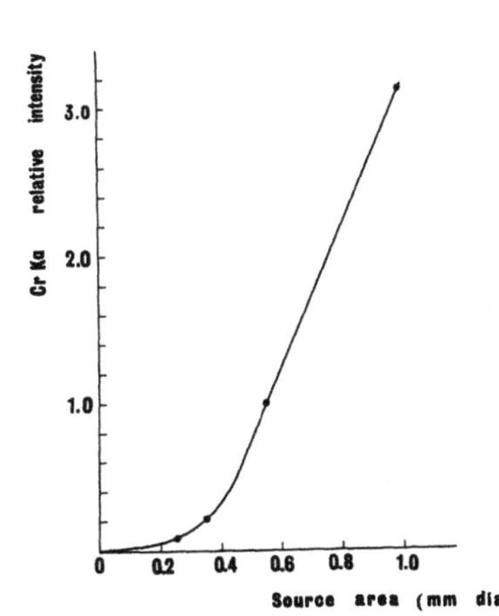

Figure 4. Relative intensities of fluorescent X-rays Cr $K\alpha$ *versus* irradiated source area on pure chromium.

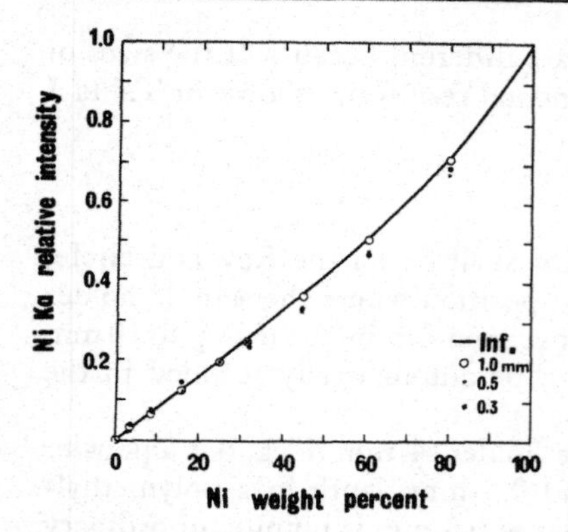

Figure 5. Relative intensities of fluorescent Ni *Kα* X-rays emitted from nickel–iron binary alloys with the use of various slits. The solid line shows the theoretical working curve for an infinite source area of the samples, and the open circles are for slits of 1.0-mm diameter, the crosses are for 0.5-mm-diameter slits, and the solid circles are for 0.3-mm-diameter slits.

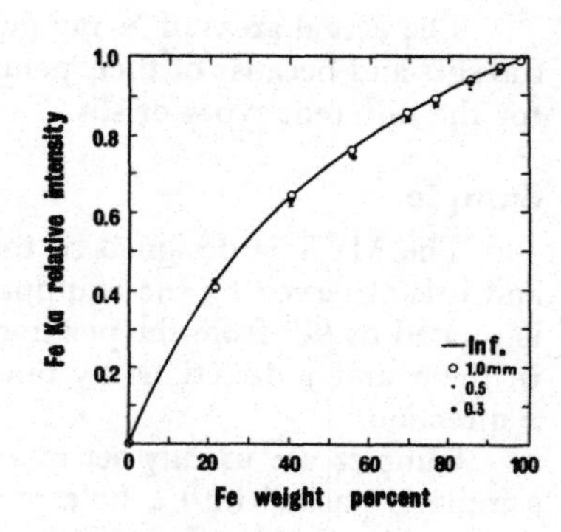

Figure 6. Relative intensities of fluorescent Fe *Kα* X-rays emitted from nickel–iron binary alloys with the use of various slits. The solid curve shows the theoretical working curve for an infinite source area of samples, and the open circles are for slits of 1.0-mm diameter, the crosses are for 0.5-mm-diameter slits, and the solid circles are for 0.3-mm-diameter slits.

samples used were nickel–iron binary alloys, and three sorts of slits, 0.3, 0.5, and 1.0 mm in diameter, were tested. The tungsten tube was operated at 30 kVP, 5 mA. The relative intensities of Ni *Kα* and Fe *Kα* are shown in Figures 5 and 6, respectively. According to the results, the working curves were coincident with those of an infinite sample when the diameter of the slit was larger than 1 mm.

For examples of the working curves of powder samples, mixtures of Fe_2O_3 and Cr_2O_3 powder were examined and the results are shown in Figures 7 and 8, respectively. The samples, each weighing 1 mg, were packed into the powder sample holder.

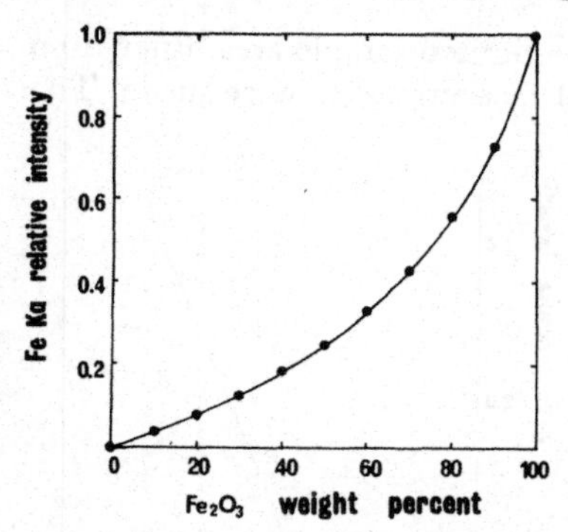

Figure 7. Relative intensities of the fluorescent X-rays Fe *Kα* from the mixtures of Fe_2O_3 and Cr_2O_3; the solid circles are experimental results, and the curve is the theoretical result. Experimental conditions were 30 kVP, 5 mA, and a slit of 1-mm diameter.

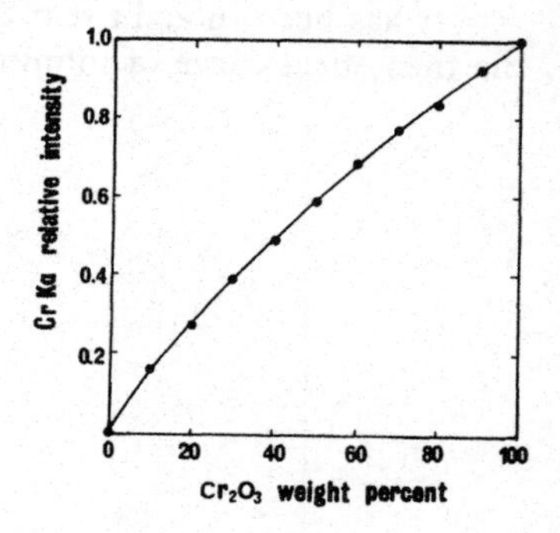

Figure 8. Relative intensities of the fluorescent X-ray Cr *Kα* from the mixtures of Fe_2O_3 and Cr_2O_3; the solid circles are experimental results, and the curve is the theoretical result. Experimental conditions were 30 kVP, 5 mA, and a slit of 1-mm diameter.

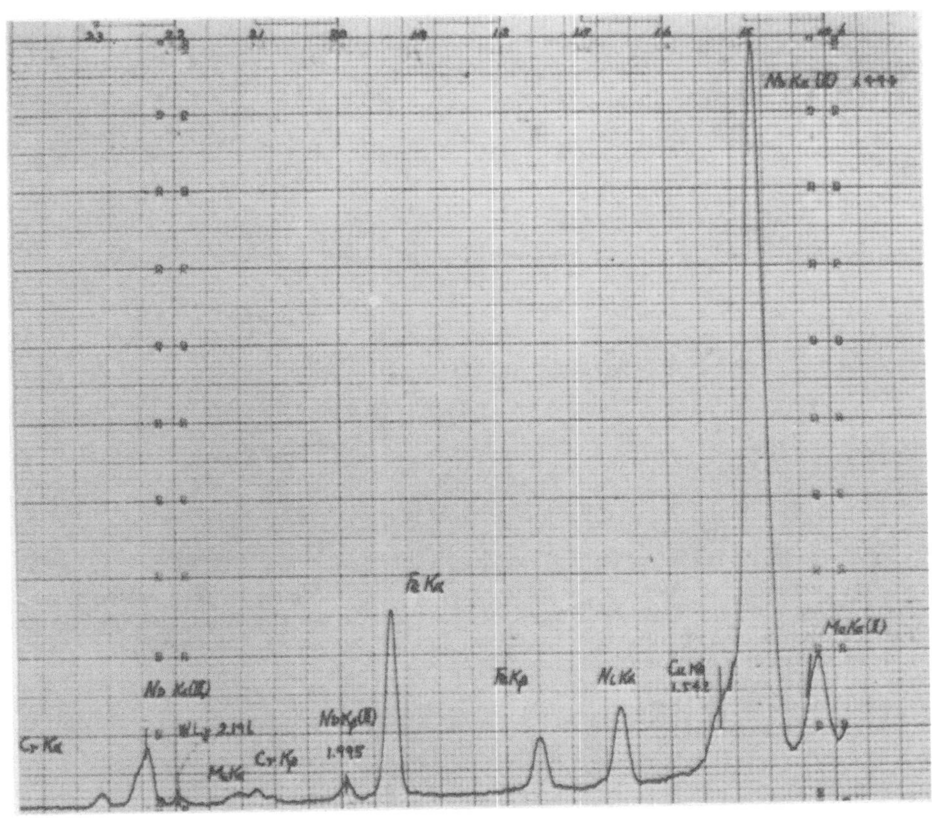

Figure 9. An X-ray spectrographic chart recording of the extracts from a stainless steel containing 2.2% molybdenum and 0.8% niobium. Experimental conditions were 45 kVP, 10 mA, and the full scale of the chart is 1000 cps.

APPLICATION

Extracts from a Stainless Steel

Figure 9 is the X-ray spectrographic chart recording of extracts from a stainless steel whose composition was 16.2% chromium, 14.8% nickel, 2.2% molybdenum, and 0.8% niobium. The sample was extracted by aqueous $FeCl_3$, and it was put into the powder sample holder. Operating conditions were 45 kVP and 10 mA, with an LiF crystal, krypton Multitron, and 1000 cps full scale. The spectral lines of niobium and molybdenum were of the second-order of Nb *Kα* and Mo *Kα*. The extracts were identified as NbC by X-ray powder diffraction.

Analysis of Chromizing Steel

The X-ray intensities of Fe *Kα* and Cr *Kα* from chromizing steel are shown in Figure 10, a slit of 0.1-mm width having been used and a scanning speed of 0.1 mm/min. Analytical results for chromium and iron are shown in Figure 11.

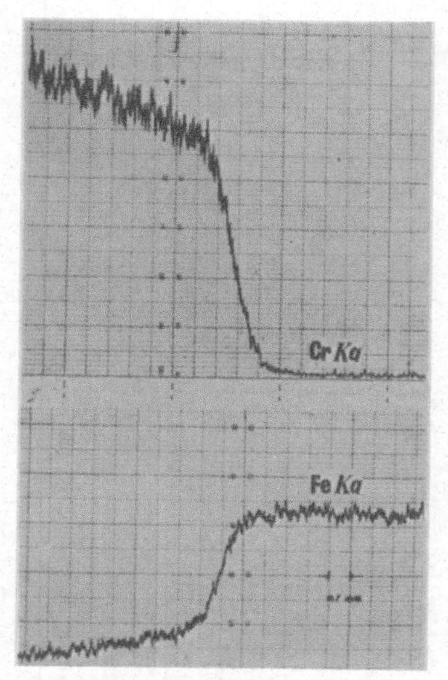

Figure 10. Intensities of the fluorescent X-rays Cr $K\alpha$ and Fe $K\alpha$ from a cross section of chromizing steel being moved at the rate of 100 μ/min.; experimental conditions were 30 kVP, 5 mA, and a slit of 0.1-mm width.

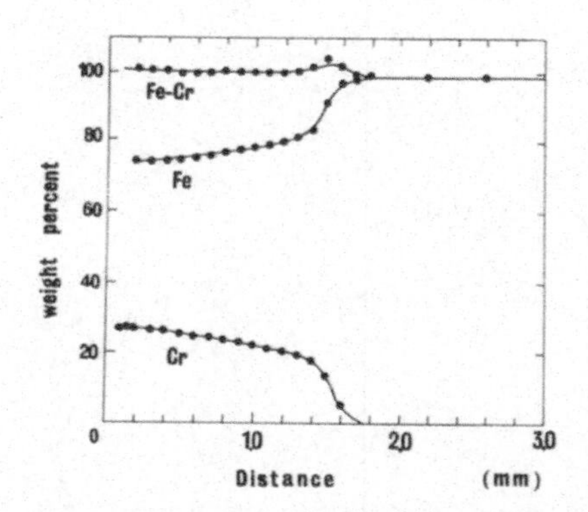

Figure 11. The analytical result for chromizing steel from the data shown in Figure 12.

Scale

Boiler-tube scale produced by vanadium attack was analyzed with the MFX. The scale which was taken from the attacked position and the neighboring positions contained vanadium of 40 and 13% and were examined by X-ray diffraction. One of the authors reported[8] that the new X-ray diffraction lines were observed and they corresponded to those of the synthetic sodium–vanadium oxide.

Fluorescent X-Ray Microanalysis

Several methods of analyzing trace elements quantitatively by means of fluorescent X-ray have been reported,[9,10] and an improvement in the sensitivity of the solution method was expected by using the MFX. The solution method was examined with the use of a 3.5-μ-thick rice-paper carrier for nickel. With the use of slits of 2 × 4.5 mm, the Ni $K\alpha$ fluorescent X-ray intensity was about 1200 cps, as shown in Figure 12, and the irradiated quantity of nickel which was soaked and dried in the rice paper was 0.25 μg. If the sensitivity is represented by counts per second per microgram, the sensitivity is nearly 5000 cps/μg for Ni $K\alpha$. It is expected that the MFX can be used for the microanalysis of trace elements by the ion-exchange method.[11]

CONCLUSION

An MFX having a vacuum type of spectrometer with a curved crystal has been designed and gives sufficient fluorescent X-ray intensity for practical use. A spectral

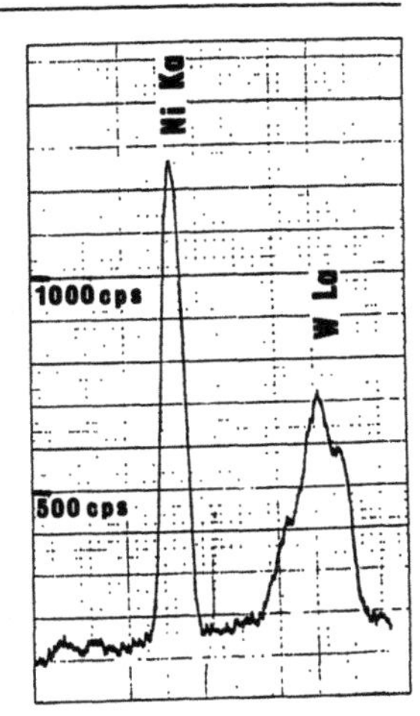

Figure 12. Chart recording of fluorescent X-rays Ni *Kα* and scattered X-rays W *L* from rice paper containing 0.25 μg of nickel in a 2 × 4.5 mm irradiated area. Experimental conditions are 45 kVP and 40 mA.

range of the MFX is 1.20 to 9.94 Å, with the use of LiF and ADP analyzing crystals, for the analysis of all elements above aluminum in atomic number. If a slit larger than 1 mm in diameter is used, the working curve corresponds to that of an infinite surface area. In the case of a powder sample, quantitative analysis is possible by using a holder of 1-mm diameter and 0.3-mm depth. Application of the MFX to fluorescent X-ray microanalysis has been examined, and the sensitivity is enough for practical use.

ACKNOWLEDGMENT

The authors express their heartfelt appreciation to Dr. M. Sumitomo, Director of Central Research Laboratories, Sumitomo Metal Industries, who has always encouraged them in the work presented here. They also wish to thank Dr. Y. Tachibana, Director of the Scientific Division, Shimadzu Seisaku-sho, Ltd., for his advice and support and Dr. H. Tsukiyama who helped them design the MFX.

REFERENCES

1. J. W. Thatcher and W. J. Campbell, *Fluorescent X-Ray Spectrographic Pròbe–Design and Applications, U.S. Bur. Mines Rept. Invest.* 5500, 1959.
2. I. A. Adler, J. Axelrod, and J. J. R. Branco, "Further Application of the Intermediate X-Ray Probe," in: W. M. Mueller (ed.), *Advances in X-Ray Analysis, Vol. 2,* Plenum Press, New York, 1960, p. 167.
3. J. Despujols, H. Roulet, and G. Senemaud, "X-Ray Fluorescence Analysis with a Focused Primary Beam," in: H. H. Pattee, V. E. Cosslett, and A. Engstrom (eds.), *X-Ray Optics and X-Ray Microanalysis,* Academic Press, New York, 1963, p. 445.
4. J. V. P. Long and H. O. E. Rockert, "X-Ray Fluorescence Microanalysis and the Determination of Potassium in Nerve Cells," in: H. H. Pattee, J. E. Cosslett, and A. Engstrom (eds.), *X-Ray Optics and X-Ray Microanalysis,* Academic Press, New York, 1963, p. 513.
5. D. C. Miller, "Norelco Pinhole Attachment," in: W. M. Mueller (ed.), *Advances in X-Ray Analysis, Vol. 5,* Plenum Press, New York, 1961, p. 513.
6. W. J. Campbell, "Fluorescent X-Ray Spectrographic Analysis of Trace Elements, Including Thin Films," *ASTM STP No.* 349, p. 48, 1964.
7. P. K. Lund, "X-Ray Fluorescence Spectroscopy in Biology and Medicine," in: H. H. Pattee, V. E. Cosslett, and A. Engstrom (eds.), *X-Ray Optics and X-Ray Microanalysis,* Academic Press, New York, 1963, p. 523.
8. T. Shiraiwa and F. Matsuno, "Identification of Scales of Heavy Oil-Fired Boilers by X-Ray Diffraction Method," *Tetsu To Hagane* **52**: 1592–1594, 1966.
9. H. G. Pfeiffer and P. D. Zemany, "Trace Analysis by X-Ray Emission Spectrography," *Nature* **174**: 397, 1954.

10. H. A. Liebhafsky, H. G. Pfeiffer, E. H. Winslow, and P. D. Zemany, *X-Ray Absorption and Emission in Analytical Chemistry*, John Wiley & Sons, New York, 1960.
11. L. S. Birks, "Comparison of X-Ray Fluorescence and Electron Probe Methods: Future Trends," *ASTM STP No.* 349, p. 151, 1964.

USE OF PRIMARY FILTERS IN X-RAY SPECTROGRAPHY: A NEW METHOD FOR TRACE ANALYSIS*

S. Caticha-Ellis,† Ariel Ramos, and Luis Saravia‡

Instituto de Fisica
Facultad de Ingenieria y Agrimensura
Montevideo, Uruguay

ABSTRACT

A method to improve the detectability of trace elements by X-ray fluorescent spectrography is described. The method consists of using appropriate filters in the primary, or exciting, beam. The effects of using filters in the primary beam on the peak-to-background ratio R of a fluorescent line have been analyzed on theoretical grounds. In fact,

$$R = \frac{I_f(\lambda_0)}{I_1(\lambda_0) + I_2 + I_f(\lambda_0)}$$

where $I_f(\lambda_0)$ is the intensity of the analytic fluorescent line, $I_1(\lambda_0)$ is the background intensity due to coherent and Compton scattering of the primary radiation by the specimen, and I_2 is the background intensity due to scattering of the fluorescent radiation by the analyzing crystal. Analytical expressions were derived for $I_1(\lambda_0)$, I_2, and $I_f(\lambda_0)$, from which it has been concluded:

1. The ratio $I_1(\lambda_0)/I_f(\lambda_0)$ decreases when the filter used has its absorption edges at wavelengths longer than λ_0.
2. The ratio $I_2/I_f(\lambda_0)$ can be separated into two parts which vary in opposite ways. The influence of these two parts on the value of R is discussed in the text.

It is then shown that the method should work well at short wavelengths and less well at longer wavelengths. The method was tested in the difficult case where overlapping of the analytical line with a characteristic line of the tube occurred, *i.e.*, in the determination of traces of selenium by using tungsten radiation. The analytic line Se $K\alpha$ has a wavelength of 1.106 Å, while W $L\gamma_1$ occurs at 1.098 Å. There is a marked effect of the filter thickness on the detectability; an optimum thickness appears to exist for each case. In the analysis of selenium, the best filter thickness (which can be selected by mere inspection of the diagrams reproduced in the text) increased the detectability of selenium traces by an order of magnitude. Finally, from statistical considerations, the quantity $t\sigma^2$ is proposed as an index of the effectiveness of the filter: the smaller $t\sigma^2$ is, the better the filter is. Here σ is the standard deviation of the intensity of the analytic line and t is the total counting time spent on the measure of the analytic line and background. In order to study the dependence of the index $t\sigma^2$ on the filter thickness, measurements were made on samples of sugar containing known concentrations of strontium. Then $t\sigma^2$ was plotted against the thickness of the filter for each concentration; these curves do show a minimum. Thus, an optimum filter thickness exists in each case.

---

* The section "Theory" was written by S. Caticha-Ellis; the section "Experimental" by Arie[l] Ramos, Luis Saravia, and S. Caticha-Ellis.

† Present address: School of Physics, Georgia Institute of Technology, Atlanta, Georgia.
‡ Present address: Department of Physics, Northwestern University, Evanston, Illinois.

THEORY

Introduction

When an X-ray beam traverses a filter made of a single element, the spectrum of the emergent beam changes as shown in Figure 1, where curve A shows qualitatively the spectrum of the primary X-ray beam (generated by electron bombardment in an X-ray tube), curve B shows the spectrum of the beam after traversing the filter, and curve C shows the absorption coefficient curve which has been superposed on the same figure for the sake of clarity. The last curve can be approximated by a λ^3 curve in an interval of λ not containing an absorption discontinuity.[1]

Let $I_0(\lambda)$ be the intensity of the primary beam as a function of λ (curve A), and $I_0(\lambda)$ (curve B) the intensity of the filtered beam. Then

$$I_0'(\lambda) = I_0(\lambda) \exp(-\mu\rho t) \tag{1}$$

where $\mu = \mu(\lambda)$ is the mass absorption coefficient, ρ the density, and t the thickness of the filter. The product ρt is the surface density of the filter.

The filtered X-ray beam is used in this method to excite the sample in an X-ray fluorescent spectrograph.

Figure 2 shows schematically the primary filter in position between the X-ray tube and the sample holder in an X-ray fluorescent spectrograph; actually, the filter is located on the window of the X-ray tube.

Let λ_a be the wavelength of the absorption edge of the filter element (Figure 1); $\lambda_0 < \lambda_a$, the wavelength of the characteristic radiation of the element to be analyzed; and

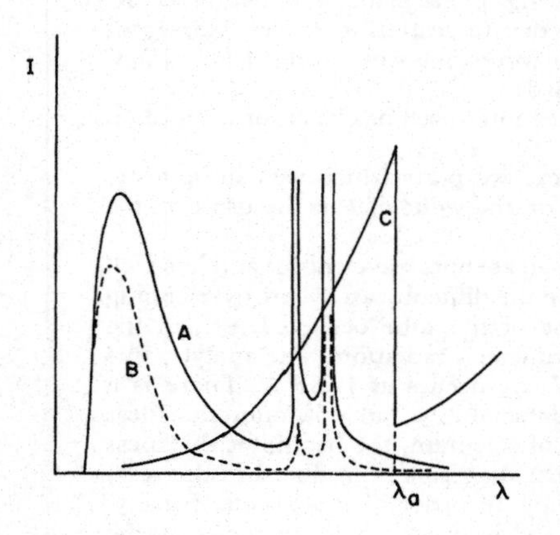

Figure 1. Curve A is the spectrum of the primary X-ray beam $I_0(\lambda)$; curve B, the spectrum of the filtered beam $I_0'(\lambda)$; and curve C, the mass absorption coefficient of the filter.

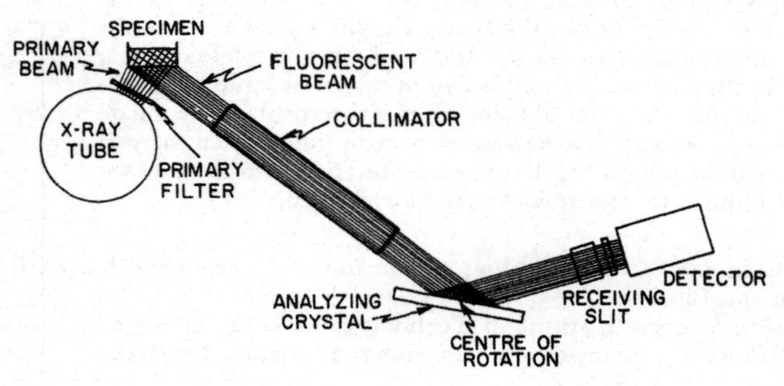

Figure 2. Scheme of the X-ray fluorescent spectrograph, showing the primary filter in position.

λ_q, the wavelength of the corresponding absorption edge. Excitation of the fluorescent line λ_0 is accomplished by radiation with $\lambda \leq \lambda_q$; the wavelengths longer than λ_q contribute to the background by scattering from the sample. When such a filter is used, the part of the primary beam which excites the analytic line decreases less than the part which only contributes to the background. Thus, we may expect to obtain a higher peak-to-background ratio for those peaks appearing in the region of wavelengths shorter than the absorption edge of the filter used.

The qualitative arguments given above indicate that it should be possible to improve the peak-to-background ratio of a given line in X-ray fluorescent analysis and thus the detectability of a given element by making adequate use of filters in the primary X-ray beam.

Theory

Peak to Background Ratio. The background intensity is mainly due to:

1. Coherent and Compton scattering of the primary radiation from the specimen, denoted hereafter by $I_1(\lambda)$.
2. Scattering of the fluorescent radiation from the analyzing crystal, I_2.

According to Birks,[2] both sources have approximately the same importance.

If there is no primary filter and $I_0(\lambda)$ is the incident radiation, the intensity originated by coherent or Compton scattering in the sample can be expressed as

$$I_1(\lambda) = k(\lambda)I_0(\lambda) \tag{2}$$

where $k(\lambda)$ is a factor which depends on the specimen and on instrumental factors.

The intensity of a fluorescent line of an element A of the sample can be written

$$I_f(\lambda) = \int_0^{\lambda_q} h_{q,A}(\lambda,\lambda')I_0(\lambda')\,d\lambda' \tag{3}$$

where $h_{q,A}(\lambda, \lambda')$ is a conversion function for radiation of wavelength λ' to λ, which takes into account all factors affecting fluorescent intensities, *i.e.*, physical factors such as fluorescent yields, absorption and interelement effects in the sample, *etc.*, and instrumental factors such as diffraction efficiency and absorption of the analyzing crystal, quantum efficiency of the detector, *etc.* Here λ_q stands for the wavelength of the absorption edge of the electron shell responsible for the radiation being detected (*i.e.*, $q = K, L, M, \ldots$).

The function $h_{q,A}(\lambda, \lambda')$ depends on many parameters and correction factors, so that its exact determination is usually not practicable. Of course, $h_{q,A}(\lambda, \lambda') = 0$ for $\lambda' > \lambda_q$.

The part of the background caused by scattering of the fluorescent radiations from the analyzing crystal may be expressed as

$$I_2 = \sum_A \sum_q k'(\lambda) \int_0^{\lambda_q} h_{q,A}(\lambda, \lambda')I_0(\lambda')\,d\lambda' \tag{4}$$

where $k'(\lambda)$ is a factor depending on the analyzing crystal; Σ_q indicates a summation over the different electronic levels of the atoms, and Σ_A indicates a summation over the different types of atoms in the specimen.

I_2 will vary in general with the position of the detector, *i.e.*, $I_2 = I_2(2\theta)$. However, in this case, no correspondence should be assigned between angles 2θ and wavelengths as in the case of Bragg reflection from the analyzing crystal. Actually, as equation (4) indicates, I_2 is composed of all fluorescent radiations originated in the sample and scattered by the analyzing crystal.

The peak-to-background ratio for a given fluorescent line λ_0 is given by

$$R = \frac{I_f(\lambda_0)}{I_1(\lambda_0) + I_2 + I_f(\lambda_0)} \tag{5}$$

or by

$$\frac{1}{R} = \frac{I_1(\lambda_0)}{I_f(\lambda_0)} + \frac{I_2}{I_f(\lambda_0)} + 1 \tag{5a}$$

In order to increase R, we can operate on one or both of the ratios

$$\frac{I_1(\lambda_0)}{I_f(\lambda_0)} \quad \text{and} \quad \frac{I_2}{I_f(\lambda_0)}$$

We shall presently show that, by introducing a convenient primary filter, the ratio $I_1(\lambda_0)/I_f(\lambda_0)$ will decrease.

Effect of a Primary Filter on the Peak-to-Background Ratio. If a filter is now introduced in the primary beam, the exciting beam will have the distribution of intensities given by equation (1). The intensity $I_1'(\lambda)$ scattered by the sample will then be

$$I_1'(\lambda) = k(\lambda)e^{-\mu(\lambda)\rho t}I_0(\lambda) \tag{6}$$

and the intensity of a given fluorescent line can be expressed as

$$I_f'(\lambda) = \int_0^{\lambda_q} h_{q,A}(\lambda, \lambda')e^{-\mu(\lambda')\rho t}I_0(\lambda')\,d\lambda' \tag{7}$$

Let λ_0 be the wavelength of the analytic line. If, *e.g.*, $q = K$ of some given element A_0, then, by choosing a primary filter whose absorption edge has a wavelength $\lambda_a > \lambda_0$, the ratio $I_f'(\lambda_0)/I_1'(\lambda_0)$ becomes

$$\frac{I_f'(\lambda_0)}{I_1'(\lambda_0)} = \frac{\int_0^{A_0\lambda_k} h_{K,A_0}(\lambda_0, \lambda')e^{-\mu(\lambda')\rho t}I_0(\lambda')\,d\lambda'}{k(\lambda_0)e^{-\mu(\lambda_0)\rho t}I_0(\lambda_0)} = \frac{\int_0^{A_0\lambda_k} h_{K,A_0}(\lambda_0, \lambda')e^{+[-\mu(\lambda')+\mu(\lambda_0)]\rho t}I_0(\lambda')\,d\lambda'}{k(\lambda_0)I_0(\lambda_0)} \tag{8}$$

But $\mu(\lambda_0) > \mu(\lambda')$ since excitation only takes place for $\lambda' \leq \lambda_k < \lambda_0$. Then,

$$e^{[-\mu(\lambda')+\mu(\lambda_0)]\rho t} > 1$$

holds. Hence,

$$\frac{I_f'(\lambda_0)}{I_1'(\lambda_0)} > \frac{\int_0^{A_0\lambda_k} h_{K,A_0}(\lambda_0, \lambda')I_0(\lambda')\,d\lambda'}{k(\lambda_0)I_0(\lambda_0)} = \frac{I_f(\lambda_0)}{I_1(\lambda_0)} \tag{9}$$

The second member in the inequality (9) is the ratio of fluorescent radiation to the radiation scattered from the sample when no primary filter is used. Therefore, the use of a primary filter increases the ratio of a given fluorescent radiation to radiation scattered from the sample, irrespective of the wavelength analyzed.

Let us now study how the ratio $I_2/I_f(\lambda)$ is affected by the filter

$$\frac{I_2'}{I_f'(\lambda_0)} = \frac{\sum_A \sum_q k'(\lambda_0) \int_0^{\lambda_q} h_{q,A}(\lambda_0, \lambda') I_0(\lambda') e^{-\mu(\lambda')\rho t} \, d\lambda'}{\int_0^{A_0\lambda_k} h_{K,A_0}(\lambda_0, \lambda') I_0(\lambda') e^{-\mu(\lambda')\rho t} \, d\lambda'} \tag{10}$$

Equation (10) shows that the ratio $I_2'/I_f'(\lambda_0)$ is very complicated. In order to assert whether this ratio is increased or decreased by the primary filter, we shall classify the integrals of the numerator [equation (10)] in two groups, those with $\lambda_q \leq \lambda_K$ and those with $\lambda_q > \lambda_K$. From mean value theorems, we may conclude that the first group will increase its value, while the second will decrease.

In other words, the background caused by scattering of fluorescent radiation from the analyzing crystal may be considered formed by two parts; one, originated by wavelengths shorter than the absorption edge λ_K related to the analytical line, tends to decrease the peak-to-background ratio when a primary filter is used; the other, composed of wavelengths longer than λ_K, tends to increase this ratio.

The equivalent of equation (5a) in the case of the primary filter is

$$\frac{1}{R'} = \frac{I_1'(\lambda)}{I_f'(\lambda)} + \frac{I_{2l}'}{I_f'(\lambda)} + \frac{I_{2s}'}{I_f'(\lambda)} + 1 \tag{11}$$

where the two parts mentioned in the preceding paragraph have been explicitly introduced, and the subscripts l and s remind us that these terms are due to wavelengths respectively longer and shorter than λ_K.

Then, in equation (11), both terms $I_1'(\lambda)/I_f'(\lambda)$ and $I_{2l}'/I_f'(\lambda)$ decrease when a filter is introduced, while the term $I_{2s}'/I_f'(\lambda)$ increases.

It is clear that, if a relatively short wavelength is being analyzed, comparatively few fluorescent radiations will contribute to I_{2s}', which appears in the term likely to decrease the peak-to-background ratio; but, if a relatively long wavelength is analyzed, the contribution to I_{2s}' will be larger than that to I_{2l}'. Thus, $I_2'/I_f'(\lambda)$ as a whole will decrease at short wavelengths and will increase at long wavelengths. In principle, it may be expected that the method will work well at short and less well at longer wavelengths.

The principles just discussed have been successfully applied in one case which is difficult to analyze by conventional methods, *i.e.*, the case where one characteristic radiation of the X-ray tube overlaps the line one wishes to detect. Of course, a tube with a different target could have been used, but our aim was to show that the method worked well even under very severe conditions.

EXPERIMENTAL

Application of the Method

The method discussed above was applied to the determination of traces of selenium in an organic matrix (sugar). At the same time, criteria for the determination of the optimum thickness of the filter were evolved. The composition of the filter is irrelevant

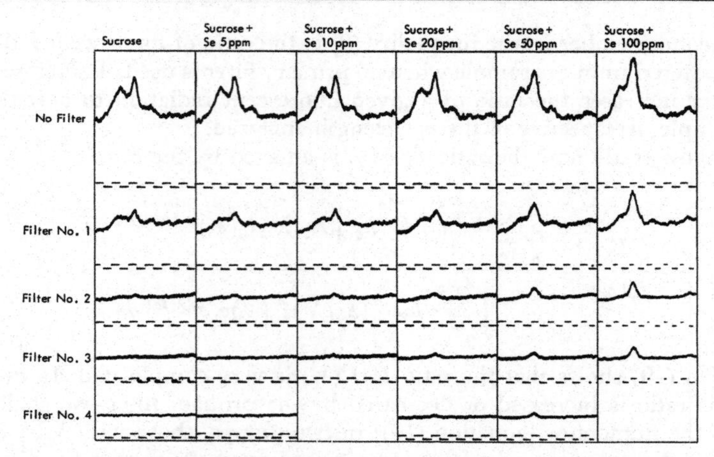

Figure 3. Patterns of samples of sucrose with varying amounts of selenium analyzed with different filters. Peaks in the pattern are of the Se $K\alpha$ line and of the W $L\gamma_1$ of the tube.

provided that its absorption edges are at wavelengths longer than λ_0. In our case, $\lambda_0 \simeq \lambda_t$, where λ_t is one of the characteristic lines of the tube target. The analytic line was Se $K\alpha$ ($\lambda_0 = 1.106$ Å), which corresponds to an angle $2\theta = 31.88°$ for an LiF analyzing crystal. As a tungsten tube was used, the Se $K\alpha$ line was overlapped by W $L\gamma_1$ (W $L\gamma_1 = 1.098$ Å and $2\theta = 31.64°$ for LiF). Filters of 7.3, 17.6, 25.0, and 33.7 mg/cm^2 of surface densities were made by superposing from one to four foils of nickel. The concentrations of selenium in the analyzed samples were 0, 5, 10, 20, 50, and 100 ppm by weight.

These samples were analyzed with the different filters by simple scanning with the spectrograph and registering the patterns which are reproduced in Figure 3. The diagrams appearing in the first row were obtained without filter. They show that even for a concentration of 100 ppm the presence of selenium may only be inferred from the rise in the height of the W $L\gamma_1$ peak. However, 10 ppm of selenium can be clearly detected by using filter No. 2 (17.6 mg/cm^2), and even the 5-ppm sample exhibits some indication of selenium being present. The detectability of selenium has thus been increased by about ten times by using the correct filter. By inspection of Figure 3, it can be stated that the filter of 17.6 mg/cm^2 is the best suited to this case. This fact can be explained as follows: Too thin a filter does not sufficiently decrease the background to enhance the peak-to-background ratio appreciably, while too thick a filter decreases excitation by a large factor and, although the background due to scattering of long wavelengths in the sample decreases more than the short wavelength range which is responsible for excitation, the rest of the background, which has a different origin, is not modified. However, when quantitative determinations of traces are needed, more rigorous criteria must be used. This is discussed in the next paragraph.

Evaluation of the Filter Quality

Let n_1 be the total measured intensity (peak plus background) in pulses per second (pps), and n_2, the background intensity. The intensity n of the analytic line in pulses per second is

$$n = n_1 - n_2 \tag{12}$$

The values of n_1 and n_2 are obtained accumulating N_1 and N_2 pulses, respectively, during intervals t_1 and t_2 sec, so that

$$n_1 = \frac{N_1}{t_1} \tag{13}$$

and

$$n_2 = \frac{N_2}{t_2} \tag{14}$$

The relative standard deviations of N_1 and N_2 are

$$\sigma_1 = \frac{1}{\sqrt{N_1}} \tag{15}$$

and

$$\sigma_2 = \frac{1}{\sqrt{N_2}} \tag{16}$$

and the relative standard deviation of the indirect measure n is

$$\sigma = \frac{\sqrt{\sigma_1{}^2 n_1{}^2 + \sigma_2{}^2 n_2{}^2}}{n_1 - n_2} \tag{17}$$

The total counting time $t = t_1 + t_2$ can be expressed as

$$t = \frac{N_1}{n_1} + \frac{N_2}{n_2} = \frac{1}{\sigma_1{}^2 n_1} + \frac{1}{\sigma_2{}^2 n_2} \tag{18}$$

It is desirable that the absolute errors of the counting rates n_1 and n_2 are equal.[3] The same is true of the standard deviations

$$\sigma_1 n_1 = \sigma_2 n_2 \tag{19}$$

Substituting equation (19) in equation (17), we get

$$\sigma^2 = \frac{2\sigma_1{}^2 n_1{}^2}{(n_1 - n_2)^2} = \frac{2\sigma_1{}^2 n_1{}^2}{n^2} = \frac{2\sigma_2{}^2 n_2{}^2}{n^2} \tag{20}$$

Hence,

$$\sigma_1{}^2 = \frac{\sigma^2 n^2}{2n_1{}^2}$$

and

$$\sigma_2{}^2 = \frac{\sigma^2 n^2}{2n_2{}^2} \tag{21}$$

Substitution for σ_1 and σ_2 in equation (18) yields

$$t = \frac{2(n_1 + n_2)}{\sigma^2 n^2}$$

or

$$t\sigma^2 = \frac{2(n_1 + n_2)}{(n_1 - n_2)^2} \tag{22}$$

The value $t\sigma_2$ can be taken as an index of the quality of the filter for the present purpose since σ and t are the relevant parameters to be considered when evaluating the measurement. The smaller $t\sigma_2$ is, the better the filter is. The values of n_1 and n_2 depend on, among other factors, the thickness of the filter. Then, if one wishes to calculate $t\sigma_2$ analytically, one should be able to calculate the variations of n_1 and n_2 as a function of the thickness of the filter. This is very difficult to calculate since such effects as fluorescent yields, matrix or interelement effects, *etc.*, should be exactly allowed for. Thus the index just defined was empirically determined for different filter thicknesses.

Several standards of sucrose with known traces of strontium were prepared to this effect; the concentrations used were 5, 10, 20 and 30 ppm. Counting was done on the Sr $K\alpha$ line ($\lambda = 0.877$ Å) by using the nickel filters previously described. The calibration curves obtained are depicted in Figure 4 for the different filters; the values of $t\sigma^2$ have been

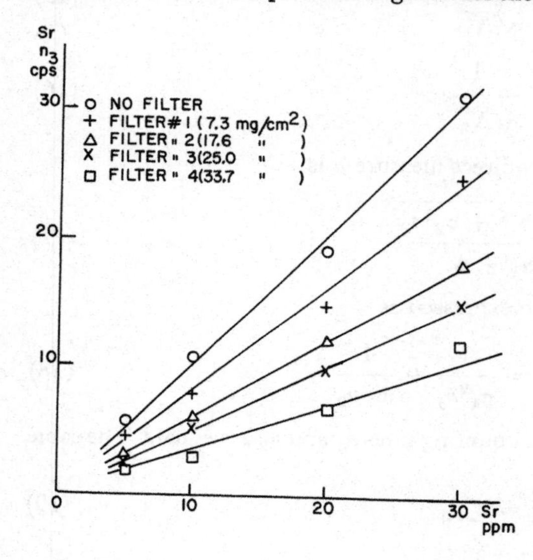

Figure 4. Calibration curves for nickel filters and the Sr $K\alpha$ line.

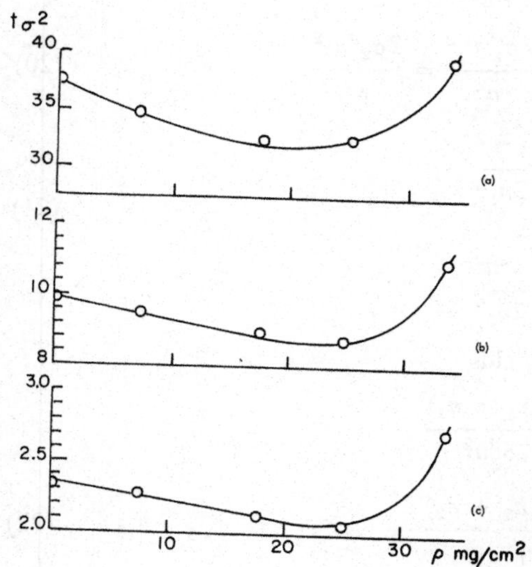

Figure 5. Filter quality index *versus* filter density for three different concentrations of strontium: (a) 5 ppm; (b) 10 ppm; and (c) 20 ppm.

plotted in Figure 5 as a function of the filter density for each of the concentrations used. Figure 5 shows that a minimum value of $t\sigma^2$ does exist for each sample and that this value approximately sets the optimum thickness of the filter.

The efficiency of the filters is strongly dependent on the wavelength measured, as predicted before; at long wavelengths, the intensity emitted by the tube is very small, and the contribution to the background by diffusion is correspondingly small. Thus, very little is to be gained by the use of the present method at the long wavelengths. This was indeed the case in an analysis for cobalt ($\lambda = 1.787$ Å, $2\theta = 52.81°$ for LiF) where very little enhancement could be observed.

In conclusion, it can be said that the method outlined in this paper works well in the analysis of elements involving moderate wavelengths, while, for longer wavelengths, other sources of background (*i.e.* scattering of fluorescent radiation from the analyzing crystal) are mainly responsible for a low peak-to-background ratio; height-pulse discrimination can then be used advantageously, or both methods can be used together to further improve the detectability of a given element.

Other aspects not treated here, such as the influence of the tube voltage and the matrix effects, remain for further study.

REFERENCES

1. A. H. Compton and S. K. Allison, *X-Rays in Theory and Experiment*, D. Van Nostrand Co., Inc., Princeton, 1935.
2. L. S. Birks, *X-Ray Spectrochemical Analysis*, Interscience Publishers, New York, 1959.
3. W. Parrish, *Philips Tech. Rev.* **17**: 206, 1956.

PRECISION AND ACCURACY OF SILICATE ANALYSES BY X-RAY FLUORESCENCE

A. K. Baird and E. E. Welday

Department of Geology
Pomona College
Claremont, California

ABSTRACT

The relative merits of different methods of sample preparation, of different instrumental operating procedures, and of different calibration techniques for major and minor elemental analyses of silicates must be judged by adequate estimates of precision and accuracy. Sources of imprecision in the X-ray method include electronic deviations, counting statistics, and sample collection and preparation. The variations attributable to these sources set limiting values to any X-ray method. Because most X-ray methods are comparative and thus dependent upon standards, the accuracy is more difficult to establish. Sources of inaccuracy in calibration methods include biases introduced by instrumental design or operation, matrix and particle size effects reflected as scatter or calibration line curvature, and an inadequate knowledge of the chemistry of the standard or standards used. With an adequate estimate of the precision, a *practical* measure of the accuracy of the X-ray fluorescence method can be obtained by the predictions of the compositions of standard silicates—standards whose compositions are *assumed* to be established by repeated wet-chemical analyses. This practical approach to an assessment of errors suggests to us that the comparative X-ray methods are highly precise. Within the limits of precision and with appropriate standards, accurate predictions of the compositions of silicate rocks can be made without applying empirically or theoretically derived correction factors. In this paper tests of precision and accuracy will be described which have been used to assess variabilities in quantitative analysis from the sample collection, through the sample preparation, to the calibration with standards.

INTRODUCTION

A large and rapidly expanding literature exists on the subject of quantifying the X-ray fluorescence (or emission) method of analysis. It is an understatement to say that there is little agreement among the various workers who have contributed to this literature, in part due to valid differences in the aims of their various analytical programs. In our opinion, however, some writers have an unfortunate tendency to claim or imply universal applicability of their particular procedures when, in fact, they are only dealing with a restricted compositional range of materials. Conversely, other quantification methods have been attacked by some writers and claimed to be invalid by the selection of extreme analytical cases far removed from practical analytical problems. In addition, some analysts still have an inadequate appreciation of the meaning of precision and accuracy (and difference between them), what aspects of their methods affect each measure of error, and what practical limitations are placed on their analytical results by incomplete assessments of errors.

In this paper, a relatively narrow field of quantitative X-ray fluorescence will be considered, namely, the sampling and analysis of silicate rocks, ranging in composition

Table I. Ranges of Weight Percent of Oxides Considered in this Paper for Silicate Analyses

Oxide	Weight percent	
	Minimum	Maximum
Na_2O	0.5	6.5
MgO	0.1	10.0
Al_2O_3	11.0	22.0
SiO_2	45.0	75.0
K_2O	0.2	5.0
CaO	0.7	12.0
TiO_2	0.1	2.3
Fe_2O_3*	1.4	16.0

* Total iron as Fe_2O_3.

from alkali granite to gabbro, for the major and minor constituents only (Table I indicates the oxide ranges).

DEFINITIONS AND QUALIFICATIONS

A major source of argument between workers stems from inadequate definitions of what is being discussed and what limitations exist. We present the following definitions and qualifications to be followed throughout the remaining sections of this paper.

Precision

We accept the definition of the Advisory Board of *Analytical Chemistry*:[1]

> Measurements which relate to the variation among test results themselves—i.e., the scatter or dispersion of a series of test results, without assumption of any prior information. The following measures apply:
>
> *Variance.* The sum of squares of deviations of the test results from the mean [symbol: $\bar{x}$] of the series after divisions by one less than the total number of test results. [symbol: s^2].
>
> *Standard Deviation.* The square root of the variance. [symbol: s].
>
> *Relative Standard Deviation.* The standard deviation of a series of test results as a percentage of the mean of this series.
>
> *Range.* The difference in magnitude between the highest test result and the lowest test result in a series.

Accuracy

We accept the definition of the Advisory Board of *Analytical Chemistry*:[1] "Measurements which relate to the difference between the average test results and the true result when the latter is known or assumed."

Specimen

A volume (variable) of rock which contains an assemblage of minerals adequate for the assignment of a common rock name. That is, material consisting of a fragment of only one or two of the five or six minerals commonly defining a rock or occurring in a rock is not an adequate specimen. The volume of a specimen varies with the grain size of the individual minerals which make up the rock, and an adequate minimum volume must be determined empirically.

Sample

A composite of several specimens, each selected by unbiased and predetermined procedures, to represent the population of rock from which they are collected. For this discussion, the sample represents specimens from an outcrop of a few square feet, or the population of an area of rock exposure of up to 600 by 600 ft; the number of specimens per sample varies from rock type to rock type and scale to scale, and the minimum number at any scale in any rock type must be determined empirically.

Instrumentation

Table II lists all instrumental parameters used for the analyses of sodium, magnesium, aluminum, silicon, phosphorus, potassium, calcium, titanium, manganese, and iron in our laboratory. (In this paper the very minor elements phosphorus and manganese are not discussed.) We emphasize that three different excitation sources are used, each with appropriate optic path conditions, and that the X-ray emission results so obtained on a given sample for a given element are not necessarily comparable with results from other equipment and parameters in intensity, primary absorption, and secondary enhancement, among other factors. Descriptions of this equipment and its use are published elsewhere.[2,3]

The X-Ray Method

We believe that quantitative X-ray fluorescence analysis of silicates is a comparative technique requiring reference to standards whose compositions are known or assumed (see definition of accuracy, above).

Analytical Goals

To determine the major and minor elemental compositions of silicate rocks (within the compositional ranges indicated in Table I) to levels of precision and accuracy consistent with (1) the inherent variability of the rock body in question, (2) the cost of sampling

Table II. X-Ray Instrumental Parameters and Operating Conditions for Silicate Analyses

Element	Na Mg	Al Si P	K Ca Ti Mn Fe
Excitation	Al *K* (demountable)	Ag *L* (demountable)	Tungsten (sealed FA 60)
Power, kV	10	12	Maximum, 40 kV
mA	150	150	40 mA
X-ray tube window	6-μ Aluminum foil	25-μ Beryllium foil	1500-μ Beryllium
Primary collimation	4 × 1 × 0.5 in.	4 × 0.6 × 0.035 in.	4 × 0.6 × 0.015 in.
Crystal	Gypsum	EDDT	LiF
Flow-counter window	6-μ Aluminum foil		Aluminized Mylar
Flow-counter collimation	0.5 × 1 × 0.011 in.		
Counter voltage	As appropriate to bring pulse distribution to 20-V base line. Signal preamplified by Tennelec 100B and appropriate pulse discrimination used.		
Counter gas	P-10 gas at 0.5 cfh (air)		

the body, (3) the cost of sample preparation for X-ray analysis, (4) the capability of available instrumentation, and (5) the number and quality of available rock standards. It is obvious that compromises are necessary and that a balance of effort and expense should be achieved. For example, efforts to achieve high analytical precision on individual samples are unwarranted if the rock is highly variable; analyses of lower precision on more samples are clearly more valuable and, in some cases, more easily and cheaply obtained.

General *Versus* Special Laboratories

The goals outlined above place our laboratory in the class of special application; we are not required to handle any and all specimens submitted. The quantification methods we use, which are evaluated in this paper, must be considered in this light; however, we believe that most academic and many industrial silicate analytical problems are of the special type (*e.g.*, studies of individual rock masses and process control).

COMPROMISES FOR SOURCES OF IMPRECISION AND INACCURACY

Sources of error (random, gross, and bias) which affect the precision and accuracy of X-ray results are complexly interrelated, and it is not possible to isolate imprecision from inaccuracy specifically in all aspects of sampling and analysis. In the discussions to follow, several major effects are noted and separated where possible. In no sense is the list to be considered exhaustive.

SAMPLE COLLECTION

No two specimens collected for X-ray analysis from a single rock body (or industrial stockpile) are exactly alike. Because the usual reason for performing an analysis is to enable one to distinguish chemical differences, it is imperative that analytical efforts (primarily, attainment of a given level of precision) be balanced with collection procedures. The following equation illustrates the compromises necessary:

$$s_s^2 = \frac{s_a^2}{n_a n_o n_r} + \frac{s_r^2}{n_o n_r} + \frac{s_o^2}{n_o}$$

where s_s^2 is the effective variance for a given element within 600- by 600-ft subareas; s_a^2 is the analytical variance based on one analytical determination per specimen; n_a is the actual number of determinations per specimen; s_r^2 is the variance of rocks within a single outcrop of rock; n_r is the number of rock specimens collected per outcrop; s_o^2 is the variance of outcrops within a subarea; and n_o is the number of outcrops sampled per subarea.

(Note that the equation above applies equally well to the industrial example by substituting the words *stockpile* for subarea, *power shovel scoop* for outcrop, and perhaps *grab sample* for specimen.)

Meaningful chemical differences between subareas can be obtained only if s_s^2 is small and this can often be achieved most economically by increasing n_r and n_o rather than attempting to minimize s_a^2. Usually, several elements per sample are to be determined; those elements which are most variable or most important for a particular study will control the optimum method of collecting. Similarly, those elements which present the greatest challenge in analytical precision (s_a^2) will control how much effort is placed in preparing samples for analysis.

SAMPLE PREPARATION

To reduce scatter or dispersion of values (mainly imprecision) and to minimize (not eliminate) absorption or enhancement (mainly inaccuracy), samples are fine ground, or fine ground and fused, or fine ground and fused with an added oxide of an absorbing heavy element. As listed, these three approaches represent an increasing order of difficulty and possibility for gross error, but they also result in an increase in precision and sometimes accuracy, except for the lightest elements (sodium, magnesium, and aluminum) where a decrease in usable X-ray signal results from dilution and absorption.

Under the qualifications listed above, we *compromise* on the fusion technique without heavy absorber (called the *moderate-dilution method*) in order:

1. To minimize grain size-to-intensity effects and to avoid mineralogical grinding biases with mica-bearing rocks. Fusion appears to be desirable because the resultant glass can be easily ground in commercially available impact mills to a uniform fine size, which is required for reasonable precision.[4-7] Unfused mixtures of minerals (with varying physical properties) are relatively intensity sensitive to length of grinding time,[7,8] and, in the case of rocks containing micaceous minerals, demonstrable biases are present.[9,10]

2. To obtain the maximum possible signal, at the highest possible signal-to-noise ratio, especially for the light elements, in a single prepared sample which can be used for all elemental determinations, sodium through iron. Heavy absorber techniques[11] have been shown to permit extension of the compositional range of analysis (more nearly linear calibration curves over greater ranges) and to reduce matrix effects in the analyses of diverse substances. The method is not used in our special application because undesirably low signals and high signal-to-noise ratios result for the lightest elements sodium and magnesium. In addition, the relatively small improvement in calibration linearity over the compositional ranges we consider does not warrant the extra expense of preparing a second heavy-absorber sample for the heavier light elements.

INSTRUMENTAL OPERATION

Obvious sources of imprecision and gross error resulting from short-term electronic instability in either X-ray generation or counting and recording of fluorescent X-ray photons and from long-term instability or "drift" resulting from the use of demountable X-ray sources, the targets of which become contaminated over periods of hours, are not considered in detail here. Tests for these errors can be and are built into the running procedure, and computer evaluation of the X-ray emissions of standard samples in the run permit corrections for long-term drift.[3,12]

The statistics of counting random pulses from an X-ray detector require further compromises in practical instrumental operation. (Note that dead-time corrections are, perhaps lamentably, not a problem in light-element analyses.) If signal-to-noise ratios of approximately 10 : 1 or greater can be achieved, then counting times of 1 min or less can be used and remain within a theoretical precision of 1% relative standard deviation. If a low signal or a low signal-to-noise ratio must be dealt with because of inadequate excitation sources, low elemental concentration, and high absorption in the sample and X-ray optical path, then prohibitively long counting times result. We believe that a practical compromise between the type of sample preparation used and the allowable counting error, given the optimum instrumental conditions, should not exceed 100 sec of fixed timing per sample analysis in routine work. With more than 100 sec per sample, the possible volume of analytical work is seriously hampered. These considerations are of greatest importance in the routine analysis of sodium, magnesium, and aluminum and are the primary considerations in rejecting the heavy-absorber method.

CALIBRATION METHODS: ACCURACY

If, from the light elements composing the rocks in question, measured X-ray emission intensities are obtained that have precisions consistent with the predictable counting error, consistent with a known (measured) error in sample preparation, and consistent with required levels of sampling variability, three general approaches are possible for conversions to weight per cent of cation or oxide:

1. Forming some ratio of net X-ray intensities from unknowns to intensities from a standard or standards whose compositions are known or assumed without corrections for matrix effects. Most commonly this is done with calibration curves, either hand-drawn or calculated by the method of least squares deviation.
2. Application of empirically derived correction factors for absorption and enhancement (*i.e.* total matrix effects) to emission intensities before deriving a ratio.
3. Direct calculation of composition from X-ray intensities, knowing or assuming basic parameters of elemental interactions in the energy region of soft X-rays and making all other necessary corrections commonly associated with the X-ray spectrographic technique, in samples composed dominantly of very light elements.

Method 1 has, as a disadvantage, the lack of any corrections for the effects of absorption and enhancement (not corrected for previously by sample preparation), which can cause an unreal scattering of X-ray results or a departure from a linear, proportional relationship between intensity and composition. Method 2 has been shown to have great value in relatively simple and closed systems of compositional variation (*e.g.*, metal alloys) and is now being used successfully in mineralogical analyses by electron microprobe. In bulk-rock analyses, however, the number of compositional variables is large, and often the limits of variation of single components are not well defined. In such cases, derived empirical corrections could be used erroneously, outside their applicable range, without the analysts' knowledge. Both methods 1 and 2 depend upon standards, and, because comparative techniques are used, the X-ray results cannot be of higher accuracy than the standards employed. Method 3, at present, suffers (in our opinion) from a lack of accurate values for absorption coefficients in the light-element region and, to a lesser extent, a lack of information on the interactions of soft primary X-rays with the sample.

In most discussions of calibration techniques using working curves (method 1), an unstated assumption appears to be made that the matrix effects constitute the single source of inaccuracy (not imprecision). Such assumptions may be justified in studies where two end members of divergent composition are mixed in different proportions. In practical silicate rock analysis, however, complex unknowns are not referred to lines or curves drawn between two compositionally divergent end members; an attempt is made to use several standards close in overall composition (and thus close in matrix characteristics) to the unknowns and to reduce interpolations to a minimum. This technique requires a library of standards and, unfortunately, neither the compositional diversity nor the analytical quality is entirely adequate. Problems encountered are illustrated diagrammatically in Figure 1, and a specific case for alumina is shown in Figure 2. The problems fall into two classes:

1. *X-ray emission and absorption biases.* Alumina results for W-1 and G-1 reported by several X-ray laboratories and studied in detail by Czamanske *et al.*[13] are an extreme example. These standards have had more wet-chemical analytical determinations reported for them than any other rocks, and the preferred alumina values are not in serious question. While close in alumina content, they have very different matrices, and the absorption effects are pronounced and result in a calibration curve of reverse slope (Figure 2). Other elements and other rocks must exhibit the same problem, probably to a much lesser

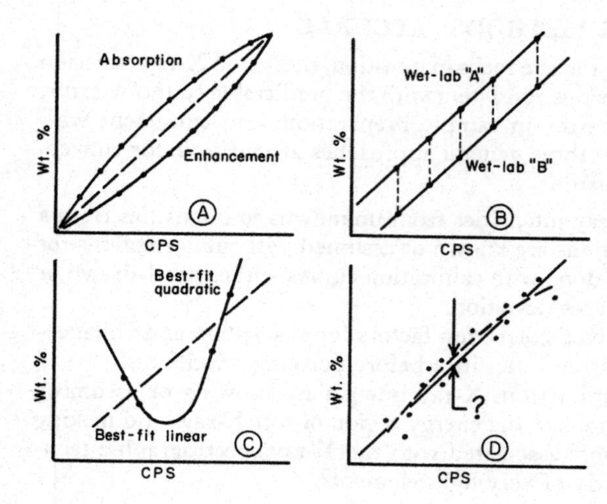

Figure 1. Diagrammatic plots to illustrate some of the calibration problems with quantitative X-ray analysis of silicates. (a) Calibration line curvature due to matrix effects of absorption and enhancement, assuming no wet-chemical error and high X-ray precision; (b) offset, parallel calibration lines due to interlaboratory wet-chemical biases, assuming high X-ray precision and no X-ray matrix effects; (c) least-square fits to insufficient number of data points exhibiting scatter from any cause (danger of mathematically "ill-behaved" quadratic fits of high correlation); (d) scatter of data points due to causes illustrated in (a) and (b). Is a quadratic a significant improvement over a line, and, if so, what magnitude of errors results from accepting the line?

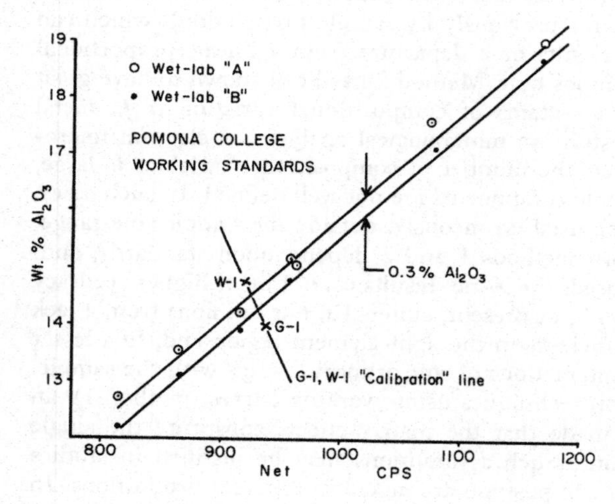

Figure 2. Calibration of seven Pomona College working standards, each analyzed by two wet laboratories for Al_2O_3. Superimposed W-1 and G-1 X-ray calibration of negative or reverse slope.

extent, and, when several different rocks are plotted on the same calibration, some scatter from strict linearity or from smooth curvature is bound to result [Figure 1(d)].

2. *Chemical and sample biases*. The cooperative studies of the standard rocks W-1 and G-1[14,15] have served to emphasize the relatively low interlaboratory precision of wet-chemical silicate analysis; but, despite this problem, any comparative X-ray technique is largely dependent upon these standards (and the newer U.S. Geological Survey standards, the Tanganyika T-1, and the C.A.A.S. syenite). In addition, we have assembled a library of 30 rocks, many of which have been wet-analyzed by three or four respected analysts in this country, Great Britain, and Norway. Others of the 30 have been correlated by X-ray analysis with all available wet-analyzed rocks and the resultant data are used as secondary working standards. As shown above, the two standards with the best established chemical values (W-1 and G-1) cannot be used together in any practical comparative X-ray calibration; thus, the other available standard materials are critically important.

What confidence can we presently place in the assigned or preferred values for available standards? Several lines of evidence suggest a relatively low confidence for some elements in some standards:

1. When G-1 and W-1 are used as unknowns and other standards of appropriate matrix are used to predict their composition, we derive lower values of SiO_2 for each, lower total FeO, MgO, and CaO for G-1, and higher K_2O for G-1 than reported as preferred values by Fleischer and Stevens.[16] It is interesting to note that the sense and approximate magnitudes of these differences are the same as reported by Ingamells and Suhr,[17] who used special spectrographic techniques and variations of standard methods of chemical analysis (Table III).

2. Splits of several of our rock standards have been analyzed by different chemists. We find, for certain pairs of laboratories, two calibration lines of the same slope but different intercept. For example, two such lines for alumina differ by 0.3 wt.% (Figure 2), even though the correlation coefficients between X-ray intensity and chemical values for each laboratory are 0.97 and 0.99. We obviously have no means for judging which, if either, set of wet values is accurate.

3. It can be shown that different samples of the same standard rock yield different X-ray emission intensities for some elements. This was first suspected when one of three samples of W-1 available to us yielded discrepant results compared with the other two. In order to isolate suspected splitting errors from preparation, including analytical errors, six samples of C.A.A.S. syenite were obtained and four preparations of each were analyzed. Significant (90% confidence) differences can be demonstrated between samples for Na_2O and Al_2O_3 compared with the analytical error (Table IV). Results of a second test, using the 30 Pomona College working standards, are summarized in Table V. For each of the first 16, four preparations were analyzed from one impact-milled split; for each of the remaining 14, one preparation each from four impact-milled splits was analyzed. The mean variance for each element in each group can be compared. The first group yields a pooled measure of analytical error; and the second, a pooled measure of variance between splits including analytical error. For CaO, MgO, and K_2O, there is a significant increase in

Table III. Predicted X-Ray Values and Preferred Chemical Values for Rock Standard G-1 (Compositions are Presented on a Moisture-Free Basis)

	Weight percent of oxide		
Oxide	X-ray predictions*	Ingamells and Suhr[17]	Fleischer and Stevens[16]
Na_2O	3.38	3.30	3.34
MgO	0.28	0.35	0.41
Al_2O_3	14.49	14.14	14.20
SiO_2	72.76	72.81	72.93
K_2O	5.52	5.54	5.50
CaO	1.29	1.37	1.41
TiO_2	0.27	0.26	0.27
Fe_2O_3†	1.80	1.90	1.94

* The $\bar{x}$ of 16 determinations, Pomona College.
† Total iron calculated as Fe_2O_3.

Table IV. Homogeneity of Six C.A.A.S. Syenite Samples for Na_2O and Al_2O_3

	Variance (s^2)			
	Total	Within samples	Between samples	F
Na_2O	0.0074	0.0054	0.0020	2.54
Al_2O_3	0.0030	0.0018	0.0012	3.72

Table V. Mean Variances of Pomona College Standard Rocks 1 to 30

For 1 to 16, Four Preparations Each from One Impact-Milled Split; for 17 to 30, One Preparation Each from Four Splits, Each Impact-Milled

	Variance (s^2)		
	Nos. 1–16	Nos. 17–30	F
MgO	0.00372	0.00730	1.96
K_2O	0.00049	0.00074	1.50
CaO	0.00066	0.00218	3.29

variance for the second group. In both tests, the magnitude of chemical differences between standard samples is small but real and should be considered when evaluating the accuracy of X-ray analysis.

TESTS OF ANALYTICAL AND SAMPLING PRECISION

The X-ray analyst is in a preferred position compared with the wet analyst in evaluations of precision: Replication is much cheaper and easier to perform. In most of our routine work, every field sample has been prepared and analyzed in duplicate, which provides us with both high-confidence estimates of precision and checks for gross error. Similarly, field samples are replicated on each succeedingly larger scale to provide measures of rock variability by scale and to permit the construction of maps of areal chemical variability. An example from a recent study of the Lakeview Mountain tonalite, southern California, is shown in Table VI (a complete description of this study is available elsewhere;[18] other similar studies have also been published[19–21]). Column 2, Table VI, indicates the measured analytical precisions for elements sodium through iron, and subsequent columns show the components of additional chemical variation due to inherent rock variabilities, on the sampling scales indicated at the head of each column. For all elements, the analytical errors are less than those reported[14] for wet-analytical precision for rocks of similar composition and are approximately an order of magnitude lower than the inherent variability of rocks collected only 5 ft apart. Except for calcium and silicon, the outcrop scale represents the level of highest variability for the Lakeview Mountain tonalite. In fact, for sodium, all demonstrable chemical variation is found at this level. For magnesium, silicon, potassium, and calcium, only, will it be possible to construct contour maps showing *valid* areal changes in chemistry (at 95% confidence).

Table VI. Summary of the Chemical Composition and Variability of the Lakeview Mountain Tonalite, Southern California

Element	Analytical precision,* s in wt.%	Variability† of rocks within an outcrop, s in wt.%	Variability‡ of outcrops within a subarea, s in wt.%	Variability‡ of subareas over 10 mile², s in wt.%	Composition over 10 mile²,‡ $\bar{x}$, in wt.% $\pm 2s$
Sodium	0.04	0.17	n.s.§	n.s.§	2.33 ± 0.05
Magnesium	0.04	0.24	0.24	0.38	1.99 ± 0.16
Aluminum	0.06	0.34	0.35	n.s.§	8.99 ± 0.13
Silicon	0.09	0.32	n.s.§	0.85	27.66 ± 0.32
Potassium	0.01	0.20	n.s.§	0.29	1.13 ± 0.12
Calcium	0.02	0.23	n.s.§	0.53	4.75 ± 0.20
Titanium	0.003	0.030	0.052	n.s.§	0.48 ± 0.015
Iron	0.04	0.47	0.64	n.s.§	5.17 ± 0.18
Degrees of freedom	174	30	30	29	

* Based on two determinations per sample.
† Thirty outcrops each with two specimens 5 ft apart.
‡ Thirty subareas each with two specimens 300 ft apart.
§ n.s. = component of variability not significant at 95% confidence.

We conclude from this test and many other similar studies, that the *precisions* of silicate analyses by X-ray fluorescence spectrography are considerably better than are required for our application. We emphasize that the majority of our work is of the type reported in Table VI: detection of relatively small chemical differences in discrete and relatively homogeneous rock masses. In all such studies, arguments about the *accuracies* of analyses are essentially irrelevant. However, if different rock bodies are to be compared with each other or our results compared with results of other analysts, accuracies must be considered.

TESTS OF ANALYTICAL ACCURACY

In using calibration curves, principal detectable inaccuracies as discussed above are: curvatures due primarily to matrix effects, scatter due primarily to differing matrix effects in different rocks and to inaccurate assigned chemical values for some standards, and incorrect intercept values due primarily to biases in chemical values from different laboratories.

The overall magnitude of these inaccuracies has been studied by using least-square techniques to best-fit lines and quadratic curves to net emission intensities *versus* assigned chemical compositions for each element in the 30 Pomona College working standards. For the line and curve for each element, correlation coefficients were computed to provide measures of the degree of scatter. Each quadratic curve was tested for statistically significant (at 95% confidence) improvement in fit over the corresponding line. Results show (Table VII) that the worst scatter occurs for sodium and aluminum, and that the fits for data for sodium, silicon, and potassium are *statistically* significantly improved by the use of a quadratic curve. Plots for these latter elements are shown in Figures 3 to 5.

Though quadratic curves for sodium, silicon, and potassium are statistically significant, the *practical* analytical significance is assessed in Table VIII. We make the following conclusions:

1. The improvements in the correlation coefficients are very small, even though quadratic curves are statistical improvements.

2. The maximum deviations in predicted values between linear and quadratic calibrations are of the same magnitudes as the analytical precision errors of the X-ray method,

Table VII. Tests for Quadratic Calibration Curves and Scatter in Thirty Pomona College Standard Rocks

Element	Linear correlation coefficient	Quadratic correlation coefficient	*F* values		Quadratic significant at 95% Conf.?
			Measured	Tabled	
Sodium	0.9827	0.9846	4.84	4.16	yes
Magnesium	0.9988	0.9988	0.31	4.16	no
Aluminum	0.9878	0.9876	0.65	4.24	no
Silicon	0.9949	0.9965	14.83	4.19	yes
Potassium	0.9986	0.9987	5.45	3.98	yes
Calcium	0.9991	0.9991	1.44	4.19	no
Titanium	0.9971	0.9971	0.94	4.16	no
Iron	0.9976	0.9977	2.56	4.24	no

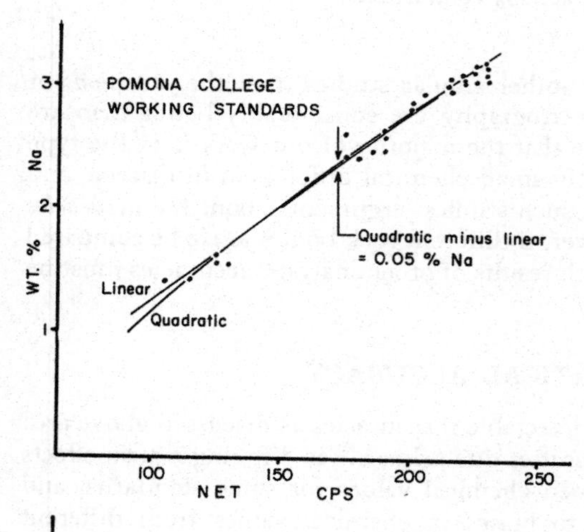

Figure 3. Calibration (linear and quadratic least-squares fits) of 29 Pomona College working standards for sodium.

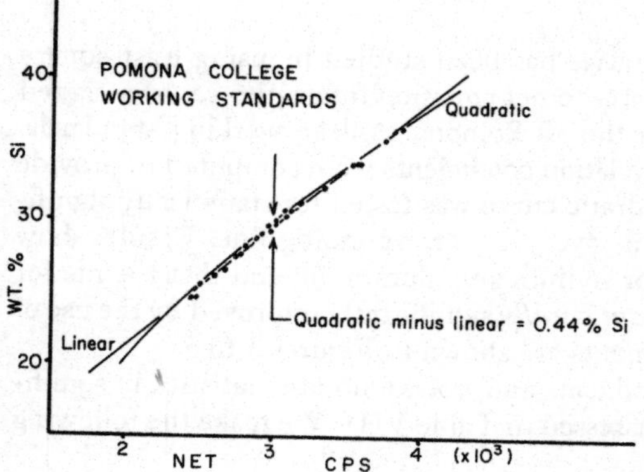

Figure 4. Calibration (linear and quadratic least-squares fits) of 24 Pomona College working standards for silicon.

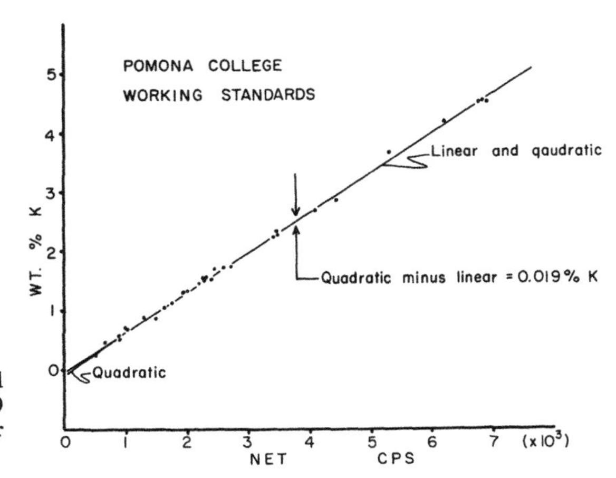

Figure 5. Calibration (linear and quadratic least-squares fits) of 30 Pomona College working standards for potassium.

Table VIII. Practical Analytical Significance of Quadratic Calibration Curves for Sodium, Silicon, and Potassium in Pomona College Standard Rocks

	Cation		
	Sodium	Silicon	Potassium
Linear correlation coefficient	0.9827	0.9949	0.9986
Quadratic correlation coefficient	0.9846	0.9965	0.9987
Max. deviation* (quadratic minus linear), wt.% cation	0.05	0.45	0.02
X-ray counting precision, *s* in wt.% cation	0.02	0.07	<0.01
X-ray analytical precision,† *s* in wt.% cation	0.06	0.30	0.02
Interlaboratory wet-chemical precision,‡ *s* in wt.% cation	0.21	0.20	0.46

* Computed at midrange of data.
† Average values for 3000 replicate analyses performed at Pomona College, 1960–1967.
‡ Fairbairn *et al.*,[14] p. 37; converted to cation percentages.

which in turn are an order of magnitude better than required for this type of silicate analysis.

3. In routine analytical runs, all standards are never used in the same calibration (standards of appropriate similarity in matrix to the unknowns in question are selected and used); thus, narrower ranges of calibration are routinely considered, and the deviations noted in conclusion 2 must, therefore, be larger than routinely encountered.

4. The uncertainties (lack of accuracy) in the values assigned to the rocks by wet-analytical methods are too large to judge the practical significance of the nonlinearity. For one example, in the calibration for silicon (Figure 4), the two points at the low-concentration end of the calibration must have a highly weighted control on the curve since the computation used minimizes the *squares* of the deviations.

A fundamental test of the accuracy (not precision) of any instrumental method of analysis is to compare instrumental results with the best established standards. For silicate-rock analysis, this means W-1 and G-1; a comparison for G-1 has been made in Table III, and we note resonably good agreement. We emphasize, however, that W-1 and G-1 generally lie at the compositional extremes of most of the elemental ranges we

Table IX. Predicted X-Ray Values and Preferred Chemical Values for Rock Standard T-1

	Oxide (wt.%)	
Oxide	X-ray predictions	Preferred values[22]
SiO_2	62.8(0)	62.65
Al_2O_3	16.6(3)	16.52
Fe_2O_3*	5.88	6.03
MgO	1.96	1.89
CaO	5.16	5.19
Na_2O	4.44	4.39
K_2O	1.26	1.23
TiO_2	0.57	0.59
P_2O_5	0.12	0.14
Loss on ignition	1.50	1.53

* Total iron calculated as Fe_2O_3.

are concerned with. The standard geochemical sample T-1 was specifically selected as a standard of intermediate composition between W-1 and G-1. Our X-ray predictions of the composition of T-1 are compared with the preferred chemical values in Table IX. We believe the agreement is well within the capabilities of the X-ray instrumentation and the present knowledge of the real composition of this standard.[22]

CONCLUSIONS

Within the compositional ranges indicated in Table I, silicate analyses performed by X-ray emission spectrography on samples prepared by the moderate-dilution fusion method are highly precise compared with wet-analytical results and are approximately an order of magnitude better (in precision) than is required to study major and minor element chemical variations in rock bodies. This high precision (plus the rapidity and low cost of the method) makes X-ray spectrography a powerful tool for petrologic research.

Though the accuracies of the silicate analyses are more difficult to assess and cannot be better than the primary standards used for comparison, several lines of evidence suggest that *prudent* working-curve calibrations provide analytical answers consistent with conventional methods when standards of similar matrices are used without application of empirically or theoretically derived correction factors. Prudent, in the preceding sentence, is underlined: Use of poorly established standards, standards of grossly different matrices on the same calibration, and gross extrapolations beyond the range of standards, obviously cannot be tolerated.

ACKNOWLEDGMENTS

The research program within which our methods of X-ray analysis have been developed has been supported by the National Science Foundation (grants G-19075, GP-1336, and GA-686 to A. K. Baird and D. B. McIntyre). We are indebted to B. L. Henke for the application of soft X-ray sources of his design to our research. K. W. Baird has aided us greatly in our analytical program, and D. B. McIntyre has written all the computer programs for calibration, significance tests, and analysis of variance reported here.

REFERENCES

1. Advisory Board, "Guide for Measures of Precision and Accuracy," *Anal. Chem.* **35**: 2262, 1963.
2. B. L. Henke, "Sodium and Magnesium Fluorescence Analysis—Part I: Method," in: W. M. Mueller and M. J. Fay (eds.), *Advances in X-Ray Analysis, Vol. 6*, Plenum Press, New York, 1963, p. 361.
3. A. K. Baird, D. B. McIntyre, and E. E. Welday, "Soft and Very Soft Fluorescence Analysis: Spectrographic and Electronic Modifications for Optimum, Automated Results," *Develop. Appl. Spectry.* **4**: 3, 1965.
4. F. Bernstein, "Application of X-Ray Fluorescence Analysis to Process Control," in: W. M. Mueller (ed.), *Advances in X-Ray Analysis, Vol. 5*, Plenum Press, New York, 1962, pp. 486–499.
5. F. Claisse and C. Samson, "Heterogeneity Effects in X-Ray Analysis," in: W. M. Mueller (ed.), *Advances in X-Ray Analysis, Vol. 5*, Plenum Press, New York, 1962, pp. 325–354.
6. E. L. Gunn, "The Effect of Particles on Surface Irregularities on the X-Ray Fluorescent Intensity of Selected Substances," in: W. M. Mueller (ed.), *Advances in X-Ray Analysis, Vol. 4*, Plenum Press, New York, 1961, pp. 382–400.
7. K. W. Madlem, "Matrix and Particle Size Effects in Analyses of Light Elements, Zinc Through Oxygen, by Soft X-Ray Spectrometry," in: G. R. Mallett, M. J. Fay, and W. M. Mueller (eds.), *Advances in X-Ray Analysis, Vol. 9*, Plenum Press, New York, 1966, pp. 441–455.
8. E. E. Welday, A. K. Baird, D. B. McIntyre, and K. W. Madlem, "Silicate Sample Preparation for Light-Element Analysis by X-Ray Spectrography," *Am. Mineralogist* **49**: 889–903, 1964.
9. A. Volborth, "Biotite Mica Effect in X-Ray Spectrographic Analysis of Pressed Rock Powders," *Am. Mineralogist* **49**: 634, 1964.
10. A. K. Baird, D. A. Copeland, D. B. McIntyre, and E. E. Welday, "Note on 'Biotite Mica Effect in X-Ray Spectrographic Analysis of Pressed Rock Powders,' by A. Volborth," *Am. Mineralogist* **50**: 792, 1965.
11. H. J. Rose, I. Adler, and F. J. Flanagan, "Use of La_2O_3 as a Heavy Absorber in the X-Ray Fluorescence Analysis of Silica Rocks," *U.S. Geol. Surv. Profess. Papers* 450-B, p. 80, 1962.
12. D. B. McIntyre, "Fortran II Programs for X-Ray Spectrography," *Technical Report* 13, Department of Geology, Pomona College, 1964.
13. G. K. Czamanske, J. Hower, and C. Millard, "Non-proportional, Non-linear Results from X-Ray Emission Techniques Involving Moderate-dilution Rock Fusion," *Geochim. Cosmochim. Acta* **30**: 745, 1966.
14. H. W. Fairbairn and others, "A Cooperative Investigation of Precision and Accuracy in Chemical, Spectrochemical and Model Analysis of Silicate Rocks," *U.S. Geol. Surv. Bull.* **980**: 71, 1951.
15. R. E. Stevens and others, "Second Report on a Cooperative Investigation of the Composition of Two Silicate Rocks," *U.S. Geol. Surv. Bull.* **113**: 126, 1960.
16. M. Fleischer and R. E. Stevens, "Summary of New Data on Rock Samples G-1 and W-1," *Geochim. Cosmochim. Acta* **26**: 525, 1962.
17. C. O. Ingamells and N. H. Suhr, "Chemical and Spectrochemical Analysis of Standard Silicate Samples," *Geochim. Cosmochim. Acta* **27**: 897, 1963.
18. A. K. Baird, D. B. McIntyre, E. E. Welday, and D. M. Morton, "A Test of Chemical Variability and Field Sampling Methods, Lakeview Mountain Tonalite, Southern California Batholith," *Calif. Div. Mines and Geol. Shorter Contributions Spec. Rept.* **92**: 11–19, 1967.
19. J. F. Richmond, "Chemical Variation in Quartz Monzonite from Cactus Flat, San Bernardino Mountains, California," *Am. J. Sci.* **263**: 53, 1965.
20. A. K. Baird, D. B. McIntyre, E. E. Welday, and K. W. Madlem, "Chemical Variation in a Granitic Pluton and Its Surrounding Rocks," *Science* **146**: 258, 1964.
21. A. K. Baird, D. B. McIntyre, and E. E. Welday, "Geochemical and Structural Studies in Batholithic Rocks of Southern California: Part II, Sampling of the Rattlesnake Mountain Pluton for Chemical Composition, Variability, and Trend Analysis," *Bull. Geol. Soc. Am.* **78**: 191, 1967.
22. Standard Geochemical Sample T-1, Supplement No. 1, Geological Survey Division, Ministry of Commerce and Industry, United Republic of Tanganyika and Zanzibar, 1963.

DISCUSSION

Chairman H. J. Rose (U.S. Geological Survey): I should like to comment here that we do not use the fusion heavy-absorber method for sodium or magnesium.

A. K. Baird: My reply is that we want to make one specimen only which we can use for the analyses of all elements, sodium through iron. We should like to avoid making two specimens.

Chairman H. J. Rose: I don't blame you, but we do use other techniques for that.

K. F. J. Heinrich (National Bureau of Standards): I think you have misunderstood the purpose of these methods. The corrections are planned and devised for people who need them, not for those who don't need them.

A. K. Baird: I agree. I think it's also fair to say that there are many people who don't know whether they need them or not. There are too many people who think they have to calculate corrections when they do not need to.

APPLICATIONS OF COMPUTERIZED STATISTICAL TECHNIQUES IN QUANTITATIVE X-RAY ANALYSIS

Betty J. Mitchell

Chemicals and Plastics Division
Union Carbide Corporation
Charleston, West Virginia

ABSTRACT

The place of statistical techniques and the digital computer in analytical chemistry is clearly exemplified in the X-ray spectrographic laboratory. As the statistician is obsolete who uses only a desk calculator for his calculations in this era of rapidly changing technology and the need for increased productivity, so the X-ray man is severely handicapped who ignores the place of statistical techniques and associated digital computation in his work. For his calculations, the statistical X-ray analyst may choose from among the following: a small specialized computer designed as part of his X-ray equipment, a medium-size computer located in the laboratory building with the capacity to handle all the problems generated by the various analytical specialists, a time-sharing computer station which he shares with other analytical or testing sections of his main laboratory, or batch processing of his data by his company's computer center. Advanced types of desk calculators are now available which will permit rapid completion of some types of complex calculations, especially if they are set up in a systematic fashion. Careful study of an individual laboratory's problems will provide economic justification for the most appropriate type of computer calculation; the choice is dependent on many factors. This paper describes the advantages and disadvantages of the various computer systems for the X-ray spectrographer and the numerous statistical techniques which he may employ as an aid in his work. Statistics provide him with the tools to design his experiments, interpret his data, control the process which supplies his samples, improve sampling procedures, calibrate his X-ray equipment, perform theoretical studies, and calculate, correct, average, and evaluate his analyses. Digital computation provides him with an extraordinarily rapid means of making these calculations; in some cases, they are virtually impossible without computerization. Examples of analyses are described which are made practicable only by this combined statistical computerized approach. The accuracy and precision of computerized analyses are higher than for results obtained by the usual X-ray methods; analyst error is reduced to a minimum. The economics of present-day industrial life, especially applied to the analytical laboratory as a service organization, demand the most efficient possible operation of all analytical equipment. The X-ray laboratory presents an outstanding example of the need for statistical analysis and its associated digital-computer computation.

INTRODUCTION

The purpose of this paper is to bring together in an organized fashion information concerning (1) the role of the digital computer in the analytical laboratory, (2) the use of general statistical techniques in the evaluation of X-ray or other analytical data and, (3) the use of specific computerized statistical–mathematical techniques in quantitative X-ray analysis.

This threefold goal is an important one in view of the rapid growth of the digital computer as a remarkable tool in all fields of endeavor and because of the failure on the part of many analytical chemists and, in particular, many X-ray spectrographers to make use of the equally remarkable tools of statistics which have been available for many years.

Part of this failure has been due to the scarcity of what may be termed "practical" statisticians and "nonostrichlike" spectrographers and further, to the lack of communication between those who do exist. The term nonostrichlike spectrographer can be applied to the individual who realizes that the so-called "correct answer" or "highly accurate number" which is frequently assigned to a single sample of an unknown may be meaningless until the complete analytical picture is considered. This individual has come to realize that the correctness of his assumptions is basic to true success in analytical studies.

In analytical work, many assumptions are made which should first be recognized as assumptions and then questioned every day of the week. For example, does the single bottle of sample or piece of material which is being analyzed truly represent the drum, or batch, or heat, or mineral field from which it was taken? Is the sample homogeneous? Are the single or even replicate samples which are removed from the bottle representative of the material therein? Is the drum, or batch, or heat from which the sample was removed representative of normal production? Is the laboratory striving to obtain an accuracy or precision in its analysis which is inconsistent with the realistic requirement? Which step in the sample preparation procedure is least precise and open to improvement? Is the simplest, most rapid, and the most accurate method used for calculating quantitative results from the measurements? Is the best possible analytical method in use?

The term practical statistician refers to the statistician who thinks realistically of inexpensive experimental and sampling design and allows himself some latitude in the interpretation of analytical measurements—in other words, the statistician who understands the day-to-day problems of the spectrographer and is willing to compromise somewhat with theoretical statistics in order to come up with practical solutions to the problems of handling analytical measurements. The practical statistician also avoids the use of sesquipedalian terms like *leptokurtic*, *homoscedasticity*, and *interpenetrating replicate subsamples*,[1] which may or may not impress the spectrographer, but which certainly confuse him. Furthermore, he is willing to make the spectrographer an equal partner in their endeavor to understand and interpret numerical data.

THE DIGITAL COMPUTER

The digital computer enters this proposed picture of harmony between spectrographer and statistician because statistical analysis would be very tedious and time-consuming, indeed, without its facility, and, in some instances, the statistical analysis or the analytical calculations would be impossible by slide rule or desk calculator.

Additional descriptions of terms would be advantageous at this point. Certain meanings are associated with the words *digital* and *computer* which distinguish the term *digital computer* from all other types of computers. According to Webster, to compute is to determine by calculation, and a computer is a device that handles numbers. Obviously, this definition includes the abacus, mechanical adding machines, and even a man's fingers and toes. The electronic automatic digital computer, however, is a device that can perform arithmetic operations at a tremendous rate of speed and make simple logical decisions according to instructions it has been given. Human control is unnecessary when it is running. On the other hand, the human operator of a mechanical desk calculator must feed numbers into the machine and select the proper arithmetic operation at the start of every calculation. The digital in digital computer tells us that the computer is a calculator that solves a problem by actually doing arithmetic in much the same way a person would "by hand"; the computer represents quantity with integers. (An analog computer operates on the principle of creating a physical analogy of the mathematical problem to be solved.)

The technology of computers is changing so rapidly that statements made about it today may not be quite true tomorrow. In fact, the very choice of how to compute is a difficult one, and the analyst must think as far ahead as possible to make the best decision. There is no single right decision because no single type of computer is the only one which will serve the purposes of the laboratory. The individual laboratory's problems and its economic position dictate the best choice.

For his calculations, the statistically oriented X-ray analyst may choose from among the following: (1) a small specialized computer designed as part of his X-ray equipment, (2) batch processing of his data by his company's large-capacity computer center, (3) a medium-size computer located in the laboratory building with capacity to handle all the problems generated by the various analytical specialists, (4) a time-sharing computer station which may be shared with other analytical or testing sections in the laboratory, and (5) a desk calculator.

The first alternative, the small specialized computer, is linked directly to the X-ray spectrometer and serves as part of the recording console which records, digitizes, corrects, and prints out the resulting analysis. This type of calculation is ideal where medium or large-scale computers are not available, where immediate results from the X-ray analysis are essential, or for the routine repetitive type of analyses. It has the distinct advantage of being available exclusively to the X-ray spectrographer, who makes the choice of the required calculations. His work may proceed without interruption by others seeking computer time. His calculations are programmed into the computer console by the instrument manufacturer, and the program may be modified as dictated by subsequent events. If immediate analytical results are necessary, as in process control, the small specialized computer is a good choice.

Most large companies have centralized, "giant brain" computer facilities intended to serve their various groups with the least overall expense. These computers operate on a batch-processing plan, i.e., jobs are sent to a computer center where they are batched with many other jobs and computed according to priority. The required *on* time of the computer for a specific problem is very short, but the elapsed time between transmission of the raw data to the computer center and the return of the calculated results may be many hours or even several days. If slow response is not a handicap to the spectrographer, this type of calculation is the cheapest and most efficient for the individual user. The availability of experienced company programmers to write programs for more complicated problems may be an advantage to the spectrographer.

Savitzky[2] offers a comprehensive review of the advantages of various computers in the analytical laboratory. He suggests that two developments in the field of computers promise a solution to the problem of lack of timely response from the large computer center. One is the development of small or medium-size computers which are relatively inexpensive but fast and powerful enough to become basic laboratory tools. This type of computer is located in the laboratory building and serves several or all analytical groups. It can be programmed for predetermined calculations by the computer manufacturer; calculations which are developed subsequent to its installation may be added to the program with the services of an experienced programmer.

The time-sharing system is a recent development which offers timeliness of response. Just a year or so ago, time sharing was a novelty. Today, it is the fastest-growing segment of the already booming computer industry. Basically, it involves the sharing of a central computer among many input-output stations. The simplest such station is the teletype machine or desk-top keyboard located in the individual laboratory. These stations at any time can be dialed directly into the computer. At a remote station, the user appears to "own" the computer. Data from as many as 50 remote stations can be submitted to the computer over telephone lines as soon as they are ready, and answers are returned almost immediately. Although each station gets total computer service, the cost per month is a fraction of total computer costs. The computer can handle many stations at the same time because of the speed with which it can switch from one to another. It does not actually handle the stations simultaneously but in turn. It deals with a typical statement from a sender in a few microseconds and then switches to the next station. Each sender receives a reply from the computer within a few seconds as he works through his program, and he gets the impression that he has the computer all to himself. In addition to this advantage of overall input-output speed, time sharing has other advantages which the X-ray spectrographer, as well as other specialists, should consider in evaluating their computer needs. A company or research organization that needs extra computer capacity or one that does not have a large computer can contract for a time-sharing station for about half the monthly cost of an analytical assistant. Also, a user can go through his program step by step with constant checks from the computer. He can easily alter his program and check and correct his data without being highly trained in computer methods. Users with little computer experience can program their own calculations, or they can use appropriate general-purpose programs which have been stored in the computer library for immediate access. For the more computer-sophisticated and for the professional programmer, the conversational approach of time sharing provides a useful debugging tool and enables errors in complex programs to be tracked down quickly. The time-sharing computer may be classified as medium in size as, at present, it does not have the capacity of large-size computers; it may be used, however, as a preprocessor for data to be sent later to the larger, general-purpose computer. Time sharing on large computers is in the development stage. In the near future a laboratory may have time-sharing stations which operate from the company's own computer center.

For some calculations, it is not economically justifiable to use computer time; for these, a desk calculator is the appropriate tool. Its role should not be neglected in this age of electronic computers. In cost, it is probably the least expensive in purchase price of the various calculating choices. Changes in desk calculators have kept pace with changes in digital computers. It is probably safe to say that the mechanical desk calculator is obsolete in that it is slow, oversize, noisy, and limited in its performance compared with some electronic desk calculators which are now available. There are electronic desk units which can instantaneously compute logarithmic and trigonometric functions and for which keyboard operations may be automated by simple programming devices. Certainly, in many laboratories, the flexibility and speed of these desk-top calculators are adequate for simple statistical calculations and for many complex ones as well if they are set up in a systematic fashion.

A realistic choice between the various types of calculators and computers depends upon sufficient understanding of the laboratory environment, the number and type of calculations, the characteristics of the data, and the need for timely response. These factors will, in fact, largely determine how the work can most efficiently be divided between the local small computer, the giant-brain computer, and the time-shared computer. Once the laboratory has access to general-purpose digital computers, however, its horizon and its usefulness are broadened. This is true not only of the X-ray laboratory but of any laboratory which has to do calculations.

APPLICATIONS OF GENERAL STATISTICAL TECHNIQUES TO PROBLEMS IN X-RAY ANALYSIS

A large number of standard statistical techniques may be useful to the X-ray analyst or to anyone specializing in the chemical or physical analysis of samples. Needless to say, any statistical technique requires manipulation of numbers in the fashion necessary to develop the "answer." Any manipulation of numbers ranging from the summation of two numbers to a multiple regression analysis can be programmed for the digital computer. Laboratory needs determine whether computer handling of a statistical analysis is necessary or beneficial. A description of various statistical techniques, all of which have been programmed for computer calculation, follows. Examples of the application of these techniques, taken from X-ray spectrographic data, are included. The techniques described are analysis of variance, experimental design, quality control by control charts, and regression analysis.

Analysis of Variance and Experimental Design

Analysis of variance is recognised by statisticians as the most powerful procedure in the field of experimental statistics. The technique is particularly useful when applied to problems in X-ray and other analytical laboratory procedures.

In the X-ray laboratory, the sampling and preparation of the unknown, whether it be received as a powder, liquid, or solid-metal specimen, is as important in the overall analysis as the conditions used for measurement of the fluorescent intensities or the calculation of the final numbers from the measured intensities. Sampling and preparation of the unknown may involve a number of steps, each of which contributes variability or bias to the analysis. There are many variations in the statistical method for determining variability or bias. Basically, what is required is a practical experiment aimed at determining, first, which factors in the experiment have a significant effect upon the result and, secondly, what the relative effects of these factors are upon the result. The technique used for discerning these factors and their effects is called *analysis of variance.*

The two major uses of analysis of variance may appear similar but they are not. In the first, the factors which are significant are usually not known, and the analysis sorts out those that are important. In the second case, the analysis makes it possible to estimate the relative importance of several sources of variability that are thought to be significant. Estimation of sampling and analytical errors is a common example of the second usage. Generally, more work is required to measure several known sources of variance than to choose the significant items in a list of factors.

The design of an experiment used to make an analysis of variance is vastly important to the success of the technique, and it often requires as much or more time to plan the procedure than to analyze the data which result from that procedure. Once the design of the data collection is fixed, so is the form of the statistical analysis. The reverse of this procedure, in which a set of analytical data is gathered and then, if possible, a suitable statistical form is fitted to it, usually has disappointing results. Realistic conclusions require valid data collected according to a sound plan. An understanding of the wide variety of plans is the province of the statistician, and his advice should be sought. If a statistician is not available for consultation, the statistically minded analyst should broaden his fund of knowledge to include the techniques of experimental design and analysis of variance by study of any of the many excellent textbooks which are available to him.[3-8] Many large computers and time-sharing systems have standard programs for handling the data generated from an analysis of variance. Smaller computers require the programming of these calculations by trained programmers.

Some of the fundamental principles which must be considered in the development of a good statistical experiment are:[9] (1) there must be a clearly defined objective; (2) as far as possible, the effects of the controlled variable factors (the conditions, procedures, *etc.*, under study which are being deliberately varied in a controlled fashion) should not be obscured by other variables; (3) as far as possible, the results should be neither consciously nor unconsciously biased by the design of the experiment or by the experimenter; (4) the experiment should provide some measure of precision; and (5) the experiment should be conducted with sufficient care to accomplish its purpose.

Three examples of designed experiments that will be described to illustrate applications of analysis of variance to problems in X-ray analysis are: (1) a factorial experiment to determine which steps in sample preparation have a significant effect upon the results, (2) a nested design to estimate the variance of each of the steps in sample preparation, and (3) a nested design to estimate the variance of each of the steps in a sampling procedure and to separate analytical error from sampling error.

Example of Factorial Experiment to Determine the Significance of Each Step in Sample Preparation. Suppose that we are interested in investigating the effect of each of several steps in the preparation of a powder sample for X-ray analysis. A portion of a sample is removed from a bottle of material which, for the purpose of this study, we assume to be homogeneous. The usual procedure is to grind the subsample to a fine powder, mix it, weigh a portion of it, and briquette it for presentation to the X-ray instrument. We may have prior knowledge or suspect that the length of the grinding and mixing periods, the amount of subsample used to form briquettes, and the pressure applied for briquetting are variables which cause a bias in the analysis. Table I presents a scheme for determining whether or not these factors are important. This scheme is called a *factorial experiment*—in this case, a $2n$ factorial, where $n = 4$ factors. Each of the factors A, B, C, and D is studied at each of two levels (specific values), *i.e.*, a short and a long grinding period, a short and a long mixing period, small and large amounts of subsample, and low and high briquetting pressures. By convention, the zero subscript used in Table I refers to the lower level, and the subscript 1 to the higher level. The symbols (1), a, b, ab, *etc.*, represent various combinations of levels. In following this scheme, it would be necessary to prepare 16 briquettes according to the various combinations of conditions. An estimate of the experimental error could be made from the factorial design or, better yet, by the complete replication of the entire experiment.

Table I. Factorial Design for Determining the Effects of Four Steps in Sample Preparation, Each Step at Two Levels

		A_0		A_1	
		B_0	B_1	B_0	B_1
C_0	D_0	(1)	b	a	ab
	D_1	d	bd	ad	abd
C_1	D_0	c	bc	ac	abc
	D_1	cd	bcd	acd	$abcd$

Factors: A, grinding time; B, mixing time; C, weight of powder; and D, briquetting pressure.

Table II. Nickel Content of Powder Material, Measured for Factorial Experiment

		Per cent of nickel			
		A_0		A_1	
		B_0	B_1	B_0	B_1
C_0	D_0	12.5	12.3	12.2	12.2
	D_1	12.4	12.3	12.0	12.1
C_1	D_0	12.6	12.5	12.2	12.3
	D_1	12.5	12.6	12.4	12.4

Table II lists the determinations of the nickel content of briquettes prepared from a powdered sample of a light material according to this four-factor design in which factor A is the grinding time, factor B is the mixing time, factor C is the weight of powder, and factor D is the briquetting pressure. Based on previous experience with other similar materials, it had been suspected that grinding time was an important factor; the significance of other factors in the preparation of this particular material was unknown. Analysis of the tabulated data confirmed the significance of grinding time, factor A, and indicated that factor C, the weight of the powder used in preparing the briquette, was also important. In this experiment, the critical numbers for 95 and 99% significance are 0.926 and 1.46, respecively. (The 95 and 99% significance mean a 5 and 1% chance of making an error in the decision.) A factor with a numerical effect greater than these critical numbers is significant. The numerical effect which was calculated for factor A was 1.9, and, for factor C, it was 1.3. If the decision had previously been made to judge significance according to the 99% critical value, the weight of the powder would not qualify. For the X-ray analyst, however, this indication of significance at the 95% level is sufficient to suggest that a careful control of this variable should be made.

It was interesting to note that the percent of nickel indicated by the X-ray analysis was lower in samples of the finer particle size (those which were ground for the longer period). This result was contrary to the commonly expected increase in the fluorescent intensity of nickel with decrease in particle size, but it was consistent with our previous experience, which showed that a metal heavier than its matrix tends to coat the matrix particles. Reduction in the overall particle size of the material resulted in greater dilution of the heavier metal on the surface of the particles by the matrix ,which reduced fluorescent intensity.

Example of Nested Design for Estimating the Variance of Each Step in Sample Preparation. If an X-ray or another analyst desires to improve the overall precision of his measurement, he has two alternatives. He may either run additional complete analyses and average them, or he may improve the analytical procedure; the choice is governed by the economics of the situation. The cost of improving precision through additional analyses is easily estimated. The cost of improving the analytical procedure is not known.

Figure 1 diagrams a nested sampling design which would aid in determining the cost of improving the analytical procedure. It can be employed in conjunction with analysis of variance to separate the components of variation and to point out where the major variations occur. Matthias[10] describes the use of a nested subsampling design (Figure 1)

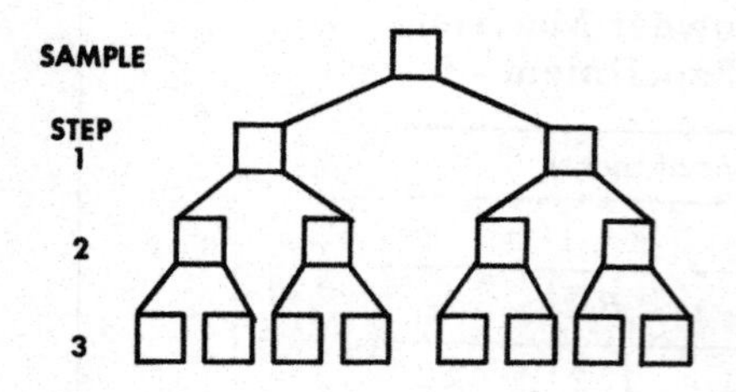

Figure 1. Nested sampling design, three-step preparation procedure.

in control laboratory problems which could be used profitably in the X-ray laboratory. According to this scheme, the first step in a preparation procedure is performed on a pair of subsamples from a homogeneous sample. The second step is performed in duplicate on each of the portions from step 1; and the third step, in duplicate on each portion from step 2. If the procedure involved a fourth step, as did the one described in the previous section for preparing a light matrix sample for nickel measurement, a fourth step would be added to the nested scheme, and performed in duplicate on each portion from step 3. The total number of briquettes prepared would be 8 in the case of the three-step plan and 16 in the case of the four-step procedure, for each pair of subsamples used. The estimates of experimental error which are required for calculation of the analysis of variance are automatically built into this design.

The intent of the nested sampling study with this stepwise design is to determine the variation in a standard procedure which is already in use. Step 1 would be performed at a constant grinding time; step 2, at a constant mixing time, *etc.* Even under controlled conditions, one of the steps may contribute more variability to the analysis than the others, and, if so, it would be advantageous to modify the preparation procedure and to perform this step in replicate or to improve it, if possible.

Changes in a nested scheme are possible so as to effect a reduction in the cost of the experiment. This is almost essential because a single-"unit" scheme, as in Figure 1, should be replicated if it is to estimate all sources of variation with reasonable sensitivity. That is, the sequence of operations should be performed on several pairs of subsamples. For example, if five pairs of subsamples were taken, a total of 80 briquettes would be required for a four-step procedure. Bainbridge[11] describes staggered nested designs which offer the advantage of a reduction in the number of required measurements. However, computer calculation of the variance components is necessary because of their complexity.

Example of Nested Design for Estimating the Variance of Each Step in a Sampling Procedure. The most important single point which could be made in this paper is that none of this statistical work in the refinement of the analytical method is worthwhile if the sample itself is not representative of the source material from which it was taken. The *only* way to detect and estimate sampling error and to separate it from the analytical error is by an analysis of variance, based on data from a properly designed experiment, from which the variance components are determined.

To aid in the understanding of the practical application of these techniques, assume a laboratory routinely determines the nickel content of composites of several batches of material. Each batch is normally loaded into drums; samples are taken from them in some fashion, composited, and submitted to the laboratory for analysis. A subsample is taken from the bottle of composited material, and duplicate analyses are performed on it. The question is: How good is the analytical number obtained by duplicate analyses, and does it really indicate the true nickel content of the batches which were sampled?

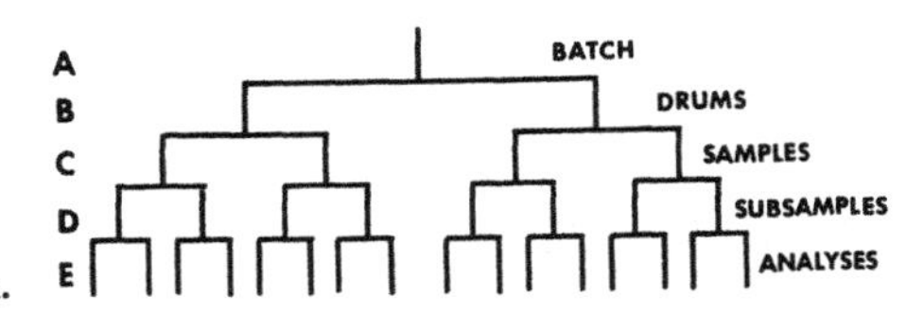

Figure 2. Five-factor nested design.

If the standard laboratory procedure were to make a single nickel determination on a single subsample from a single sample taken from a single drum from a single batch, the analysis could be considered as a precise measure of the nickel content of the batch only if there were no variations in any of the sampling steps.[12] But the single sample from each of these "universes" is only one of the many which could have been taken, and variation must be assumed to exist in each universe as well as in the analytical one. Unfortunately there is no inexpensive way to estimate the variation in the steps. They can be estimated only after a statistically designed experiment, such as the one outlined in Figure 2, has been carried out.

We can perform a sampling study of the nickel-containing material by sampling N_a batches, taking N_c samples from each of N_b drums of each batch, taking N_d subsamples from each sample, and making N_e determinations of nickel on each subsample. Obviously, the expense of the analyses can become prohibitive unless the most efficient design is chosen. Figure 2 presents a *single unit* of a five factor nested sampling design which is a highly inefficient one but is the best one if the calculations have to be done without the aid of a computer. If two units of the five factor-design are to be used, the experimental pattern is established by sampling each factor in duplicate, beginning with the first, factor A. Two portions are taken from each of two A's; two portions, factor C, are taken from each B subsample; *etc.* The scheme (Figure 2) is a sampling outline which could be followed in a study of the components of variance. It can be considered a diagram showing the analytical measurement as the end point of a chain of samples which begins with the batch of material.

The results of the analyses from a replicated unit design are subjected to the analysis of variance, or ANOVA as it is abbreviated by statisticians. A computer can be programmed to make the analysis of variance and to calculate the variance components. The results will indicate which of the steps in the sampling process contribute a significant variability and will estimate the variances in each of the sampling steps. The V's shown in Table III represent these variances (the squares of the standard deviations) for each factor. As shown, their sum gives the variance of the average of all the analyses. If the unit design in Figure 2 had been replicated with N_a, N_b, *etc.*, each equal to 2, the denominators

Table III. Results Calculated from Nested Design for Planning Economical Sampling

$$V(\bar{X}) = \frac{V(A)}{N_a} + \frac{V(B)}{N_{ab}} + \frac{V(C)}{N_{abc}} + \frac{V(D)}{N_{abcd}} + \frac{V(E)}{N_{abcde}}$$

$V(A) =$ calculated variance for factor A, *etc.*

$N_a =$ number of samples of factor A, *etc.*

for each variance term would be 2, 4, 8, 16, and 32, respectively. A total of 32 analyses would be required (16 for each unit in the design).

If our sampling study of the nickel-containing material is typical of many sampling studies which have been performed at Union Carbide's Metals and Chemicals Divisions, the variation in the factors *A*, *B*, *C*, and *D* will be many times larger than the analytical variation, factor *E*. According to these results, it is not practical to refine the analysis unless the variability in the sampling factors is also reduced. In many situations, it is impossible to reduce the variability of the product being analyzed, but it remains desirable to improve overall precision by replicating and averaging one or more of the sampling steps. The measured content of a composite of several batches of material will depend on the *N*'s *i.e.*, the number of samples taken from each step, with the assumption that the variability of the manufacturing process remains the same as it was during the sampling study. The total variance of the average, $V(\bar{X})$, becomes smaller as the *N* is increased for an individual step. The best sampling procedure can be determined by minimizing $V(\bar{X})$. The more drums which are sampled and analyzed, the less the variability is that the drums contribute to the total; the more samples taken from each drum, the smaller this term becomes; etc. It is wise, then, to take multiple samples at any point of significantly high variability. Depending on $V(E)$, the variance of the analyses, there may or may not be a need to improve the analytical technique or to run repetitive analyses.

The economics of sampling and testing is discussed by Davies,[4] who develops a formula for determining the most economical sampling scheme.

Quality Control by Control Chart

Statistical analysis can show us how to improve the preparation of a sample for analysis and how to decrease the variability in the sampling of the material being analyzed. At this point, we may still ask what the value of our analysis really is. If analytical numbers are obtained periodically on a manufactured product, the quality of which is indicated by the level of its metal content, the evaluation of these numbers is still another statistical task and can be facilitated by the control chart technique. Although the concept of quality control by application of control chart techniques is not new, there are many industries which have not adopted it simply because of unfamiliarity with its potential value.

Much of the value of analytical data being generated in the laboratory is wasted. These data could be accumulated by the analyst for the establishment of control on the production of the material being analyzed. This can be done if the production process is reasonably continuous and if analyses can be made in a regular fashion.[13] Consider that a given number of samples are taken from the process at more or less regular intervals and analyzed for their metal content. If the manufacturing process is operating under normal conditions, fluctuations will occur in the average metal content of each set of samples. The fluctuations will be distributed in a random statistical pattern, the old familiar bell-shaped or normal curve which is illustrated in the left portion of Figure 3.

If enough sets of samples (a minimum of 30) are analyzed, it is possible to estimate the mean and standard deviation of the set averages $\bar{X}$. A chart can be drawn with a vertical scale calibrated for $\bar{X}$ and the horizontal scale marked to designate time or order of production. If horizontal lines are drawn through $\bar{\bar{X}}$, the estimated grand mean of $\bar{X}$, and through each of the extreme values on the upper and lower tails of the distribution of $\bar{X}$, the result is a control chart for $\bar{X}$. These upper and lower control limits are three standard deviations above and below the grand mean and include 99.73% of the data.

Once the control chart is established, it can be used as a "watchdog" on the future manufacturing operation. Sets of samples are analyzed periodically and the average of

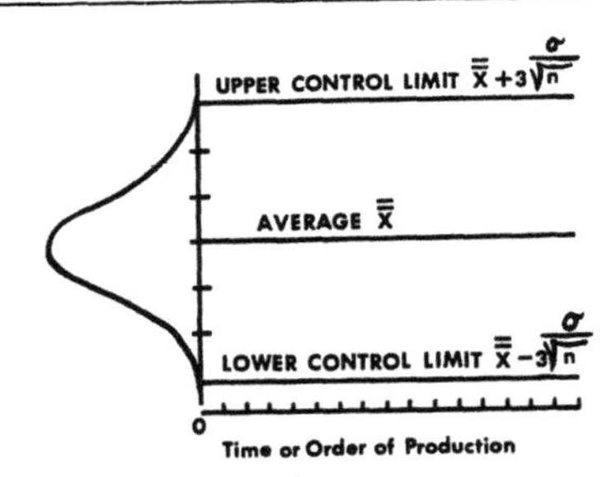

Figure 3. Illustration of the theoretical basis for a control chart.

each set is plotted on the control chart. If these values fall within the established control limits and show no cycles nor runs above or below average nor runs up or down, then it can be said that the manufacturing process is in a state of statistical control at a designated level of probability. If the data do not conform to a pattern of random variation within the control limits, which is frequently the case in the beginning, then departures from the pattern should be investigated and the reason for the variation—the "assignable cause," as the statisticians call it—should be tracked down. If an assignable cause of variation is found to be detrimental to the product, an effort should be made to eliminate it. In this way, the variability of the product is kept under control, and the quality of the product is assured at a constant level. Conversely, if the variation indicates an improvement in the product, an effort should be made to extend and perpetuate the cause effecting the improvement. In this way, the process may be improved. Statistical control, therefore, is a method of focusing attention on the presence of assignable causes for exceptional operation, both good and poor, and tends to prevent deterioration of the quality of the product and to suggest the existence of a condition which might improve it.

In addition to controlling the level of the critical metal in a product, the variability of the metal content may be kept under control by the preparation and use of a control chart for ranges, *i.e.*, a chart of the ranges of the numbers in each set of analyses.

A computer program may be used to calculate the control limits for averages and ranges from the analytical data taken to set up the control charts. A computer may also be programmed to evaluate the analyses as they are made on subsequent production and to indicate when the critical analytical number is "out of control."

The long- and short-term benefits which result from statistical handling of laboratory data are often quite impressive, both from an economic and a technological standpoint. For example, use of statistical quality control coupled with high-speed X-ray analysis for investigation of certain plant-produced catalysts has brought large improvement in this segment of Union Carbide's business. As a matter of fact, scientific approaches of this type are slowly but steadily transforming the subject of catalysis from an art to a very exact science.

The Role of Regression Techniques in the X-Ray Laboratory[14]

Quantitative determinations in X-ray fluorescence analysis, as in other types of instrumental techniques, require calibration of the instrument with standards of known composition and inverse estimation of unknown quantities from the established calibration. In many cases, simple methods which relate the measured X-ray fluorescent intensity of an element in a set of standards to its known concentration are applicable. The relationship between the two variables, intensity and concentration, can be expressed graphically by plotting the data and drawing a straight line through the points, and there appears to be no need for a statistical analysis. Not all calibration points, however, will fall perfectly in a straight-line pattern, even when the analyst is certain that the relationship is a linear one. A least-squares determination of the slope and intercept may be helpful in drawing the best line to fit the data; and, of course, if the calculations of an unknown are subsequently to be made by a computer, the equation for that line must be known. For the simple linear example, a single independent variable X, the intensity of the element in question, is used to predict the dependent variable Y, or concentration (see Table IV).

It would be well to point out an apparent inconsistency between mathematical notation and the physics of an analysis requiring a calibration curve. When a linear (or other) equation is used to determine the concentration of an element in an unknown, the X, or fluorescent intensity, could be considered the independent variable, and Y, or concentration, the dependent variable. On the other hand, in the relationship between intensity and concentration in the physical sense, the concentration is the independent variable and the intensity is dependent upon it. This is true when one is measuring the standards in order to establish the calibration curve. The choice in labeling axes on a calibration curve is the analyst's, and a designation of the axes which avoids confusion is the best choice.

In the regression equations in this paper and my previous one on the subject,[14] fluorescent intensities are used as independent variables and concentrations as dependent variables when the best equations for calculation of concentrations are to be determined. This usage could be considered incorrect mathematically, as matrix inversion is the accepted way to convert from the concentration–intensity relationship of independent to dependent variable to that of dependent–independent variable. However, because errors in intensity readings and in concentration values are both small and the method of analysis sought was an "empirically correct" one, this inconsistency may be overlooked in favor of the proved, practical advantages of the less rigorous procedure.

Not all laboratory calibrations are linear, and many exhibit a curvature, either concave or convex, which is hyperbolic or parabolic in shape. An example is a two-element matrix in which one element highly absorbs or enhances the fluorescence of the other. Under these conditions, a quadratic equation best describes the calibration (see Table V). The calibration curve may be "eyeballed," but the constants in an equation for the curve are best determined by calculation, which can be done conveniently with computer help.

There are analytical situations in which linear or quadratic calibrations are not adequate. In these cases, the relationship between concentrations and intensity is very complex, and the true shape of the calibration, *e.g.*, may be similar to that of a telephone receiver or a series of hilly surfaces. Figure 4 illustrates a nonlinear or nonquadratic calibration which cannot be represented adequately by a two-dimensional plot. Figure 4 shows uncorrected X-ray fluorescent intensities plotted against concentration of the element of interest in a complex matrix (in this case, the intensity of Mn $K\alpha$ in a slag matrix of the oxides of silicon, magnesium, aluminum, calcium, barium, iron, and minor impurities).

Table IV. Straight-Line Calibration Curve, Linear Equation

$Y = a_0 + a_1 x$

$Y =$ the dependent variable representing concentration

$X =$ the independent variable representing fluorescent intensity

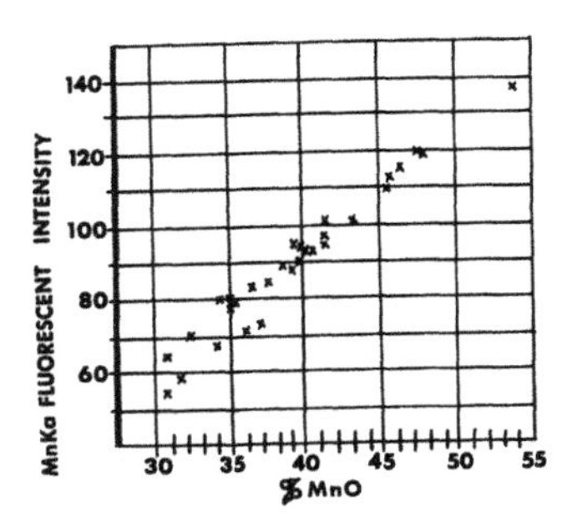

Figure 4. Manganese fluorescent intensity–concentration relationship for slag samples.

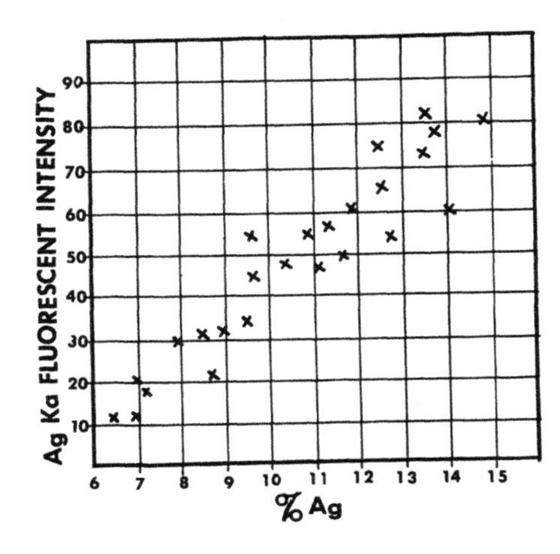

Figure 5. Silver fluorescent intensity–concentration relationship.

Table V. Concave or Convex Calibration Curve, Quadratic Equation

$Y = a_0 + a_1x + a_2x^2$

$Y =$ percentage of element A in matrix material B

$X =$ fluorescent intensity of element A in an absorbing or enhancing matrix B

An even more frustrating situation is shown in Figure 5. The matrix in this figure is a light powdered material, a commonly used catalytic material in the petrochemical industry, and the element for which we are attempting to plot a calibration is silver. Prior to the application of the technique of multiple regression analysis, all attempts to untangle the data by theoretical corrections or by elaborate preparation of the powder sample had failed. (A solution analysis was satisfactory but very time-consuming.) In these two situations, one, an unusually complex sample and, the other, a material with an unusual physical or crystalline structure, the use of a multiple-regression analysis for development of a satisfactory mathematical relationship between concentration and intensity was found to offer the simplest approach. In fact, in the second sample, that of silver in a light matrix, it is the only approach that worked satisfactorily on a powdered sample.

Simple regression relates the dependent variable to a single independent variable. Multiple regression relates the dependent variable to several independent variables.

Table VI. *N*-Dimensional Calibration Curve, Polynomials of Higher Order

$$Y = a_0 + a_1x_1 + a_2x_2 + a_3x_3 + a_4(x_1)^2 + a_5(x_2)^2 + a_6(x_3)^2 + a_7x_1x_2 + a_8x_1x_3 + \cdots + a_n(x_1)_3 + \cdots$$

Y = percentage element A in matrix of B, C, *etc.*
X_1 = fluorescent intensity element A
X_2 = fluorescent intensity element B
X_3 = fluorescent intensity element C, *etc.*

Multiple linear regression uses only linear variables, while a nonlinear regression uses higher-order transformations of the variables (with linear coefficients). Table VI lists a polynomial of higher order which can be used as a model for a multiple-nonlinear-regression analysis.

The appropriate model for analysis of X-ray data is empirically determined. If the analyst has prior knowledge of the effects of matrix elements on the element being analyzed, he may include intensity measurements of these elements in the analysis. This technique is satisfactory for slag samples in which each element has a specific effect on manganese intensity, some having a more severe effect than others. For the silver catalytic material, element intensities alone are not adequate. In the development of a satisfactory relationship between silver concentration and sample fluorescence, many measurements were made at various emission lines of the matrix metals and various angles of scattered radiation. The use of computer-calculated correlation coefficients which indicated the degree of correlation of the various measurements with silver intensity was very helpful in selecting the variables to be included in the regression analysis. Table VII lists the correlation coefficients between the per cent of silver and the measured fluorescent intensities of a series of standard samples. A number of variables, including the background read at 0.45 Å and its reciprocal, were more highly correlated with the per cent of silver than was the fluorescent intensity of Ag $K\alpha$ itself. Correction of the Ag $K\alpha$ line by subtraction of or ratioing to background measurements resulted in high correlations, the later correction giving somewhat higher coefficients than the first. Inclusion of the measurement at 0.45 Å in the regression resulted in the successful calibration of six types of catalytic materials of varying silver concentrations and matrix with a single equation. Although the manufacturer of the X-ray instrument on which this analysis was finally refined and established as a routine procedure may have thought we were mistaken in "wasting" an X-ray spectrometer on background measurements, we were completely justified in our decision. The regression technique, given the appropriate variables, was the only solution possible for this severe analytical problem.

Several other comments on the use of regression techniques are appropriate. The regression equation is no more accurate than the data on which it is based; if they are subject to a large experimental error, the equation will predict inaccurately. Precise measurement of standards and accurate knowledge of their metals content are essential. For example, we recently have been able to reduce the number of measurements made on our silver catalytic material from three to two by improving the instrumental precision of measurement of the line and background intensities. Also, chemical analyses in duplicate were performed on all standards, as analytic inaccuracies add to any error which may be inherent in the regression analysis. Preparation of the samples and the standards prior to analysis was carefully optimized in order to present a highly uniform surface to the X-ray beam. The final calibration equation was checked intermittently by chemical analysis of unknowns to ensure continuing accuracy.

Table VII. Correlation Coefficients Between Per Cent of Silver and Matrix Fluorescent Intensities

Description of variable, fluorescent intensity	Correlation coefficient
Ag $K\alpha$	0.943
Background at 0.45 Å	−0.965
Background at 0.63 Å	−0.856
Calculated background at 0.56 Å	−0.958
Al $K\alpha$ matrix	0.433
Ag $K\alpha$ −background at 0.45 Å	0.968
Ag $K\alpha$ −background at 0.63 Å	0.961
Ag $K\alpha$ −calc. background at 0.56 Å	0.962
Ag $K\alpha$/background at 0.45 Å	0.975
Ag $K\alpha$/background at 0.65 Å	0.976
Ag $K\alpha$/calc. background at 0.56 Å	0.976
Reciprocal of Ag $K\alpha$	−0.811
Reciprocal of background at 0.45 Å	0.987

Calibration with this technique must be based on as large a number of standards as is practical. It will fail to predict unknown concentrations accurately if the calibration base does not include all meaningful combinations of the measured variables as established by regression analysis. Extrapolation beyond the range of the standards can lead to highly erroneous results.

The main value of this technique is that it can bring order out of the chaos of a mass of data and may be the only practical solution to a difficult X-ray problem, as was the case in our analysis of the silver catalyst. Use of the technique is now saving over $11,000 per year in analytical cost plus an added saving of $15,000 per year due to time-shared communications with computer facilities and rearrangement of the laboratory.

APPLICATION OF COMPUTERIZED MATHEMATICAL–STATISTICAL TECHNIQUES TO SPECIFIC PROBLEMS IN X-RAY ANALYSIS

The extent to which the X-ray laboratory can effectively employ the digital computer is limited only by the imagination of the spectrographer. There are very few, if any, X-ray laboratories which would not benefit from the services of a digital computer. Control laboratories and research laboratories are alike in this need for automated calculation. The X-ray spectrographer is remiss if he does not seek computer assistance, and management is remiss if it does not provide the service in some form.

Possible applications are so numerous that only a few of the more obvious ones can be mentioned here. It is impossible to provide a comprehensive list because each X-ray analyst has his own way of tackling a problem in X-ray analysis; but, whatever his personal method might be for handling numerical data, he will benefit from automation. If nothing else, possible human error in transposition of data can be eliminated. Data transposition from an X-ray instrument to a time-shared or other independently located computer can be automated by digitizing the instrumental output and automatically recording it on punch tape or cards which are compatible with the computer input requirements. Direct transmission of digitized readings to the independent computer is also possible; obviously, automatic data transposition is one of the functions of the small specialized X-ray computer.

Practical Applications of Computerization

The accuracy of an X-ray determination is frequently improved by measurement of scattered X-radiation adjacent to the analytical line. In some cases, the measurement of coherent or incoherent scattering is beneficial in compensating for physical inhomogeneities.[15] The background measured adjacent to the line or the background at the line after interpolation from background measurements on both sides may be subtracted from or ratioed to the line measurement itself. The background correction may be the first step in a sequence of corrections, such as correction of intensities for detector nonlinearity, calibration shifts, matrix effects, and correction to 100% of total analysis. Any or all corrections that may be useful in refining analyses can be programmed for computer calculation.

An interesting correction which could be made is for an actual line interference such as that of Cr $K\beta$ with Mn $K\alpha$. If the computer has sufficient information in storage and is given proper instructions, it can make any type of correction. To calculate a correction for line interference, the computer would need prior quantitative information which would establish the relationship between the magnitude of the interference and the concentration of the interfering element.

Calibration shifts are of several types. A calibration curve may change in slope, shape, or intercept (the level of background intensity). Given precise readings on a minimum number of standards, the computer program can easily correct the data from unknowns for whichever change has occured, prior to determination of concentrations from the calibration.

Corrections for matrix effects can be made in numerous ways, and each analyst has a favorite technique which he believes to be best suited for his purposes. He may program his favorite method along with several other matrix correction techniques and, by computerization, compare the several techniques. In this unbiased fashion, he can discern which approach is best. For example, different regression equations can be evaluated by the analysis of a set of unknown samples according to each equation; the equation which produces the most accurate analysis is best.

The calculation of final concentrations in an X-ray analysis may involve adjustment of the actual analysis for changes due to the preparation of the sample. If a sample has been concentrated or a separation has been performed or a solution has been prepared from a powder or the sample has been subjected to any other of the many procedures followed in sample preparation, an appropriate computer program can compensate for the changes and adjust percentages accordingly.

If regression equations are the favorite method of matrix correction, as they are in our laboratory, the measured standard and sample intensities are fed directly to the computer, which then corrects for day-to-day calibration fluctuations, substitutes corrected intensities in the appropriate equations, corrects for sample dilutions, averages results from duplicate subsamples, averages results from sets of samples, calculates ranges within sets of samples, and compares and plots the final numbers relative to predetermined control-limit values for the material being analyzed.

Application of Computerization to Theoretical Problems

Strictly speaking, there are no "theoretical problems" in quantitative X-ray fluorescence analysis. There are only various theories which have been developed to help solve the basic practical problem of how to relate X-ray fluorescent intensity accurately to element concentration. Each theory describes some kind of mathematical relationship, which varies from the very simple to the very complex. And any mathematical relationship will benefit from the assistance of a digital computer for its application. It is not within the scope of this paper to list these numerous theoretical studies; only a few recent ones will be mentioned.

The use of multiple-regression analysis to relate fluorescent intensity to element concentration empirically was discussed in a previous section as a method for predicting concentration from fluorescent intensity. The inverse relationship between intensity and concentration was suggested[14] as a method for predicting fluorescent intensities from composition. This method makes possible a systematic study of matrix effects and a better understanding of absorption–enhancement phenomena.

The output from a computer program for calculating the regression analysis may include a tabulation of the coefficients of correlation between each of the input variables. If it does not, a program may be written that is intended to calculate only these coefficients. The coefficients vary from 0 for no correlation to 1.0 for perfect positive correlation and from 0 to −1.0 for perfect negative correlations. They are useful in the interpretation of the significance of matrix effects. Table VII lists the correlation coefficients between a set of independent variables (measured fluorescent intensities and their transformations) and silver concentration. The intensity variables which have a significant effect on the concentrations of the element of interest are easily picked out from a tabulation of this kind (see the section "The Role of Regression Techniques").

Correlation coefficients can also be calculated between the measured intensity of the element of interest, as a dependent variable, and either percentages or fluorescent intensities of the various elements in the sample matrix, as independent variables. An interesting study could be made of the comparative numerical effects of matrix concentrations and matrix fluorescent intensities on the fluorescent intensity of the analytical element. As previously mentioned, prior determination of mathematical correlations is an aid in the choice of the independent variables to be included in a regression analysis, whether one is considering intensity or concentration as the dependent variable.

R. J. Traill and G. R. Lachance[16] have developed a simple relation between relative X-ray intensities and mass concentrations of the elements in a multicomponent system. A correction constant α_{AB} is a quantitative measure of the effect of matrix element B on element A. Various α values are related to the concentration of A according to the equation in Table VIII. If the concentrations of elements B and C are unknown, the measured relative intensities of A, B, and C are substituted in a set of equations which are solved for the concentrations of A, B, and C. The use of computer assistance in the solution of the series of simultaneous equations is indicated.

Because it is not always possible to obtain standards for the actual measurement of the α's, an equation for their calculation from theoretical considerations was developed by

Table VIII. Alpha Type of X-Ray Analysis

$$R_A = \frac{\text{measured intensity}}{\text{intensity of } 100\%\ A}$$

$$\alpha_{AB} = \text{effect of matrix } B \text{ on element } A$$

$$\alpha_{AB} = \frac{C_A - R_A}{R_A C_B}$$

$$C_A = R_A + R_A \alpha_{AB} C_B + R_A \alpha_{AC} C_C$$

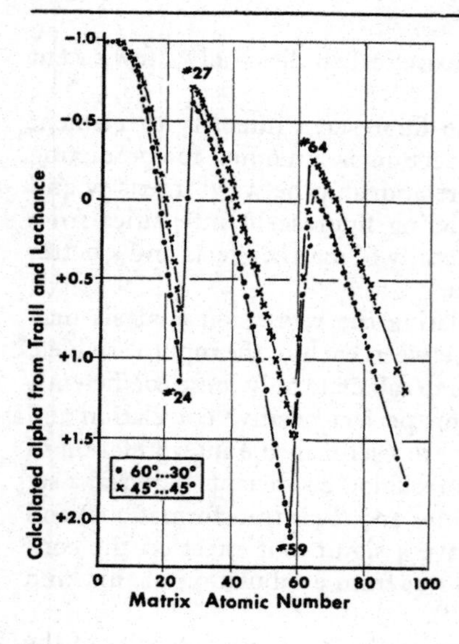

Figure 6. Fe $K\alpha$ alpha–atomic number pattern.

Table IX. Calculation of Theoretical Alpha Value*

$$\alpha_{AB} = \frac{(\mu_1 \operatorname{cosec} \phi_1 + \mu_2 \operatorname{cosec} \phi_2)B}{(\mu_1 \operatorname{cosec} \phi_1 + \mu_2 \operatorname{cosec} \phi_2)A} - 1$$

* After Lachance and Traill.[16]

Traill and Lachance (Table IX). The variables in the equation are the mass absorption coefficients of the element and the other matrix materials and the cosecants of the angles of the incident and emergent X-ray beams. The consistency of theoretical and empirical α values may hinge on the condition of sample excitation. A factor for the integrated energy of the exciting radiation was included in the theoretical calculations made by Shiraiwa.[17]

Among the numerous programs which have been written by Union Carbide's Statistical Group in Charleston for the use of our inorganic analytical laboratory was one designed to calculate the theoretical α's of Traill and Lachance and to plot them against the atomic numbers of the matrices. Figure 6 shows this plot for the Fe $K\alpha$ line at 1.94 Å, calculated at incident and emergent angle combinations of 60°, 30° and 45°, 45°. The pattern compares with a similar plot of relative intensity (absorption–enhancement index) *versus* atomic number,[18] as in Figure 7.

Expansion of matrix correction techniques leads to the possibility of the ultimate in computer programs—a universal solution to the problem of matrix effects in X-ray spectrography and electron probe microanalysis. The computer input should consist of complete and accurate quantitative information on absorption and enhancement effects, for example, the α type of numbers for the effects of all elements on X-ray emission lines at constant conditions of excitation and instrumental optics. Calibration of the X-ray instrument would involve only the measurement of backgrounds at the analytical lines and the 100% standard of each element in the same physical form as the unknown sample,

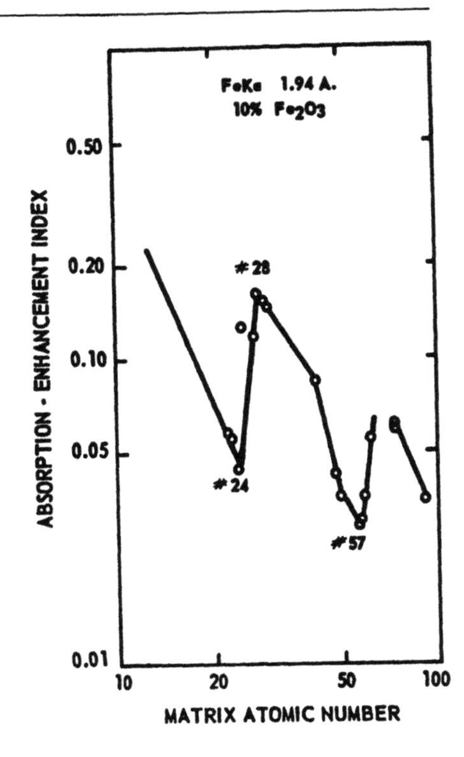

Figure 7. Fe $K\alpha$ intensity–atomic number pattern.

i.e., powder or metal. These fluorescent intensities, the fluorescent intensities of the unknowns, pertinent information concerning elements like carbon which are not normally analyzed by X-ray fluorescence, and any necessary sample weights would be fed to the computer for its calculations. The computer would calculate the true percentages of the sample components by using one of various techniques, such as the solution of series of simultaneous equations.

CONCLUSIONS

The applications of computer-assisted statistical and mathematical techniques in quantitative X-ray analysis are limited only by the imagination of the spectrographer. Many of these same techniques are valuable tools in other analytical disciplines. Among the general statistical methods which are useful in the interpretation of X-ray data are analysis of variance, experimental design, quality control charts, and regression analysis.

Typical applications are for the evaluation of the various steps in sample preparation, the determination of variability in sampling procedures, process improvement by referral of analytical numbers to control chart probability limits, and the development of analytical methods based on regression analysis of calibration data. Many computerized statistical techniques make significant economic saving possible by the reduction of analytical and plant operating costs and by the improvement of product quality.

Additional savings are made possible with computer assistance in X-ray instrumental-calibration procedures and the calculation of concentrations after matrix and background corrections. Computerized methods offer other advantages of increased speed and dependability.

The study of theoretical problems in quantitative X-ray analysis is facilitated by digital computer calculation. A universal X-ray method is now possible, based on computer evaluation of all pertinent sample data.

Statistical and other mathematical manipulations are not a substitute for good analytical technique and will not reveal anything which is not already in the data. Statistical procedures are simply tools the X-ray spectrographer can use to get the most out of his experiments.

The choice among the various types of computers depends upon an adequate understanding of the laboratory situation, the number and type of calculations, the characteristics of the data, and the need for timely response. Once the laboratory has access to general-purpose computers, its horizon and its usefulness are significantly broadened.

ACKNOWLEDGMENTS

The author thanks the Union Carbide Corporation, Chemicals and Plastics Division, for its permission to publish this work. She is indebted to F. L. Boggs, F. N. Hopper, J. E. Kellam, and C. J. Kennett for their generous assistance in reviewing this work.

REFERENCES

1. M. J. Slonim, "Sampling in a Nutshell," *Am. Statistical Assoc. J.*, **52**: 143, June 1957.
2. A. Savitzky, "Computers Reduce Data for Better Quality Control," *Chem. Eng.* **73**: 99, 1966.
3. C. A. Bennett and N. L. Franklin, *Statistical Analysis in Chemistry and the Chemical Industry*, John Wiley & Sons, Inc., New York, 1954.
4. Owen L. Davies, *The Design and Analysis of Industrial Experiments*, Hafner Publishing Co., New York, 1954.
5. R. L. Anderson and T. A. Bancroft, *Statistical Theory in Research*, McGraw-Hill Book Company, New York, 1952.
6. K. A. Brownlee, *Statistical Theory and Methodology in Science and Engineering*, John Wiley & Sons, Inc., New York, 1960.
7. W. J. Dixon and F. J. Massey, *Introduction to Statistical Analysis*, McGraw-Hill Book Company, New York, 1951.
8. R. A. Fisher, *The Design of Experiments*, Hafner Publishing Co., New York, 1960.
9. M. G. Natrella, "Experimental Statistics," *Natl. Bur. Std. (U.S.) Handbook* **91,** 1963.
10. R. H. Matthias, "Use of Subsampling in Control Laboratory Problems," *Anal. Chem.* **29**: 1046, 1957.
11. T. R. Bainbridge, "Staggered, Nested Designs for Estimating Variance Components," *Industrial Quality Control*, p. 12, July 1965.
12. F. N. Hopper, "Notes for Instruction in Statistical Control," Unpublished manuscript, Union Carbide Corp., 1962.
13. A. J. Duncan, "Quality Control and Industrial Statistics," Richard D. Irwin, Inc., Homewood, Illinois, 1953.
14. B. J. Mitchell and F. N. Hopper, "Digital Computer Calculation and Correction of Matrix Effects in X-Ray Spectroscopy," *Appl. Spectry.* **20**: 172, 1966.
15. G. Andermann and J. W. Kemp, "Scattered X-Rays as Internal Standards in X-Ray Emission Spectroscopy," *Anal. Chem.* **30**: 1306, 1958.
16. R. J. Traill and G. R. Lachance, "A New Approach to X-Ray Spectrochemical Analysis," *Can. Dept. Mines Tech. Surv. Geol. Surv. Can., Paper* **64-57**, 1965.
17. T. Shiraiwa, "Application of Theoretical Calculations to X-Ray Fluorescent Analysis," in: Pittsburg Conf. on Anal. Chemistry and Applied Spectroscopy, Paper No. 284, March 1967.
18. B. J. Mitchell, "A Study of the Complete Pattern of Calibration Curves for a Variety of Matrices," 12th Annual Symposium on Spectroscopy, Chicago, Illinois, May 1961.

DISCUSSION

F. Bernstein (General Electric Co.): In the experiment you described on the nickel samples, I believe one of the conclusions was that the weight of sample in preparing the briquette was a significant factor in affecting the precision or accuracy of analysis.

B. J. Mitchell: Yes, that's right.

F. Bernstein: Could you elaborate on the order of magnitude of variation you are talking about?

B. J. Mitchell: In this case, perhaps it was smaller than that we normally worry about. The difference in grinding times, of course, caused the biggest error, a matter of 0.3 or 0.4%. The difference in the weight of sample we used to make the briquette resulted in, perhaps, near 0.1 to 0.2%. In other words, it was necessary to control the weight of the sample, which was birquetted; otherwise, the 0.1 to 0.2% error would be added to the analysis.

F. Bernstein: Control it to what degree? To 1 mg, to 0.1 mg?

B. J. Mitchell: To within a couple of tenths of a gram.

X-RAY FLUORESCENCE ANALYSIS OF A MANGANESE ORE

W. D. Egan and F. A. Achey

Homer Research Laboratories
Bethlehem Steel Corporation
Bethlehem, Pennsylvania

ABSTRACT

To replace time-consuming chemical analysis, a procedure was developed for applying X-ray fluorescence to the analysis of a manganese ore. This X-ray fluorescence method is rapid; the total time for both sample preparation and analysis is ½ hr. The method is also simple enough for routine laboratory use. The components determined, the concentration ranges, and the agreement between chemical and X-ray analysis in terms of standard deviation are:

Component	Concentration range, %	Agreement standard deviation, % concentration
Manganese	14–38	±0.40
Iron	2– 8	0.12
SiO_2	2–11	0.29
CaO	1–24	0.45

The agreements between chemical and X-ray analysis noted for manganese and iron are obtained by correcting for CaO concentration when determining the manganese and for manganese and SiO_2 concentrations when determining the iron. The corrections employ empirical equations developed by a multiple regression technique.

Sample preparation is reduced to a minimum because it consists of only two steps, pulverization of the manganese ore followed by briquetting. Four minutes of grinding with Boraxo as the grinding aid gives sufficient uniformity to minimize the effect of particle size variation. The ground mix of manganese ore and Boraxo is pressed into 1.25-in. briquettes for analysis.

INTRODUCTION

Wet-chemical determinations of manganese, iron, SiO_2, and CaO in Mexican Molango manganese ores take from 4 to 7 hr, depending on ore solubility. In the interests of finding a faster yet accurate method of analyzing large numbers of these ores, Bethlehem's Research Department studied the possibility of using X-ray fluorescence.

The use of X-ray fluorescence to analyze various components in different types of manganese ore had been reported by previous investigators.[1–3] Sample preparation techniques used in those studies were fusion,[1] solution,[2] and pulverization.[3] Although fusion and solution sample preparation techniques may provide a basis for more accurate analysis, they tend to be slower than pulverization. Since it was desired to keep the overall time of analysis as short as possible, one of the objectives of our work was to incorporate a rapid pulverization technique into the analytical procedure. The X-ray analyses of samples prepared with a rapid pulverization technique were in good agreement with the corresponding chemical analyses.

EXPERIMENTAL

Ore Samples

Twenty-eight samples of the manganese ore were selected from core drillings taken at the mine site. All samples were analyzed for manganese, iron, SiO_2, and CaO by wet-chemical methods. Of these samples, 17 were used for instrument calibration and the remaining 11 served to verify the agreement between chemical and X-ray analyses determined with the calibration samples. Table I lists the chemical analysis of the 17 calibration samples. The calibration samples were selected to cover the concentration ranges for manganese, iron, SiO_2, and CaO that were expected in the area to be mined. Concentration ranges covered by the calibration samples were

Manganese 14–38%	SiO_2 2–11%
Iron 2–8%	CaO 1–24%

For the most part, the ore consisted of a fine-grained calcium manganese carbonate. Since the mineralogy did not vary greatly among the ore samples, it was felt that fine grinding would give sufficient sample homogeneity to provide good X-ray analytical results. Test results showed that this assumption was correct.

Sample Preparation

Before the manganese ore samples were received at the Homer Research Laboratories they were ground to −100 mesh. These samples contained less than 1% moisture.

Table I. Chemical Analysis of Manganese Ore Samples Used for X-Ray Calibration

	Chemical analysis, %			
Sample No.	Manganese	Iron	SiO_2	CaO
1	20.79	2.77	4.85	20.28
3	26.77	4.04	6.19	12.23
4	32.40	7.82	8.91	1.04
6	27.31	4.67	4.89	10.50
7	20.89	5.53	8.35	15.15
8	29.06	6.14	6.08	6.89
9	17.10	3.95	3.55	23.78
11	29.41	4.02	7.84	10.82
12	28.09	6.24	10.11	3.39
14	37.81	5.18	7.77	0.79
15	14.59	3.93	10.84	22.18
16	25.50	5.20	2.46	13.65
18	29.30	2.87	8.25	8.32
19	26.03	4.88	6.41	12.73
21	25.22	2.61	3.82	15.70
22	20.72	4.52	2.96	19.43
23	30.86	3.22	9.17	3.13

Laboratory preparation of the samples for X-ray analysis consisted of:

1. Combining 15.0 g of the −100-mesh sample with 0.6 g of Boraxo, a grinding aid.
2. Grinding the mixture for 4 min in a Shatterbox disc grinder.
3. Pressing the mix into a 1.25-in.-diameter steel cap by applying a force of 40,000 lb for 15 sec.

In laboratory tests to determine the relation between X-ray intensity and grinding time, it was found that the silicon showed a larger variation in X-ray intensity with changes in grinding time than the manganese, iron, or calcium. Since the silicon X-ray intensity reached a constant level at 4 min and additional grinding time did not cause significant X-ray intensity changes (Figure 1), the grinding time was fixed at 4 min for all samples.

The average particle size of the manganese ore samples ground for 4 min with Boraxo as a grinding aid was microscopically determined to be less than 10 μ. The maximum particle size observed was approximately 20 μ.

X-Ray Analyzer Conditions

An Applied Research Laboratories Vacuum X-ray Quantometer was used to make all X-ray measurements. The Machlett OEG-60 platinum-target X-ray tube was operated at 50 kV and 35 mA. The X-ray analysis of a sample consisted of duplicate 2-min counting times.

Spectrometer conditions used for the determination of the elements manganese, iron, silicon, and calcium are listed in Table II.

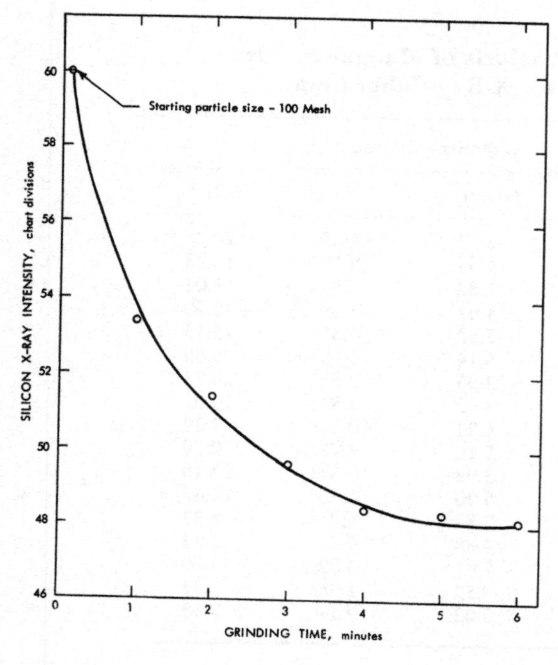

Figure 1. Change in silicon X-ray intensity with grinding time.

Table II. Spectrometer Conditions for the X-Ray Analysis of Manganese Ore

Element	Line	Wavelength, Å	Crystal	Atmosphere	Detector
Manganese	$K\alpha$	2.103	LiF	Vacuum	Proportional (Multitron)
Iron	$K\beta$	1.756	LiF	Vacuum	Proportional (Multitron)
Silicon	$K\alpha$	7.126	EDDT	Vacuum	Geiger (Exatron)
Calcium	$K\alpha$	3.360	LiF	Vacuum	Proportional (Multitron)

RESULTS

Reproducibility

To determine the reproducibility, *i.e.*, precision, for the X-ray analysis of our manganese ore, 16 of the ore samples were divided into two parts when received at the laboratory. One of the parts was prepared and analyzed by X-ray fluorescence. The other part was prepared and analyzed 2 months later. With this procedure, differences in the X-ray analysis reflected errors due to variations in sample preparation and any instrument instability.

The differences between the X-ray analyses of the two sample parts were used to calculate[4] the reproducibility of the X-ray method. In Table III, the reproducibility for the X-ray determination of each component is expressed as standard deviation and coefficient of variation. The fact that these values are small shows the good X-ray analysis reproducibility obtained by using samples prepared with the pulverization technique.

X-Ray Analyzer Calibration

The X-ray equipment was calibrated for the determination of manganese, iron, SiO_2, and CaO with 17 chemically analyzed manganese ore samples.

The agreement between chemical analyses and X-ray intensities showed that matrix changes did not have a significant effect on X-ray determinations of SiO_2 and CaO. The X-ray determinations could be obtained directly from the measured X-ray intensities, as shown by the calibration curves in Figures 2 and 3. These curves represent equations derived by the least-squares method.

Table III. Reproducibility of X-Ray Analysis

		Reproducibility	
Component	Concentration range, %	Standard deviation	Coefficient of variation
Manganese	14–38	±0.21	0.9
Iron	2–8	0.06	1.2
SiO_2	2–11	0.15	2.3
CaO	1–24	0.08	0.7

Tests showed that the measured X-ray intensities for manganese and iron vary with changes in the composition of the matrix. It was therefore necessary to correct the manganese and iron X-ray intensities for the matrix effect. The corrections were made by using empirical equations developed by a multiple-regression technique.[5,6] The data

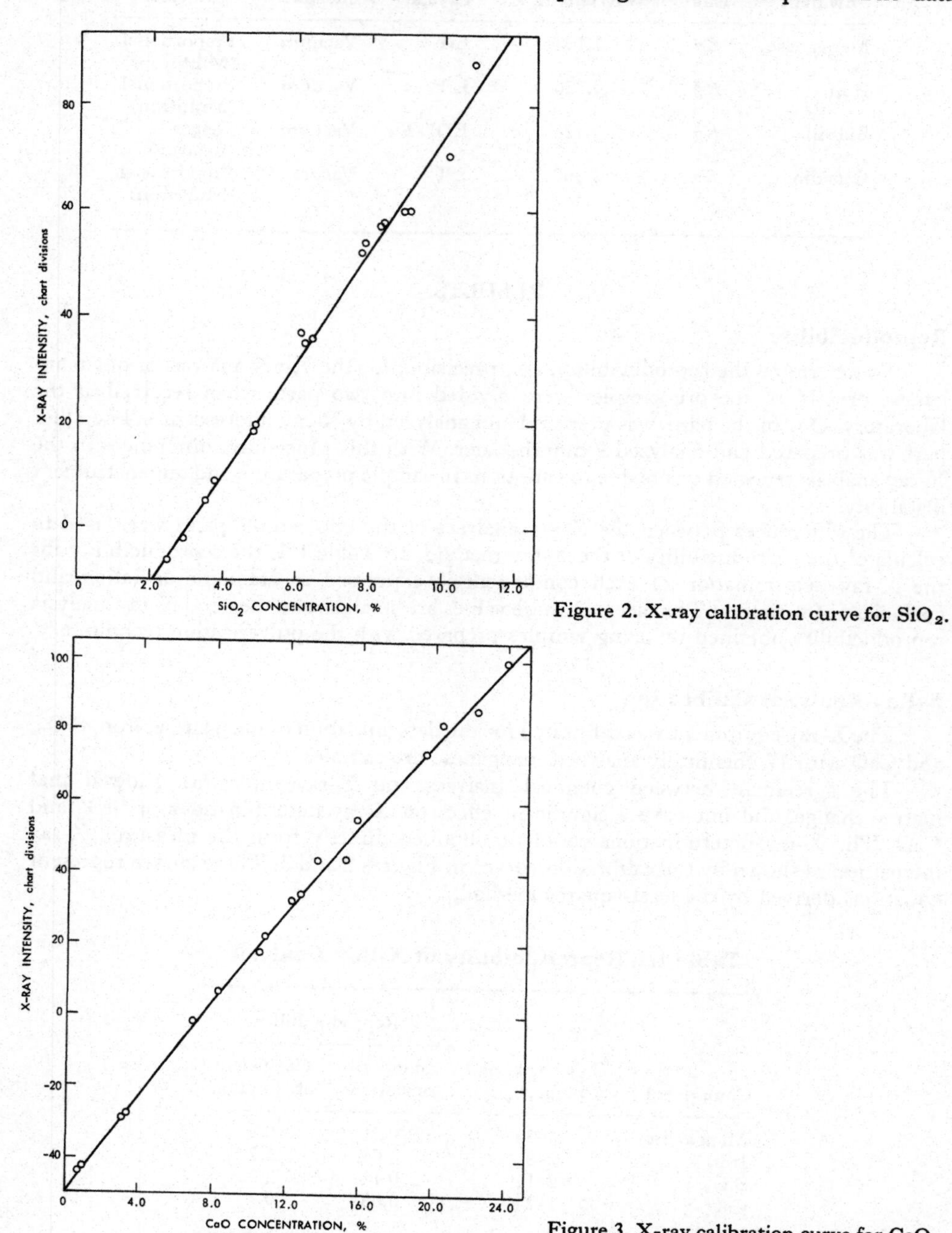

Figure 2. X-ray calibration curve for SiO_2.

Figure 3. X-ray calibration curve for CaO.

Table IV. Agreement Between Chemical and X-Ray Analyses of Calibration Samples

Component	Concentration range, %	Agreement, % concentration	
		Standard deviation	2 (SD)*
Manganese	14–38	±0.40	±0.80
Iron	2–8	0.12	0.24
SiO_2	2–11	0.29	0.58
CaO	1–24	0.45	0.90

* Approximate 95% confidence limits.

analysis showed that CaO interferes with the direct determination of manganese and that manganese and SiO_2 interfere with the direct determination of iron. The equations developed to determine corrected percent concentrations of manganese and iron from the measured X-ray intensities (xi) are:

Manganese determination

$$\% \text{ Mn} = 20.39 + 0.222 (\text{Mn}_{xi}) + 0.187 (\text{Ca}_{xi}) - 0.00064 (\text{Ca}_{xi})^2 - 5.24 (\log_{10} \text{Ca}_{xi}) \tag{1}$$

Iron determination

$$\% \text{ Fe} = 0.88 + 0.065 (\text{Fe}_{xi}) + 0.012 (\text{Mn}_{xi}) - 0.000087 (\text{Si}_{xi})^2 \tag{2}$$

By applying these equations to determine the percent concentrations of manganese and iron, the difference (expressed as standard deviation) between the chemical and X-ray determinations was reduced from 0.95 to 0.40% for manganese and from 0.44 to 0.12% for iron.

With the use of the calibration samples, the agreement between chemical and X-ray determinations of manganese, iron, SiO_2, and CaO was calculated by the least-squares method. This agreement is shown in Table IV and was used to predict the expected performance of an X-ray analysis of manganese ore.

Table V. Observed Agreement Between Chemical and X-Ray Analyses of Manganese Ore Samples

Sample No.	Chemical analysis, %				Difference between chemical and X-ray analyses, % concentration			
	Manganese	Iron	SiO_2	CaO	Manganese	Iron	SiO_2	CaO
24	17.78	4.73	3.20	21.36	0.00	0.26	0.06	0.49
25	25.13	4.97	2.42	14.25	0.38	0.15	0.08	0.72
26	23.79	3.30	2.98	17.20	0.06	0.03	0.05	0.61
27	28.16	3.49	7.84	8.95	0.01	0.01	0.16	0.10
28	27.27	4.29	3.62	12.06	0.18	0.19	0.26	0.65
29	24.60	4.64	3.83	14.28	0.28	0.15	0.30	0.62
30	26.55	5.01	11.63	7.01	0.70	0.21	0.28	0.28
31	28.09	6.23	10.14	3.26	0.34	0.24	0.56	0.01
32	29.45	3.42	9.40	5.38	0.13	0.04	0.20	0.01
33	22.50	5.22	2.16	16.29	0.31	0.15	0.07	0.61
34	29.18	4.05	8.85	7.62	0.35	0.03	0.01	0.02

Agreement Between Chemical and X-Ray Analyses

To verify the predicted 95% confidence limits shown in Table IV, 11 additional ore samples from the same mine were analyzed. Table V gives the chemical analyses together with the differences in percent concentration between the chemical and X-ray analyses for these samples. Matrix corrections were applied for the X-ray determinations of manganese and iron. The X-ray and chemical analyses agree, with one exception, within the predicted approximate 95% confidence limits.

CONCLUSIONS

X-ray fluorescence has been shown to be a rapid and accurate means of analyzing manganese ore samples. The fast analysis time, $\frac{1}{2}$ hr, was made possible by the application of a sample preparation technique based on pulverization. Pulverization is suitable because there are only slight variations in the mineralogy of the ore samples and because a relatively small particle size can be produced in a short grinding time.

Matrix corrections are not required for the X-ray determinations of SiO_2 and CaO in the manganese ore. In the case of manganese and iron, simple matrix correction procedures can be applied manually or with a small digital computer to improve the accuracy of the X-ray determinations significantly.

ACKNOWLEDGMENTS

The authors wish to thank F. H. Ruch, B. S. Mikofsky, and G. F. Heck of the Homer Research Laboratories for their assistance.

REFERENCES

1. J. Hanon and R. Winand, "Analysis of a Manganese Ore by X-ray Fluorescence. Calibration of the Method by Synthetic Mixtures," *Mem. Sci. Rev. Met.* **62**: 45, 1965.
2. B. J. Mitchell and H. J. O'Hear, "General X-ray Spectrographic Solution Method for Analysis of Iron-, Chromium-, and/or Manganese-Bearing Materials," *Anal. Chem.* **34**:1620, 1962.
3. H. F. Carl and W. J. Campbell, "The Fluorescent X-ray Spectrographic Analysis of Minerals," *ASTM Spec. Tech. Publ.* **157**: 63, 1954.
4. H. J. Rose, I. Adler, and F. J. Flanagan, "Suggested Method for Spectrochemical Analysis of Rocks and Minerals Using an X-ray Spectrometer," in: *Methods for Emission Spectrochemical Analysis*, ASTM, Philadelphia, 1964, p. 814.
5. B. J. Alley and R. H. Myers, "Corrections for Matrix Effects in X-ray Fluorescence Analysis Using Multiple Regression Methods," *Anal. Chem.* **37**: 1685, 1965.
6. B. J. Mitchell and F. N. Hopper, "Digital Computer Calculation and Correction of Matrix Effects in X-ray Spectroscopy," *Appl. Spectry.* **20**: 172, 1966.

DISCUSSION

M. Berman (U.S. Bureau of Mines): I noticed on most of your element determinations that, if your intensities were zero, you ended up with usually 2 or 3% of an element present, from either your regression formulas or just extrapolating the curve to zero intensity. Is there any significance to this?

W. D. Egan: Our intensity readout is not in counts, and we can have a negative instrument reading. The readout is set from −100 to +200 chart divisions on the strip chart recorder. Thus, for the lower concentrations, the instrument intensity reading would be a negative number.

M. Berman: Also, another thing I noticed, you had a logarithmic term in the determination of manganese, logarithmic in the calcium. If you ended up with no calcium intensity, or pretty low, you could be in an awful lot of trouble. Was there any reason why you had to use a logarithmic term?

W. D. Egan: The logarithmic term was used because manganese determinations corrected for calcium with an exponential equation were better than the uncorrected determinations. Including the log term in the regression analysis improved the standard error of the manganese determination by about 35%.

M. Berman: But there was no theoretical reason why?

W. D. Egan: No.

B. S. Sanderson (National Lead Co.): I wonder why you didn't take your 11 additional samples and plug them back into your regression equation. Actually, all you have proven is that there was homogeneity at variance in your method, and you would get better results, I think, if you used all of your standards.

W. D. Egan: The only reason we kept the 11 samples aside to analyze later was to verify the empirical equations.

B. S. Sanderson: After doing that, what you have proved is that you have homogeneity of variance, and then this allows you to go back and use those as standards.

W. D. Egan: I don't believe this would have changed the results.

S. D. Rasberry (National Bureau of Standards): Can you tell me why the Fe $K\beta$ line was selected in preference to the Fe $K\alpha$?

W. D. Egan: Because of the proximity of the Mn $K\beta$ peak to the Fe $K\alpha$ line and the high concentration of manganese in some of the samples.

TOTAL NONDESTRUCTIVE ANALYSIS OF CAAS SYENITE

A. Volborth, B. P. Fabbi, and H. A. Vincent

Nevada Mining Analytical Laboratory
Mackay School of Mines
University of Nevada
Reno, Nevada

ABSTRACT

Determination of 35 trace and 8 major elements in a granitic rock by nondestructive X-ray emission and fast-neutron activation methods has been accomplished. Trace elements arsenic, barium, cadmium, cerium, cobalt, chromium, cesium, copper, gallium, gadolinium, germanium, hafnium, mercury, indium, lanthanum, molybdenum, manganese, niobium, neodymium, nickel, rubidium, antimony, scandium, samarium, tin, strontium, tantalum, thorium, titanium, vanadium, yttrium, ytterbium, zinc, and zirconium and major elements aluminum, calcium, iron, potassium, and silicon have been estimated in the vacuum X-ray spectrograph. Phosphorus has been excited by a chromium-target tube. Major elements magnesium, sodium, and oxygen have been estimated by the Henke aluminum-target X-ray spectrograph. Major elements oxygen and silicon have also been determined by fast-neutron activation. Detection limits for traces are in our system in the range of 1 to 100 ppm with standard deviations of about 1 to 10 ppm. Precision in the determination of major constituents is better than 1% in relative standard deviation. The data are reported in parts per million for the trace elements and for the major elements as well. It is suggested that this type of reporting of geochemical data be preferred to the conventional reporting of major constituents as oxides.

INTRODUCTION

Conventionally an analyst determines only a few of the total number of constituents in a substance. A glance through the literature on X-ray emission and fast-neutron activation techniques, however, shows a few attempts at total analysis of complex substances.[1–5] An analyst reporting a total silicate rock analysis for all major as well as trace elements is faced with the 100 $\pm$ 1% limitation acceptable under normal circumstances of a gravimetric analysis. In practical nondestructive work, this requirement is frequently difficult to achieve. A first attempt at a total yield in major elements will often result in figures nearer 96%. This is due mainly to the fact that the trace and minor elements, which often constitute over 2% of the total,[3] and the major element oxygen,[6] which escapes detection if driven off as water or by fluxing or in vacuum chambers of X-ray spectrographs, are thus not determined. Since, conventionally, total oxygen is not determined and appears as equivalent oxygen in the total, and since the major multivalent element in rocks, iron, can be nondestructively determined satisfactorily only as total iron disregarding the actual Fe^{+3} : Fe^{+2} ratio in the substance, the total sum of a nondestructive analysis can have little significance and may not be used as an indication of the quality of analytical work unless oxygen is estimated independently. Substitution of fluorine, chlorine, and sulfur for oxygen in minerals further complicates this picture.

Therefore, we have attempted to develop an analytical system which would permit rapid, nondestructive analyses of rocks, ores, cement, or complex inorganic mixtures. For this, we have selected a combination of X-ray emission and fast-neutron activation methods. (Methods for individual elements and specific interferences, such as sample preparation, contamination, spectral, compositional, mineralogic and enhancement effects, have been described before.[1,2,5,7-11]) It may be pointed out that this combination includes only truly nondestructive methods and that no fluxing or dilution or heavy absorber techniques are used. The method is based on analyses of two portions of the same sample, each prepared differently which enables us to establish the concentrations of specific contaminants and to correct for these.[8] (No mathematical corrections for absorption, enhancement, or matrix differences have been performed. Therefore, the data given presents only first quantitative approximations of a total composition of a granitic rock.) In order to be able to make the necessary interelement corrections of a matrix, the approximate composition of this matrix must be known. Data obtained by our methods can therefore be regarded as a first step in the difficult task of providing a total nondestructive analysis of a complex substance. In a textbook (in press), one of us has recently compiled the integrated methods used at this laboratory.[12]

INSTRUMENTAL

The Syenite Rock-1 standard prepared by the Canadian Association for Applied Spectroscopy[13] was analyzed by the combined X-ray spectroscopic and fast-neutron activation techniques. The major elements aluminum, calcium, iron, potassium, and silicon were determined in a Norelco vacuum X-ray spectrograph with tungsten- and chromium-target tubes. Elements magnesium, sodium, and oxygen were also estimated by using the Henke aluminum-target demountable vacuum X-ray spectrograph built for us by Norelco. Gypsum and lead stearate crystals were used as dispersion media. In addition, elements oxygen and silicon were determined with the use of the corresponding (n, p) reactions induced by 14-MeV neutrons produced in the Kaman model A-1001 drift tube fast-neutron generator. A dual transfer system built for us by the Technical Measurement Corporation coupled with a dual detection system were used. Trace elements arsenic, barium, cadmium, cerium, cobalt, chromium, cesium, copper, gallium, gadolinium,

Table I. Major and Minor Elements in Syenite Rock 1

Element	Parts per million, X-rays	Standard deviation, ppm	Parts per million, fast neutrons	Parts per million, Webber's[13] data
Oxygen	458,400	± 5200	447,100 ± 1000	444,700
Silicon	286,000	± 1300	278,200 ± 1600	277,100
Aluminum	49,700	± 500		51,700
Iron	56,900	± 100		58,200
Calcium	66,500	± 100		73,800
Magnesium	24,000	± 800		24,500
Potassium	22,900	± 300		22,800
Sodium	23,300	± 200		24,000
Manganese	3,270	± 30		3,100
Titanium	2,400	± 50		2,940
Phosphorus	620	± 10		870
Zirconium	1,570	± 20		3,048
Total	995,560		976,460	986,758

germanium, hafnium, mercury, indium, lanthanum, molybdenum, manganese, niobium, neodymium, nickel, phosphorus, rubidium, antimony, scandium, samarium, tin, strontium, tantalum, thorium, titanium, vanadium, yttrium, ytterbium, zinc, and zirconium were all determined in the standard Norelco vacuum X-ray spectrograph with LiF or EDDT crystals. Standards were prepared in our laboratory by mixing known amounts of the trace element oxides (in most cases) in a vibration type mill with a 20% Al_2O_3 + 80% SiO_2 matrix.[11] Plain rock powders were pressed into pellets for X-ray and packed in plastic containers for neutron activation work. United States Geological Survey standard rock G-1 and, when feasible, rock W-1 have been used to check the accuracy of the calibration curves used.

RESULTS

Results for major elements and minor elements manganese, titanium, phosphorus, and zirconium are grouped in Table I. The second column gives the X-ray data and our

Table II. Trace Elements in Syenite Rock 1

Element	Parts per million, X-rays	Standard deviation	Parts per million, Webber's[13] data, means
Antimony	<75	—	—
Arsenic	30	1	10.4
Barium	350	10	273
Cadmium	<10	—	1.5
Cerium	362	4	593
Cesium	1.5	0.5	0.9
Chromium	47	1	58
Cobalt	18	1	19
Copper	19	1	25
Gadolinium	137	2	150
Gallium	12	1	18
Germanium	<1	—	<10
Hafnium	<10	—	<100
Indium	<30	—	<10
Lanthanum	220	10	243
Mercury	<200	—	<10,000
Molybdenum	<50	—	4.1
Nickel	45	1	42
Neodymium	266	5	325
Niobium	142	2	146
Rubidium	295	3	204
Samarium	~10	1	<100
Scandium	23	1	14
Strontium	105	1	320
Tantalum	<100	—	<200
Thorium	545	6	1,324
Tin	<55	—	12
Vanadium	66	1	87
Ytterbium	37	1	73
Yttrium	280	3	447
Zinc	175	2	200
Total traces, not including figures marked	< 3185.5		

precision in terms of standard deviation, the third gives the neutron activation estimates, and the fourth lists Webber's[13] reported means recalculated to parts per million.

Table II gives data on the 31 trace elements with standard deviations of our system. The elements of which the concentration limits were below our detection limits have not been considered in the sum total of the trace elements.

The total for the X-ray fluorescence analysis is 998,745.5 ppm. The total for X-ray fluorescence analyses where fast-neutron activation data for oxygen and silicon have been substituted for the X-ray data is 979,645.5. These figures represent sums of the major and trace elements combined taken from Tables I and II.

DISCUSSION AND CONCLUSIONS

These data indicate that, when a total X-ray spectrographic and fast-neutron activation analysis is performed comparing intensities of irradiated elements in rocks of similar physicochemical composition with carefully prepared or pre-existing standards, one may expect fully acceptable results compared with data that can be obtained by conventional, spectroscopic, gravimetric, or volumetric techniques. Because of the effort involved in preparation of standards, only one rock has been analyzed. Therefore, it may be too early to make any generalizations; however, it is encouraging that a multitude of elements are determined satisfactorily and that our data are comparable to those previously reported.

The application of this rapid method may prove exceptionally adaptable in geochemistry where precision of data is often more valuable than single absolute results. However, it must be emphasized that the analysis presented represents only a first step in the determination of a total composition of a rock. Since oxygen percentage given by neutron activation is believed to be accurate, the authors venture to propose that the sum totals given in this analysis may have more significance as indications of accuracy than those in a conventional rock analysis. The problem of total water in a rock and the problem of ferrous–ferric iron ratios in nondestructive analysis may in this case also be closer to a quantitative solution.

REFERENCES

1. A. Volborth, "Total Instrumental Analysis of Rocks," *Nevada Bur. Mines Rept.* **6**:Al–72, B1–12, 1963.
2. A. Volborth, "Hard, Soft, and Ultrasoft X-Radiation Combined with Fast-Neutron Activation, Applied to the Total Analysis of Rocks," *Materials Science and Technology for Advanced Application, Vol. II*, American Society for Metals, San Francisco, 1964, pp. 117–142.
3. A. Volborth, "Rock Analysis for Geochemical Studies by Combining X-ray Fluorescence and Neutron Activation Techniques," in: *22nd International Geological Congress, Rept., India*, 1964 (in press).
4. A. K. Baird and B. L. Henke, "Oxygen Determination in Silicates and Total Major Elemental Analysis of Rocks by Soft X-ray Spectrometry," *Anal. Chem.* **37**: 727–729, 1965.
5. A. Volborth and B. P. Fabbi, "X-Ray Spectrographic Determination of 38 Trace Elements in a Granite and Standards G-1 and W-1," *XIII Colloquium Spectroscopicum Internationale, Ottawa, Canada, Abstracts of Papers*, p. 120, 1967.
6. A. Volborth, "On the Distribution and Role of Oxygen in the Geochemistry of the Earth's Crust," Première Reunion de L'Association Internationale de Géochime et de Cosmochime, Paris, Sect. **5**: 33–34, 1967.
7. A. Volborth, "Biotite Mica Effect in X-ray Spectrographic Analysis of Pressed Rock Powders," *Am. Mineralogist* **49**: 634–643, 1964.
8. A. Volborth, "Dual Grinding and X-Ray Analysis of All Major Oxides in Rocks to Obtain True Composition," *Appl. Spectry.* **19**: 1–7, 1965.
9. H. A. Vincent and A. Volborth, "Fast-Neutron Activation Analysis for Silicon and Phosphorus in Rocks and Meteorites," Soc. of Appl. Spectry., 18th Mid-America Symposium on Spectroscopy, Chicago, Abstracts, No. 33, 1967.

10. H. A. Vincent and A. Volborth, "High Precision Determination of Silicon in Geological Samples by Fast-Neutron Activation Analysis," *Trans. Am. Nucl. Soc.* **10**(1): 26, 1967.
11. A. Volborth and H. A. Vincent, "Accurate Determination of Total Oxygen in Six New U.S. Geological Survey Rock Standards by Fast-Neutron Activation," *Trans. Am. Nucl. Soc.* **10**(1): 27–28, 1967.
12. A. Volborth, *Elemental Analysis in Geochemistry*, Elsevier Publishing Co., Amsterdam, 1967 (in press).
13. G. R. Webber, "Second Report of Analytical Data for CAAS Syenite and Sulfide Standards," *Geocinm. Cosmochim. Acta* **29**: 229–248, 1965.

DISCUSSION

M. Berman (U.S. Bureau of Mines): How long does it take you to analyze a rock by this method? Weeks, days, minutes?

A. Volborth: You imply to analyze for some 50 elements that I have just reported?

M. Berman: Yes. If you were given a rock now and asked to analyze it for these elements, how long would it take to do this?

A. Volborth: We can do that in approximately 2 weeks if we drop everything else. When it is a question of getting a rare sample analyzed, we have to put all our energies into this. We have three men working, and I should say a careful estimate is 2 weeks. I know we can do it in 2 weeks, but this is dropping everything else.

E. L. Gunn (Esso Research and Engineering): I should like to ask Dr. Volborth: What do you use as your reference or standardization materials in the determination of oxygen both for neutron activation and for fluorescence?

A. Volborth: As reference in this case, we are using pure quartz, specifically, pure quartz provided by Johnson-Matthey and Co. We have also compared with some industrially available quartz powder which gives us the same results. We have carefully calibrated known oxides like titanium oxide, aluminum oxide, also calcium carbonate and quartz. We have these secondary standards prepared in our laboratory. They give us the same gamma count ratio per theoretical oxygen percent as the other samples. We have five of these basic standards prepared in our laboratory. They all give us the same count per percentage of oxygen within approximately 0.05%. These data are slightly better than those we report. We cannot know the amount of true oxygen in any of these substances, but the fact that we have taken five different pure substances and compared oxides and calcium carbonate and have received the same intensities after, of course, igniting these and keeping them in a desiccator and packing them immediately into the containers, is encouraging. We believe that these substances are stoichiometric in terms of the major component oxygen.

V. Lazar (U.S. Plant, Soil, and Nutrition Laboratory): You implied 2 weeks total elapsed time. Would that mean a total of 240 man-hours for each sample?

A. Volborth: For each complex sample?

V. Lazar: Yes.

A. Volborth: I should say approximately, yes, 240 man-hours. But, for industrial work, the speed can be considerably increased.

V. Lazar: One other question. Would you consider that 12 samples of a very complex nature, almost as complex as this where there is a large amount of organic matter, could be handled by your technique?

A. Volborth: We have not tried that. It depends on what type of organic sample it is, how much water or other volatiles are present. Our problem in providing complete analysis includes taking in account the amount of hygroscopic water in the sample and its effect on the accurate oxygen analysis. We have been able to predict approximately amounts of water (moisture), upon the basis of a complete analysis. We have done this on the new U.S. Geological Survey standard rocks. We have two samples within these, both magnesium-rich (PCC-1 and DTS-1), with approximately 40% magnesium oxide. One of them has much more moisture than the other. As you know, water contains 88.88% oxygen, so, considering this, we can determine the differences in absorbed water if a total analysis of similar materials is made.

A. P. Langheinrich (Kennecott Copper): Why do you show on some of your tables four or more significant figures when you also show that the third figure is usually very much in doubt?

A. Volborth: This is as our computer's output. I do not pretend that the three last figures are of much significance. You are right. The problem is that, when you do report major elements and you have recalculated them and you report trace elements in the same column, you just leave these last three meaningless figures in the major element like this: 488,330 ppm. The 330 ppm in oxygen would not have any meaning whatsoever. The only reason it is reported is that we also sum up the trace elements at levels of 10 and 100 ppm. That's the only reason we did it, but we do not imply, I am glad you asked this question, that the major elements are reported with this accuracy. The precision on the major elements was given on each of the results.

K. F. J. Heinrich (National Bureau of Standards): Am I right in assuming that you have driven your precision to such a high point that there seem to be significant differences in your subsamples in oxygen? The question then arises: Will similar sampling problems, *i.e.*, subsampling in the standards, not be then the ultimate limit of the precision that you hope to obtain?

A. Volborth: That's the way it looks at the present moment. We have discovered this on G-1 and W-1 and on the syenite. As Professor Baird mentioned, the Pomona team has also discovered that the different samples in different bottles are not as homogeneous and cannot be reproduced as well as within a sample. Therefore, there is no reason for us to go further. We have reported the data, the precision, as we get it. We do not imply that this has any significance as such in the analysis of that particular sample or bottle.

X-RAY FLUORESCENCE OF SUSPENDED PARTICLES IN A LIQUID HYDROCARBON

E. L. Gunn

Esso Research and Engineering Company
Research and Development Division
Baytown, Texas

ABSTRACT

In the application of X-ray fluorescence to the inspection of petroleum or petrochemical products, the measured element may not be ideally dispersed in the organic substrate but may exist as suspended particles. A question naturally arising is how small the particles must be to approximate the fluorescent intensity level of the true solution. The fluorescent emissions of suspensions of particles in a viscous hydrocarbon were measured and the intensities compared with true solutions of zinc, iron, and silicon, respectively. Particle segregates were classified by microsieve and photomicrographic inspection. Iron or zinc particles of 8-μ size yield 65% of the intensity of the solution state. Increase of particle size to 70 μ reduces intensity by several fold. Silicon, dispersed as porous silicon dioxide particles of 28-μ size, yields 80% of the intensity of the solution form. Reliable measurement of a suspension can be made provided it is stable and a standard closely simulating it is available for reference or if proper intensity corrections are applied for a known particle size in suspension. Comparisons of experiment with theory for the particle size–intensity effect agree only moderately well. This is attributed to uncertainties in the parameters relating the two. Theory indicates that zinc or iron in suspended form should be less than 1 μ in size to approximate the intensity of the elements in solution. Additional research aspects of particles in suspension are suggested.

INTRODUCTION

The application of X-ray fluorescence to the inspection of petroleum or petrochemical products is favorable for the majority of samples received in the laboratory. The element sought frequently is in minor concentration and is ideally dispersed as solution in a low-atomic-number organic carrier. The X-ray intensity is high and the measurements precise. In certain systems, the sought element, although present in minor to trace amounts, is not ideally dispersed in the substrate. The substance is not miscible as a liquid or solid solution but is dispersed as finely divided entities or phases.

The influence of such particles in the X-ray analysis of powders and inhomogeneous solids has long been recognized. For selected powder systems, the effect of particle size has been evaluated.[1–6,8] Methods of preliminary sample treatment which eliminate or minimize particle effect and inhomogeneity have been proposed for both solids[4] and slurries.[7]

Examples which illustrate inhomogeneity in organic carriers include used and contaminated lubricants; oils, plastics, and waxes containing dispersed additives; and various fluids having suspended matter in them. The X-ray analysis of such systems is uncertain. A question arises logically as to how fine the dispersion of a specific substance must be to approach reliable measurement as a truly homogeneous solution. The present effort is concerned with properties of a few selected dispersions of suspended particles in a liquid hydrocarbon. Measurement of elements in such particles often is of prime importance in these laboratories. Hence, X-ray measurement of these suspensions was carried out to determine limitations of the X-ray fluorescence technique in such analyses. Points of difference between the above-mentioned powder systems and these suspensions are that powders are mixtures of one or more species of solid phase particles, whereas the suspensions have solid particles dispersed in a continuous low-atomic-number liquid phase with no voids.

EXPERIMENTAL

Preparation of Standards

The X-ray fluorescent properties of suspensions can best be assessed by measuring systems which contain particles of known composition, size, shape, and uniformity of dispersion. Selection of the particles and the dispersion medium for a synthetic composition is relatively simple; control of size, shape, and dispersion is considerably more problematic. The segregation and accumulation of solid particles uniform in both size and symmetry and also ideally distributed in hydrocarbon suspension would indeed be difficult to achieve. Even though these significant limitations exist and are recognized, dispersions of a somewhat less perfect type can be prepared without great difficulty. In turn, these can be employed in providing some meaningful information about the X-ray properties of solids suspended in organic matter.

The use of hydrocarbon oil in the analysis of thorium and uranium ores dispersed as slurries has been described.[7] A viscous hydrocarbon oil was selected as the dispersing medium for the present particle study. The oil flowed very slowly at room temperature, and its kinematic viscosity at 99°C was 650 SSU. The viscosity was reduced by heating to 150°C as an aid to stirring in the added solid particles. After heating, either a magnetic stirrer or a Waring blender was found to be effective for mixing. During the period of cooling to room temperature, the contents were mixed manually. Each dispersion was stirred immediately before X-ray measurement.

Metals chosen for dispersion were iron and zinc. Baker and Adamson reagent-grade metal powder of each was used. Micro sieves were used to segregate particles into size ranges above 20 μ. Microscopic estimation of average particle size was also made, which will be discussed in a subsequent section. The weights of oil and powder were carefully predetermined before dispersion to provide approximately 0.1% of iron or zinc concentration.

Silicon was dispersed as a finely divided diatomaceous earth powder. The amount added was 1% as elemental silicon. The material available was a relatively pure form of silicon dioxide having a high specific surface, which showed the material to be very porous.

The foregoing three elements represent a broad range in wavelength of characteristic K radiation from the respective standards.

In addition to the particle dispersions, a homogeneous solution of each element was also prepared. Zinc and iron naphthenates, respectively, were added to the viscous oil so that it would contain an exact known amount of the element.

Particle Size Estimation

The segregation of particles into size ranges by microsieve segregation has been described.[5] The average of nominal size values of sieves was taken as the value of particles segregated between them. As an example, those particles passing a sieve designated 43 to

47 μ but not passing one designated 28 to 32 were assigned a value of 37 μ. The outside nominal range, thus, is 28 to 42 μ, which represents a significant variation in particle size.

Photomicrographs, also, were used in characterizing particles as to size and shape. These represented an image magnification of $\times$ 77. The examination of a reference scale under the same magnification was used as a means of calibrating the photographic print, as well as the occular reticle in the eyepiece of the microscope. Both were used in estimating an average size for a number of suspended particles in the oil. At the magnification power used, the field depth is small. A particle out of focus gave a distorted and enlarged image. Only those particles in focus were measured.

Photomicrographs

Typical photomicrographs which illustrate dispersions of iron, zinc, and silicon dioxide in the viscous oil are shown in Figures 1 to 9. The sample area irradiated by the primary X-ray beam in fluorescence measurement was approximately 300 times that represented in a photograph.

Figures 1 to 6 exemplify iron-particle dispersions. Conclusions regarding the dispersed particles are evident from inspecting these figures.

The particle size range within a sample varies somewhat and can be greater than a factor of 2.

Figure 1. Iron suspension, 8 μ.

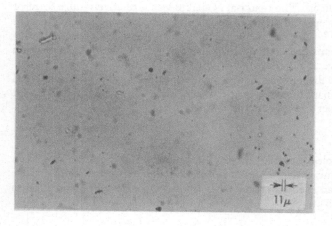

Figure 2. Iron suspension, 14 μ.

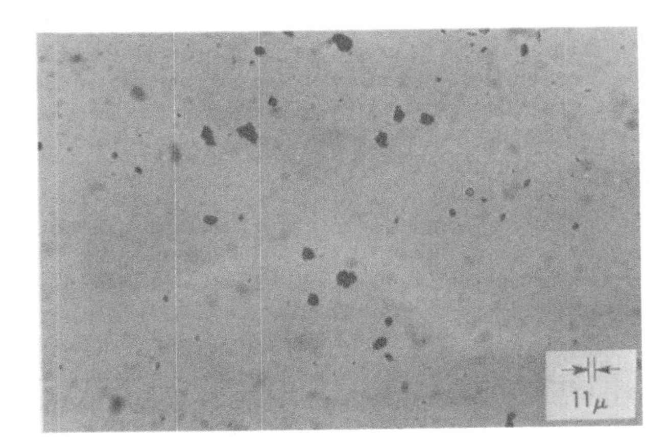

Figure 3. Iron suspension, 25 μ.

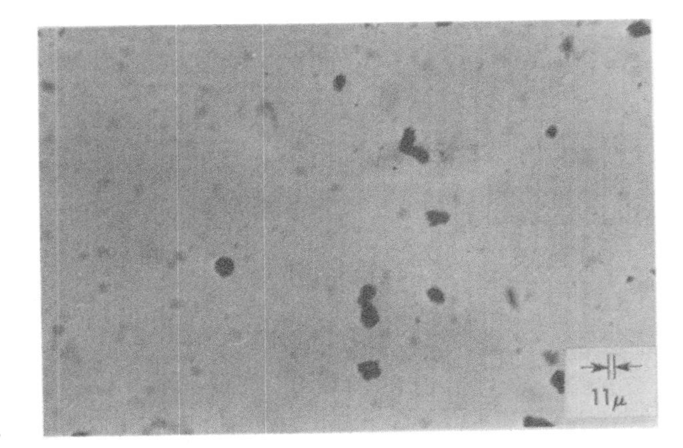

Figure 4. Iron suspension, 37 μ.

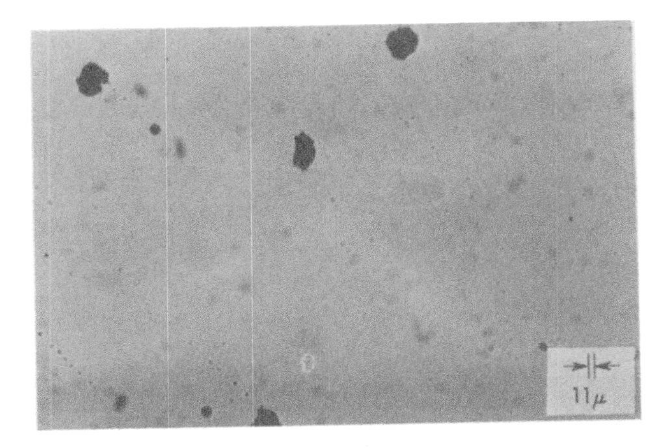

Figure 5. Iron suspension, 52 μ.

There is no regular particle symmetry or shape; this varies from particle to particle. For example, a particle may be three times as long as it is thick and have an irregular cross section. Two measurements, one transverse to the other, were made on the photograph to obtain a single size estimate for a particle. A number of particles were measured in this manner, and the arithmetic average was taken as characterizing the size of the particles in suspension.

Spatial dispersion in the suspensions is not uniform in the sense that particles are equidistant one from the other. This becomes quite evident with large particles, as shown in Figures 5 and 6.

Two suspensions of zinc particles are shown in Figures 7 and 8. They are similar in character to those of iron, and the same conclusions apply as for iron suspensions.

Figure 9 illustrates the dispersion of silicon dioxide. The powdered form used could not be conveniently segregated in the same manner as were the powdered metals; however, a particle segregation was made by floatation in isopropyl alcohol. The range in particle size of silica was found to be greater than that of the segregated metals, whether segregated from alcohol or unsegregated.

It should be pointed out that the synthetic suspensions described above do not differ drastically from those observed in plastics, oils, or other organic substrates which

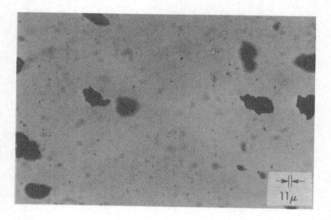

Figure 6. Iron suspension, 67 μ.

Figure 7. Zinc suspension, 7 μ.

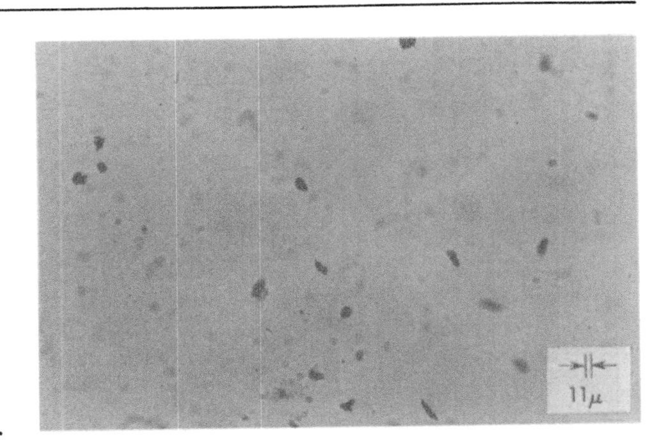

Figure 8. Zinc suspension, 28 μ.

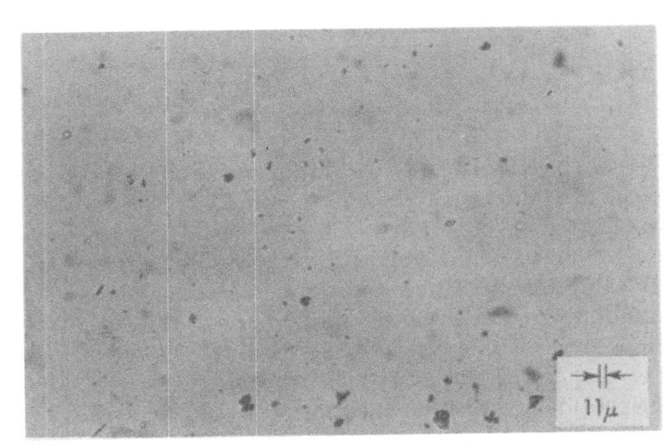

Figure 9. Silicon dioxide suspension, 18 μ.

occasionally are submitted to the laboratory for analysis. These usually have a greater variety in the composition of suspended forms and in size ranges than the present synthetics, however.

X-Ray Measurement

An X-ray spectrometer with tungsten-target radiation was used in the measurements. Scintillation-counter measurement was used for iron and zinc suspensions; and flow proportional, for silicon. The analyzing crystals used were, respectively, LiF and EDDT. Peak measurement of the element was corrected for background in each case.

Iron. Measurements of the suspensions and solution of iron were all normalized and expressed as intensity in counts per second for exactly 0.1% of the element. In this manner, the intensity of each suspension could be compared directly with that of the homogeneous solution. Each intensity, then, was expressed as a fraction of that of the solution. Inspection of the data suggested an exponential relationship between intensity and particle size; a semilogarithmic plot was prepared which was linear, as shown in Figure 10. The curve relating particle size to intensity is connected by a dotted line with the solution intensity point. This is not to suggest that the particle curve should be extrapolated to the solution point, but to emphasize the lack of experimental data in this region.

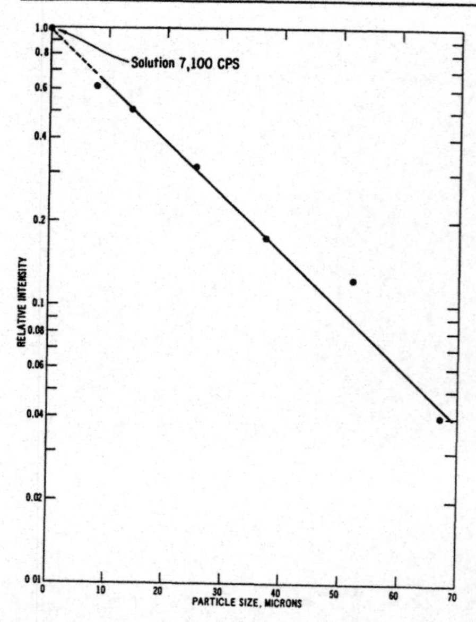

Figure 10. The effect of particle size on the intensity of iron.

A particle size of 10 μ exhibits about 60% of the intensity of a true solution. At a size of 65 μ, the intensity is only about 5%. The results demonstrate that the fluorescent intensity from suspended iron particles is very sensitive to change in particle size in the range shown. They signify further that suspensions should not be compared with solutions in X-ray analysis unless the particle size characteristics are known and proper corrections can be applied. Measurement of different preparations of the same suspended material did show that analysis can be performed with confidence provided the size and properties of the suspended matter remain unchanged and the matter occurs as dilute amounts in suspension. Numerous analyses made on suspensions in this laboratory have confirmed this conclusion.

Zinc. Figure 11 exhibits the results for the measurement of zinc. Small particles of 7-μ size gave 65% the intensity of the homogeneous solution. Intensity dropped very rapidly with size increase, in contrast to the corresponding decrease for iron. Sizes greater than 30 μ, namely, 37, 52, and 67 μ, gave very low Zn *Kα* intensity measurement. These results indicate a more critical relationship between intensity and particle size than was observed for iron.

Silicon. The silicon dioxide dispersion, containing particles of 28-μ average size, were measured in reference to a standard solution of dimethyl silicone. The particle intensity of Si *Kα* was 80% that of the silicon in solution. This is greater than for the equivalent size of either iron or zinc, for which the emissions were about 40 and 20%, respectively.

The suspension prepared by direct dispersion of the silicon dioxide in the oil gave the same intensity as the suspension in which the silicon dioxide was segregated by isopropyl alcohol floatation. Furthermore, no difference between these suspensions in particle size average was revealed by microscopic examination.

It is recognized that silicon was measured in a combined form and not as a free element as were the metals. Furthermore, and probably of greater significance, the structure of a metal particle is much different from that of the silicon dioxide particle. The diatomaceous earth has a large surface area, which indicates that each dispersed particle has

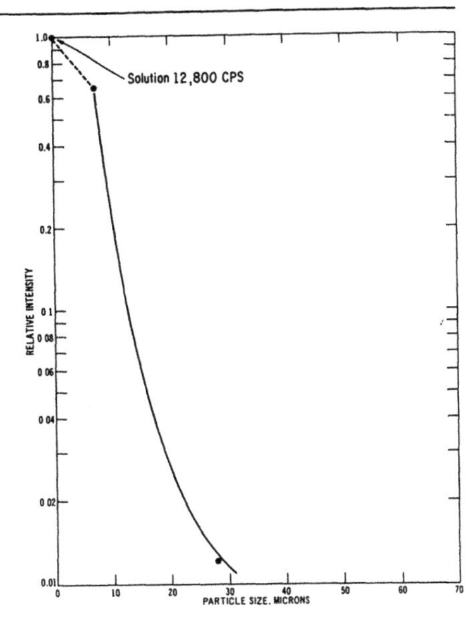

Figure 11. The effect of particle size on the intensity of zinc.

many submicroscopic pores and interstices in its structure, whereas the metal particles are much more compact. These differences in composition and structure may account for the greater relative emission by the silicon dioxide suspension.

No estimation was made of the particle size reduction necessary to approach more closely the intensity of the silicone solution. The nonavailability of properly segregated sizes of less than an average of 28 μ prevented this estimate's being made.

Replicate measurements of the foregoing suspensions of iron, zinc, and silica indicated the suspensions to be stable during the course of measurement. This signifies that reliable measurement of suspensions of a given particle size can be made provided the particle dispersion remains stable and a standard (or standards) of similar particle size is used for reference. As pointed out previously, when a different known particle size is measured, correction for particle size difference can be applied.

DISCUSSION OF RESULTS

Experimental Expression

Although all the points of Figure 10 do not fit a straight line precisely, the plot immediately suggests a simple exponential relationship. The form of the expression is

$$\frac{I_p}{I_s} = e^{-km} \tag{1}$$

where I_p and I_s are the particle and solution intensities, respectively; m is the micron size of the particle; and k is a constant derived from the slope of the line. For convenience in interpretation, we may write

$$m_0 = \frac{1}{k} \tag{2}$$

where m_0 has the dimension of microns.

Thus,

$$\frac{I_p}{I_s} = e^{-m/m_0} \tag{3}$$

The derived value of m_0 is 22 μ. If the iron particles in suspension are 22-μ size, the relative intensity is $1/e$ times the solution intensity, or 0.368. Similarly, if the particle intensity is one-half that of the solution, the calculated particle size is 15 μ.

Because of the paucity of data for zinc and silicon, expressions relating intensity to particle size were not developed.

Dispersion States

Suspensions of 10-μ metal particles of iron and zinc give about 60% the intensity of these metals in solution. It is of interest to compare the state of dispersion of these suspensions with the solutions.

Assume the suspension contained dispersed 10-μ spheres and the solution contained atoms dispersed in discrete molecular form. The number of separate atoms in solution per unit of volume is approximately 10^{13} times that of the separate spheres in suspension. The mean distance between spherical particles is approximately 10^4 times that between atoms. These parameters emphasize the vast magnitude of difference in state of dispersion between the nominal suspension and true solution.

Theoretical

Consideration of a fundamental relationship between X-ray fluorescent intensity and the size of particles in suspension was made for two reasons: first, to establish a relationship in basic physical principles and, second, to provide a possible explanation for the difference in behavior between iron and zinc particles.

Two considerations which are basic are total particle surface and the X-ray absorption by particles.

The fluorescent intensity emitted by the measured element is a direct function of the excitation area presented to the analytical beam. By assuming a symmetrical uniform particle such as a sphere, arithmetic calculations can readily be performed. To illustrate the arithmetic involved, doubling the diameter of a sphere increases its volume and weight eightfold and its area fourfold; the number of spheres per unit of mass is decreased by one-eighth. The specific surface—area per unit of mass—of uniform spherical particles can easily be calculated. However, as pointed out in a foregoing section, it must be recognized that particles in the present suspensions actually are neither symmetrical nor uniform in size.

The X-ray fluorescence emitted will be lessened by internal absorption loss within the particle. The loss decreases as the particle size is reduced, the extent of which is dependent upon the absorption coefficients of the particle for both incident and emerging X-rays. Both Bernstein[2] and Gunn[6] have related intensity and particle size by the well-known expression

$$\frac{I_x}{I_\infty} = 1 - e^{-apx} \tag{4}$$

where I_x is the intensity for a particle of x thickness; I_∞, the intensity for infinite thickness; p, the density; and a, an absorption parameter. The last may be written

$$a = \mu_i \csc \theta_i + \mu_e \csc \theta_e \tag{5}$$

where μ_i and μ_e are absorption coefficients for incident and emergent X-rays and θ_i and θ_e are the incident and emergent angles, respectively.

Substitutions in equation (4) show that, although the fluorescent emission per unit of mass increases as the value of x decreases, the intensity per single particle decreases because there is less mass to emit.

In considering the relationship of symmetry and absorption of particles, Claisse[4] has pointed out that, for a sphere, the effective absorption distance is two-thirds the diameter. This has been employed in the present considerations.

The critical thickness x_c of a particle was calculated by arbitrarily assigning the ratio I_c/I_∞ a value of 0.99. The uncertainty in this calculation resides mainly in the value used for μ_i, the absorption coefficient of the particle for the incident X-rays. Effective values of 250 and 220 were taken for iron and zinc, respectively, in the calculations. The accepted value of μ_e for the emitted X-rays is in each case exact, of course. The intensity for a single particle size relative to that for critical thickness was calculated.

The product of surface and intensity ratio I_x/I_c may be used to predict the intensity behavior with change in particle size. With decrease in particle size, two effects result. The area per particle decreases; the specific surface increases. The emission per single particle decreases, but the emission per unit of mass increases. The total area of a given mass of particles of thickness x can be expressed relative to that of a similar mass of particles having critical thicknesses x_c. Likewise, the intensity ratio I_x/I_c can be expressed so as to relate the intensity of a particle of thickness x to that of a particle of critical thickness.

Although the intensity ratio per particle decreases as the total particle area increases, the net result is that their product actually produces an intensity increase of several fold. This is because the rate of increase in area is greater than the concurrent rate of decrease in particle intensity. A plot of this product *versus* particle size produces a sigmoid type of curve.[2-4] The higher intensity ratio levels out at a particle size of 0.1 μ and the lower intensity ratio at 200 to 300 μ.

Figure 12 shows the experimental and calculated curves relating intensity to particle size for iron. The calculated curve was normalized to the experimental curve by reference to the calculated intensity-ratio value for 0.1 μ, the value assigned being unity. Although the symmetries of the two curves are somewhat similar, the calculated curve shows a greater upward concavity in the particle range shown than does the experimental curve. At 60 μ, the experimental intensity is equal to that of theory, whereas at 20 μ it is only 60%. Possible explanations for this difference are the following: The value of μ_i selected

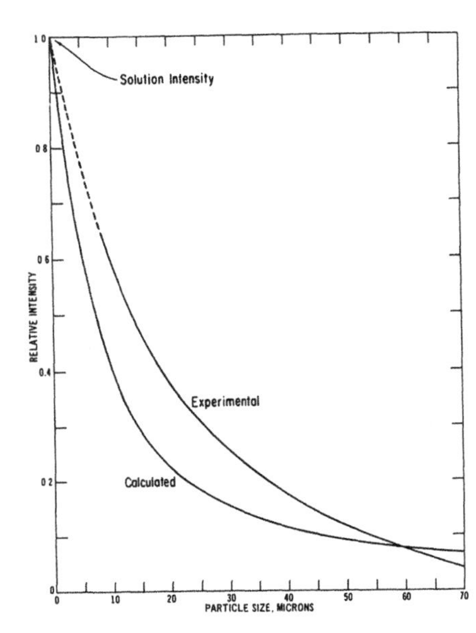

Figure 12. The comparison of the observed and calculated effects of particle size on the intensity of iron.

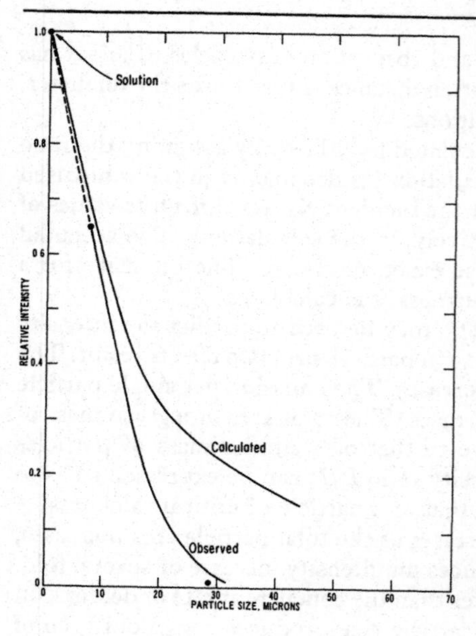

Figure 13. The comparison of the observed and calculated effects of particle size on the intensity of zinc.

for calculating the experimental curve, if too large, would increase the concavity; X-rays generated from particles in suspension have depths of origin which differ from that of particles lying at the geometric surface of the sample cell; the nominal particle size of the experimental particles may not describe their distribution within the segregated range limits very well; the shape and structure among individual experimental particles is not uniform and their interaction with X-rays differs somewhat from that of uniform spheres. The last two nonideal factors in particular may account for the curves of experiment and theory's not being in better agreement. In reference to particles, it is Madlem's[7] conclusion that many of the effects attributed to the matrix are in reality due to grain size and shape heterogeneity.

Figure 13 shows experimental and calculated curves for zinc. Although there are only two experimental points shown for the zinc particles, the experimental curve indicates greater intensity response to particle size changes than does the calculated. This is a reverse in behavior compared with the curves for iron. The possible explanations for the difference between theory and experiment proposed for iron also may apply in some measure to zinc, but in a reverse manner.

The experimental data do not include measurements of particles 1 μ or less in size. However, the theoretical curves of Figures 12 and 13 indicate that, to approximate the intensity of the elements in solution, the particles must be less than 1 μ in size.

CONCLUSIONS

From information gained in the present limited project, certain conclusions may be drawn regarding the X-ray fluorescent behavior of particles in hydrocarbon suspension.

Iron and zinc particles 8 μ in size yield about 65% the intensity of a true solution. An increase in size up to 70 μ results in a decrease in fluorescent intensity by at least several fold, *e.g.*, 15 or more.

Silicon dispersed as a highly porous silicon dioxide form of 28-μ particle size yields 80% the intensity of an organosilicon form in solution.

Reliable measurement of a suspension can be made provided it is stable and a standard closely simulating the unknown is available for reference. It also may be measured reliably if proper correction for a known particle size can be applied.

Comparisons of observed with theoretical calculation of the intensity–particle size relationship for iron and zinc agree only moderately well. The nonideal properties of actual particles and the uncertainty in parameters used in calculation may explain the lack of better agreement. Theory indicates that particles in suspension should be less than 1 μ in size to produce intensity approaching that of the element in true solution.

The present project on suspension measurement should be regarded only as an initial effort. More information on the X-ray properties of suspended systems is needed. These would include more elements and compounds, other concentration levels, sharper segregation and smaller particle sizes, additional dispersants, liquid- as well as solid-particle suspensions, and possible variations in the analytical measurements. Such information would greatly assist in defining the limits of X-ray fluorescence in analyzing dispersed systems.

REFERENCES

1. F. Bernstein, "Applications of X-Ray Fluorescence Analysis to Process Control," in: W. M. Mueller (ed.), *Advances in X-Ray Analysis*, *Vol. 5*, Plenum Press, New York, 1962, p. 486.
2. F. Bernstein, "Particle Size and Mineralogical Effects in Mining Applications," in: W. M. Mueller and M. Fay (eds.), *Advances in X-Ray Analysis*, *Vol. 6*, Plenum Press, New York, 1963, p. 436.
3. F. Bernstein, "Mineralogical Problems in X-Ray Emission Analysis of Sylvite Concentrates," in: W. M. Mueller, G. R. Mallett, and M. J. Fay (eds.), *Advances in X-Ray Analysis*, *Vol. 7*, Plenum Press, New York, 1964, p. 555.
4. F. Claisse and C. Samson, "Heterogeneity Effects in X-Ray Analysis," in: W. M. Mueller (ed.), *Advances in X-Ray Analysis*, *Vol. 5*, Plenum Press, New York, 1962, p. 335.
5. E. L. Gunn, "The Effect of Particles and Surface Irregularities in the X-Ray Fluorescent Intensity of Selected Substances," in: W. M. Mueller (ed.), *Advances in X-Ray Analysis*, *Vol. 4*, Plenum Press, New York, 1961, p. 382.
6. E. L. Gunn, "Practical Methods of Solving Absorption and Enhancement Problems in X-Ray Emission Spectrography," in: J. E. Forretti and E. Lanterman (eds.), *Developments in Applied Spectroscopy*, *Vol. 3*, Plenum Press, New York, 1964, p. 69.
7. K. W. Madlem, "Matrix and Particle Size Effects in Analyses of Light Elements, Zinc through Oxygen, by Soft-X-Ray Spectrometry," in: W. M. Mueller, M. J. Fay, and G. R. Mallett (eds.), *Advances in X-Ray Analysis*, *Vol. 9*, Plenum Press, New York, 1966, p. 441.
8. C. R. Hudgens and G. Pish, "X-Ray Fluorescence Emission Analysis of Slurries," in: L. R. Pearson and E. L. Grove (eds.), *Developments in Applied Spectroscopy*, *Vol. 5*, Plenum Press, New York, 1966, p. 25.

DISCUSSION

M. Berman (U.S. Bureau of Mines): Have you thought of taking advantage of this size effect and using it sort of backward, *i.e.*, if you know how much iron, *e.g.*, is present in your suspension, this is used as a means of determining particle size?

E. L. Gunn: No, I haven't. Perhaps you have a point.

M. Berman: We have thought about doing this on coal maybe by taking two measurements on every sample you're dealing with, one as it comes to you and the other after you have reduced the particle size to that which you need to give you the true value of the element present.

E. L. Gunn: I think my situation might be a little more artificial than yours. You see, I was working only with synthetics. Not that we haven't had actual samples with dispersed particles in them, but I haven't been faced with the practical reality which I think you have.

F. Bernstein (General Electric Co.): I was curious about what you use as a solution point.

E. L. Gunn: To make the true solution?

F. Bernstein: Right.

E. L. Gunn: Iron and zinc naphthenates, and the dimethyl silicone was a General Electric product.

F. Bernstein: Thank you. Also, I was a little curious—we seem to have lost the higher-level plateau that we see on looking at ground powders, and I was wondering if you would have an explanation. I've been thinking about it, and I don't seem to have any explanation for it, and I was wondering if you do.

E. L. Gunn: Oh, you mean, use smaller particles than I did so as to approximate this plateau.

F. Bernstein: Somewhere in the 1 to 10-μ range, I should expect, similar to what we've done on powders where we've seen more of a leveling.

E. L. Gunn: Actually, this sigmoid type of curve which I described began to fall off at about 70 μ, but, at 20 and along in that region, the curve is not as steep. I don't know where the inflection is; I think that, more generally, one or two are in that region where you found them. I think that's what I would have found also, had I used smaller particles.

THE EFFECT OF SURFACE ROUGHNESS IN POLYMERS ON X-RAY FLUORESCENCE INTENSITY MEASUREMENTS

J. Gianelos and C. E. Wilkes

The B. F. Goodrich Company Research Center
Brecksville, Ohio

ABSTRACT

We sought to determine how seriously surface roughness affects X-ray intensity measurements in polymers. Fourteen elements ranging from lead to silicon were added singly to fourteen batches of trans-1,4-polyisoprene. Smooth pressings of each batch were made, and intensity readings were taken (I_0). Reproducibly rough surfaces were made from these by molding a square wire-mesh pattern into them, with the use of Tyler standard sieve screens. The amount of roughness was controlled by using screens of very fine to very coarse mesh. We studied the change in the X-ray intensity of the rough surfaces *versus* the smooth $[(I/I_0) \times 100]$ with respect to: (1) the degree of roughness, (2) concentration of the added element, (3) emitted wavelength of the added element, (4) X-ray tube target material, and (5) correction for matrix effects on the intensity. We found that, at wavelengths emitted below 1 Å, intensity differences are small, regardless of which factors were varied. At wavelengths emitted above 1 Å, however, we found large differences. The intensity changes are highly dependent on roughness. Also, they become greater at the longer emitted wavelengths and with increasing concentration of added elements. Beginning with Ti $K\alpha$, losses are much higher with the use of chromium primary radiation than with tungsten. A technique of milling polyethylene into polymers with rough surfaces to provide a smooth surface is discussed.

INTRODUCTION

X-ray fluorescence is becoming more widely used for the analysis of inorganic elements in organic polymers. Surface quality is an important variable in these analyses. Some polymers exhibit nonreproducible surface wrinkling shortly after molding. The purpose of this controlled study was to determine how this surface wrinkling, or roughness, affects X-ray fluorescence intensities.

SUMMARY

The results of our study are:

1. Surface roughness always results in a loss of X-ray intensity.
2. The amount of loss increases with increasing roughness and with increasing concentration of the inorganic element.
3. The amount of loss is dependent upon the wavelength of the emitted radiation measured.
4. Losses are relatively small when the emitted radiation measured is less than 1 Å.

5. Losses are similar at short wavelengths for both tungsten- and chromium-target X-ray tubes. At long wavelengths, beginning with Ti *Kα* (2.75 Å), losses are much greater with chromium-target tubes.
6. Surface roughness does not significantly change the ratio of coherent- to Compton-scattered primary radiation.

We also found that the adverse effects of surface roughness can be reduced or eliminated by milling 10 to 20% of polyethylene into the polymer. The polyethylene imparts enough "body" to keep the surface smooth.

EXPERIMENTAL

The polymer used throughout this study was synthetic trans-1,4-polyisoprene (TPI). It was chosen for three reasons. First, when molded, it gives a lasting smooth surface. Second, it is semirigid and will retain the form of anything impressed into it. Third, it mills well, and most inorganic materials added to it disperse easily and homogeneously.

We added 14 different inorganic elements separately into batches of TPI by milling. The added elements were lead, erbium, gadolinium, antimony, copper, nickel, iron, chromium, titanium, calcium, potassium, chlorine, sulfur, and silicon. All were added at a fixed 5% concentration, as oxide, carbonate, or organometallic compounds. Since the anion part of the compound contained no element heavier than oxygen, the light element character of the polymer matrix was preserved, and possible interelement effects were minimized.

Six smooth-surface pellets, 1.5 in. in diameter by 0.16 in. thick, were made from each batch. Each pellet was counted to obtain I_0, the intensity with a smooth surface, and to ensure that the added inorganic was homogeneously dispersed.

Following counting, the smooth surface was left undisturbed on one pellet. On the remaining five pellets, the smooth surface was reproducibly roughened by impressing a standard sieve screen (W. S. Tyler Co.) into it. The amount of roughness was controlled by using sieve screens of very fine to very coarse mesh. This simulated to some extent naturally occurring surface wrinkling in polymers.

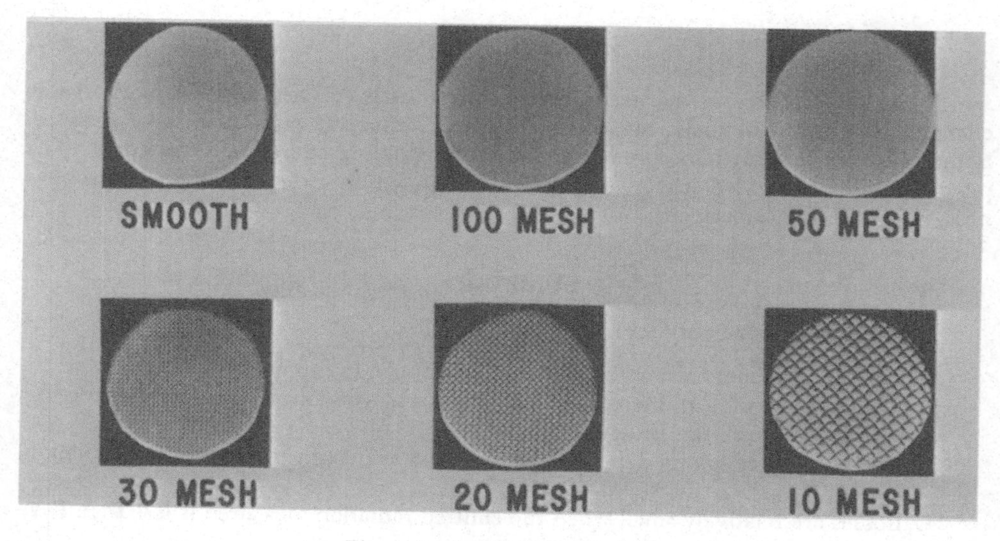

Figure 1. A typical set of pellets.

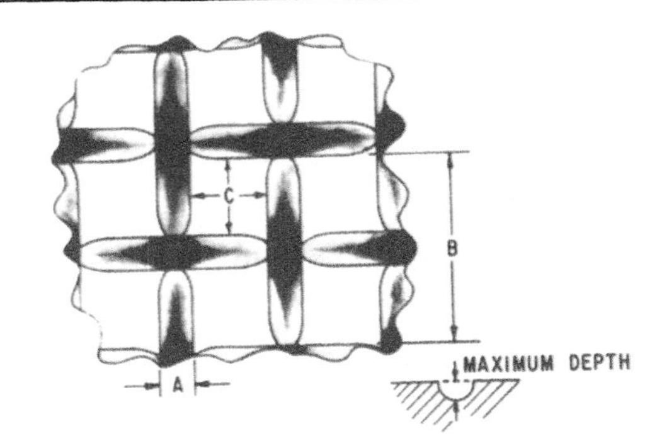

Figure 2. Impression dimensions.

Table I. Data on Symmetrical Impressions* Produced by Various Sieve Screens

Mesh	Maximum depth of groove (in.)	*A* (in.)	*B* (in.)	*C* (in.)	% Undisturbed flat surface area
10	0.024	0.023	0.180	0.077	59
20	0.021	0.013	0.090	0.037	54
30	0.020	0.009	0.059	0.021	45
50	0.010	0.006	0.032	0.014	51
100	0.008	0.003	0.016	0.006	43

* See Figure 2 for impression dimensions.

Figure 1 shows the surface of a typical set of pellets, one with a smooth surface and the others with their surface impressed by 100-, 50-, 30-, 20-, or 10-mesh screens. Each pellet was recounted to obtain I, the intensity with a reproducibly roughened surface. For each group of pellets, we calculated $100 - [(I/I_0) \times 100]$, the percent of intensity loss of the rough- compared with that of the smooth-surfaced pellet.

Figure 2 and Table I show the dimensions of the symmetrical impressions produced by the various sieve screens, measured by microscopy. The undisturbed flat surface area only changes from 43 to 59% for the different screens. The most significant difference in the various impressions is the increase in the maximum depth of the grooves from 0.008 to 0.024 in. because of the larger wire size used with the coarser screens.

RESULTS AND DISCUSSION

Figure 3 shows the overall reproducibility of our experimental procedure from weighing and milling to molding and impressing the pellets. The data appear in clusters of three points because three separate identical batches were made, each containing 5% of titanium. Data for both tungsten and chromium excitation are shown. With the roughest surfaces, the reproducibility approaches $\pm 1\%$. The amount of intensity loss increases with roughness and is greatest for the 10-mesh surface. The intensity loss is also much greater with chromium excitation than with tungsten.

Figure 4 shows the loss of X-ray intensity with wavelength. All added inorganic elements are present at 5% concentration, and all pellets have a 30-mesh surface impression. The K, L, and M spectral lines are intermixed. Data are shown for both chromium and tungsten excitation.

Below 1 Å, the intensity loss with either excitation is about 1%. Part of this relatively small loss is probably due to the slight increase in effective coupling distance, since parts of the pellet are depressed away from the X-ray tube. The intensity loss gradually increases with increasing wavelength, and, beginning with Ti $K\alpha$ at 2.75 Å, the curves separate, with the intensity losses being much greater with chromium excitation.

This is shown again in Figure 5, which is the same as Figure 4 except that the data are for 100-mesh and 10-mesh pellets. The lower curves, for 10-mesh, separate more abruptly in this figure. The data suggest that the greater intensity loss beginning with Ti $K\alpha$ is associated with strong line excitation by Cr $K\alpha$. The Cr $K\alpha$ does not penetrate deeply into the pellet, and the mass absorption coefficients of elements in this wavelength region for Cr $K\alpha$ are relatively high. Therefore, the surface contributes a larger percent of the total radiation counted. Since the surface is rough and absorbs more of the fluorescence radiation, the intensity loss is greater. The small loss as the curves approach silicon is attributed to lower mass absorption coefficients at these wavelengths.

Figure 6 shows the effect of concentration *versus* intensity loss. At the Sb $K\alpha$ line (0.47 Å), there was essentially no difference in intensity loss at concentrations of 0.5, 5, and 9%. All plot together at about 1% intensity loss. However, the same pellets at the Sb $L\alpha$ line (3.44 Å) show increasing intensity loss with increasing concentration.

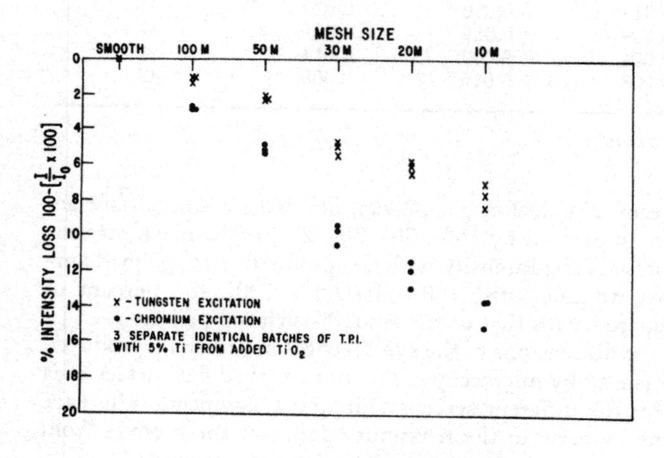

Figure 3. Experimental procedure reproducibility.

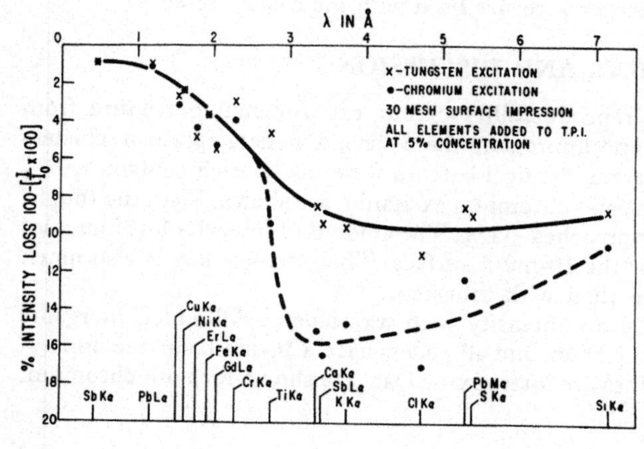

Figure 4. Intensity loss *versus* wavelength; 30-mesh impression.

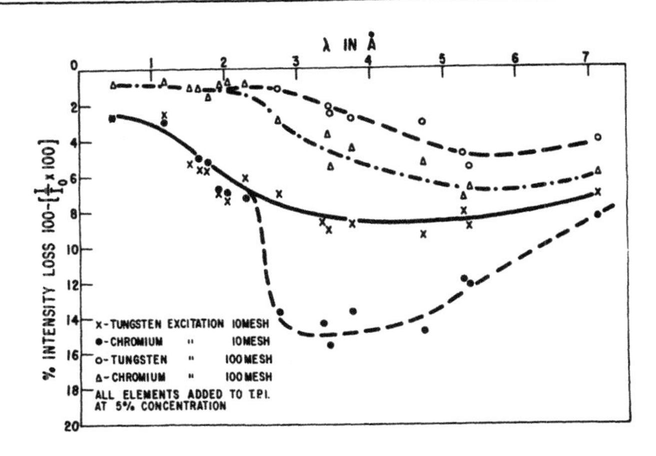

Figure 5. Intensity loss *versus* wavelength; 10- and 100-mesh impressions.

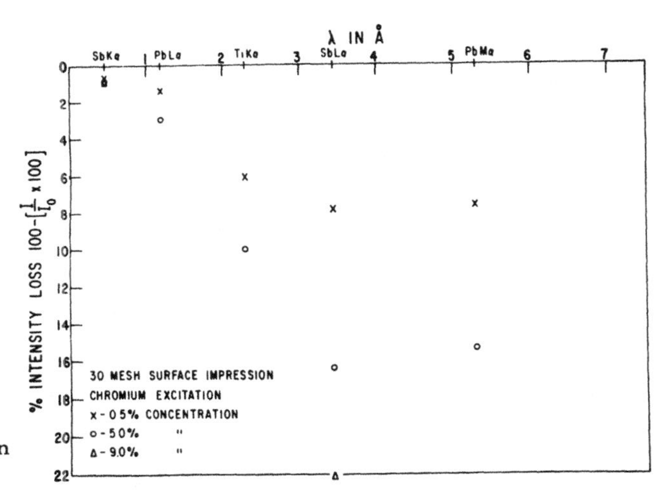

Figure 6. Effect of concentration on intensity loss.

The intensity loss is 7.8% for 0.5% antimony and increases to 22.1% for 9.0% antimony. In practical terms, this means an analyst would have reported only 7.0% on the 9.0% antimony pellet by counting the $L\alpha$ line and comparing with smooth-surfaced standards. By counting the $K\alpha$ line, the analyst would have obtained 8.9%.

We believe that the differences in critical depth account for these losses. Polymers are normally very transparent to X-rays, and radiation is emitted from a considerable depth in them. At longer wavelengths, the critical depth is reduced. Inorganics, however, are very strong absorbers of X-rays, and the critical depth of most inorganic powders and metals is quite shallow. As the concentration of an inorganic increases in a polymer, the critical depth of the polymer is reduced further. At long wavelengths, the net result is a greater intensity loss as the concentration of the inorganic increases in a rough-surfaced polymer.

In discussing experimental reproducibility above, we stated that the greatest intensity loss for titanium was for pellets having a 10-mesh surface impression. This was true for all other elements which emitted radiation below about 3 Å. Above 3 Å, however, the 30-mesh pellet showed the greatest intensity loss. This is shown in Figure 7 where

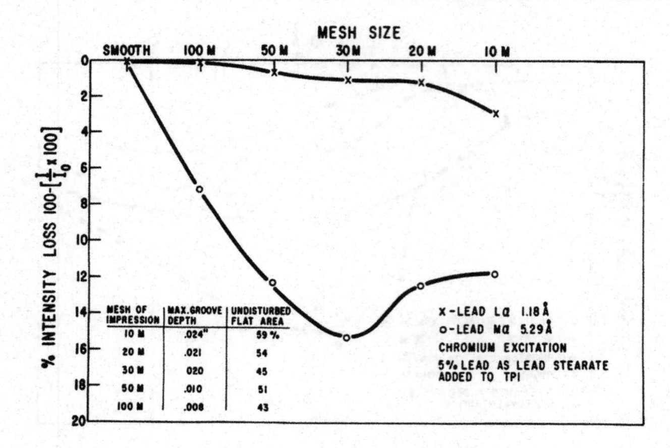

Figure 7. Comparative intensity loss for Pb $L\alpha$ and Pb $M\alpha$.

intensity loss is plotted for Pb $L\alpha$ and $M\alpha$ radiation emitted from the same pellets. The inset table taken from Table I helps explain why this occurs. With Pb $L\alpha$, the loss increases with the increasing depth of the impression. With Pb $M\alpha$, the loss is dependent both on groove depth and on the amount of undisturbed flat-surface area. The 30-mesh impression has one-third less flat-surface area than the 10-mesh with only slightly less groove depth. With critical depth reduced at long wavelengths, the pellet with less flat-surface area would be expected to show a greater loss. The 100-mesh pellet, which has the least surface area of all but also has the shallowest groove depth, shows a large intensity loss.

Some workers[1] in the polymer field measure the coherent- and Compton-scattered primary radiation from a polymer sample and use the ratio of these to correct for minor matrix differences. We investigated whether surface roughness affected the ratio obtained and found that it did not. The intensity of both the coherent- and Compton-scattered primary radiation is reduced proportionally, so the ratio of the two remains unchanged.

Having learned that surface roughness can cause serious X-ray intensity errors, we became concerned with eliminating or reducing them. We found that polyethylene can be readily blended into most polymers by hot milling. If 10 to 20% by weight is milled in, enough "body" can be imparted to the polymer to keep its surface smooth and wrinkle-free. By correcting for dilution and, if necessary, by correcting for matrix changes also, good analytical results can be obtained.

Figure 8 shows four pellets of a different commercial polymer which exhibited natural surface wrinkling. We added 1% titanium as TiO_2. Pellet 1 was freshly molded and exhibited no surface wrinkling. The surface of this pellet progressively wrinkled on standing, and pellet 2 shows the same material after standing 1 day; the surface had become wrinkled and moderately rough. Pellet 3 contained 20% polyethylene and had stood 1 day; its surface appeared as smooth as freshly molded pellet 1. Pellet 4 was milled, unmolded polymer and had a very rough wrinkled surface. With enough time, the surface of pellets 1 and 2 would eventually revert to this surface.

Table II shows the percent of titanium determined on all four samples. Very good agreement was obtained between the freshly molded pellet and the one with added polyethylene. The low results for the two rough-surfaced pellets were proportional to the amount of roughness.

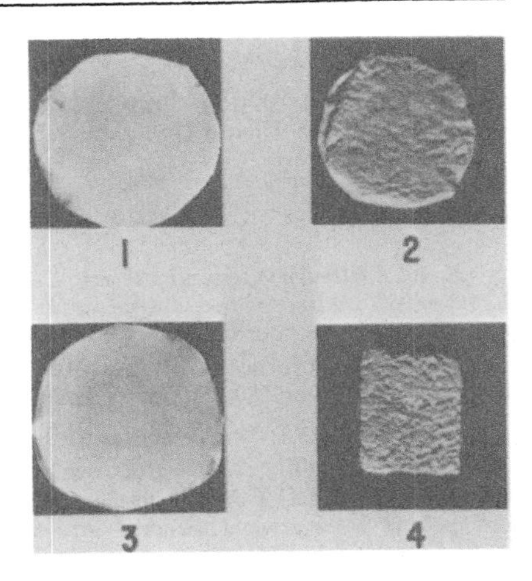

Figure 8. A commercial polymer exhibiting surface wrinkling.

Table II. Analytical Data on Four Samples of a Commercial Polymer Exhibiting Surface Wrinkling

Pellet	Description	% Ti	% Error
1	Freshly molded	1.00	—
2	After standing 1 day	0.96	−4.0
3	20% Polyethylene added	0.99	−1.0
4	Milled unmolded	0.93	−7.0

CONCLUSIONS

The results indicate that surface roughness in polymers causes X-ray fluorescence intensity losses. This can be serious at long wavelengths and at high concentrations. Fortunately, the majority of polymers do not exhibit surface wrinkling. When it does occur, the individual case must be considered in deciding whether the probable error justifies the additional time needed to mill in polyethylene. Practically speaking, if a heavy element such as lead were being determined at trace levels in a polymer, adding polyethylene would not be necessary. If light elements such as chlorine were being determined at a high concentration, the smoothness and reproducibility of the surface would be of paramount importance. Any measure such as adding polyethylene to provide a smooth surface would improve the accuracy of the results greatly.

The amount and kind of naturally occurring surface wrinkling is not reproducible in polymers. It varies with time and is influenced by such factors as average molecular weight, molecular weight distribution, and degree of branching. Our procedure for simulating surface wrinkling by impressing sieve screens into a smooth-surfaced polymer is, at best, an approximation of actual conditions. However, we believe this study has given us a good feeling of actual conditions and the probable size of errors to be expected.

ACKNOWLEDGMENT

We wish to express gratitude to Messrs. C. J. Carman, H. L. Shaw, and R. W. Smith for their assistance in this study.

REFERENCE

1. C. J. Carman, "X-Ray Fluorescent Determination of Major Constituents in Multielement Matrices by the Use of Coherent to Incoherent Scattering Ratios," *Develop. Appl. Spectry.* **5**: 45–58, 1966.

DISCUSSION

E. W. Millburn (General Electric Co.): On your artificially wrinkled samples, did you explore the effect of rotation of the sample about the normal to the surface and what effect this might have on intensity, since you did have orthogonal grooves in it?

J. Gianelos: I'm not quite sure what axis you're rotating. Since you have a square impression, if you rotate it 90 or 180°, you have basically the same impression.

E. W. Millburn: I'm referring to intermediate positions.

J. Gianelos: Oh, yes, we did. But we didn't go to any lengths exploring this. This work was done on a GE XRD-6 X-ray, and its sample holder has a rectangular opening about $\frac{3}{4}$ by 1 in. The pellets were always inserted into the holder reproducibly by making the impression lines parallel to the edges of the opening. In setting up the experiment, we tried some pellets with the lines running diagonally but found the data wasn't as reproducible because it's more difficult to align the pellet if you're not completely parallel to something or unless you make reference marks. In a paper published recently in *Analytical Chemistry* [**39** (6): May, 1967, p. 26A], Hasler used grind marks in steels in different rotations and got some interesting results, but we didn't explore this aspect.

E. W. Millburn: Thank you.

F. Bernstein (General Electric Co.): You mentioned that the coherent and Compton scatter both were reduced when the surface roughness increased and their ratio didn't change, but I was wondering whether you have explored the possibility of using either one as a measure of surface roughness to correct your other values.

J. Gianelos: We haven't considered this, but since we have all the data, this would be quite easy to do in the future.

K. Lawrence (Naval Ammunition Depot): What was the form of the material that was added to the samples?

J. Gianelos: They were added as very, very fine powders. In general, the organometallics added, such as stearates, essentially dissolve in the polymer. With oxides and carbonates that might not dissolve, we used material that passed through a 325- or 400-mesh screen. In some cases, we ground the powders further, for perhaps 15 min or $\frac{1}{2}$ hr, in a Spex mill. We felt the extra grinding was important for soft elements, such as silicon, where we wanted to be sure we weren't seeing particle size effects.

PRODUCTION CONTROL OF GOLD AND RHODIUM PLATING THICKNESS ON VERY SMALL SAMPLES BY X-RAY SPECTROSCOPY

George H. Glade

International Business Machines Corporation, Components Division
Essex Junction, Vermont

ABSTRACT

Manufacture of reed switches, critical components in present-day data-processing support devices, requires a means of accurate, rapid analysis of elements used in plating the levers of the switch. Because of greatly reduced feedback time, X-ray spectroscopy has replaced metallographic sectioning and optical measurement as a plating-thickness control method. While 6 hr were required to obtain thickness data for a given sample size by sectioning, X-ray spectroscopy requires only 2 hr, which permits better control of the plating operating. X-ray spectroscopy is now used routinely to control both gold and rhodium plating thicknesses in the 20- to 100-μin. (1×10^{-6}) thickness range. The large number of samples prevents long count duration, while the small sample size (0.110 by 0.035 in.) reduces the precision of the analysis. However, the precision of the X-ray and optical methods is approximately the same, 8% variance. X-ray accuracy is comparable to that of sectioning since the standards are obtained by sectioning. Simplicity of operation is required since relatively untrained operators are used. An aperture system is used to reduce background. The rhodium thickness measurement is obtained from gross rhodium intensity. Attenuation of gross nickel intensity from the base material was found to be a better measure of gold thickness intensity. Calibration for both gold and rhodium is performed by using the same wide detector conditions. The choice of analysis is made by changing only the 2θ angle, thus avoiding the time required for recalibration when changing analysis.

INTRODUCTION

Reed switches are used extensively in data-processing support equipment and in the electronics and communications industries as well. The heart of the reed switch is the lever tip, which is the electrical contact and which is successively plated with gold and rhodium for electrical and wear characteristics. Accurate control of the plating thickness is required to ensure long life and trouble-free operation.

DISCUSSION

For more than 20 years, X-ray fluorescence has been used as a method of plating thickness control. However, the small size of the lever tips presents unique problems. The flat area on the lever tip suitable for X-ray measurement is 0.110 in. by 0.033 in. The amount of plated material on the tip is very small, which reduces the precision of the analysis. This could be compensated for by long count durations except that effective process control demands that a large number of samples be analyzed. To optimize conditions, very good sample masking is required to define the sample position reproducibly and to limit background.

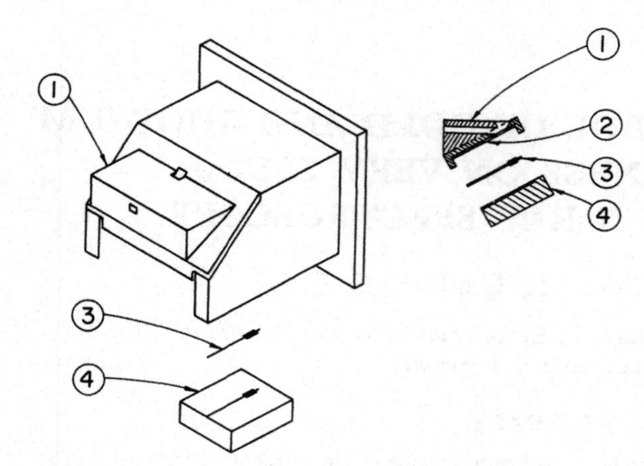

Figure 1. Modified Lloyd aperture as mounted on a General Electric sample drawer and in cross section. (1) Aperture; (2) mask; (3) sample; (4) fixture with cavity to hold sample.

The well-known Heinrich Probe[1] was an immediate consideration for the examination of a small selected area but was rejected because of its physical delicacy and slow sample alignment. Our needs called for a rugged, foolproof device which a relatively untrained worker could operate reliably and reproducibly.

Aperture System

Such a device is found in a modification of the aperture system developed by J. C. Lloyd, International Business Machines Corporation, East Fishkill, New York.[2] It limits both the primary and secondary X-ray beams. It is shown in Figure 1, mounted on the General Electric SPG-4 spectrometer sample drawer and in cross section.

The lever is placed in a cavity in a tool steel fixture. The cavity, which conforms to the dimensions of the lever, is created by electric discharge machining.

The fixture holding the lever tip is aligned by the boxlike underside of the aperture system so as to place the lever tip beneath an 0.080- by 0.025-in. opening in a 0.010-in.-thick copper mask.

The mask opening was made smaller than the available sample area to compensate for variations in sample position caused by necessary mechanical clearances.

The primary X-ray beam is limited by the opening in the top of the aperture system. The large, flat top surface of the aperture system significantly reduces the background due to primary radiation scattered from the sloping surface of a General Electric sample mask. Our original aperture had a comparatively narrow top. Widening the top reduced the background by a factor of 7.

The exit tunnel of the aperture system limits the area from which secondary radiation is measured and reduces scattered radiation.

As with the Heinrich Probe, the middle blades were removed from the beam tunnel of the spectrometer. A Heinrich Probe adjustable-slit collimator is used in place of the Soller slit. The line profile is no wider than that obtained with a 0.010-in. Soller slit.

Measurement of Gold Thickness

The levers are stamped from nickel–iron wire and are first plated with gold. Control samples for gold plating thickness are taken following the gold plating. Gold $L\alpha_1$ I, $\lambda = 1.28$ Å, is without interference except from first-order tungsten from the primary tube, which cannot be eliminated by discrimination. Our instrument is equipped with an EA-75 Dual Target primary X-ray tube, which permits gold analysis with chromium excitation. Under these conditions, Au $L\alpha_1$ I is interference-free.

Experiments showed that measurement of the attenuation of nickel radiation from the body of the lever is a more sensitive measure of gold plating thickness than is the direct measurement of gold intensity. Nickel $K\alpha$ I, $\lambda = 1.66$ Å, is interference-free.

Instrumental conditions for gold and nickel analysis are given in Table I.

Table I. Instrumental Conditions for the Analysis of Gold, Nickel, and Rhodium and Broad-Range Conditions for Nickel and Rhodium Combined

Element:	Gold	Nickel	Rhodium	Broad range (Ni and Rh)
Primary tube			EA-75 Dual Target	
Target	Chromium	Tungsten	Tungsten	Tungsten
kV/mA	50/50	60/55	60/55	60/55
Analyzing crystal	LiF	LiF	LiF	LiF
Path	Air	Air	Air	Air
Detector		General Electric	SPG-4 scintillation	
Detector voltage	1060	1050	1050	1050
Amplifier gain	16 × 0.88	16 × 0.88	16 × 0.88	16 × 0.88
Analytical line	$L\alpha_1$ I	$K\alpha$ I	$K\alpha$ I	$K\alpha$ I/$K\alpha$ I
Degrees 2θ	36.96	48.66	17.56	48.66/17.56
E_L, V	4	3	10	3
E_U, V	30	21	60	60
Counting duration, sec	40	40	40	40

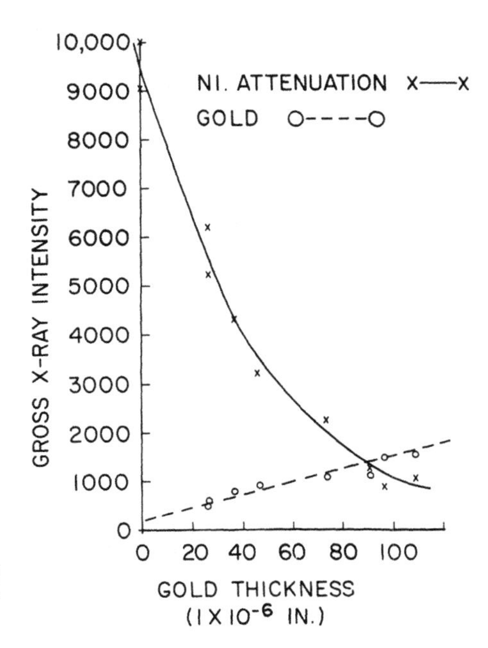

Figure 2. Typical working curves for gold plating thickness by gold X-ray intensities and by attenuation of nickel X-ray intensity.

Typical working curves for gold plating thickness for both the direct measurement of gold intensity and nickel attenuation are shown in Figure 2. It should be noted that, because of the flattening of the slope of the nickel attenuation curve, nickel attenuation cannot be used as a measure of gold plating thickness above 100 μin. Above 100-μin. thickness, gold intensity must be used instead.

Initially, gold and nickel background measurements were made, and net X-ray intensities were used. Background was found to be predominately scattered radiation from the Lloyd aperture and to be very constant. Experiments showed that the use of a background correction increased neither the precision nor the accuracy of the analysis, and its use was dropped in favor of the ability to analyze the greater number of samples needed for process control.

Measurement of Rhodium Thickness

After gold plating, levers are rhodium plated. Rhodium thickness control samples are taken after the rhodium plating.

Rhodium *Kα* I, $\lambda = 0.615$ Å, has no interferences from nickel, iron, gold, and copper. The only interference is W *Kα* III, which is eliminated by discrimination.

Instrumental conditions for the measurement of rhodium are given in Table I. A typical working curve for rhodium plating thickness is shown in Figure 3.

The effect of various thicknesses of gold plating under the rhodium plating was investigated with specially prepared batches of levers. Figure 4 is a plot of gross rhodium intensity *versus* rhodium plating duration for samples having three different levels of

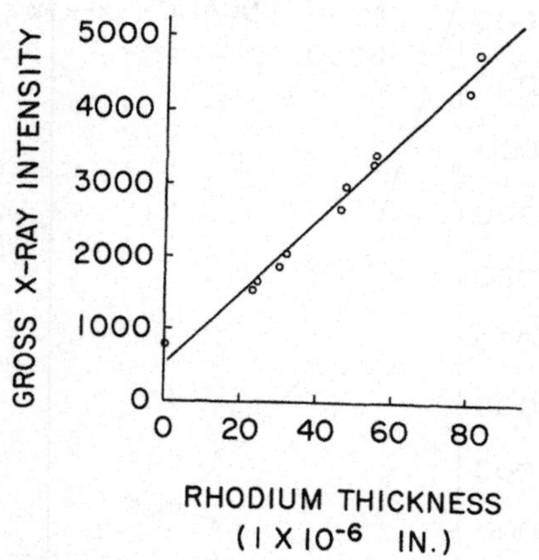

Figure 3. Typical working curve for rhodium plating thickness (with broad-range conditions).

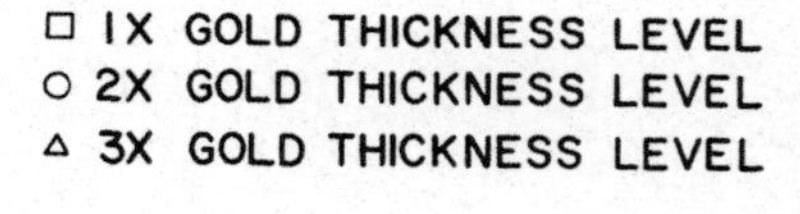

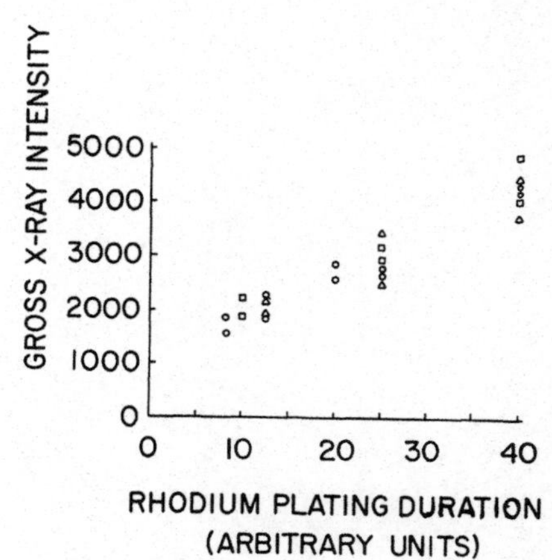

Figure 4. The effect of various underlying gold thicknesses on rhodium X-ray intensity; plotted on the basis of plating time, which is proportional to plating thickness.

gold thickness beneath the rhodium. (Plating thickness is proportional to plating duration.) Various thicknesses of gold plating beneath the rhodium had no effect on rhodium intensity.

A rhodium background correction was also used initially but was dropped for the same reasons as the gold and nickel background corrections.

Preparation of Standards

Metallographic sectioning and optical measurement provide the only absolute method of measuring plating thickness. Since this method is destructive, it cannot be used to create standards. Standards for both gold and rhodium analysis were made in the following manner:

Batches of levers were plated for different durations to produce a range of plating thicknesses. Plating thickness varied from side to side on the levers. The standard levers were identified by side, and each side was considered a separate standard. A dual set, two of each thickness, was selected, and X-ray data were taken on them. The plating thickness of one lever of each pair was then measured metallographically. It was known that plating thickness varied over the surface of the lever and (that a single-point metallographic thickness determination was subject to some uncertainty. Therefore, plating thickness was measured at nine different points on the flat surface of the lever tip, and the values were averaged. The average metallographic plating thicknesses were plotted against the average X-ray intensities to obtain a calibration curve. The average X-ray intensities for the second members of the lever pairs were read into the calibration curve to obtain the plating thicknesses on the second levers, which then became standards.

Broad-Range Discriminator Conditions

Both gold plating thickness and rhodium plating thickness samples are received for measurement three times each shift. It is necessary to perform the analysis without taking time to measure the standards. Further, operators are not trained to adjust the parameters of the instrument. To enable both gold and rhodium thickness analysis to be performed interchangeably, without recalibration, wide-range discriminator conditions are used. These conditions are given in Table I. The choice of analysis is made by operator adjustment of the spectrometer to the correct 2θ angle. Since none of the parameters are changed, the working curves remain valid.

Both the gold and the rhodium standards are measured twice each shift.

The only effect of the wide-range detector conditions is a slight increase in the background; 10 counts/sec for nickel, 14 counts/sec for rhodium. At the 50-μin. thickness level, the signal-to-noise ratio when using the broad-range condition is 10:1 for gold measured by using nickel attenuation and 4:1 for rhodium.

RESULTS

The precision of the gold thickness analysis was studied during the preparation of standards. A group of 36 lever sides having a gold-plating-thickness range of 26 to 106 μin. and an average thickness of 64 μin. was studied. The variance of the X-ray data was 4.0%. The variance of the metallographic data was 8.3%.

The precision of the rhodium thickness analysis was studied by using 10 samples having a rhodium thickness range of 40 to 80 μin. and an average thickness of 60 μin. The variance of the X-ray data was 6.1%. The variance of the metallographic data on the same 10 samples was 7.4%.

Table II. Comparison of X-Ray Measurements of Rhodium Plating Thickness Made at Two Different Times on the Same Process Control Samples

Sample	Rhodium thickness* as measured on: 3/14	3/30	Δ*
1. Side 1	65	63	−2
Side 2	66	68	+2
2. Side 1	63	65	+2
Side 2	57	61	+4
3. Side 1	72	68	−4
Side 2	76	65	−11
4. Side 1	76	78	+2
Side 2	63	60	−3

* Rhodium thickness in microinches.

On one occasion, it was necessary to reanalyze four process control samples 2 weeks after initial measurement. Table II is a comparison of the two sets of thickness measurements. The thickness range of the samples was 60 to 78 μin. The average thickness was 67 μin. The range of the differences between the two measurements on the same lever side was -11 to $+4$ μin. The average difference was 3.8 μin.

CONCLUSION

Prior to the development of the X-ray method described above, process control was performed by metallography using a single-point measurement. Seven hours were required for the measurement, and sampling was performed once each shift. The X-ray measurement, which was developed under the auspices of the Burlington Quality Assurance function, has the same, or slightly better, precision and requires only 2 hr to examine samples from all plating lines. Sampling is now performed three times per shift, resulting in better control of the product.

REFERENCES

1. K. J. F. Heinrich, "X-Ray Probe with Collimation of the Secondary Beam," in: W. M. Mueller (ed.), *Advances in X-Ray Analysis, Vol. 5*, Plenum Press, New York, 1962, p. 516.
2. J. C. Lloyd, Department 255, International Business Machines Corp., Fishkill, New York, private communication.

A STUDY OF X-RAY FLUORESCENCE METHOD WITH VACUUM AND AIR-PATH SPECTROGRAPHS FOR THE DETERMINATION OF FILM THICKNESSES OF II–VI COMPOUNDS

Frank L. Chan

Aerospace Research Laboratories
Wright–Patterson Air Force Base, Ohio

ABSTRACT

Films of II–VI compounds have been prepared by vacuum technique for the determination of their thicknesses. Some difficulties have been experienced in preparing good quality films. The position of the substrate in the vacuum chamber, the rate of deposition, and the temperature of sublimation are some of the factors influencing the quality of these films. For instance, films of cadmium sulfide could be prepared in the usual yellow or orange color or in colors ranging from brown to black, depending on the conditions enumerated. For energy conversion, a film of proper thickness is one of the requirements for aerospace application. Among the various methods used for the determination of the thickness of these films, X-ray fluorescence can be performed rapidly and nondestructively. After determination of thickness by the X-ray fluorescence method, the samples can be used for other determinations and for energy conversion without their efficiency's having been affected. Other physical and chemical methods have been worked out. Comparison of these methods with the X-ray fluorescence method is made. Procedures and results are presented.

INTRODUCTION

After the discoveries of the photovoltaic effect of cadmium sulfide and other II–VI compounds by the Aerospace Research Laboratories some 14 years ago,[1] considerable efforts were made to use these materials for the fabrication of solar cells for aerospace applications. Generally speaking, for the energy conversion by using silicon or germanium, it is necessary to use single crystals. By using II–VI compounds, however, such as cadmium sulfide for fabrication of solar cells, it is not necessary to use single crystals. Films of this compound can and are being used for this purpose. They are now in outer space for energy conversion. The use of II–VI compounds for this purpose has certain advantages over use of the conventional ones. Although the conversion efficiency is 5.4% in the case of cadmium sulfide, pound for pound cadmium sulfide is more efficient than silicon and germanium. Cadmium sulfide can be fabricated into rolls, and these rolls can be spread out in space for energy conversion.

One of the requirements for the fabrication of cells of the II–VI compounds is proper thickness. With the present state-of-the-art, cadmium sulfide must be of certain optimum thickness. Efforts are being made to reduce this thickness without reducing the overall efficiency.

Among the methods used for measuring the film thickness, X-ray fluorescence is perhaps the best. It is well known that this method is nondestructive in nature. Films, after their thicknesses have been determined, can be used for other determinations and applications. They can also be preserved for future reference. The constituents of these compounds, because of their characteristic wavelengths evidenced by results obtained, can be utilized for very thin films to comparatively thick films suitable for aerospace applications.[2] With the X-ray instruments available, soft X-rays such as $S\,K\alpha$, Si $K\alpha$ when present in the substrate, Cd $L\beta_1$, and Te $L\alpha_1$ are being analyzed for this purpose with the vacuum spectrometer that will shortly be described. Very high count rates have been obtained with the analyzing crystal PET, prepared by this author in the Aerospace Research Laboratories.[3,4] Hard X-rays such as Cd $K\alpha$ and Te $K\alpha$ are also being used for this purpose with the air-path spectrometer.

Since the X-ray fluorescence method is not an absolute method, the thickness of films determined by this method is correlated with those determined by chemical and physical methods. The physical methods used are optical, neutron activation, and laser microprobe.

For dealing with thin films, which is a special situation of limited quantity in X-ray fluorescence analysis, Birks[5] has derived an equation (for details and nomenclature, see the original source)

$$I_i = \frac{I_{pi}\rho_i}{\sum_j (a_{ij}\rho_i)} \left\{ 1 - \exp\left[-\left(\sum_j a_{ij}\rho_j \right) t \right] \right\}$$

such that, for a thin layer with small t, the exponential may be approximated by the first two terms of the series expansion, which reduces the above equation to

$$I_i = I_{pi}\rho_i t$$

Without knowing ρ_i, the measured intensity is simply related to element i in grams per square centimeter.

More recently, Addink[6] derived the usual equation (for details and nomenclature see the original source)

$$dI_A = kI_{\text{prim}} \exp(-Qm)\, dm\, C_A$$

After integration and for the case of infinite thickness, $e^{-Qm} = 0$, the final equation

$$I_A = kI_{\text{prim}} \frac{C_A}{Q}$$

enabled him to make two conclusions, namely:

1. When Q remains constant for various concentrations of the element A, B, *etc.*, I_A and C_A are linearly proportional.
2. When e^{-Qm} is unity, Qm is approximately equal to zero and, therefore, there is a linear relationship of I_A to C_A.

Therefore, for thin films, $C = KI$, where K equals the proportionality factor called the *micro K factor* and C equals the concentration of the element to be analyzed. The validity of these derivations has been experimentally verified by Liebhafsky and Zemany[7] and by Rhodin.[8]

In the case of II–VI compounds in thin films, the derivations given above are applicable.

It is the purpose of this paper to show that:

1. Thin films of II–VI compounds have been successfully vacuum deposited (among other materials) on polymeric materials such as Mylar, polypropylene, and Kapton. Other uses of Mylar for X-ray fluorescence analysis can be cited.[9]

2. As many as eight layers of Mylar have little effect on the X-ray fluorescence intensity of elements present in the II–VI compounds provided wavelengths of these elements are less than 1.4 Å, *i.e.*, Zn $K\alpha$.

3. The linearity region of thin films of II–VI compounds can easily be determined by using multiple layers of Mylar-coated films, as described in this paper.

The thickness of film on a specified substrate can quickly be determined by a procedure described in this paper. By this procedure the X-ray fluorescence intensity of film of unknown thickness is measured along with the known thickness of the same II–VI compound on Mylar placed over the same substrate as that of the unknown. In this manner the effect of the substrate is eliminated.

INSTRUMENTATION

X-ray fluorescence intensity of films of II–VI compounds and of elements present in the substrate is measured by two instruments made by the General Electric X-ray Department. The reason for using two instruments is that the vacuum spectrograph has dual targets, one of chromium and one of tungsten. In the air-path spectrograph, there is a molybdenum target. Furthermore, a variety of elements is involved in the determination of film thickness of the II–VI compounds. Elements such as sulfur with atomic number 16 and mercury with atomic number 80 are routinely analyzed. The number of elements involved is further increased by analyzing those existing in the different substrates used for the deposition of the film. The selection of one of these targets is deemed necessary because of possible interferences by spectra lines originating from the target material. A brief description is as follows.

Vacuum Spectrometer

The spectrometer in principle is the same as other types of spectrometers in that the intensity of secondary X-rays (the fluorescence X-ray) produced by the films of II–VI compounds is determined by a detector after they are diffracted by an analyzing crystal. The analyzing crystal and the detectors are enclosed along with the samples in a compartment so that the whole spectrometer proper can be operated in the absence or presence of helium or air.

At the present moment the vacuum spectrometer can accommodate two analyzing crystals. (Provisions will soon be made to accommodate four crystals.) These crystals can be turned to the reflecting position with a remote-control mechanism. The PET and LiF are located in the crystal holder and are kept dry by a device described previously by this author.[2] A total of four samples can be loaded into the instrument at one time. These samples with the present setup were located below the target tube and held in a position 30° to the horizontal. The size of the opening for the samples to be exposed to the primary X-ray beam is 1.9 by 2.6 cm. To prevent the slipping of the sample in the holder because of the constant revolving of the sample holder by the remote-control mechanism, a soft organic polymeric material is placed in between the sample and a spring located on the lower part of the holder. The No. 7 counter tube normally is operated at 1620 V with the pulse-height selector usually set at $E = 2$ V and $\Delta E = 4$ V. A cover measuring 10.575 in. at the top can be opened for removing or changing any parts inside the vacuum chamber such as soller slits, counter tubes, or analyzing crystals.

Air-Path Spectrometer

Besides the SPG-3 spectrometer, an air-path or helium-path SPG-2 has also been installed as shown in Figure 1. These instruments were connected in such a manner that both units use a common X-ray generator and the SPG-4 detector to reduce the cost of operation. The SPG-2 is provided with a modified Heinrich probe which can be used to scan a sample for homogeneity and is therefore supplementing the vacuum spectrometer. It is planned that the homogeneity of some of the films will be determined by this method. This spectrometer has a molybdenum target which is useful for the determination of films containing selenium. It also provides a two-crystal changer operated by a remote-control mechanism. The detector in this spectrometer is a scintillation counter operating at a high efficiency over a wide range of elements.

The entire setup consists of the two spectrometers described, the X-ray generator, and XRD-6 high-voltage assembly, and an SPG-4 detector. A special cable and a transformer were installed for the entire setup to obtain electricity from the main supply line. Both instruments are capable of operating at a maximum load of 75 kVP and 50 mA or 37.5 kVP and 20 mA.

The goniometer can be used for measuring 2θ angles accurately. These 2θ angles can be scanned manually or automatically with increasing or decreasing angles. The worm gear can be quickly disengaged, and the protractor turned to any desired position.

The SPG-4 detector has an amplifier, a pulse-height selector, a ratemeter, a scaler–timer combination, a strip chart recorder, and a digital computer.

EXPERIMENTAL

Preparation of Films of II–VI Compounds

Figure 2 shows the vacuum deposition apparatus. It is a modified model CVE 14 supplied by the Consolidated Vacuum Corporation. This apparatus has a mechanical pump, diffusion pump, baffle, valve, manifold, baseplate, and ionization gage. An 18-in.-diameter bell jar, with guard and hoist is located above the baseplate. Substrates for the deposition of film are placed inside the jar before evacuation. Evacuation is accomplished

Figure 1. Some of the instruments used for X-ray fluorescence analysis.

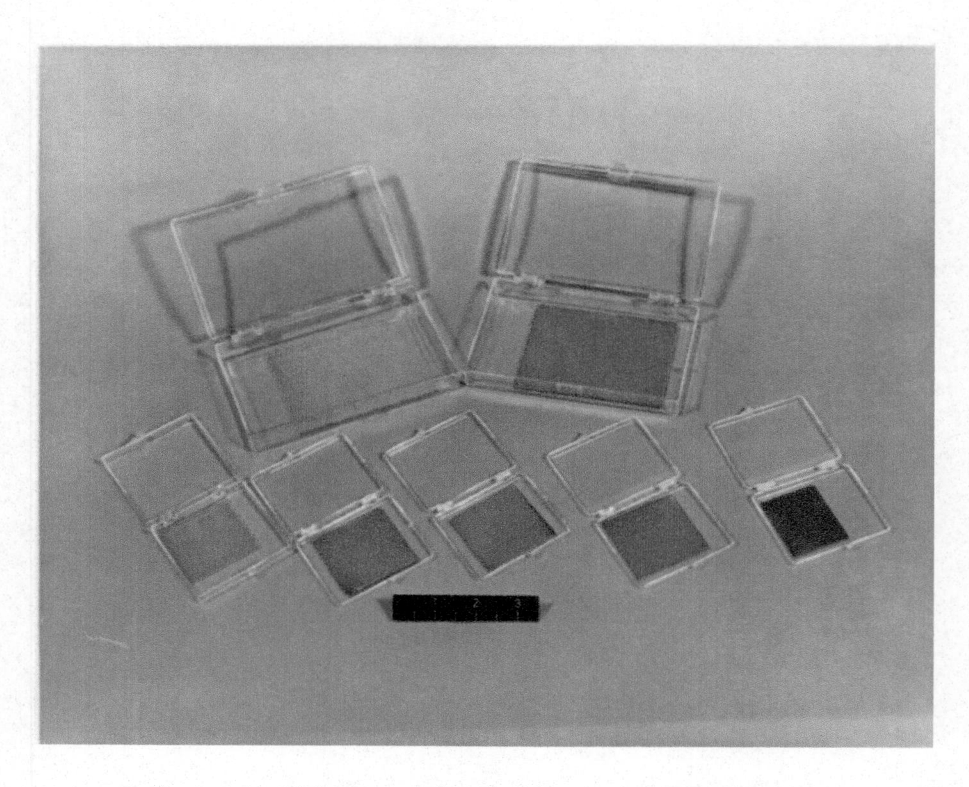

Figure 3. A few of the vacuum-deposited films on a variety of substrates.

Figure 2. Apparatus for vacuum deposition of films of II–VI compounds.

in two steps. In the first step, air is removed by a mechanical pump until the chamber pressure is about 50 to 100 mtorr. In the second step, the diffusion pump is turned on and normally reduces the chamber pressure to 8 to 9 μtorr. The heating of the II–VI compound located in a fused-quartz boat or molybdenum boat commences after the vacuum is reached. Electrical connections from a Variac to the boat are introduced through the baseplate.

Deposition of the II–VI compound was carried out by placing the substrate to be coated in a certain part of the vacuum chamber. When substrates are cut to size, they are measured and weighed before they are placed in the chamber. For most II–VI compounds, when a film of about 2 μ thick is used, it is not necessary to heat the substrate. A device for heating the substrate is provided by installing an electrical connection through the baseplate. When thin films are to be coated, the substrates of proper size are placed in the chamber in pairs, one behind the other, so that only one size is coated.

Materials for the vacuum deposition are the purest that can be obtained for this purpose. Except HgSe, the materials are resublimed before being used. Each material is provided with a boat for sublimation to avoid contamination, and a new charge is placed in the boat for each run. A few of these vacuum-deposited films on a variety of substrates are shown in Figure 3.

Procedures

Preliminary examination was carried out by inspecting the evenness of the coating and decomposition of the compound. Unevenness is detected by the appearance of fringes; the greater the number of fringes per centimeter, the greater is the unevenness.

The size of a sample suitable for the vacuum spectrograph is 2 by 2 in. or less and for the air-path spectrograph 1.75 by 1.875 in. Thickness of the sample can be 0.125 in. or more.

The sample is placed in the holder (four samples can be loaded in the vacuum spectrograph for each run). When a vacuum spectrograph is used, the chamber is evacuated when vacuum is required. The goniometer is turned to the proper setting, and the intensity of the spectral line in which one is interested is recorded. Depending on other elements present in the film, the background count is from 1 to 3° from the peak position in increasing and decreasing 2θ angles.

To correlate the net count per second with the thickness of films, several methods were used. The first method involved weighing the substrate before and after the deposition of film. The second method involved the use of the neutron-activation method. This is done by irradiating either the film on the substrate or an extract of the film in the event of interference by the material present in the substrate. The radioactivity of the element of interest is statistically determined. The third method involved the use of laser microprobe based on the intensity of the emission line of the element of interest.

Chemical methods have been used in determining the thickness of these films, such as the conversion of sulfide into sulfate and precipitation of the latter with barium chloride as the sulfate, or a direct conversion of the sulfide to sulfate that is weighed as such. Optical methods have also been used for thicker samples.

RESULTS AND DISCUSSION

The conditions for the determination of film thickness based on the count rate of various elements existing in the II–VI compounds are briefly listed in Table I. The combination of different components shown in this table is merely a suggestion. Other

Table I. Summary of Conditions for the Determination of Film Thickness of II–VI Compounds by X-Ray Fluorescence Method

Element	Spectral line	2θ	Analyzing crystal	Target	Counter	PHS*	Remarks
Sulfur	$K\alpha$	75.76	PET	Chromium	Flow Proportional		Vacuum
Selenium	$K\alpha$	31.89	LiF	Chromium Molybdenum	Flow Proportional	In presence of cadmium	
Tellurium	$K\alpha$	12.91	LiF	Tungsten	Scintillation		
	$L\alpha_1$	109.54	LiF	Molybdenum	Flow Proportional		Vacuum
Zinc	$K\alpha$	41.80	LiF	Tungsten Chromium	Scintillation Flow Proportional		
Cadmium	$K\alpha$	15.31	LiF	Tungsten	Scintillation		
	$L\beta_1$	136.35	LiF	Molybdenum	Flow Proportional		Vacuum
Mercury	$L\alpha_1$	35.90	LiF	Tungsten Molybdenum	Scintillation Flow Proportional		

* In this and following tables, PHS denotes pulse-height selector.

combinations can be used depending on circumstances and the possible breakdown or malfunctioning of some equipment. Figure 1 shows the general layout of the instrument. In the vacuum spectrograph, there is a dual-target tube of tungsten and chromium and, in the air-path spectrograph, a molybdenum-target tube as described under instrumentation. Although these target tubes are interchangeable to fit particular determinations with interference from elements present in the sample or substrate, the present arrangement appears optimum and a determination of any element listed could be carried out without much change in the setup. Scintillation counters of course could be adapted for a large number of elements ranging from Se $K\alpha$ down to Hg $L\alpha_1$ spectra. The flow proportional counter, on the other hand, is most suitable for long-wavelength spectra with the tail end at about 0.4 Å. When there are a large number of samples with different elements to deal with, as in the study of II–VI compounds, it is necessary to use less optimum conditions, or lower count rates, as circumstances demand. For instance, in the investigation of the effect of substrates on the count rate of such compounds as ZnSe, CdSe, CdTe, and HgSe, as shown in Table II, all the data were obtained without interruption from the molybdenum target installed on the air-path spectrograph. By using the white radiation of a molybdenum target for the Cd $K\alpha$ and Te $K\alpha$ spectra, a satisfactory count rate has been obtained. Likewise, the white radiation of a chromium target was at times used for Zn $K\alpha$. By using this radiation to excite Zn $K\alpha$ with no vacuum in the vacuum spectrograph and a flow proportional counter operated at 1680 V, a count rate of 3899 counts/sec (net) was obtained compared to 14,209 counts/sec (net) when a tungsten target was used. Depending on circumstances, a moderate count rate is preferred to a higher rate, especially when very thick films are dealt with. Numerous other examples could be cited.

Figure 4 shows a strip chart of an actual tracing of the spectra peaks from a film of mercuric selenide deposited on an aluminum substrate. The 2θ angles are in degrees, and the intensities are as shown. These peaks have been identified and confirmed. Their identity is indicated above each peak. Determination of the thickness of mercuric selenide

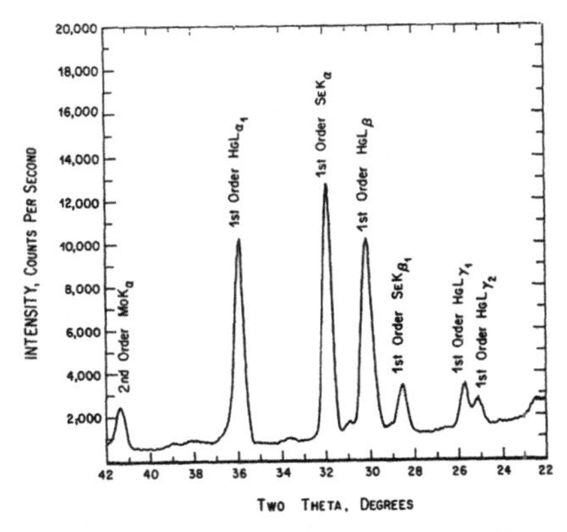

Figure 4. Recorder chart showing actual tracing of spectra peaks with the use of a sample of thin film of mercuric selenide on aluminum substrate. Conditions: 60 kVP; 50 mA; no vacuum used in spectrometer; molybdenum target; no PHS; analyzing crystal, LiF; scintillation counter.

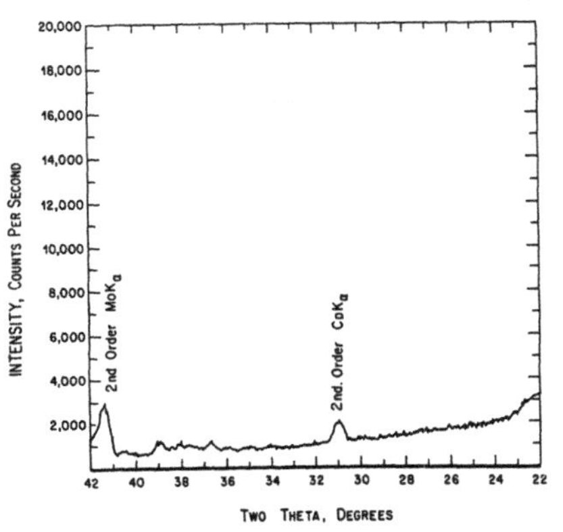

Figure 5. Recorder chart showing the actual tracing of a sample of thin film of cadmium sulfide on aluminum substrate. Conditions: 60 kVP; 50 mA; no vacuum used in spectrometer; molybdenum target; PHS in; analyzing crystal, LiF; scintillation counter.

film can be made by using the Se $K\alpha$ and Hg $L\beta_1$ spectra. Molybdenum from the target material is in regions outside this section of the strip chart. An identical tracing was carried out with a cadmium sulfide, which is shown in Figure 5. The second-order Cd $K\alpha$ appearing in the middle of this figure is indeed very near the Se $K\alpha$ spectra with lithium fluoride as the analyzing crystal. The interference of cadmium in a determination of the intensity of Se $K\alpha$ can be overcome by the use of a pulse-height selector and careful calibration of the goniometer. By comparing Figure 4 on Hg Se and Figure 5 on CdS, the noninterference of second-order Cd $K\alpha$ is shown.

The use of a molybdenum target resulted in the appearance of a second-order molybdenum peak shown on the extreme left of Figure 5. In the determination of the intensity of Zn $K\alpha$ in zinc sulfide, the second-order Mo $K\alpha$ peak is close to the Zn $K\alpha$ spectra. Accurate adjustment of the goniometer is necessary to prevent this interference. The use of a chromium or a tungsten target will also eliminate such interference.

The presence of three distinct regions in the X-ray intensity *versus* thickness of metallic film have been studied by previous workers, notably by Liebhafsky, Pfeiffer, Winslow, and Zemany[10] and others.[7,11] From the experimental results obtained by the procedure devised by this author using multiple layers of Mylar and the air-path spectrograph, linear curves were obtained by plotting intensity in counts per second (net) *versus* thickness in microns as shown in Figure 7. Other results obtained indicate that, for film thickness determined by (1) S $K\alpha$ and Cd $L\beta_1$, (2) Se $K\alpha$, Zn $K\alpha$, and Hg $L\alpha_1$, and (3) Cd $K\alpha$ and Te $K\alpha$, the linear range is estimated as (1) less than 1 μ, (2) greater than 1 μ, and (3) greater than 20 μ (see Figure 6). Because of many factors influencing the linearity, the relation between thickness and mass absorption coefficient is only qualitative.

As mentioned by Liebhafsky and Zemany, substrate spectra can also be used to determine the film thickness.[7] In plating copper and nickel on iron, the thickness has been successfully determined by Birks, Brooks, and Friedman.[11] In the case of II–VI compounds, it has been shown that silicon in glass or fused quartz can be used to determine the thickness of the compound. The decrease in intensity is a function of thickness.[2]

Table II shows typical counting statistics of Zn $K\alpha$ and Se $K\alpha$. The counts were

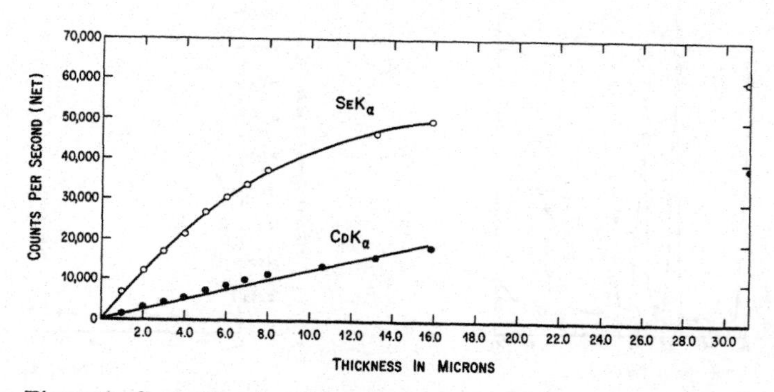

Figure 6. Curve showing the appearance of Se $K\alpha$ and Cd $K\alpha$ by the successive addition of Mylar films coated with cadmium selenide of known thickness, the linear and logarithm regions for Se $K\alpha$ as shown. Conditions: 60 kVP; 10 mA; molybdenum target; no vacuum used in spectrometer; analyzing crystal, LiF; scintillation counter; PHS in for Se $K\alpha$ and out for Cd $K\alpha$.

Table II. Typical X-Ray Fluorescence Determination of Multiple Layers of Mylar Film Coated with II–VI Compounds*

Film thickness by another method, μ	Zn $K\alpha$		Se $K\alpha$	
	Average background counts, counts/sec	Net counts, counts/sec	Average background counts, counts/sec	Net counts, counts/sec
0.28	710	3,890	692	8,847
0.56	734	7,470	806	16,888
0.83	749	10,581	898	23,913
1.101	771	14,475	1020	32,C25
1.39	778	17,780	1081	38,902

* The number of layers, plain or coated, of Mylar is the same (five layers). For plots of these results, see Figure 7. Conditions: Molybdenum target, 60 kVP, 50 mA; PHS out; scintillation counter; LiF analyzing crystal.

repeated five times and averaged as shown in this table. The film thickness has been determined by another method. For a plot of these results see Figure 7.

In the course of an investigation of films of II–VI compounds having somewhat uniform variation in thickness, it was soon found that it is possible to coat sheets of

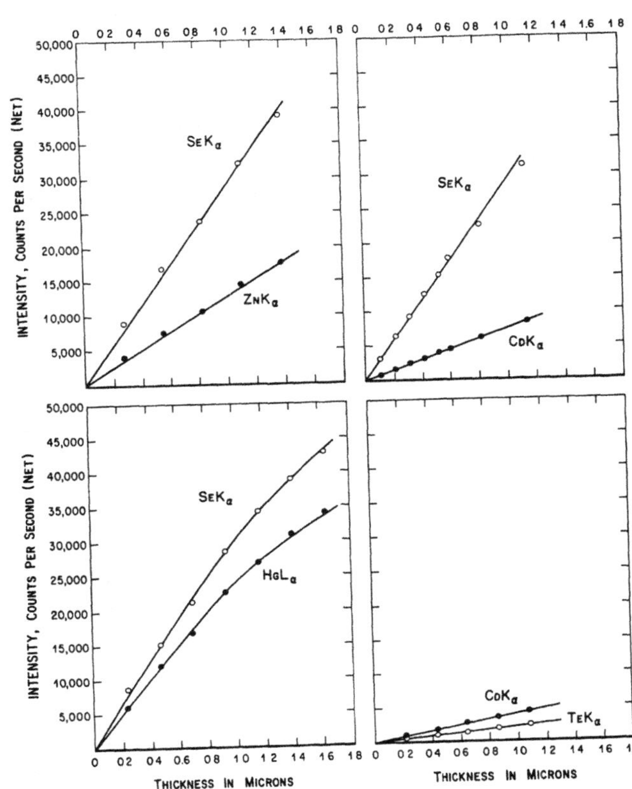

Figure 7. Curves showing the intensity of various spectra present in the II–VI compounds *versus* thickness of the film. Conditions: PHS in for Se $K\alpha$ in CdS and PHS out for all other spectra; no vacuum used in spectrometer; 60 kVP; 50 mA; molybdenum target; 0.01-in. sollar slit; analyzing crystal, LiF; scintillation counter.

Table III. Effect of Mylar Overlay on Count Rate of Cd $K\alpha$ and Se $K\alpha$*

Cd $K\alpha$		Se $K\alpha$	
Absence of Mylar overlay, counts/sec, net	Presence of 8 Mylar overlays, counts/sec, net	Absence of Mylar overlay, counts/sec, net	Presence of 8 Mylar overlays, counts/sec, net
38,610	38,555	60,152	60,171
38,841	38,528	60,029	60,134
38,725	38,452	60,128	60,063
38,805	38,384	60,062	60,009
38,827	38,507	60,010	60,024
Avg. 38,762	38,485	60,076	60,076
% Decrease in count rate	0.71		nil

* Conditions: Molybdenum target; size of mode 1 by 0.575 in.; 60 kVP; 10 mA; PHS out for Cd $K\alpha$; PHS in for Se $K\alpha$; scintillation counter; LiF as analyzing crystal.

Mylar, 15 by 15 cm square, of controlled thickness. These sheets were, after determining the film thickness, cut into the size suitable for the vacuum and air-path spectrographs. Curves having somewhat uniform increase in thickness are shown. When wavelengths shorter than 1.4 Å were used for the determination of the thickness of the film, the presence of as many as eight sheets of Mylar films on the surface of the film gave very slight change in count rate. Actual counts per second in the presence and absence of Mylar film are listed in Table III.

Because of the finding described, the curves shown in Figure 7 have been constructed. In these curves the number of Mylar films, plain or coated with II–VI compounds, in each series was the same.

Figure 8 shows the intensity *versus* thickness of HgSe in millimicrons. As in the case of CdS reported earlier,[2] the intensity of Se $K\alpha$ and Hg $L\alpha_1$ are shown in this curve. The instrument is, in this case, operated at its peak under actual working conditions.

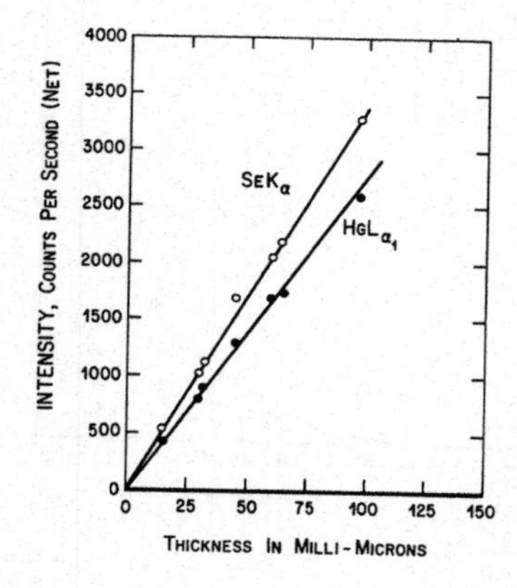

Figure 8. Typical curve showing the intensity of mercuric selenide *versus* thickness in millimicrons. Conditions: molybdenum target, 0.01-in. sollar slit; PHS out; 60 kVP; 50 mA; no vacuum in spectrometer; analyzing crystal, LiF; scintillation counter.

Table IV. Effect of Substrate on Intensity of Film of II–VI Compounds*
(Results based on nickel substrate as unity)

II–VI compounds on substrate	Zinc selenide, 0.63 μ		Cadmium selenide, 0.06 μ		Mercuric selenide, 0.47 μ	
	Zn $K\alpha$	Se $K\alpha$	Cd $K\alpha$	Se $K\alpha$	Hg $K\alpha$	Se $K\alpha$
Soft glass	0.99	1.02	1.03	1.00	1.01	0.98
Pyrex	0.97	1.02	1.01	1.01	1.02	1.01
Quartz	1.01	0.98	1.02	1.00	1.01	1.00
Aluminum	1.04	1.00	1.00	1.00	1.00	1.02
Nickel	1.00 (counts/sec, net, 6877)	1.00 (counts/sec, net, 17,730)	1.00 (counts/sec, net, 4487)	1.00 (counts/sec, net, 18,653)	1.00 (counts/sec, net, 12,977)	1.00 (counts/sec, net, 16,549)
Copper	0.90	1.01	0.99	0.98	1.01	0.98
Zirconium	0.97	1.15	1.00	1.08	1.17	1.16

* Conditions: Molybdenum target; 60 kVP; 10 mA; PHS in Se $K\alpha$, in cadmium selenide, all others PHS out; scintillation counter; LiF as analyzing crystal.

The effect of substrates is shown in Table IV. With the use of a molybdenum target and the instrument operated at 60 kVP and 50 mA and use of the scintillation counter, the count was high. With nickel substrate as an example, the selenium counts amounted to 16,549 counts/sec (net) for HgSe to 18,653 counts/sec (net) for cadmium selenide with mercury selenide in between. Because of the second-order Cd $K\alpha$ with a 2θ angle of 31.08°, which is somewhat close to the Se $K\alpha$ first order, a pulse-height selector was used to prevent interference. The recorder chart (Figure 5) has shown that there is negligible interference caused by the presence of Cd $K\alpha$ (second order). The count rate for Cd $K\alpha$ and Hg $L\alpha_1$ under identical operating conditions amounts to 4487 counts/sec (net) for Cd $K\alpha$ and 12,777 counts/sec (net) for Hg $L\alpha_1$. All of these count rates are within the linear range of the instrument used. Among the various substrates tried, quartz, Pyrex, and ordinary soft glass show high background counts for Cd $K\alpha$ compared with nickel as a substrate. However, the net counts for Cd $K\alpha$ are the same as those with nickel as a substrate. Zirconium as a substrate gave high count rates for Se $K\alpha$ in CdSe, for Se $K\alpha$ and Zn $K\alpha$ in ZnSe, and for Se $K\alpha$ in HgSe; very little difference in count rate occurs with other substrates. By using a flow proportional counter with low efficiency for cadmium, the effect appears to be reversed. Because of the effect of substrates on the count rate of II–VI compounds, it is necessary to use a number of standards for each substrate for the determination of the thicknesses of the II–VI compounds. The use of Mylar overlay coated with these compounds for the determination of film thickness eliminates the necessity of preparing these standards.

Glang *et al.*[12] pointed out in their study of vacuum evaporation of CdTe that, to avoid condensation of free tellurium, the substrate temperature has to be greater than 150°C. The ratio of the count rate of X-ray fluorescence of the constituents in II–VI compound is one of the means used to characterize the stoichiometric relationship between the constituents in the compound. Other chemical methods for sulfide have previously been described.[13] The film of HgSe has been carefully examined for free mercury and possible decomposition. There are a number of chemical methods which could be used as described by Feigl.[14] Although the starting material of HgSe shows the

Table V. Determination of Thickness of II–VI Compounds by the Use of Mylar Overlay Coated with Known Thickness of Same Material

II–VI compound	Substrate	Thickness of coating: By other method, μ	Thickness of coating: By X-ray fluorescence, μ	Thickness determined by
ZnSe*	Aluminum	0.34	0.33	Zn $K\alpha$
		0.34	0.38	Se $K\alpha$
CdSe*	Quartz	0.12	0.15	Se $K\alpha$
HgSe†	Glass	0.32	0.31	Se $K\alpha$
		0.32	0.34	Hg $K\alpha$
CdS*	Pyrex	0.95	0.90	Cd $K\alpha$

* Conditions: 60 kVP; 50 mA; chromium target; No. 7 proportional counter; PHS out; no vacuum in spectrometer; analyzing crystal, LiF.
† Same conditions as in Table II.

presence of traces of free mercury by one of these methods, the palladium method, the deposited film of HgSe was found to be free from elemental mercury.

For analysis by the X-ray fluorescence method, it is necessary to have a set of standards prepared in the same way and containing material of the same composition as that of the unknown. This is necessary because fluorescence X-ray is influenced by such phenomena as enhancement and absorption. Because Mylar does not interfere with wavelengths smaller than 1.4 Å, Mylar film coated with known thicknesses of II–VI compounds could be used to determine the thickness of the unknown film. The count rate of the Mylar film with known thickness on the same substrate as that of the unknown will enable one to calculate the thickness of the latter film. Some of the results obtained are shown in Table V.

SUMMARY AND CONCLUSION

1. Operating conditions have been given for the determination of the film thickness of II–VI compounds, based on the spectra lines of these compounds.
2. Neutron activation and laser microprobe for the intensity of emission lines of II–VI compounds have been used to supplement the study of film thickness by the X-ray fluorescence method.
3. When a substrate such as aluminum was used for the vacuum deposition of mercuric sulfide, there appeared to be an appreciable decomposition, as evidenced by the ratio of the count rates. Samples for the determination of thickness as reported in this paper gave no trace of free mercury by the palladium chloride method.
4. In general, linearity of X-ray fluorescence and film thickness is governed by the mass absorption coefficient of the fluorescence, primary X-ray, and the II–VI compounds under investigation. From experimental evidence of film thickness determined by S $K\alpha$ or Cd $L\beta_1$, the linear range is estimated to be less than 1 μ with chromium radiation; for Se $K\alpha$, Zn $K\alpha$, or Hg $L\alpha_1$, greater than 1 μ with molybdenum as the target; and for Cd $K\alpha$ or Te $K\alpha$, greater than 20 μ with tungsten radiation.
5. Besides various substrates used for vacuum deposition of II–VI compounds, the use of an organic film of polymeric material such as Mylar for the coating of II–VI compounds has been successful.

6. Coating of films of II–VI compounds on polymeric material, when the thickness has been determined by other methods, can be used to determine the thickness of an unknown film of the same compound provided the polymeric material does not seriously affect the count rate of the spectral line of the II–VI compound films.

7. It has been found that, for spectra lines with wavelengths shorter than 1.4 Å, such polymeric film does not seriously affect the count rate. Thus, such films can be placed over a substrate of the same material that an unknown film is on and the thickness of the unknown film can thus be determined. Standards normally prepared for a particular substrate for the determination of an unknown film may be dispensed with when this procedure is used.

ACKNOWLEDGMENT

The author would like to thank the following gentlemen and research organization for assistance: Mr. D. C. Reynolds for active interest in this work; Captain Max F. Thompson of Nuclear Reactor, AFIT, for neutron activation analysis; Mr. S. F. Brokeshoulder of Aerospace Medical Research Laboratories for the emission analysis of cadmium with the laser microprobe; Mr. Roy Hayslett for assistance with the vacuum deposition of films; Clevite Electronic Research Division for furnishing a number of samples; and Dr. H. W. Pickett of the General Electric X-Ray Department for his comments on X-ray fluorescence analysis.

REFERENCES

1. D. C. Reynolds, G. Leies, L. L. Antes, and R. E. Marburger, "Photovoltaic Effect in Cadmium Sulfide," *Phys. Rev.* **96**: 533, 1954.
2. Frank L. Chan, "Determination of Film Thickness of Cadmium Sulfide Used in Energy Conversion with Soft and Hard X-Rays," paper presented at the 18th Annual Mid-America Symposium on Spectroscopy, May 15–18, 1967, Chicago, Ill.
3. Frank L. Chan, "A Study of Silicon Determination in Organo–Silicon Compounds by X-ray Fluorescence with Vacuum Spectrograph," in: P. W. Shallis (ed.), *Proceedings of the SAC Conference, Nottingham*, W. Haffer and Sons, Ltd., Cambridge, England, 1965, p. 89.
4. Frank L. Chan, "An Apparatus for the Analysis of Liquid Samples by the X-ray Fluorescence Method with Vacuum Spectrograph," in: L. R. Pearson and E. L. Grove (eds.), *Developments in Applied Spectroscopy, Vol. 5*, Plenum Press, New York, 1966.
5. L. S. Birks, *X-Ray Spectrochemical Analysis*, Interscience Publishers, Inc., New York, 1959, p. 63.
6. N. W. H. Addink, "Optical and X-Ray Spectroscopy," in: W. Wayne Meinke and Bourdon F. Scribner (eds.), *Trace Characterization, Natl. Bur. Std.* (*U.S.*) *Monograph* **100**: 121, 1967.
7. H. A. Liebhafsky and P. D. Zemany, "Film Thickness by X-ray Emission Spectrography," *Anal. Chem.* **28**: 455, 1956.
8. T. N. Rhodin, "Chemical Analysis of Thin Films by X-ray Emission Spectrograph," *Anal. Chem.* **27**: 1857, 1955.
9. Frank L. Chan, "Detection Confirmation and Determination of Trace Amounts of Selenium by X-ray Methods," in: W. M. Mueller, G. R. Mallett, and M. J. Fay (eds.), *Advances in X-Ray Institute, Vol. 7*, Plenum Press, New York, 1964.
10. H. A. Liebhafsky, H. G. Pfeiffer, E. H. Winslow, and P. D. Zemany, *X-Ray Absorption and Emission in Analytical Chemistry*, John Wiley and Sons, Inc., New York, 1960.
11. L. S. Birks, E. J. Brooks, and H. Friedman, "Fluorescent X-Ray Spectroscopy," *Anal. Chem.* **25**: 692, 1953.
12. Reinhard Glang, John G. Kren, and William J. Patrick, "Vacuum Evaporation of Cadmium Telluride," *J. Electrochem. Soc.* **110**: 407, 1963.
13. Frank L. Chan, "Spot Test Detection of Sulfide in Minerals," *Anal. Chim. Acta* **37**: 391, 1967.
14. Fritz Feigl, *Spot Tests in Inorganic Analysis*, Elsevier Publishing Co., Amsterdam, 1958, p. 364.

RARE-EARTH ANALYSES BY X-RAY-EXCITED OPTICAL FLUORESCENCE

W. E. Burke and D. L. Wood

Bell Telephone Laboratories, Incorporated
Murray Hill, New Jersey

ABSTRACT

X-ray excitation causes rare-earth impurities in yttrium oxide and gadolinium oxide to emit intense and highly characteristic optical line fluorescence, which enables their analyses at low concentrations. The limits of detection for praseodymium, neodymium, samarium, europium, gadolinium, terbium, dysprosium, holmium, erbium, thulium, and ytterbium in these two oxides range from 1 to 100 parts per billion (ppb). In other rare-earth oxides which have been investigated, the fluorescent intensities are greatly reduced. Successful analyses can be made only by dilution in high-purity Y_2O_3. This dilution raises the detection limits for rare earths in these other oxides to the part per million range. X-rays from the chromium target of a dual-target X-ray tube are about two times more efficient in exciting rare-earth optical fluorescence than are the tungsten-target X-rays, even though the total energy output of the chromium target is only about one-third that of the tungsten target. With either target material, the rare-earth intensities vary linearly with the X-ray tube current, but a plot of intensity *versus* the square of the accelerating potential is not linear; it drops off at higher voltages.

INTRODUCTION

When suitably excited, solid-state fluorescent materials emit optical radiation which is often characteristic of the activator ion. This is demonstrated by the trivalent rare-earth ions, which are easily excited to give off their highly individual sharp line spectra. When X-rays impinge on Y_2O_3 and Gd_2O_3 crystalline powders, the rare-earth impurity ions are so efficiently excited that their characteristic spectral lines can readily be detected at very low levels. Typical rare-earth spectra are shown in Figures 1 to 4. Each spectrum was recorded with the rare earth at a concentration of 100 ppm in Y_2O_3.

Low, Makovsky, and Yatsiv,[1,2] were the first to propose that X-ray-excited optical fluorescence be used to determine rare-earth impurities in crystals, and Linares *et al.*[3] reported the detection of europium, gadolinium, terbium, and dysprosium in Y_2O_3 by X-ray-excited optical fluorescence. Walters *et al.*,[4] evidently concurrently with our work, also extended Linares' procedure and detected samarium, europium, gadolinium, terbium, dysprosium, erbium, and thulium.

Direct bombardment of the sample with high-speed electrons is also a good method of excitation. Terol and Aguino[5] successfully used cathodoluminescence to determine rare-earth concentrations in yttrium oxide and lanthanum oxide matrices. Another communication by these same authors[6] gives the detection limits for rare-earth elements in various lanthanum oxy-compounds when subjected to cathodoluminescence.

Karjakin *et al.*[7] have also successfully used ultraviolet radiation as an excitation mechanism for the quantitative detection of rare earths by fluorescence.

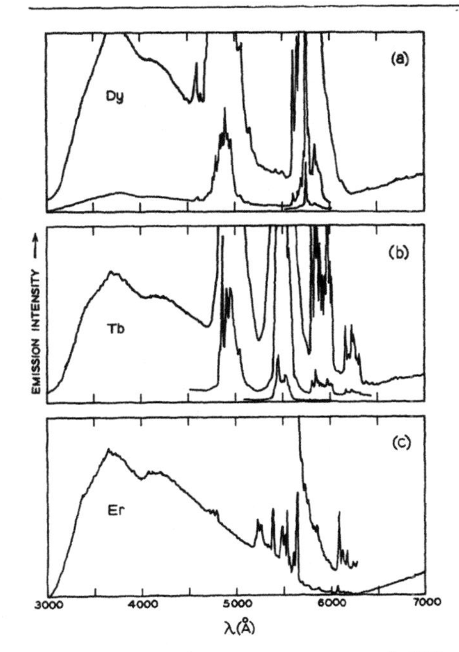

Figure 1. Luminescence spectra of 100 ppm (a) dysprosium, (b) terbium, and (c) erbium in Y_2O_3.

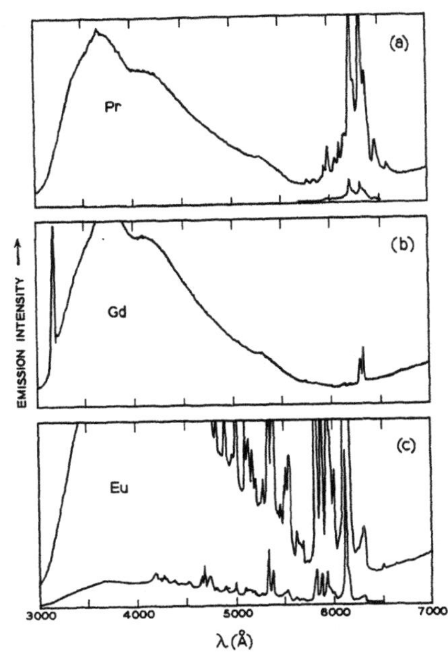

Figure 2. Luminescence spectra of 100 ppm (a) praseodymium, (b) gadolinium, and (c) europium in Y_2O_3.

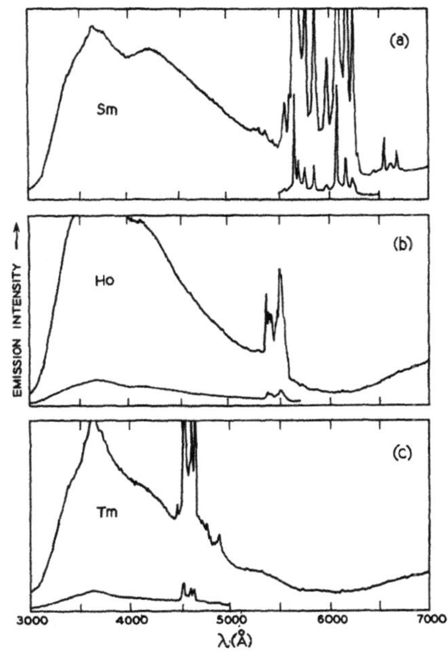

Figure 3. Luminescence spectra of 100 ppm (a) samarium, (b) holmium, and (c) thulium in Y_2O_3.

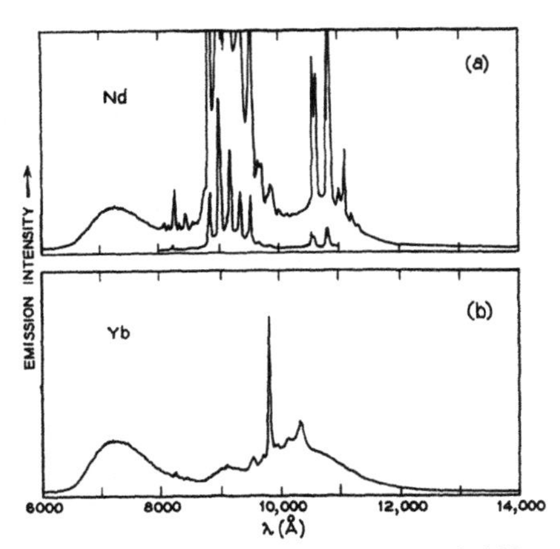

Figure 4. Luminescence spectra of 100 ppm (a) neodymium and (b) ytterbium in Y_2O_3.

Our communication describes a further extension of Linares' procedure which enables the quantitative analyses of these and other rare earths in both Y_2O_3 and Gd_2O_3 with the following detection limits in parts per billion: praseodymium, 100; neodymium, 1; samarium, 10; europium, 10; gadolinium, 100; terbium, 1; dysprosium, 1; holmium, 100; erbium, 100; thulium, 10; and ytterbium, 100. Yttrium, lanthanum, cerium, promethium, and lutetium were not sought by this procedure because yttrium, lanthanum, and lutetium do not have characteristic line fluorescence, promethium is radioactive (not found in nature), and the cerium lines are emitted in the infrared region which is beyond the range of our detectors.

EXPERIMENTAL

The apparatus shown in Figure 5 is similar to that of Linares.[3] The excitation unit consists of a dual-target X-ray tube operated at 45 kV and 39 mA from a suitable power supply. The tube (General Electric EA-75) contains two electron sources and two targets, of tungsten and chromium, so that the excitation may be electrically selected with a high-voltage switch.

The powdered sample is packed into a 0.5-in.-diameter by 0.03125-in.-deep recess in an aluminum sample holder and then positioned so that its surface is about 4 cm from the X-ray window. The X-rays impinging upon the sample excite rare-earth fluorescence, which enters the monochromator and is dispersed into its component wavelengths. These optical wavelengths are then measured by the detector.

A 15,000 lines/in. grating is used with a 0.5-meter Ebert scanning monochromator. A liquid-nitrogen-cooled photomultiplier detector with S-1 cathode is used for the longer-wavelength neodymium and ytterbium radiations, and a second photomultiplier with S-11 cathode surface is used for the remaining rare earths. The photomultiplier current is measured by an electrometer amplifier and displayed on a strip chart recorder.

PROCEDURE

The sample to be analyzed (either Y_2O_3 or Gd_2O_3) is scanned "as received" with the S-11 detector (from 3000 to 13,000 Å) in order to obtain a rough estimate of the various rare-earth elements present. The wavelengths ordinarily used to detect the various elements are given in Table I. As these wavelengths have been selected to give the greatest intensity with the least interference, some of these wavelengths are not the most intense for a given rare earth, and several must still be corrected when interference from certain other rare earths occurs.

When working with the first order of a grating at the 8980-Å neodymium and 9800-Å ytterbium wavelengths detected by the S-1 surface tube, any second-order lines of half the given wavelength may also produce output signals. These interferences are prevented by using a red filter (Corning C.S. 2-58) which blocks out radiation shorter than 6200 Å.

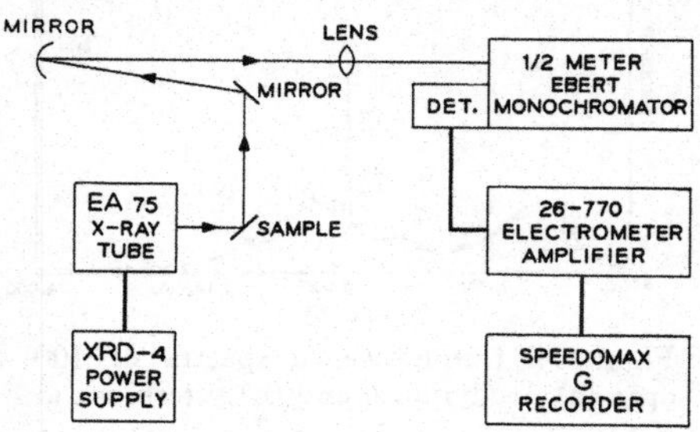

Figure 5. Schematic of apparatus for X-ray excitation of luminescence.

Table I. Rare-Earth Wavelengths for S-11 Detector*

Rare earth	Wavelength, Å
Gadolinium	3160
Thulium	4560
Erbium	5260
Holmium	5400
Terbium	5450
Dysprosium	5730
Samarium	6090
Europium	6140
Praseodymium	6480

* Use S-1 detector for neodymium and ytterbium at 8980 and 9800 Å, respectively.

After thus roughly estimating the amount of each rare earth present, stock solutions of rare earths dissolved in 19% HCl are added to the unknown sample (also dissolved in 19% HCl) to approximately double the amount of each rare-earth impurity present in the original sample; this produces additive standards. To minimize interference from overlapping peaks, neodymium, samarium, europium, holmium, thulium, and ytterbium are added to one Y_2O_3 or Gd_2O_3 aliquot; praseodymium, gadolinium, terbium, dysprosium, and erbium are added to another aliquot.

The two prepared solutions and a sample blank are evaporated to dryness on a hot plate, then transferred to platinum crucibles, and heated overnight in a furnace at 1000°C or higher, while a gentle stream of air is blown into the furnace. After cooling, the spectra of the blank and the two additive samples are recorded with the apparatus shown in Figure 5.

Dissolving in HCl, precipitating, and firing to the oxide are necessary because the rare-earth impurities must be in the lattice in order to produce the same line spectrum as the blank. A mechanical mix of Y_2O_3 and the rare-earth impurities will not give the same fluorescence as will the rare-earth ions in the Y_2O_3 lattice.

Because the absolute intensity of a given rare-earth line may depend rather strongly on the details of preparation of the sample, it is necessary to measure each rare-earth line intensity relative to that of an internal standard. It has been found that, for each group of elements mentioned in connection with the addition technique, one of the elements from the other group can be used for the internal standard, and no such standard need be added. Dysprosium is usually present in Y_2O_3 and Gd_2O_3 and has been the best for an internal standard for the group including neodymium, samarium, europium, holmium, thulium, and ytterbium. For the other group, including praseodymium, gadolinium, terbium, dysprosium, and erbium, europium has been present in the blank and is useful for an internal standard. Instead of line peak intensity, then, the ratio of peak intensity to that of the internal standard in the blank is used to calculate the amount of a given ion present in the sample. For the additive standards, the ratio of the element being estimated is corrected by subtracting the ratio of this element in the blank, and this value gives the slope of the curve of the intensity ratio *versus* concentration. The slope is then used to calculate the amount of the element in the blank, the validity of the linear relation discussed below being assumed. Mathematically, the following relationship is used. If C_{su} is the concentration of a given element of unknown concentration added to the blank to make a standard, I_{bi} is the intensity of the internal-standard line in the blank, I_{bu} is that of the unknown element in the blank, and I_{si} and I_{su} are the corresponding

Table II. Rare-Earth Analysis of Y_2O_3 and Gd_2O_3 by X-Ray-Excited Optical Fluorescence

In parts per million

	Praseo-dymium	Neodym-ium	Samar-ium	Euro-pium	Gado-linium	Terbium	Dyspro-sium	Holmium	Erbium	Thulium	Ytterbium
Y_2O_3 Sample											
1	ND*	3.2	0.13	0.21	ND	0.01	0.02	ND	ND	ND	ND
2	2	9.7	0.99	4.4	4	1.2	40	ND	21	ND	11.5
3	8	8.5	0.72	0.23	ND	2.2	10	ND	24	ND	2.1
4	ND	0.11	0.09	0.78	ND	0.03	0.03	ND	ND	ND	ND
5	ND	1.3	13	ND	1	0.5	2.4	ND	ND	ND	26
6	ND	1.1	0.72	0.17	1	0.7	36	80	ND	ND	ND
Gd_2O_3 Sample											
7	0.29	7.5	ND	3.6		0.20	7.1	ND	ND	ND	12
8	0.24	0.66	0.21	0.24		0.19	0.26	ND	ND	ND	ND
9	0.35	0.81	ND	1.2		0.30	0.18	ND	ND	ND	ND
10	0.35	0.50	ND	0.85		0.39	0.31	ND	ND	ND	0.32
11	0.43	0.51	0.16	0.68		0.27	0.22	ND	ND	ND	ND
Avg. detection limits	0.10	0.001	0.01	0.01	0.10	0.001	0.001	0.10	0.10	0.01	0.10

*ND = not detected.

intensities of the internal-standard line and unknown line in the additive standards, then the concentration of the unknown in the blank, C_{bu} is given by

$$C_{bu} = \frac{I_{bu}/I_{bi}}{\dfrac{1}{C_{su}}\left(\dfrac{I_{su}}{I_{si}} - \dfrac{I_{bu}}{I_{bi}}\right)} = \frac{I_{bu}C_{su}}{I_{bi}\left(\dfrac{I_{su}}{I_{si}} - \dfrac{I_{bu}}{I_{bi}}\right)}$$

and the concentration in the blank is determined.

RESULTS

The analyses of some samples used for making high-purity single crystals are given in Table II. Of the six Y_2O_3 samples, the first four had been specially purified, and sample 1 has less than 4 ppm total rare-earth impurities. The other two samples were commercial lots, and their contamination is higher for the rare earths. All of the Gd_2O_3 samples had been specially purified, and the level of rare-earth impurities is very low. The detection limits for the method are also given in the table, and "ND" for a given element means that element is present with a concentration less than the detection limit noted.

Utilization of this procedure presupposes straight-line calibration curves, and we have found that all of the rare earths determined by this method have straight line calibrations, at least up to 100 ppm. Figure 6 shows a typical calibration curve for dysprosium in Y_2O_3. The Y_2O_3 used for all calibration curves is sample 1 in Table II. The average error of the procedure as determined by replicate analyses is $\pm 30\%$ of the amount present. Variations of fluorescence efficiency due to unintentional differences in preparation are probably the greatest sources of error.

A comparison of the detection limits of this procedure with those of the emission spectrograph is given in Table III. Emission spectrographic data were obtained from Nash.[8] Although both atomic absorption and X-ray fluorescence are also faster methods for rare-earth analyses than is X-ray-excited optical fluorescence, their detection limits are higher. Slavin[9] in his review article on atomic absorption shows rare-earth detection

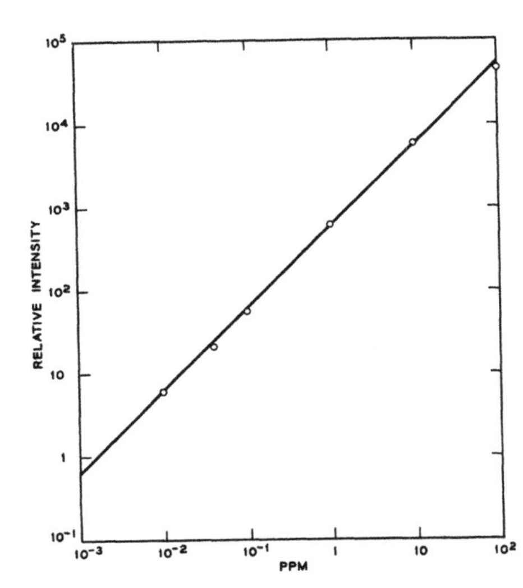

Figure 6. Calibration curve for analysis of dysprosium in Y_2O_3 by X-ray excitation of luminescence.

Table III. Detection Limits (ppb) for Rare Earths in Y_2O_3

Rare earth	Emission spectrograph	X-ray-excited optical fluorescence
Lanthanum	5,000	Probably not detectable
Cerium	40,000	Not sought
Praseodymium	25,000	100
Neodymium	20,000	1
Promethium	Radioactive, not sought	
Samarium	20,000	10
Europium	10,000	10
Gadolinium	3,000	100
Terbium	15,000	1
Dysprosium	1,000	1
Holmium	30,000	100
Erbium	5,000	100
Thulium	5,000	10
Ytterbium	500	100
Lutetium	500	Probably not detectable

limits ranging from 0.04 μg/ml of H_2O solution for ytterbium to 10 μg/ml for praseodymium. These correspond to 400 ppb of ytterbium and 100,000 ppb of praseodymium in a solid sample if a 10% solution is assumed. Although activation analysis[10] and spark-source mass spectrometry[11,12] are both capable of rare-earth analyses in the parts per billion range, both methods require more expensive instrumentation and facilities. Activation analysis also may incur separation problems when complex mixtures are analyzed.

DISCUSSION OF RESULTS

A twofold intensity increase is obtained by dissolving the Y_2O_3 in and recrystallizing it from HCl instead of HNO_3. The HCl reaction proceeds as follows:

$$Y_2O_3 + 6HCl + 3H_2O \rightarrow 2YCl_3 \cdot 6H_2O \xrightarrow{\Delta} 2YCl_3 + 6H_2O\uparrow \xrightarrow{(\frac{3}{2}O_2)} Y_2O_3 + 3Cl_2\uparrow$$

The $YCl_3 \cdot 6H_2O$ obtained by evaporating the solution is heated to at least 1000°C in a slow stream of air for 12 to 16 hours to convert it to the oxide.

As shown in Table IV, both terbium and samarium (each at 100 ppm in Y_2O_3) reached maximum intensity after having been in the furnace at 1000°C between 1 and

Table IV. Temperature and Time Effect on Terbium and Samarium Peak Intensities

	Relative peak intensity	
	100 ppm terbium, 5450 Å	100 ppm samarium, 6090 Å
Dried on hot plate	280	0
1 hr at 1000°C	17,000	4000
16 hr at 1000°C	26,500	5700
40 hr at 1000°C	26,600	5690

16 hours. A shorter time than 16 hours could probably be used, but 16 hours (overnight) is a convenient time period. The low emission intensity observed after drying only on the hot plate is due to the low fluorescent efficiency of rare-earth ions in a hydrated crystalline environment or in the presence of chloride ligands.

We compared the relative efficiency of chromium and tungsten targets for rare-earth excitation and found that the chromium-target X-rays are about two times as efficient as are the tungsten-target X-rays, although the total energy output of the chromium target is only about one-third that of the tungsten target. The chromium-target tube has more energy at longer wavelengths (>2 Å) than does the tungsten target[13] (Figure 7) because the $K\beta$ and $K\alpha$ lines of chromium occur at 2.08 and 2.29 Å, respectively, while the W $L\beta$ and $L\alpha$ lines are at 1.28 and 1.48 Å, respectively. One possible explanation for this greater efficiency of longer-wavelength X-rays may involve their lower penetrating power, which causes more of their energy to be expended at the sample surface, whereas the shorter-wavelength X-rays penetrate deeper into the sample. As the rare-earth lines utilized for this procedure are all in the ultraviolet to near-infrared range, only those optical fluorescence lines given off near the surface of the powder sample would be emitted and the remainder would be scattered before they reached the surface. Results obtained with aluminum-foil X-ray filters substantiate the above. Specifically, when the dual-target X-ray tube is used with no filter, the Cr:W intensity ratio is 2.06 for terbium radiation. When 0.08 mm aluminum foil is placed over the tube window as an absorber of the longer-wavelength X-rays, the ratio is 0.58; with 0.13-mm aluminum, the ratio is 0.55. The 0.08-mm aluminum absorbs 99.9% of all radiation longer than 2.08 Å, which eliminates the Cr $K\alpha$ and $K\beta$ lines before they reach the sample. The low ratio with an aluminum-foil filter indicates that the long-wavelength X-rays are the most effective. As chromium-target tubes are more efficient for rare-earth excitation than tungsten-target tubes are, other low-atomic-number X-ray target materials, such as titanium, should be useful.

We have also investigated the effect of X-ray-tube current and accelerating potential on rare-earth emission intensity (Figures 8 and 9). The total power[14] or integrated intensity of the X-ray beam is $I = 1.4 \times 10^{-9} iZV^2$, where I is the intensity in watts, i is the electron current in amperes, Z is the atomic number of the target material, and V is the potential difference across the tube in volts. The original sources for this formula are Ulrey[15] and Wagner and Kulenkampff.[16,17] As shown in Figure 8, the rare-earth intensity does vary linearly with the current. However, the intensity *versus* the square of the accelerating potential (Figure 9) does not result in a straight line but deviates at the higher potentials. This is to be expected since raising the voltage produces a greater

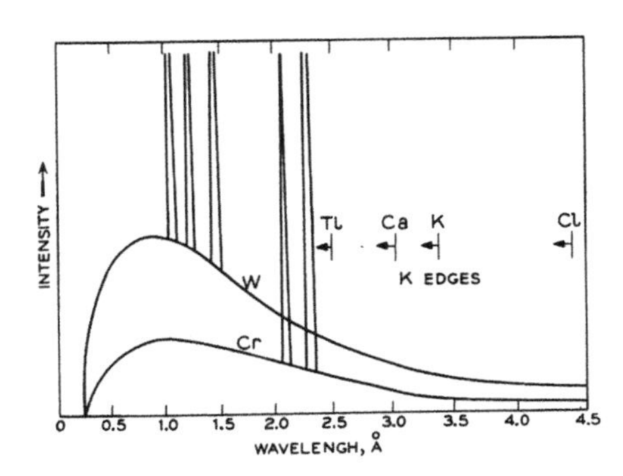

Figure 7. Typical X-ray spectra of chromium- and tungsten-target X-ray tube (from F. Bernstein[13]).

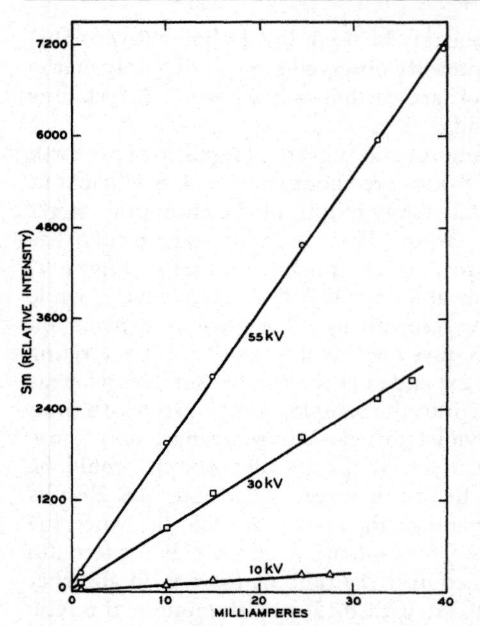

Figure 8. Emission intensity of 100 ppm samarium in Y_2O_3 *versus* chromium X-ray-tube current.

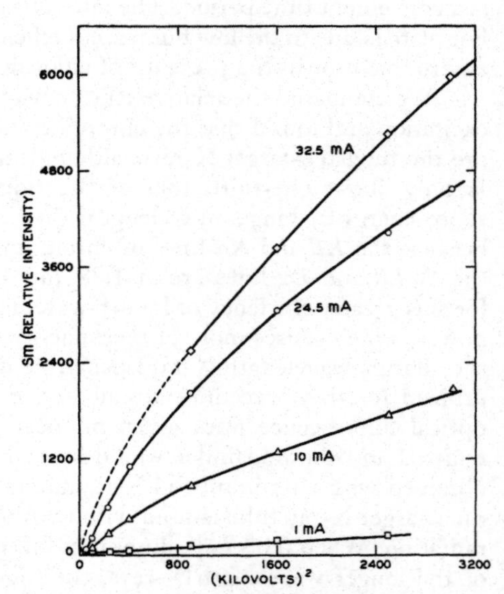

Figure 9. Emission intensity of 100 ppm samarium in Y_2O_3 *versus* square of chromium X-ray-tube accelerating potential.

proportion of short wavelength and more penetrating radiation. Because the longer wavelengths have higher efficiency for rare-earth emission, there should be a decrease of efficiency with increase in voltage, as observed.

Nd_2O_3, Eu_2O_3, Tm_2O_3, and Yb_2O_3 samples have also been analyzed for rare-earth impurities by this procedure. Severe quenching effects, however, require as much as a thousandfold dilution of each of these oxides in a Y_2O_3 matrix before a quantitative rare-earth impurity analysis can be made. This, in effect, means that the detection limits for rare-earth impurities in these rare-earth oxide hosts are raised by a factor of 1000 to the parts per million range. A preliminary examination of Pr_6O_{11}, Sm_2O_3, Tb_4O_7, Dy_2O_3, Ho_2O_3, and Er_2O_3 also indicates quenching problems in these host lattices.

ACKNOWLEDGMENTS

The authors are indebted to S. S. De Bala for designing and producing the experimental apparatus and to R. C. Linares, L. G. Van Uitert, T. Y. Kometani, D. L. Malm, and D. L. Nash for helpful discussions.

REFERENCES

1. J. Makovsky, W. Low, and S. Yatsiv, "Excitation of Optical Fluorescence Spectra of Transition Elements by Means of X-rays," *Phys. Letters* **2**: 186, 1962.
2. W. Low, J. Makovsky, and S. Yatsiv, "Fluorescence of Rare Earth Ions Upon X-ray Excitation," in: P. Grivet and M. Bloembergen (eds.), *Quantum Electronics III, Vol. II*, Columbia University Press, New York, 1964, p. 655.
3. R. C. Linares, J. B. Schroeder, and L. A. Hurlbut, "The Applications of X-ray-excited Optical Fluorescence to Analytical Chemistry," *Spectrochim. Acta* **21**: 1915, 1965.

4. R. M. Walters, J. F. Cosgrove, and D. J. Bracco, "X-ray-excited Optical Fluorescence in the Analysis of Trace Rare Earths," Paper 24 at 1966 NBS Symposium on Trace Characterization, Gaithersburg, Maryland.
5. S. Terol and F. F. Aguino, "The Determination of Rare Earth Traces in Yttrium and Lanthanum Compounds by Cathodoluminescence," *Proc. Intern. Conf. Luminescence*, Budapest, **11**: 8, 1966.
6. S. Terol and F. F. Aguino, *2. Intern. Symp. Reinstoffe in Wissenschaft und Technik, Dresden*, 1965 (to be published).
7. A. V. Karjakin, L. J. Anikina, and L. A. Filatkina, *2. Intern. Symp. Reinstoffe in Wissenschaft und Technik, Dresden*, 1965 (to be published).
8. D. L. Nash, "The Determination of Trace Amounts of Rare Earth Impurities in Yttrium Oxide," *Appl. Spectry.* **22**: 101, 1968.
9. W. Slavin, "Atomic Absorption Spectroscopy—a Critical Review," *Appl. Spectry.* **20**: 281, 1966.
10. R. C. Koch, *Activation Analysis Handbook*, Academic Press, New York, 1960.
11. A. J. Ahearn, "Spark Source Mass Spectrometric Analysis of Solids," paper presented at the 1966 NBS Symposium on Trace Characterization, Gaithersburg, Maryland.
12. D. W. Oblas, D. J. Bracco, and D. Y. Yee, 'Trace Analysis in Oxidic Matrices by Solids Mass Spectroscopy," paper 68 at 1966 NBS Symposium on Trace Characterization, Gaithersburg, Maryland.
13. F. Bernstein, "Relationship Between X-ray Tube Target Materials and X-ray Emission Intensities," in: E. N. Davis (ed.), *Developments in Applied Spectroscopy, Vol. 4*, Plenum Press, New York, 1965, p. 49.
14. H. A. Liebhafsky, H. G. Pfeiffer, E. H. Winslow, and P. D. Zemany, *X-ray Absorption and Emission in Analytical Chemistry*, John Wiley & Sons, Inc., New York, 1960, p. 8.
15. C. T. Ulrey, "An Experimental Investigation of the Energy in the Continuous X-ray Spectra of Certain Elements," *Phys. Rev.* **11**: 401, 1918.
16. E. Wagner and H. Kulenkampff, "The Intensity of Reflection of Röntgen Rays of Various Wavelengths," *Ann. Physik* **68**: 369, 1922.
17. H. Kulenkampff, "Continuous Röntgen Spectra," *Ann. Physik* **69**: 548, 1922.

DISCUSSION

E. Tomasi (Molybdenum Corp. of America): I should like to find out what the effects of non-rare-earth elements are on this particular technique.

W. E. Burke: We haven't investigated the effects non-rare-earth elements have on this procedure, but we believe the internal standard procedure we use would probably compensate for this and any other variables.

E. Tomasi: Also, are the sensitivities equal for all three matrices?

W. E. Burke: The sensitivity of any one rare earth is about the same in the three matrices (Y_2O_3, Gd_2O_3, and La_2O_3). However, the sensitivities given are average sensitivities which may show as much as a twofold variation from sample to sample. For example, inadvertent procedural variations may change the crystallite size of the sample and thus cause a sensitivity change. That is why we prefer to use an internal-standard procedure.

E. Tomasi: Thank you.

NONDISPERSIVE X-RAY FLUORESCENT SPECTROMETER

W. Barclay Jones* and Robert A. Carpenter†

Technical Measurement Corporation
Special Products Division, San Mateo, California

ABSTRACT

Recent advances in semiconductor particle detector resolutions along with new electronic circuitry associated with these detectors make possible their application in nondispersive elemental analysis. The use of radioactive sources for exciting the characteristic X-rays provides highly stable systems which can be used to accumulate data for prolonged periods. Due to the inherent stability of the detector and the excitation source, the only limitation in sensitivity is the ability to accumulate statistics above the background of scattered counts. Since this method of analysis is nondispersive, it has the capacity to determine many elements simultaneously. Solutions composed of mixtures of three or four elements were studied. The elements selected were bromine, rubidium, and strontium. These elements exhibit wide variations in mass absorption coefficients for the various characteristic X-rays emitted. The concentrations of the elements in solution varied from 10 ppm to 5% by weight. The relative intensities of the characteristic X-ray lines were compared with the concentration of the solutions to establish sensitivity curves and to study linearity of response as well. The interelemental interference was studied and the effect was evaluated for the particular elements under study. Means were developed for predicting and correcting for matrix effects.

INTRODUCTION

Many nondispersive X-ray fluorescent spectrometers have been proposed or described in recent years.[1-9] In general, they obtain energy discrimination with selective filters, differential absorption, and variation in pulse amplitudes. These methods used singly or in various combinations have been proposed for field assay of ores,[9-11] alloy identification,[12] "on stream" analysis,[13] metal-coating-thickness measurements,[14-16] mail sorting,[17,18] and many other applications.[19-22] The sensitivity differs from element to element and from matrix to matrix, but the detectable limit has been 0.01%.[23]

Recent developments[24-26] in lithium-drifted silicon detectors and associated electronic circuitry, however, permit resolutions better than 350 eV, full-width half maximum (FWHM). The separation of the $K\alpha$ energies for elements near calcium on the periodic table is about 400 eV; for manganese and iron, this separation increases to 500 eV and is over 1 keV near tin. The resolving power of these new spectrometers permits separation of $K\alpha$ X-rays of adjacent elements in the atomic scale, as illustrated in Figure 1. Detection limits have been extended beyond 10 ppm.

The spectrometer system simultaneously accumulates data on all the reradiating elements in the specimen. Figure 2 shows a spectrum obtained in a 1-min exposure of a mixture of elements from chromium to silver. Linearity of the energy scale permits positive identification of each element in the sample.

* Present address: Physics Dept., Yale University, New Haven, Connecticut.
† Present address: Nuclear Equipment Corp., 931 Terminal Way, San Carlos, California.

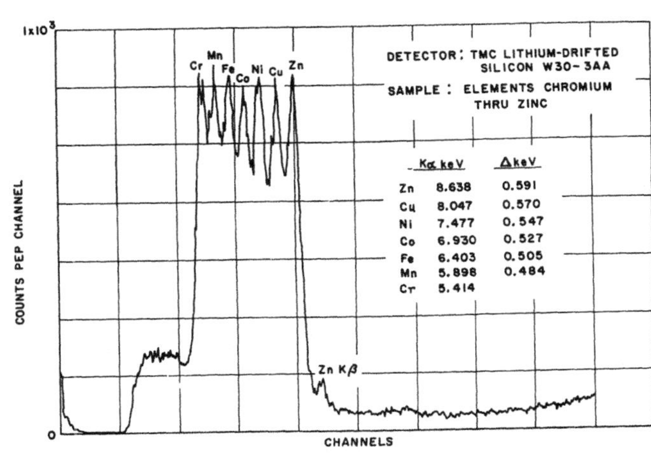

Figure 1. Photon spectrum from seven adjacent elements.

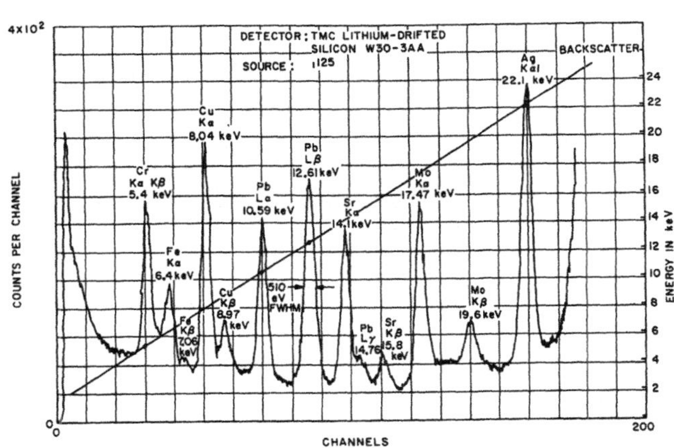

Figure 2. Spectrum of characteristic X-rays of elements used for calibration.

Quantitative analysis with this equipment is complicated by interference effects. Two sorts of interference may be distinguished, namely, superposition and fluorescent interactions. Superposition can take place only between different spectral lines as $K\beta$ of one element of $K\alpha$ of another, and is easy to detect. Fluorescent interactions may be predicted from known absorption and reradiation coefficients. Taking these factors into account permits simultaneous quantitative analysis of many elements.

APPARATUS AND PROCEDURE

Although the apparatus has been described previously,[27] we shall review its operation. The photon spectrometer is shown in Figure 3. It consists of a vacuum cryostat, liquid-nitrogen reservoir, preamplifier, ion pump, and associated power supplies. It operates in conjunction with a multichannel pulse-height analyzer (PHA) and various readout instruments. The operation of the instrument is diagrammed in Figure 4. A radioisotopic source irradiates a sample, which absorbs some or all of the radiation. Individual atoms in the sample, in turn, either reemit characteristic X-rays or scatter the primary radiation. The detector in the spectrometer absorbs both scattered and characteristic X-rays and emits electrical charge pulses proportional in amplitude to the energy absorbed. The amplifying circuit converts these pulses into voltage pulses and feeds them into the PHA. The PHA sorts and stores these pulses according to their voltages. This information is then read out on an X–Y plotter, a printed tape, or a punched tape.

Figure 3. Technical Measurement Corporation model 334 photon spectrometer.

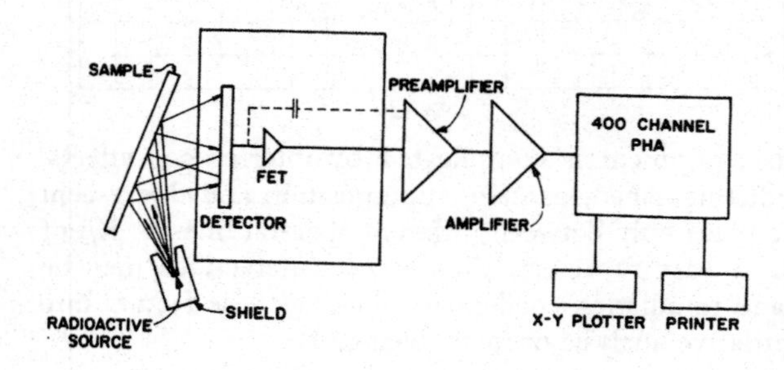

Figure 4. Functional schematic photon spectrometer.

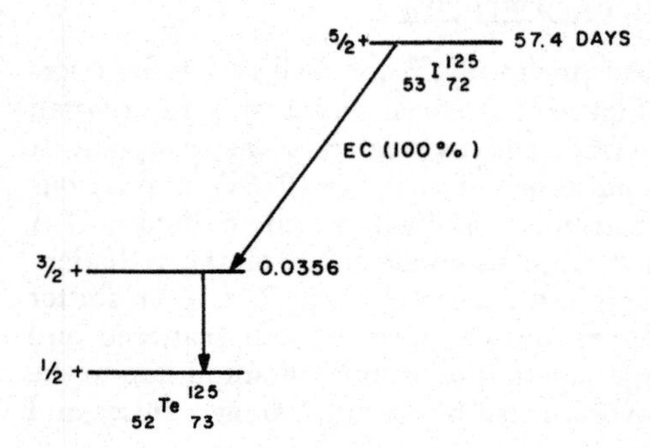

Figure 5. Decay scheme for iodine-125.

Most of the work reported in this paper was done with an iodine-125 source. Its decay scheme is shown in Figure 5; it is an electron-capture isotope which feeds a 35.6-keV-excited level of tellurium. This level is highly converted; the predominant radiation is the 27.47-keV $K\alpha$ X-ray of tellurium.[28] The half-life of iodine-125 is 57.4 days. For this experiment, we used four sources of approximately 1.5 to 2.0 mCi each, spaced equidistant around the collimating orifice. Similar arrangements have been used for this type of study by others.[29]

An iron-55 source was also used; it was fabricated by electrodeposition in an annular ring on a copper disc with a concentric 0.5-in.-diameter hole. Iron-55 is an electron-capture isotope decaying to the ground state with a 2.7-year half-life.[28] Only X-rays are emitted; the predominant radiation is at 5.9 keV.

DATA

Using an annular ring type of source presents two advantages:

1. Flux distribution is uniform over the surface of the sample.
2. The source is planar, which makes the results relatively insensitive to geometric variations.

To verify statement 1, photographic emulsions were exposed to the source. The darkening was uniform over an area of 6.5 cm^2. The area of the sample exposed to the detector was less than 1 cm^2. To verify statement 2, counting rates were determined for a 0.1% strontium chloride solution as a function of distance from the source annulus. Figure 6 shows the counting rate *versus* source position. The variations were less than 1% for 1-mm displacement, and our geometry was repeatable to better than 0.3 mm. Thus, the error from this source was negligible.

To obtain an energy calibration for the spectrometer, a mixture of chemical compounds was exposed as a single sample for 1 min, previously shown in Figure 2. The characteristic X-rays establish an energy scale from chromium through silver. In addition, this simultaneously verifies the linearity of the instrument and measures the energy resolution.

The initial work was performed with liquid samples. Solutions have the following advantages:[30]

1. No particle size effects.
2. Uniform elemental distribution.
3. Easily controlled concentrations.
4. Simplified standardization.

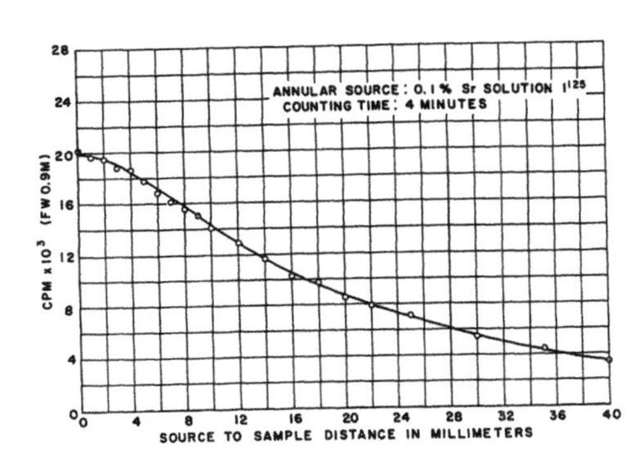

Figure 6. Counting rate for a 0.1% strontium solution *versus* distance from source.

In previous reports,[31] we established that the total integrated counting rate for any elemental species was related to the area between nine-tenths maximum limits ($A\ 0.9M$). This relation holds true in the absence of peak distortion. For this investigation, we shall use the ($A\ 0.9M$) to avoid interference between the tails of adjacent peaks. Figure 7 shows the spectrum from a solution containing 0.5% each of bromine, rubidium, and strontium. Because of overlapping, there would be large corrections on the area of each peak if the whole area were used. Integrating only between nine-tenths maximum limits eliminates this correction.

However, it is still necessary to correct for superposition. The $K\beta$ of bromine lies directly under the $K\alpha$ of rubidium. The area of the bromine $K\beta$ must be inferred by proportion from the area of the bromine $K\alpha$ or measured separately with a sample containing only bromine. Then the bromine $K\beta$ may be subtracted from the composite peak to yield the corrected $K\alpha$ rubidium peak area.

Figure 8 and Table I combine analyses of many different solution mixtures to show the effect of varying the concentration of bromine in a mixture of 0.5% rubidium and strontium. The rubidium counting rate decreases approximately 10% as the bromine concentration is increased from 0 to 0.75%. The strontium counting rate, however, decreases over 40% for this same small change in bromine concentration. Figure 9 and Table II display the effect of varying the strontium concentration from 0 to 0.75% in a mixture of 0.5% bromine and rubidium. The rubidium counting rate decreases approximately 15%, whereas the bromine counting rate is constant.

Further studies were made with the use of various steel standards. In this case, the basic matrix was iron with large variable percentages of nickel and chromium. In the initial work, we used a Bureau of Standards stainless steel, D846.[32] Three different samples were tested:

1. 1.25-in.-diameter by 0.25-in.-thick disk.
2. 1.25-in.-diameter by 0.75-in.-thick disk.
3. 0.5-in.-diameter by 0.125-in.-thick disk.

These tests verify that results are independent of sample diameter and thickness over this range of samples. All gave identical spectra within the statistical deviations expected. Figure 10 shows the superimposed spectra.

In addition, various steels were tested to determine if their spectra were distinguishable. Table III lists the steels examined and the known compositions. For many samples,

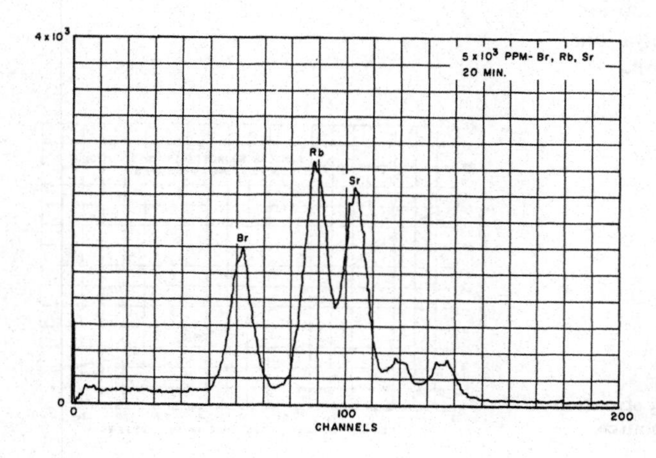

Figure 7. Spectrum from 5000 ppm (0.5%) each of bromine, rubidium, and strontium.

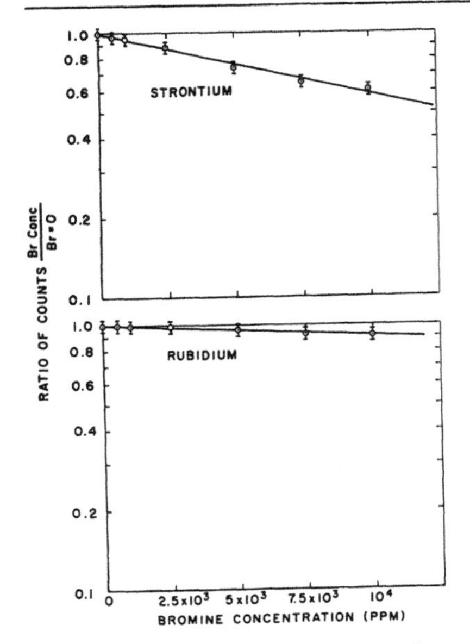

Figure 8. Effect of variations in bromine concentrations on strontium and rubidium.

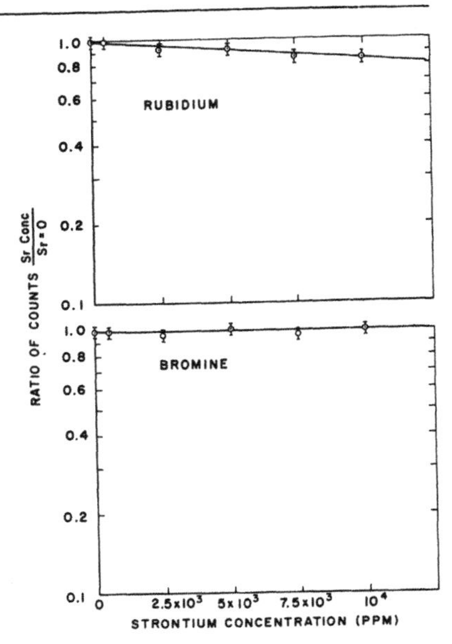

Figure 9. Effect of variations of strontium concentrations on rubidium and bromine.

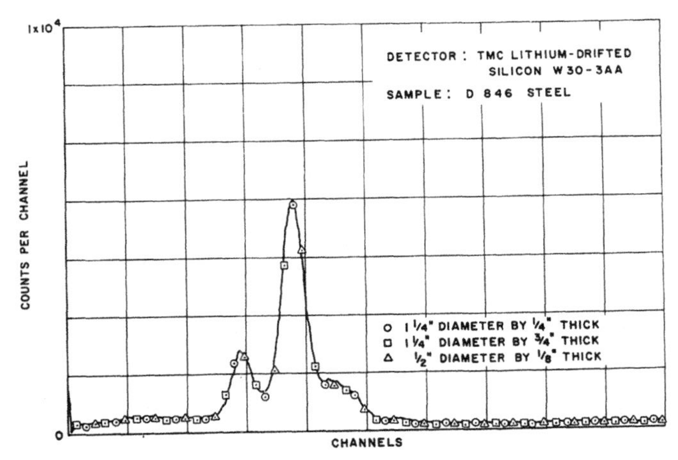

Figure 10. Photon spectrum of D846 steel showing a comparison of different sample sizes.

the statistics were inadequate to give accurate relative concentrations of all trace elements. The spectra were sufficiently distinct, however, to permit rapid identification of each sample. Compare the spectra in Figures 11, 12, 13, and 14. These analyses required only 10 min each. For molybdenum and niobium, even with the poor statistics due to the low counting rate, quantitative results were obtained to within 10% for 0.01% concentrations. The linearity of response of the system to different concentrations of molybdenum and niobium are shown in Figures 15 and 16.

Table I. Effect of Bromine Concentration on Strontium and Rubidium

Concentration: bromine, 5×10^3 ppm; rubidium, 5×10^3 ppm

Date	Count time, min	Br conc.	Rb total counts	Sr total counts	Background		Rb, cpm	Sr, cpm	Br, cpm	Correction*			Relative counting error, cpm†			Ratios, cpm	
					Rb	Sr				Rb	Sr	Br	Rb	Sr	Br	Rb $\frac{\text{Br conc.}}{\text{Br} = 0}$	Sr $\frac{\text{Br conc.}}{\text{Br} = 0}$
5-1	20	0	8083	9642	360	320	386	466	0.0	386	466	0.0	4.6	50.	0.0	1.0	1.0
5-1	20	10^2	8007	9835	360	320	382	476	6.3	382	476	6.3	4.6	5.1	1.5	1.0	1.0
4-27	10	2.5×10^2	4233	4848	185	170	405	468	17.7	377	437	16.5	6.7	7.1	2.4	1.0	0.935
5-1	20	5×10^2	7802	9338	411	320	370	446	25.0	370	446	25.0	4.5	5.0	1.8	0.96	0.96
5-1	20	10^3	8186	9131	505	320	384	441	50.9	384	441	50.9	4.7	4.9	2.1	1.0	0.95
5-3	20	2.5×10^3	8098	8216	714	320	369	395	124.0	382	409	124.0	4.7	4.6	2.9	1.0	0.88
4-26	20	5×10^3	8895	7832	1121	340	389	375	261.0	356	345	240.0	5.0	4.5	3.9	0.94	0.74
5-3	20	7.5×10^3	8098	6305	1270	320	341	300	332.0	350	310	343.0	4.9	4.1	4.3	0.92	0.67
5-3	20	10^4	8266	5837	1555	320	336	276	432.0	346	285	446.0	5.0	3.9	4.9	0.91	0.61

* Correction for decay of I^{125} source.
† Counts per minute.

Table II. Effects of Strontium Concentration on Bromine and Rubidium

Concentration: bromine, 5×10^3 ppm; rubidium, 5×10^3 ppm

Date	Count time, min	Br conc.	Br total counts	Rb total counts	Background		Br, cpm	Rb, cpm	Sr, cpm	Counting error, cpm†			Ratios, cpm	
					Br	Rb				Br	Rb	Sr	$\frac{Br}{Br\ (Sr20)}$	$\frac{Rb}{Rb\ (Sr20)}$
5-4	20	0	5020	8750	378	1015	232	387	0	3.7	5.0	0	1.0	1.0
5-4	20	5×10^2	5094	8427	378	1015	236	371	37	3.7	4.9	1.6	1.02	0.96
5-4	20	2.5×10^3	4751	7663	338	950	221	336	182	3.7	4.7	2.2	0.96	0.87
5-4	20	5×10^3	5068	7795	338	1050	236	337	342	3.7	4.8	4.2	1.02	0.87
4-26	20	5×10^3	5642	8895	414	1121	261	389	375	3.9	5.0	4.5	. .	. .*
5-4	20	7.5×10^3	4765	7511	338	960	222	323	467	3.7	4.8	5.0	0.96	0.85
	20	10^4	4962	7291	338	980	231	316	603	3.7	4.7	5.6	1.0	0.815

* Correcting for decay of I^{125} source.

† Counts per minute.

Note: We find: bromine counts per minute are 227; rubidium, 338; and strontium, 326.

Table III. Composition of Steels Examined (%)

Source*	Iron	Chromium	Nickel	Molybdenum	Niobium
845	83.5	13.31	0.28	0.92	0.11
846		18.35	9.11	0.43	0.60
847	62.0	23.72	13.26	0.059	0.03
848		9.09	0.52	0.33	0.49
849		5.48	6.62	0.15	0.31
850	71.22	2.99	24.8		0.05
1161	95+	0.13	1.73	0.30	0.011
1162	95+	0.74	0.70	0.080	0.096
1163	95+	0.26	0.39	0.12	0.195
1164	95+	0.078	0.135	0.029	0.037
1165	95+	0.004	0.026	0.005	0.001
1166	95+	0.011	0.051	0.011	0.005
1167	95+	0.036	0.088	0.021	0.29
1168	95+	0.54	1.03	0.20	0.006
Source†					
1	~95	1.55	0.29	0.39	
2	95	0.56	0.57	0.22	
3	95	0.44	0.48	0.17	
4	95	1.51	0.50	0.22	
5	95	1.36x	3.26x		
6	95	0.566x	1.26x		
7	95	0.185x	4.92x	0.042x	
8	95		2.04x	0.256x	
9	95	2.22x	3.19x		
10	95	0.19	0.13		
11		1.40x	0.21	0.07x	
12	95	1.43x	0.187x	0.34x	
13	95	0.116x	0.171x	0.06x	
14	95	0.48x	0.64x	0.08x	
15	95	0.045x	0.066x	0.008x	
16	95	5.30	0.15	1.46	
17	95	0.24	0.35	0.13	
18	95	1.43	0.13	0.030	

* Source, U.S. National Bureau of Standards.
† Source, J.B. Newkirk guarantees only concentrations marked x.

Table IV. Approximate Yields for *K* X-Rays from Alpha-Excited Sources

$E_\alpha = 5$ to 6 MeV

			X-ray yield		S_k/S_α, μCi/mCi	
Element	E_k, keV	Z	per proton × sterad	per α × sterad	$\epsilon = 0.50$	$\epsilon = 0.05$
Carbon	0.28	6	4.5×10^{-3}	1.8×10^{-2}	113.0	11.3
Aluminum	1.49	13	3.5×10^{-3}	1.4×10^{-2}	88.0	8.8
Titanium	4.51	22	3.2×10^{-4}	1.3×10^{-3}	8.2	0.82
Iron	6.40	26	1.0×10^{-4}	4.0×10^{-4}	2.5	0.25
Copper	8.05	29	4.5×10^{-5}	1.8×10^{-4}	1.1	0.11

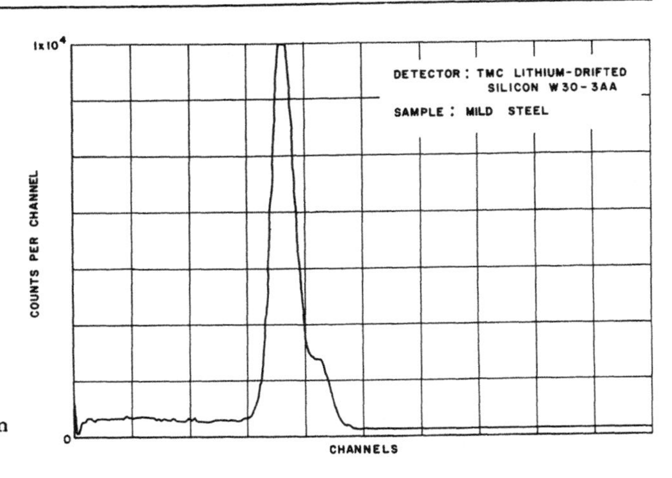

Figure 11. Photon spectrum of mild steel.

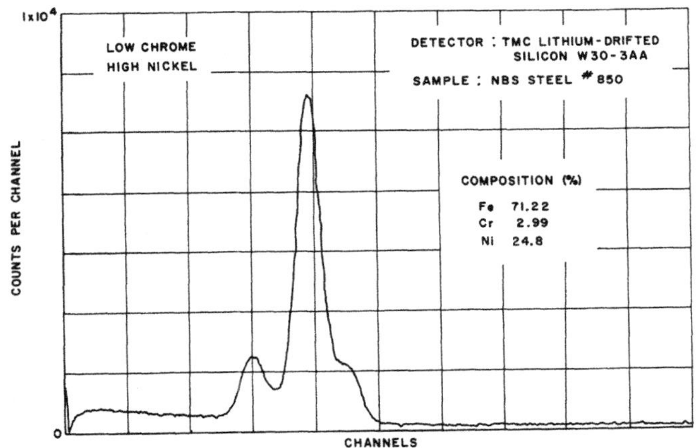

Figure 12. Photon spectrum of National Bureau of Standards steel No. 850.

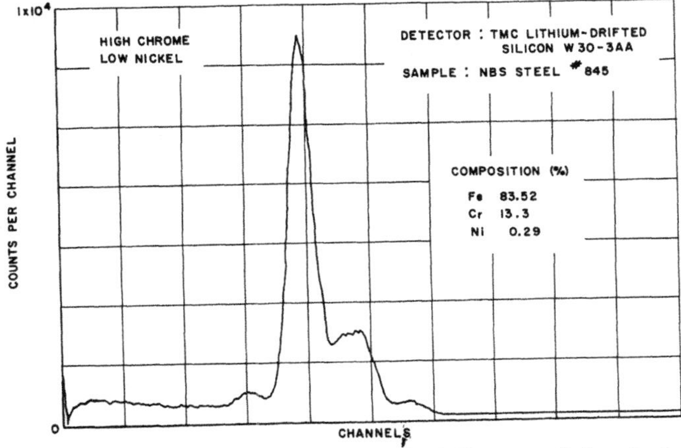

Figure 13. Photon spectrum of National Bureau of Standards steel No. 845.

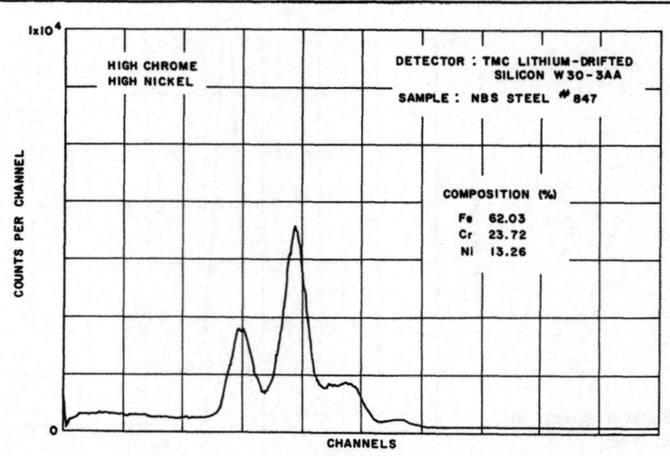

Figure 14. Photon spectrum of National Bureau of Standards steel No. 847.

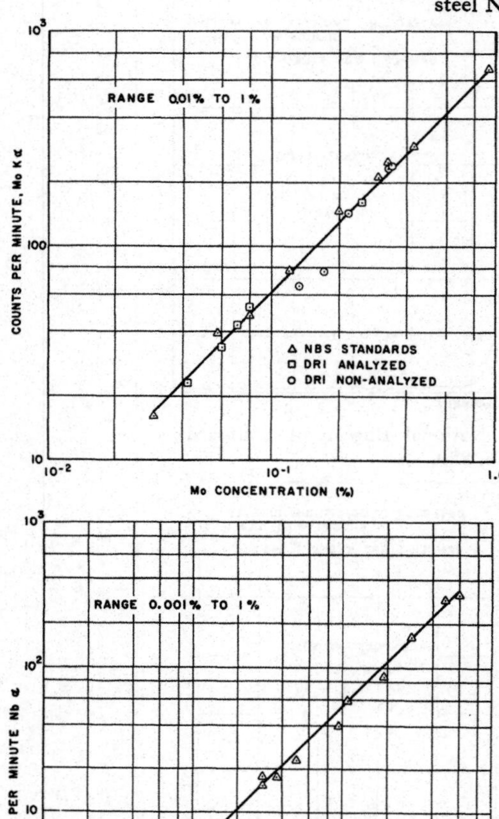

Figure 15. Molybdenum analysis in steel.

Figure 16. Niobium analysis in steel.

DISCUSSION

Portable nondispersive X-ray systems have many uses which are not feasible with conventional X-ray systems. Formerly the usefulness of nondispersive systems was limited because of the inability to identify element by element of the atomic scale. The use of filters as described by Cameron and Rhodes[8] and others can make these systems capable of distinguishing the elements; the instrument cannot, however, simultaneously distinguish element by element. The system described in this paper is truly nondispersive with adequate resolution to be useful.

Nondispersive systems view about 1% of the radiation emitted, while conventional dispersive systems have a transmission efficiency of 10^{-6}. The high transmission efficiency of the photon spectrometer permits it to use radioisotopes as sources of excitation instead of X-ray tubes. Among the many advantages of radioisotopic sources over X-ray tubes are:

1. Simplicity of design and lower cost.
2. Compact, light-weight instrumentation.
3. Elimination of variations in X-ray tube voltage and current as a source of error.

There are three types of radioisotopic sources: alpha, beta, and photon emitters. Many investigators have used alpha-emitting sources; these sources, however, have limited effectiveness in exciting heavier elements. Table IV illustrates the drastic decline in efficiency of photon yield as a function of the atomic number of the sample.[33] For exciting elements heavier than copper, an alpha-emitting source is useless.

Beta-emitting sources have been investigated as secondary emitters, *i.e.*, for exciting the characteristic X-rays of a selected target. These sources are used in absorbiometry. Beta sources are also used in producing the bremsstrahlung radiation.

Most photon-emitting sources are electron-capture isotopes which produce characteristic X-rays of the decay daughter. In addition, there may be several excited states which decay partially through internal conversion. These are also sources of characteristic X-rays of the daughter nuclide. Table V lists some of the radioisotopes that are available for use as excitation sources.[34]

In our work, we selected radioisotopes with these characteristics:

1. Photon emitters (minimal number of beta particles).
2. Essentially monoenergetic radiation.
3. Easily shielded low-energy radiation only.

We have avoided beta-particle emitters to eliminate bremsstrahlung background. Bremsstrahlung is troublesome because it is not a monotonic function of energy over the useful range. As more and more applications arise, there will be uses for sources which do not conform to these limitations.

The precision that can be obtained by a nondispersive system is limited by the capacity of the system to accumulate data. The detector amplifying system can accumulate up to 50,000 events per second without serious spectral distortion. Hence, the standard error in a 1-sec count is 0.45% if all the counts are in the peak. Any backscatter or loss of counts into any other peak will increase the standard error. To have 0.1% standard error requires a minimum counting time of 20 sec just due to the random nature of accruing information.

Other limitations include:

Source Activity. Present source strength is about 1 mCi. This can be increased by orders of magnitude before health hazards become serious.

Table V. Gamma-Ray, *K*-Capture, and Internal-Conversion Sources

Radio-isotope	Half-life	Type of decay; particle energy, keV; and abundance	γ Energy, keV; and abundance	Internal conversion factor, %	X-ray energy, keV
Fe^{55}	2.7 yr	EC, 100%		..	5.9
Cd^{109}	470 days	EC, 100%	88, 4%	96	22
I^{125}	57.4 days	EC, 100%	35.4, 7%	93	27
Am^{241}	470 yr	α5.44 MeV, 13%	27, 2.8%		
		5.48 MeV, 85%	33, 0.2%	..	11–22
			60, 40%		
			Others		
Gd^{153}	236 days	EC, 100%	70, 2%	9	
			97, 21%	8	41
			103, 25%	35	
Tm^{170}	127 days	β^-880, 22%	84, 3%	19	52, 4%
		970, 78%			7, 3%
Eu^{155}	1.7 yr	β^-154, 80%	87		
		243, 20%	105		
		Others	Others weaker	..	43
Co^{57}	270 days	EC, 100%	14, 6%	83	
			122, 88%	1	6.4
			136, 10%	1	
Dy^{159}	144.4 days	EC, 100%	0.058, 26%	83	(Tb X-rays)
			Others		

The Relative Yield of the Characteristic X-Rays of the Elements in the Sample. This will determine which source can be used. The relative amounts of the different elements of the sample will limit the accumulation of statistics for any particular element in the sample.

Interelemental Interference. This is shown by Figures 11 and 12. These results are easily accounted for by comparing the relative absorption coefficients for strontium, rubidium, and bromine. The *K* absorption edge for bromine is above the Rb $K\alpha$ X-rays, but below the Sr $K\alpha$ X-rays. Hence, bromine strongly attenuates the strontium radiation while scarcely affecting the rubidium radiation.

In the second phase, shown in Figure 12, strontium absorbs part of the rubidium radiation. Whatever bromine radiation it absorbs, however, is replaced by bromine X-rays excited by the strontium radiation. In this example, the competing processes compensate exactly, which produces a constant counting rate.

In work on stainless steels, the presence of iron or chromium limits the response of the instrument to nickel, for nondispersive and dispersive X-ray systems alike. This relationship is not unique but is evident in all matrix situations where the characteristic X-rays of one component of the sample are very highly absorbed by another component.

To illustrate this problem consider

$$N = N_0 \exp(- \Sigma \rho \mu_i r_i X) \tag{1}$$

where N_0 is the number of X-rays created at location X; μ_i, the absorption coefficient of the ith component; r_i, the concentration of the ith component; ρ, the density; X, the penetration; and N, the number of X-rays emitted.

For our case, let us consider the nickel content in a matrix of chromium, iron, copper, and molybdenum. The energy of the $K\alpha$ X-ray of nickel is 7.47 keV, just above the *K*

Table VI. Mass Absorption Coefficient for 7.47-keV Photons

Element	Atomic No.	cm^2/g
Chromium	24	350
Iron	26	420
Nickel	28	58
Copper	29	62
Molybdenum	42	190

absorption edge for iron and chromium. Table VI shows the mass absorption coefficients for the members of the matrix.[35] Iron and chromium are about six times as effective in absorbing nickel X-rays as nickel and copper. Thus, if we look for nickel X-rays emitted from any position in the sample, we must concern ourselves with the iron and chromium content.

The effective absorption by molybdenum, on the other hand, is well below that of iron and chromium. Only when there is twice as much molybdenum as iron does the molybdenum content affect the nickel X-ray yield. Consequently, in the analysis of steels, nickel–molybdenum interference is negligible.

A very small change in the iron or chromium content will change the total mass absorption per unit of thickness more than 10 times as effectively as nickel, copper, or the low-Z elements. As a consequence of this *selective absorption*, it is impossible to calculate reliably the nickel content without knowing the iron and chromium contents. On the other hand, our results indicate that it is possible to estimate the combined iron and chromium contents by the deviation of the actual nickel counting rate from that predicted on a known nickel content. Where the nickel content is not known, one can estimate the iron and chromium content only when the iron or chromium peak can be observed.

Figure 17 illustrates the variation in the Fe $K\alpha$ counting rate when the chromium content is varied. Data for this figure comes from the tests of steels listed in Table III. Figures 11, 12, 13, and 14 show representative spectra from these tests. Very small changes in the chromium concentration substantially change the counting rate for iron X-rays.

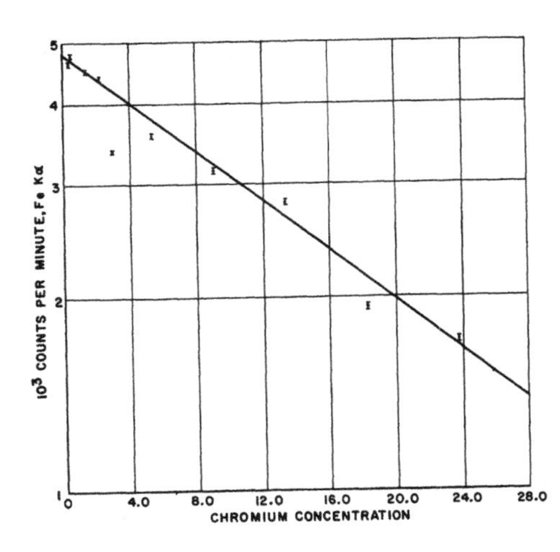

Figure 17. Effect of chromium concentration on Fe $K\alpha$ steel samples.

Up to 10% nickel, the Fe $K\alpha$ intensity is independent of nickel concentration. This effect is analogous to the condition in the bromine–strontium system. Hence, for low nickel concentration, the variation is Fe $K\alpha$ intensity may be used as a measure of chromium content.

RESULTS

1. In the bromine–rubidium–strontium system considered, the intensity of the Br $K\alpha$ line is, to a first approximation, independent of the strontium concentration. Hence, the intensity of this line is a direct measure of the bromine concentration. The bromine concentration so determined may be used to compute a correction on the Rb $K\alpha$ and Sr $K\alpha$ intensities.

Allowing for the absorption of the Sr $K\alpha$ line by bromine yields a direct measure of the strontium concentration. This concentration value supplies a further correction on the Rb $K\alpha$ intensity, which produces a measure of the rubidium concentration. Hence, so-called matrix effects may be predicted and manipulated over a large range of concentrations to yield quantitative analyses.

2. For iron alloys containing less than 10% nickel, the absorption by chromium of the Fe $K\alpha$ radiation may be used as a measure of the chromium concentration. This is done by comparing the intensities of the 6.39-keV radiation from the sample and from iron alone. This test is effective only because of the inherent stability and freedom from geometrical variations of the photon spectrometer system.

3. The analyses reported in Table III were taken over a period of 2 months, during which the spectrometer was dismantled several times and the iodine-125 source decayed through one half-life. Despite these disturbances and the lapse of time, data taken at the beginning of the period were still exactly reproducible within expected statistical deviations.

ACKNOWLEDGMENT

We wish to acknowledge the generous assistance of Dr. J. B. Newkirk, Denver Research Institute, Mr. T. C. Yao of Shell Development, Emeryville, and Prof. George Gordon, University of California, Berkeley, in providing numerous samples of steels of known composition.

REFERENCES

1. G. B. Cook, C. E. Mellish, and J. A. Payne, "Applications of Fluorescent X-ray Production by Electron Capture Isotopes," *Proc. Intern. Conf. Peaceful Uses At. Energy, 2nd* **19**: 127, 1958.
2. P. Leveque, P. Martinelli, and R. Chauvin, "Studies and Industrial Applications of Bremsstrahlung from the Beta Rays of Yttrium 90," *Proc. Intern. Conf. Peaceful Uses At. Energy, Geneva, 1955*, **15**: 142, 1955.
3. L. Reiffel, "Beta Ray Excited Low-Energy X-ray Sources," *Nucleonics* **13**: 33, 1955.
4. H. J. Dothie and B. Gale, "Nondispersive X-ray Fluorescence Absorption Edge Spectroscopy," *Spectrochim. Acta* **20**: 1735–1753, 1964.
5. G. Seibel, *Intern. J. Appl. Radiation Isotopes* **15**: 25–41, 1964.
6. S. Enomoto and C. Mori, Nondispersive X-ray Spectro-Chemical Analysis Using Nuclear Beta Ray Source, *Japan. J. Appl. Phys.* **2**: 274–280, 1963.
7. J. Niewodniczanski, "Analysis of Copper Ores in Mine Conditions by X-ray Fluorescence Induced by Isotope Sources," *Radioisotopes Instruments in Industry and Geophysics, SM* 68/10, Warsaw, 1965.
8. J. F. Cameron and J. R. Rhodes, "Filters for Energy Selection in Radioisotope X-ray Techniques," in: G. L. Clark (ed.), *Encyclopedia of X-rays and Gamma Rays*, Reinhold Publishing Corp., London and New York, 1963, pp. 387–388.
9. J. R. Rhodes, T. G. Ahier, and D. O. Poole, "Analysis of Zn and Cu Ores by Radioisotope X-ray Fluorescent," *UKAEA, Rept. AERE*-R4474, 1964.
10. T. Florkowski, B. Diunikowski, A. Kosiara, and M. Wasilewska, "Analysis of Iron, Zinc, and Copper Ores in the Field and Laboratory by Fluorescence Excited by 3H/Zr Bremsstrahlung," *Radio Chem. Method of Anal., Vol. II*, International Atomic Energy Agency, Vienna, 1965.

11. A. G. Darnley and C. C. Leamy, "The Analysis of Tin and Copper Ores Using a Portable Radioisotope X-ray Fluorescence Analyzer," *Proc. 1st AEA Symp. Radioisotope Instr. Industry Geophysics, Warsaw, 1965, Vol. 1,* International Atomic Energy Agency, Vienna, 1966.
12. S. Enomoto and T. Tanemura, "X-ray Spectroanalyser Using a Promethium-47 Beta Source," *Proc. 1st AEA Symp. Radioisotope Instr. Industry Geophysics, Warsaw. 1965, Vol. I,* International Atomic Energy Agency, Vienna, 1966.
13. K. Uchida, H. Tominga, and H. Imamura, "Light-Elements Simultaneous Analyzer by the X-ray Emission Method Using Alpha and X-ray Sources for Cement Raw Mix Control," *Proc. 1st AEA Symp. Radioisotope Instr. Industry Geophysics, Warsaw 1965 Vol. 1,* International Atomic Energy Agency, Vienna, 1966.
14. J. Cavailles and P. Martinelli, "Continuous Measurement of Galvanization Coatings by Means of X-ray Fluorescence," *Proc. 1st AEA Symp. Radioisotope Instr. Industry Geophysics, Warsaw, 1965, Vol. 1,* International Atomic Energy Agency, Vienna, 1966.
15. K. W. Ostrowski, L. Gorski, and T. Niewodneczanski, "Some Applications of Low-Energy Gamma and X-ray Sources in Poland," *Second Symp. Low-Energy X and Gamma Sources and Applications, Austin, Texas, Mar., 1967* (to be published).
16. S. Margolenas, "X-ray Fluorescence Applied to the Measurement of Zinc Coating in the Galvanizing Industry," *Second Symp. Low-Energy X and Gamma Sources and Applications, Austin, Texas, Mar., 1967* (to be published).
17. G. Robin and E. Darigny, "The Use of X-ray Fluorescence for the Automation of Mail Sorting, *Proc. 1st AEA Symp. Radioisotope Instr. Industry Geophysics, Warsaw, 1965, Vol. 1,* International Atomic Energy Agency, Vienna, 1966.
18. E. Darigny and G. Robin, "Application of X-ray Fluorescence to the Automation of Mail Sorting," *Second Symp. Low-Energy X and Gamma Sources and Applications, Austin, Texas, Mar.,* 1967 (to be published).
19. F. E. Sentfle and A. B. Tanner, "Mobile X-Ray Fluorescent Detector for Mineral Exploration," *Second Symp. Low-Energy X and Gamma Sources and Applications, Austin, Texas, Mar.,* 1967 (to be published).
20. I. Adler and J. Trombka, "Rock Analysis by Alpha Excitation of X-Rays—A Possible Lunar Probe," *Second Symp. Low-Energy X and Gamma Sources and Application, Austin, Texas, Mar.,* 1967 (to be published).
21. J. H. Patterson, A. L. Turkevich, and E. Franzgrote, "Analysis of Light Elements in Surfaces by Alpha-Particle Scattering," *Proc. Symp. Radioisotope Instr. Industry Geophysics, Warsaw,* SM 68/18, 1965.
22. A. E. Metzer, R. E. Parker, and J. I. Trombka, *IEEE Trans.* NS-B No. 1, pp. 554–561, Feb., 1966.
23. J. R. Rhodes, "Radioisotope X-ray Spectrometry," *Analyst* **91**: 683, Nov., 1966.
24. E. Elad and M. Nakamura, "High-Resolution X-ray and Electron Spectrometer," *Nucl. Instr. Methods* **41**: 161, 1966.
25. E. Elad, "A Preamplifier with 0.7 keV Resolution for Semiconductor Radiation Detectors," *Nucl. Instr. Methods* **37**: 327, 1965.
26. E. Elad and M. Nakamura, "Low-Energy Spectra Measured with 0.7 keV Resolution," *Nucl. Instr. Methods* **42**: 315, 1966.
27. W. B. Jones and R. A. Carpenter, "X-ray Fluorescent Analysis of Multiple Component Samples by a Nondispersive System," *Second Symp. Low-Energy X and Gamma Sources and Applications, Austin, Texas, Mar., 1967* (to be published).
28. Nuclear Data Sheets compiled by K. Way *et al., Natl. Acad. Sci.—Natl. Res. Councii, Publ.* 1964.
29. H. Imamura, K. Uchida, and H. Tominga, *Radioisotopes (Tokyo)* **14**: 286, 1965.
30. E. D. Pierson and R. H. Munch, "Oxide Mixture Analyses for V, Cu, Mo, Ti, Co, and Ni by Fluorescence Spectrometry," in: *Encyclopedia of X and Gamma Rays,* Reinhold Publishing Corp., New York, 1963, p. 681.
31. W. B. Jones, "Trace Element Detection by a Nondispersive X-ray Fluorescent Analysis System," Pittsburg Conference on Analytical Chemistry and Applied Spectroscopy, Mar.10, 1967.
32. *Natl. Bur. Std. (U.S.), Misc. Publ.* 260.
33. B. Sellers and C. A. Ziegler, Generation and Practical Use of Monoenergetic X-rays from Alpha-Emitting Isotopes, *Proc. 1st Symp. Low-Energy X and Gamma Sources and Applications, ORNL*–11C-5, Chicago, 1964.
34. J. F. Cameron and T. Florkowski, "Radioisotope Sources of Low-Energy Electromagnetic Radiation and Their Use in Analysis and Measurement of Coating Thickness," *Proc. Symp. Low-Energy X and Gamma Sources and Application, ORNL*–11C-5, UC-23, 1964.
35. Jean Leroux, *Encyclopedia of X and Gamma Rays,* "Absorption Coefficients (Mass), Empirical Determination Between 1 and 50 keV, calculated from Table 1," Reinhold Publishing Corp., New York, 1963, pp. 11–13.

THE INFLUENCE OF SAMPLE SELF-ABSORPTION ON WAVELENGTH SHIFTS AND SHAPE CHANGES IN THE SOFT X-RAY REGION: THE RARE-EARTH *M* SERIES

David W. Fischer and William L. Baun

Air Force Materials Laboratory (MAYA)
Wright–Patterson Air Force Base, Ohio

ABSTRACT

The $M\alpha$ and $M\beta$ emission spectra and the M_{IV} and M_V absorption spectra have been studied for the entire series of rare-earth elements. It is conclusively shown that the complicated multiplet structure observed in the emission spectra is not real emission structure but is, instead, produced by sample self-absorption. This is demonstrated by observing the emission spectra over wide variations in take-off angle and bombarding electron energies and finally by comparing the detailed structure of both the emission and absorption spectra. The M_{IV} and M_V absorption structure completely overlaps the $M\alpha$ and $M\beta$ emission lines, which are each found to have but one intensity maximum when obtained under conditions of minimum self-absorption. Some of these spectra have never been shown previously, while others have been studied in detail by several investigators. Points of agreement and disagreement with previous work are mentioned, and the wavelengths of the emission lines and absorption edges are listed for all of the rare-earth elements. It is concluded that the $4f \rightarrow 3d$ electron transitions are reversible in these elements.

INTRODUCTION

It is recognized by most X-ray spectroscopists that sample self-absorption effects exist, but it is not generally appreciated how really significant and serious these effects can sometimes be. Very little work has been done to the present on assessing the magnitude to which self-absorption can influence soft X-ray emission-line and band shapes and wavelength position. As a result, it is usually assumed, and incorrectly so, that the effects are insignificant and need not be bothered with.

A most striking example of the extreme influence of self-absorption is found in the $M\alpha$ ($4f \rightarrow 3d\frac{5}{2}$) and $M\beta$ ($4f \rightarrow 3d\frac{3}{2}$) emission lines of the rare-earth elements which fall in the 7.6 to 14.9-Å wavelength region. The rare earths are a unique series of elements in that they have an unfilled shell of electrons which are not admixed with the valence electrons. The 3*d* and 4*d* shells of the first and second transition series of elements, respectively, are also unfilled, but they are admixed to some extent with the valence band which permits the *d* electrons to become involved in chemical bonding. In the rare-earth elements, however, the unfilled 4*f* shell is truly an inner shell and the 4*f* electrons do not take part in chemical bonding.[1,2]

The $M\alpha$ and $M\beta$ emission spectra of the rare earths, therefore, arise from transitions involving the unfilled shell. Theoretically, one would expect these emission lines to be complicated multiplets because of coupling between the incomplete 4*f* shell and the singly ionized M_{IV} ($3d_{\frac{3}{2}}$) or M_V ($3d_{\frac{5}{2}}$) shell.[3] Confirmation of this general idea was apparently first shown some 40 years ago by Van der Tuuk[4] and a short time later by Lindberg.[5] Both were able to resolve the broad $M\alpha$ line into several components for most of the rare earths. The $M\beta$ line also resolved into more than one component but, in general, was not as complicated as $M\alpha$. Hirsh[6] explained the extra $M\beta$ structure as double-jump satellites and concluded that the $M\alpha$ satellites vanished at erbium, the rest of the multiple splitting being part of the main $M\alpha$ band. Apparently the theory was satisfied, and there was no further reason to question seriously the complicated multiplet structure observed for $M\alpha$. No other complete study of these rare-earth emission spectra has appeared since Lindberg's work some 36 years ago.[5]

In 1945, Rule[7] studied the $M_{IV,V}$ absorption spectra of samarium and observed a curious correspondence between wavelengths of points on his absorption curves with the emission lines of Lindberg.[5] Lindberg had, in fact, obtained the M_V absorption line for two of the rare earths (ytterbium and erbium) and considered them to correspond to the "$M\alpha_{III}$" and "$M\alpha_{IV}$" emission lines. Later Zandy[8] confirmed Rule's observation of correspondence between points on the absorption and emission spectra of samarium but was not sure just what physical significance should be attached to such correspondence. Zandy also obtained the $M_{IV,V}$ absorption spectra of europium, neodymium, and praseodymium but was unable to observe any correspondence between emission and absorption spectra. The $M_{IV,V}$ absorption spectra had also been obtained for a few other rare earths by Stewardson and Lee[9] (erbium) and Lee, Stewardson, and Wilson[10] (thulium and erbium), but no comparison was made with emission spectra. Then, in 1956, a reinvestigation of the erbium $M_{IV,V}$ absorption spectrum by Stewardson and Wilson was published.[11] The absorption spectrum was compared with the $M\alpha,\beta$ emission lines, which revealed a complete overlap in wavelength between the respective emission and absorption lines. It was also observed that the multiplet structure of the erbium emission spectrum became more distinct as higher electron-beam potentials were used. Stewardson and Wilson then concluded that the erbium emission multiplet structure of both $M\alpha$ and $M\beta$ was not real but was produced by self-absorption in the target material, the true emission being a broad single line probably accompanied by true satellites.

Our intention in this paper is to show that the multiplet structure of both $M\alpha$ and $M\beta$ observed for all the rare earths is caused entirely by self-absorption effects and to attempt to show as far as possible the true appearance of the emission spectra. This entails a detailed investigation of both emission and absorption spectra, many of which have not previously been shown in the literature.

EXPERIMENTAL TECHNIQUES

Instrumentation

The flat-crystal vacuum spectrometer used for this investigation has been described previously.[12,13] Characteristic emission spectra are produced by electron-beam bombardment of the target material, which is mounted on a four-sided, rotatable, brass anode. For absorption studies, the continum from a platinum or gold target was used. A special anode assembly was constructed for part of the self-absorption studies, and it has the capability of being rotated perpendicular to the entrance soller slit, which makes the takeoff angle continuously variable between 0 and 70°. The normal anode assembly has a fixed takeoff angle of 30°.

The detector is a thin-window flow proportional counter with an argon–methane (P-10) flow gas at a reduced pressure (120 mm Hg). All recording electronics, including the pulse-height discriminator, were commercial Picker equipment except for a low-noise Tennelec preamplifier. The resultant rate-meter scans and the curves shown in the figures have a mean deviation of $\pm 2\%$.

Several different crystals were used to disperse the radiation, a particular crystal being chosen to provide the best possible dispersion of a given wavelength interval. The following crystals were used in this study: EDDT ($2d = 8.8030$ Å) for lutetium, ytterbium, and thulium; ADP ($2d = 10.639$ Å) for erbium, holmium, dysprosium, terbium, and gadolinium; ammonium tartrate ($2d = 14.152$ Å) for europium, samarium, neodymium, praseodymium, and cerium; sucrose ($2d = 15.12$ Å) for lanthanum.

A special eight-position sample wheel was used to hold the absorption film specimens and placed directly in front of the detector slit. This wheel could be rotated to any of the eight positions from outside the vacuum. This enabled us to obtain both the emission and the absorption spectrum from the same element during the same run. Normal operating vacuum was 1 to 3×10^{-6} mm Hg.

Sample Preparation

All the rare-earth emission spectra were run, where possible, from both metal and oxide specimens. The oxides were all in the form of fine powder and were mixed into a slurry with acetone or methanol and spread directly on the anode surface. Metallic samples were prepared by vacuum evaporation of metal chunks onto aluminum foil, and some were run in the form of thin rolled sheet. A few of the metals, especially europium, were quite difficult to keep from oxidizing, and it is doubtful that the spectra obtained from them are truly characteristic of the metal. For those elements for which metallic spectra could be obtained, only the spectrum of ytterbium showed any change between metal and oxide.

Absorption specimens were prepared by vacuum evaporation of metal chunks onto aluminum foil (2.0 mg/cm^2) or onto an extremely thin formvar film. The rare-earth film thickness varied between approximately 0.05 and 0.2 mg/cm^2.

RESULTS AND DISCUSSION

The rare-earth $M\alpha$ and $M\beta$ emission spectra obtained at 10 kV are shown in Figure 1. These spectra cover the 7.6 to 14.9-Å wavelength region as one moves from lutetium down through lanthanum. It is rather difficult to compare these spectra with those which have been obtained previously since all of the earlier work was done with photographic film as the detector. Lindberg,[5] *e.g.*, includes tables of wavelength values of the $M\alpha$ and $M\beta$ components, but it is often difficult to visualize just how a spectrum looks from a table of wavelength values. It appears, however, that our wavelength positions agree quite well with those listed by Lindberg.

Since lutetium has a filled $4f$ shell, one would expect the Lu $M\alpha$ and $M\beta$ spectra to be quite similar to those of the next-higher elements hafnium, tantalum, *etc.*, which, indeed, they are. As seen in Figure 1, the lutetium spectrum consists of two sharp lines, each with some high-energy satellite structure. The next-lower element ytterbium has one $4f$ vacancy, and this causes a change in the emission spectrum. Both $M\alpha$ and $M\beta$ become broader, and their relative intensity changes significantly. Also, $M\alpha$ now shows an extra component. At thulium, the spectrum changes even more, and so on down the series as the $4f$ shell begins to empty out. Notice that, when going from lutetium down through samarium, the $M\alpha$ band shows more and more multiplet structure and becomes weaker and weaker compared with $M\beta$. The $M\beta$ band, on the other hand, does not become so complicated in structure but appears as a main $M\beta$ with a high-energy satellite which becomes more and more separated from the main component. Hirsh,[6] in fact, found that

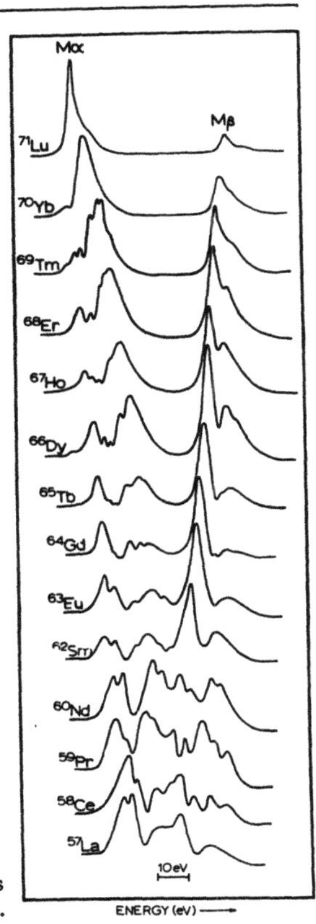

Figure 1. $M\alpha$ and $M\beta$ emission spectra of the rare-earth elements obtained with electron-beam potential of 10 kV and 30° takeoff angle.

these extra high energy peaks followed very nicely the expected Moseley relationship for $M\beta$ satellite lines. The $M\alpha$ satellites appeared to vanish at erbium, and, from there on down, the structure was all attributed to the main $M\alpha$ band. Notice that, at neodymium, the spectra again change so that, for the bottom four elements, it is difficult to pick out by sight the individual $M\alpha$ and $M\beta$ structures. From the spectra shown in Figure 1, one can follow what appears to be a nice systematic change from element to element. Theoretical predictions of multiplet structure arising from multiple $4f$ vacancies seem to be beautifully confirmed.[3]

When we first observed these spectra, we were duly impressed, but, upon closer inspection, some inconsistencies were found. Reasonably accurate values of the energy separation between the M_{IV} and M_V levels can be obtained from the $L\alpha_1$ and $L\alpha_2$ lines which were studied by Sakellaridis.[14] The $M\alpha$ and $M\beta$ lines should be separated by this same amount. In order to obtain these expected energy spreads, it is necessary to use the strongest $M\beta$ component and the longest-wavelength $M\alpha$ component, which, in some cases, is the weakest in intensity of any of the multiplets. One cannot help but notice the puzzling difference in behavior of the $M\alpha$ and $M\beta$ emission groups. The $M\beta$ group remains relatively simple, with the apparent main component and high-energy satellite. The $M\alpha$ group, on the other hand, appears to split into as many as six components. It is also difficult to explain why the $M\alpha$ emission is so much weaker in intensity than the $M\beta$ for elements such as terbium, gadolinium, europium, and samarium. For us, the answer to the problems came about rather by accident. Several of the emission spectra were obtained over a period of 3 years and, in comparing the various results from any one element, it was noticed that the spectra were not always exactly the same. Some of them showed very distinctly the multiple components of $M\alpha$ and $M\beta$, while, in others, the same components were less distinct. Also, the relative intensities of the $M\alpha$ and $M\beta$ groups did not come out the same every time. The only condition that had changed was the energy of the bombarding electrons. This led to an investigation which produced results such as those shown in Figures 2 and 3. These figures show the Dy $M\alpha$ and $M\beta$ emission spectra obtained with various electron-beam energies and takeoff angles. Notice that, as the beam voltage is increased or the takeoff angle is decreased, the multiplet structure becomes more distinct and $M\alpha$ becomes weaker compared with $M\beta$. So, then, the spectral details are dependent on the depth from which the X-rays emerge from the target material. The obvious conclusion is that the multiple components of $M\alpha$ and $M\beta$ are being produced by self-absorption effects. We then found that Stewardson and Wilson[11] had come to the same conclusion concerning erbium by comparing the $M_{IV,V}$ absorption spectra with the $M\alpha,\beta$ emission spectra. We then subjected the emission spectra of all the rest of the rare earths (except promethium) to the same variations in electron-beam voltage and takeoff angle, as shown for dysprosium in Figures 2 and 3. The results were all the same. For every element, the multiplet structure was most pronounced under conditions of high electron-beam voltage or very low takeoff angle. Under conditions of minimum self-absorption, the spectra are found to consist of a single line for both $M\alpha$ and $M\beta$, as shown in Figure 4. These spectra were all obtained with bombarding-electron energies of 200 to 300 V in excess of M_{IV} threshold from the same specimens and during the same run as the spectra shown in Figure 1. Compare these two figures, and notice the very striking difference in the spectra, especially from elements such as neodymium, praseodymium, and cerium. Stewardson and Wilson[11] speculated that the true emission would consist of single broad lines probably accompanied by true satellites. For most of the spectra shown in Figure 4, we are unable to observe a distinct satellite structure on either $M\alpha$ or $M\beta$. These spectra may still be slightly affected by self-absorption, especially in the relative

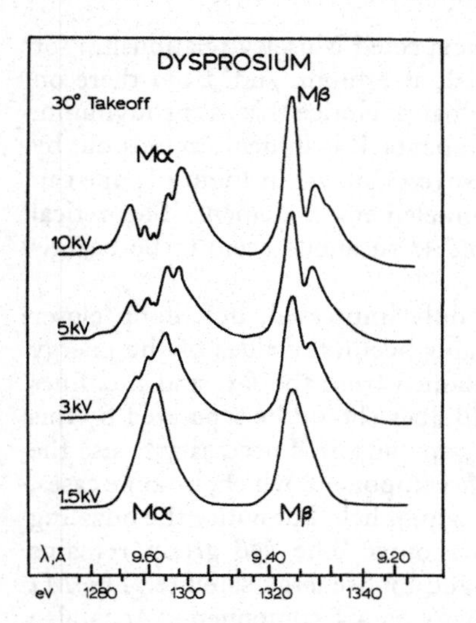

Figure 2. Dysprosium $M\alpha$ and $M\beta$ emission spectra obtained at various electron-beam energies with constant 30° takeoff.

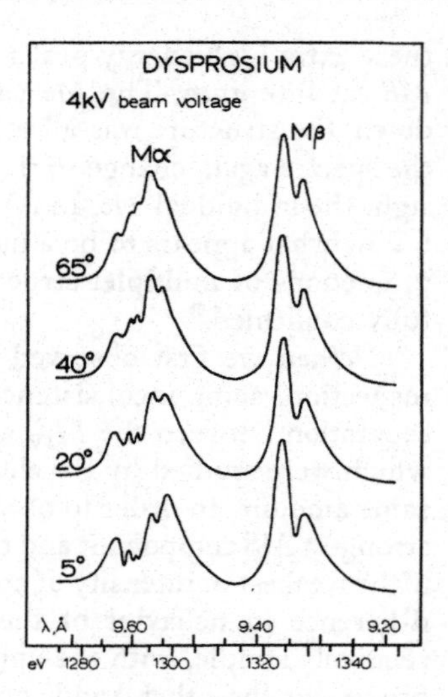

Figure 3. Dysprosium $M\alpha$ and $M\beta$ emission spectra obtained at various takeoff angles with constant 4-kV electron-beam voltage.

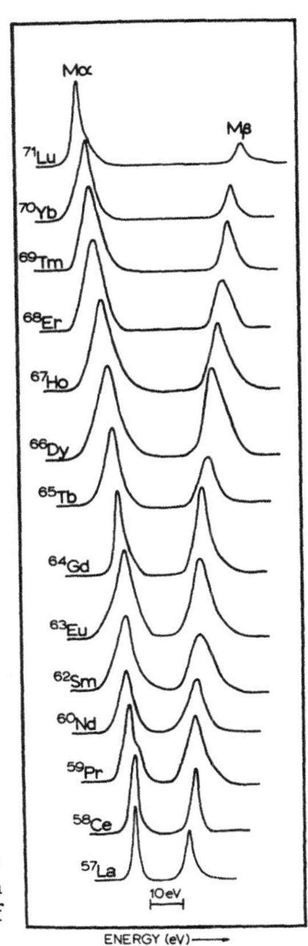

Figure 4. $M\alpha$ and $M\beta$ emission spectra of the rare-earth elements obtained under minimized self-absorption conditions. Electron-beam voltage was 200 to 300 V in excess of M_{IV} threshold with high takeoff angle.

Table I. Wavelengths of Rare-Earth $M\alpha$ and $M\beta$ Emission Lines Obtained with Minimized Self-Absorption Conditions (Oxide Targets)

Element	$M\alpha(N_{VI,VII} \rightarrow M_V)$		$M\beta(N_{VI} \rightarrow M_{IV})$	
	λ, Å	eV	λ, Å	eV
Lanthanum	14.859 ± 0.005	834.20 ± 0.08	14.573 ± 0.005	850.55 ± 0.08
Cerium	14.078	880.43	13.786	899.10
Praseodymium	13.339	929.26	13.056	949.35
Neodymium	12.667	978.54	12.399	999.71
Samarium	11.485	1079.26	11.249	1101.84
Europium	10.955	1131.50	10.736	1154.48
Gadolinium	10.469	1183.98	10.249	1209.34
Terbium	10.014	1237.79	9.787	1266.46
Dysprosium	9.589	1292.62	9.359	1324.38
Holmium	9.193	1348.28	8.959	1383.54
Erbium	8.822	1404.97	8.584	1443.93
Thulium	8.475	1462.61	8.238	1504.62
Ytterbium	8.141	1522.63	7.912	1566.59
Lutetium	7.837	1581.48	7.600	1630.90

intensities of $M\alpha$ and $M\beta$, but they should be very close to being the "true" emission spectra. It would be better to use electron-beam voltages even closer to M_{IV} threshold than we used to obtain these spectra, but, in our case, the intensity would be too low to obtain reliable line profiles. The wavelength values of the $M\alpha$ and $M\beta$ intensity maxima of Figure 4 are listed in Table I. These wavelengths are in fairly good agreement with those listed by Bearden,[15] who averaged Lindberg's multiplet component measurements[5] by using a modified Moseley diagram. When a Moseley plot is made of the values in Table I, very straight lines are obtained for both $M\alpha$ and $M\beta$.

To illustrate even more clearly that these gross distortions in $M\alpha$ and $M\beta$ are caused by self-absorption, the M_{IV} and M_V absorption spectra of all the rare earths were obtained and are shown in Figure 5. About half of these spectra have been shown in the literature,[8-11,16] and our agreement with them is quite good in most cases. All of the earlier work was done by using photographic film recording, while the spectra in Figure 5 were obtained with a flow proportional counter.

Notice in Figure 5, the progressive change in the spectra as one goes down through the rare-earth series. The M_{IV} absorption is quite weak in ytterbium and slowly increases in intensity compared with M_V as the $4f$ shell becomes more empty. As was noted by earlier investigators, the spectra themselves are quite different in appearance from the *edge* structure normally observed in X-ray absorption spectra. These might be better described as *line* absorption.

The M_{IV} and M_V edge positions are listed in Table II. A Moseley plot of these values produces two very straight lines which are only slightly displaced from those of the $M\alpha$ and $M\beta$ emission lines. These values were measured at the midpoint of the edge on the low-energy side of the absorption "line."

If the M_{IV} and M_V absorption spectra are placed in exact register with the $M\alpha$ and $M\beta$ emission spectra of Figure 1 for each of the elements, the results shown in Figures 6 to 9 are obtained. For all 13 elements, the M_{IV} absorption completely overlaps the $M\beta$ emission,

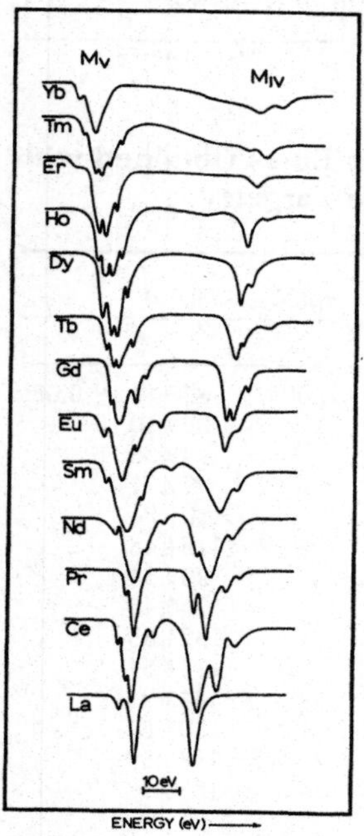

Figure 5. M_{IV} and M_V absorption spectra of the rare-earth elements.

and the M_V absorption completely overlaps the $M\alpha$ emission. Also, in every case, the absorption structure is precisely mirrored in the emission spectra. In fact, for most of these elements, one could obtain a virtually complete $M_{IV,V}$ absorption spectrum just by

Table II. M_{IV} and M_V Absorption-Edge Positions for Rare-Earth Elements

Element	M_V edge		M_{IV} edge	
	λ, Å	eV	λ, Å	eV
Lanthanum	14.875 ± 0.005	833.3 ± 0.2	14.595 ± 0.005	849.3 ± 0.2
Cerium	14.065	881.3	13.783	899.3
Praseodymium	13.350	928.4	13.080	947.6
Neodymium	12.703	975.7	12.411	995.8
Samarium	11.518	1073.0	11.254	1101.4
Europium	10.975	1129.4	10.710	1154.0
Gadolinium	10.497	1180.8	10.231	1211.6
Terbium	10.041	1234.4	9.778	1267.6
Dysprosium	9.627	1287.6	9.355	1325.0
Holmium	9.227	1343.3	8.956	1384.0
Erbium	8.845	1401.4	8.580	1444.6
Thulium	8.489	1455.9	8.229	1506.2
Ytterbium	8.155	1515.5	—	—

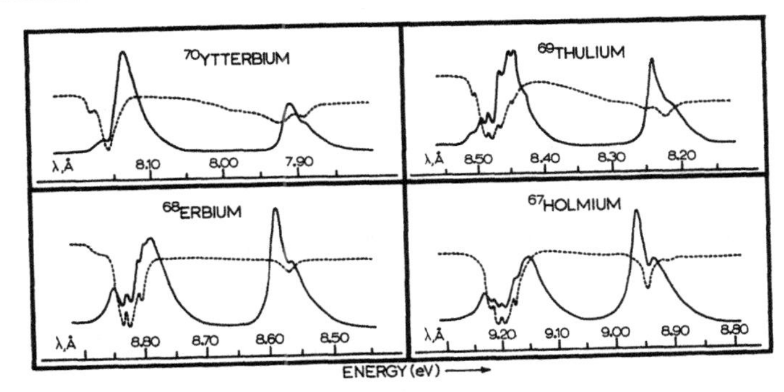

Figure 6. Correspondence between the M_{IV} and M_V absorption spectra (dashed curve) and the $M\alpha$ and $M\beta$ emission spectra at 10 kV (solid curve) for ytterbium, thulium, erbium, and holmium.

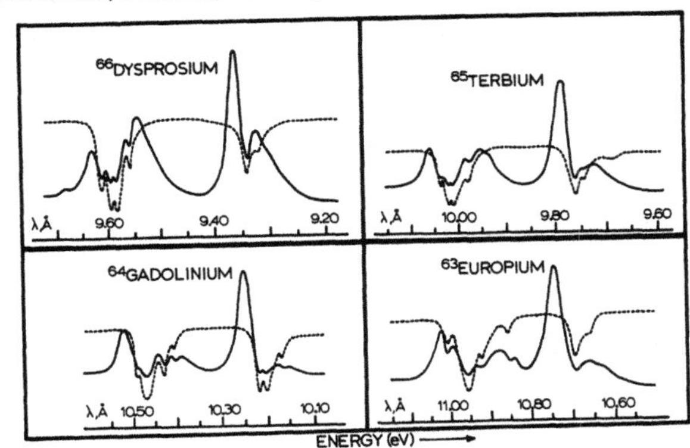

Figure 7. Correspondence between the M_{IV} and M_V absorption spectra (dashed curve) and the $M\alpha$ and $M\beta$ emission spectra at 10 kV (solid curve) for dysprosium, terbium, gadolinium, and europium.

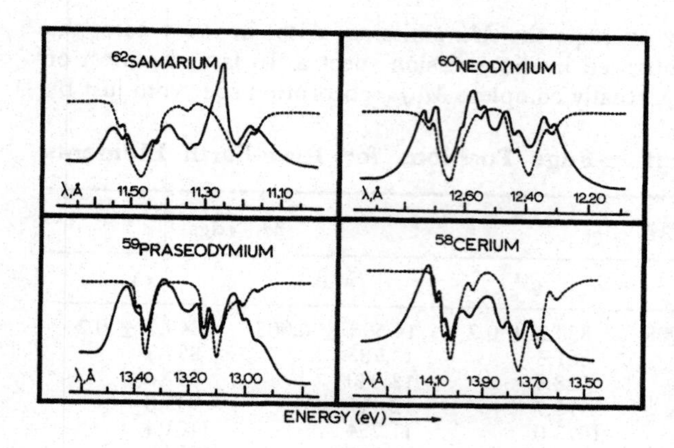

Figure 8. Correspondence between the M_{IV} and M_V absorption spectra (dashed curve) and the $M\alpha$ and $M\beta$ emission spectra at 10 kV (solid curve) for samarium, neodymium, praseodymium, and cerium.

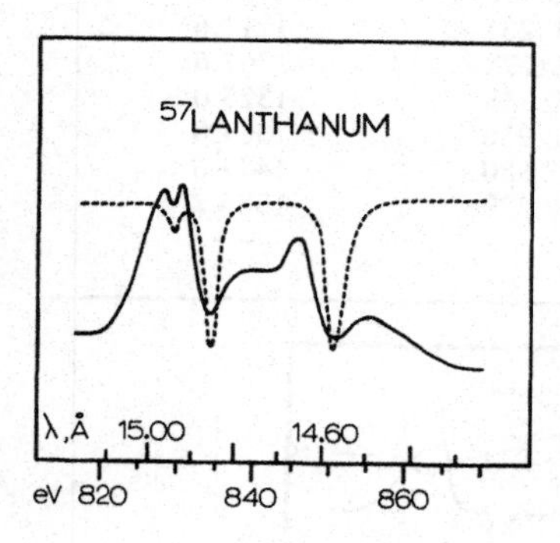

Figure 9. Correspondence between the M_{IV} and M_V absorption spectra (dashed curve) and the $M\alpha$ and $M\beta$ emission spectra at 10 kV (solid curve) for lanthanum.

studying the $M\alpha,\beta$ emission spectrum at high beam potentials. Certainly, these results show that the multiple structure observed on $M\alpha$ and $M\beta$ is not real emission structure at all. It is definitely absorption structure.

The fact that the emission and absorption spectra fall at the same energy positions indicates, as first suggested by Lindberg[5] and later by Stewardson and Wilson,[11] that the $4f \rightarrow M_{IV}$ and $4f \rightarrow M_V$ transitions are quite reversible. This is perhaps shown most strikingly in the case of lanthanum (Figure 9). In its normal state, lanthanum has no $4f$ electrons.[1,2] On this basis, one would not expect to see La $M\alpha$ or $M\beta$ emission lines. It is found, however, that these lines can be produced quite easily under electron-beam bombardment (Figures 1 and 4). We must, therefore, be injecting electrons into the $4f$ level by an absorption process. Electrons from the M_{IV} or M_V shells (and, to lesser extent, the $N_{IV,V}$ shells) are absorbed into the $4f$ shell and can then fall back into the M_{IV} or M_V vacancy. Such a reversible process is apparently unique to the rare-earth elements. All the $M_{IV,V}$ spectra shown in Figure 5 must represent absorption into the unfilled $4f$ sites and not the normal edge absorption into the continuum of states. The values listed in Table II, therefore, do not represent M_{IV} or M_V energy-level values as such but, rather, the energy just needed to "kick" them into the $4f$ shell.

Except for europium and ytterbium, neither the $M\alpha,\beta$ emission nor the $M_{IV,V}$ absorption spectra appear to be significantly influenced by chemical combination. Europium and ytterbium are apparent exceptions because they show divalent properties as metals but become trivalent when oxidized.[1,2] Williams showed that the two different valence states of europium result in significantly different $M_{IV,V}$ absorption spectra.[16] We have

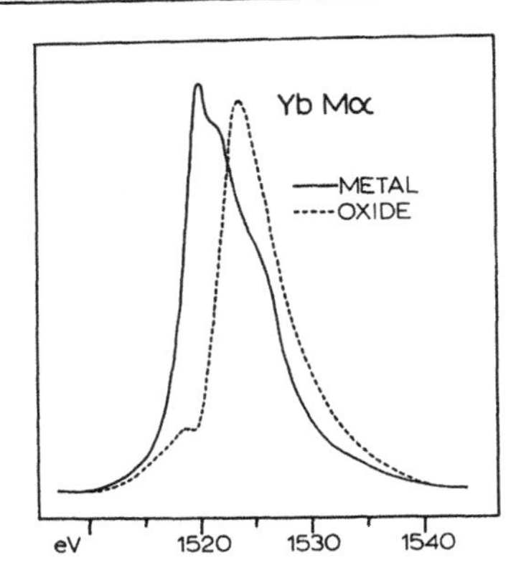

Figure 10. Differences in $M\alpha$ emission line from $+2$ and $+3$ valence states of ytterbium.

not been able to keep europium in a pure metallic state long enough to obtain either emission or absorption spectra from it because it oxidizes too rapidly, even under vacuum. We have, however, succeeded in obtaining the $M\alpha,\beta$ emission spectra from the $+2$ and $+3$ states of ytterbium, the $M\alpha$ line being shown in Figure 10. When going from metal to oxide, this line shifts to higher energy and changes considerably in shape. This change seems to confirm the general idea that the ytterbium $4f$ shell is filled in the metal but has one vacancy in the oxide.[16] The ytterbium spectra shown in Figures 1, 4, and 5 are from the oxide in order to be consistent with the rest of the series.

CONCLUSIONS

The results shown here indicate conclusively that the rare-earth $M\alpha$ and $M\beta$ emission spectra are seriously distorted by self-absorption. It is further proved that the complicated multiplet structure observed for $M\alpha$ and $M\beta$ is not true emission structure. The true spectrum consists of single lines for both $M\alpha$ and $M\beta$ with no resolved satellites for either. In a way, the general idea that the multiple emission components are caused by the unfilled $4f$ shell is correct. This multiplicity actually occurs in the absorption spectrum, however, and not in the true emission spectrum. The complete overlap of the emission and absorption spectra indicate that the $4f \rightarrow M_{IV}$ and $4f \rightarrow M_{V}$ transitions are reversible for the rare-earth elements.

The striking distortion caused by self-absorption points up the dangers of putting too much faith in emission spectra without also investigating the conditions under which the spectra were obtained. Self-absorption can also significantly affect valence-emission-band spectra[17-19] and give one a completely false impression of the actual band shapes and energy positions. Many of the disagreements in the literature concerning band shapes, band widths, etc., can be traced to the failure of the investigators to take possible self-absorption effects into consideration.[18]

REFERENCES

1. D. N. Trifinov, *The Rare Earth Elements*, Pergamon Press, New York, 1963.
2. N. E. Topp, *The Chemistry of the Rare Earth Elements*, Elsevier Publishing Co., New York, 1965.
3. I. B. Borovskii and T. M. Golovner, "Multiplicity in X-Ray Emission Spectra," *Dokl. Akad. Nauk SSSR* **88**: 233, 1953.
4. J. H. Van der Tuuk, "Höhere Multipletts in Röntgenspektrum," *Z. Physik* **44**: 737, 1927.
5. E. Lindberg, "The *M* and *N* Series. A Spectroscopic Study of X-Rays," *Nova Acta Regiae Soc. Sci. Upsaliensis* **7**: 7, 1931.
6. F. R. Hirsh, "The Satellites of *M* Series X-Ray Lines," *Phys. Rev.* **38**: 914, 1931.

7. K. C. Rule, "X-Ray Line Absorption in the M Series of ^{62}Sm," *Phys. Rev.* **68**: 246, 1945.
8. H. F. Zandy, "The M_{IV} and M_V Absorption Edges of ^{59}Pr, ^{60}Nd, ^{62}Sm and ^{63}Eu," *Proc. Phys. Soc.* (*London*) **65A**: 1015, 1952.
9. E. A. Stewardson and P. A. Lee, "The M_{IV} and M_V Absorption Edges of ^{68}Er," *Proc. Phys. Soc.* (*London*) **64A**: 318, 1951.
10. P. A. Lee, E. A. Stewardson, and J. E. Wilson, "The M_V Absorption Edges of Tm and Er and the Origin of the Er M_V Absorption Lines," *Proc. Phys. Soc.* (*London*) **65A**: 668, 1952.
11. E. A. Stewardson and J. E. Wilson, "The $M_{IV,V}$ Spectra of ^{68}Er," *Proc. Phys. Soc.* (*London*) **69A**: 93, 1956.
12. D. W. Fischer and W. L. Baun, "Diagram and Non-Diagram Lines in the K Spectra of Magnesium and Oxygen from Metallic and Anodized Magnesium," *Spectrochim. Acta* **21**: 443, 1965.
13. W. L. Baun and D. W. Fischer, "The Effect of Valence and Coordination on K Series Diagram and Non-Diagram Lines of Magnesium, Aluminum, and Silicon," in: W. M. Mueller, G. R. Mallett and M. J. Fay (eds.), *Advances in X-Ray Analysis*, *Vol. 8*, Plenum Press, New York, 1965.
14. P. Sakellaridis, "*L* Emission and Absorption Spectra of Gadolinium and Thulium," *Compt. Rend.* **236**: 1244, 1953.
15. J. A. Bearden, "X-Ray Wavelengths," *Rev. Mod. Phys.* **39**: 78, 1967.
16. K. C. Williams, "The $M_{IV,V}$ Spectra of ^{63}Eu," *Proc. Phys. Soc.* (*London*) **87**: 983, 1966.
17. H. P. Hansen and J. Herrera, "Self-Absorption in the X-Ray Spectroscopy of Valence Electrons," *Phys. Rev.* **105**: 1483, 1957.
18. R. J. Liefeld, "$L\alpha$ X-Ray Emission Lines of Nickel, Copper and Zinc," *Bull. Am. Phys. Soc.* **10**: (4): 549, 1965.
19. C. Bonnelle, "Contribution a l'étude des métaux de transition du premier groupe, du cuivre et de leurs oxydes par spectroscopie X dans le domaine de 13 a 22 Å," *Ann. Phys.* (*Paris*) **1** (7,8): 439, 1966.

DISCUSSION

H. W. Pickett (General Electric Co.): I wonder if you can make the emission sample thin enough so that you get the normal emission spectrum even though you have too high a voltage. Have you tried it?

D. W. Fischer: It's possible. We've tried this, but it seems that, if you get it thick enough so that you can see it, it is already too thick.

H. W. Pickett: I just wondered what these numbers were; 1½-kV excitation is pretty low.

D. W. Fischer: That's pretty low, right.

H. W. Pickett: It would almost have to be unsupported.

D. W. Fischer: It would have to be so thin you couldn't even see it, and I don't know how thin that is.

H. K. Herglotz (E. I. du Pont): I wonder how thick your samples were for the absorption spectra. At 14.6 Å, the absorption, of course, is extremely heavy.

D. W. Fischer: These varied all the way from, say, 0.01 to 0.1 mg/cm², which is a film you can completely see through.

H. K. Herglotz: Thank you.

THE X-RAY WAVELENGTH SCALE IN THE LONG-WAVELENGTH REGION

A. F. Burr

New Mexico State University
Las Cruces, New Mexico

ABSTRACT

Recently the X-ray wavelength scale has been put on a consistent basis by the selection of the W $K\alpha_1$ line to be the wavelength standard for the whole X-ray region of the electromagnetic spectrum. To help establish this scale, four other X-ray lines, which had been measured with a precision approaching 1 ppm, were selected as secondary standards. However, since the longest of these lines is the Cr $K\alpha_2$ line at 2.293606 Å, other lines whose wavelengths are less precisely known have been used for calibration purposes. The most interesting of these is the Al $K\alpha$ line, which plays an important part in the effort to connect the X-ray wavelength scale to the basic standard of length. Two other often-used lines are the Cu $L\alpha$ line at 13.339 Å and the O $K\alpha$ line at 23.62 Å. It is, however, necessary to take several steps to trace the accepted values of these lines back to the basic X-ray wavelength standard. Furthermore, the uncertainties introduced by the effects of the chemical combination and the large natural line widths limit the precision obtainable. It has been suggested that the first resonance line of the Ne K edge in the 14-Å region would overcome both of these difficulties. It has also been suggested that the high-frequency limit of the bremsstrahlung radiation would make a particularly convenient reference mark. This method is discussed here, and it is shown that the accuracy obtainable by this method is better than that obtainable by means of some commonly used reference lines.

THE X-RAY WAVELENGTH SCALE

Early in the history of X-ray wavelength measurements, it was noted that lines could be measured with respect to each other with greater precision than they could be related to the basic standard of length. Hence, there quickly grew up an X-ray wavelength scale which was defined by M. Siegbahn[1,2] in terms of the lattice constant of calcite crystals. As the precision of measurements improved, it became increasingly evident that all calcite crystals were not the same and that, therefore, the definition of the X-ray wavelength scale left something to be desired. Siegbahn[3] recognized the possibility of this difficulty but also recognized the greater difficulty of setting a wavelength standard with much of the X-ray wavelength region still unexplored.

There were several suggestions for a change in the definition of the X-ray wavelength scale. Merrill and DuMond[4] suggested that the Mo $K\alpha_1$ line would make a good standard. Later, DuMond[5] suggested the use of a weighted average of several wavelengths. Bergvall, Hornfeldt, and Nordling[6] noted that the Mo $K\alpha_1$ line had for some time been used as a wavelength standard in their laboratory.

After an exhaustive study of many factors, including the above suggestions, Bearden[7] redefined the X-ray wavelength scale in terms of the W $K\alpha_1$ line. He considered the wavelength region in which the primary standard should lie, the accuracy with which angles can be determined, the requirements of crystal perfection, the symmetry and width of the

Table I. X-Ray Wavelength Standards

Primary standard		
W $K\alpha_1$	0.2090100 Å*	
Secondary standards		
Line	Wavelength (Å*)	Probable error (ppm)
Ag $K\alpha_1$	0.5594075	1.3
Mo $K\alpha_1$	0.709300	1.3
Cu $K\alpha_1$	1.540562	1.2
Cr $K\alpha_2$	2.293606	1.3

standard line, the chemical and isotope shifts, and the possible use of Mössbauer lines. Bearden concluded that the best means to establish a new, consistent X-ray wavelength scale would be to define the wavelength of the W $K\alpha_1$ line to be

$$\lambda_{\mathrm{W}\,K\alpha_1} = 0.2090100 \text{ Å*}$$

where the symbol Å* is used to distinguish the new wavelength unit from the old and from the angstrom unit, which the new unit closely approaches. In order to facilitate the application of the new wavelength scale, Bearden and co-workers[8] very carefully measured, to a precision of almost 1 ppm, the wavelength of four other lines which could serve as secondary standards. These lines and their wavelengths are listed in Table I. Bearden and his co-workers [9,10] then applied themselves to the task of placing all the measured wavelengths on this new scale.

LONG-WAVELENGTH REFERENCE LINES

In the long-wavelength region, the extreme precision represented by the secondary standards is rarely needed; indeed, it is almost impossible to get. Furthermore, for most practical purposes, the new wavelength scale is indistinguishable from the angstrom scale at wavelengths greater than 5 Å. However, many measurements still make use of reference lines. Figure 1 shows some of these lines and a number of other interesting lines in the long-wavelength region. The figure covers a region from the $K\alpha$ line of aluminum at 8 Å (the most precisely measured line of the group) to an M line of potassium at almost 700 Å (the longest measured X-ray line). Note that the closest secondary standard is far off the figure to the right.

The Al $K\alpha$ Line

The Al $K\alpha$ line has been very carefully measured by a number of workers, primarily because of efforts to use it to relate the relative X-ray wavelength scale to the standard of

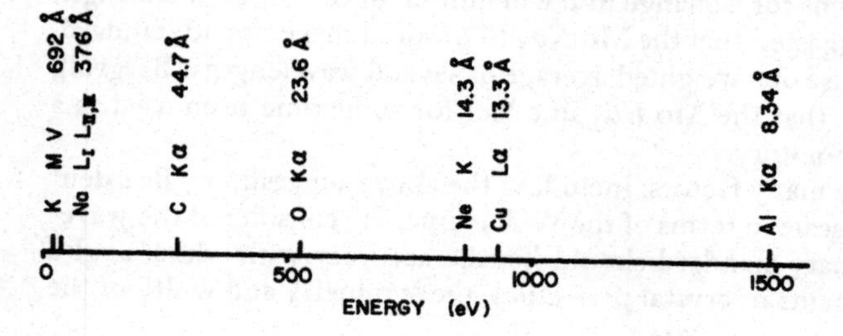

Figure 1. Some important lines from the soft X-ray region.

length. The width of the line and the effect of chemical composition on its wavelength make a precision measurement difficult. Nordfors[11] has made a very careful study of the $K\alpha$ lines of both the metal and the oxide. Tyrén[12] obtained some excellent photographs of the same lines. These photographs were recently reanalyzed by Edlén and Svensson[13] to obtain good values for the Al $K\alpha$ line on the absolute wavelength scale. This last paper also contains a large number of very accurately calculated wavelengths of highly ionized light elements which are well suited for reference use. Further work which will result in a more precise wavelength for this line is currently being done at the Johns Hopkins University.

The Cu $L\alpha$ Line

Another line which has been widely used as a reference line is the Cu $L\alpha$ line. The various values which have been assigned to this line are listed in Table II. The first line of that table contains the most probable value of the line based on all the available X-ray information. The second line contains a weighted average of all the succeeding lines. Figure 2 presents the data from the table in a graphical form where rough estimates of the probable errors involved in each value are shown so that one can obtain a better idea of the extent to which the various measurements agree with each other. The numbers above the plotted points refer to the source of each datum as given in the list of references at the end of this article.

Other Lines

A popular reference line in the longer-wavelength region is the O $K\alpha$ line since it is easy to obtain a target containing oxygen in some form. Of course, any line can be used for reference purposes if its wavelength is known with a considerably greater degree of accuracy than that of the feature which is to be measured. Convenience will then dictate

Table II. Values Assigned to the Cu $L\alpha$ Line

Year	Spectrometer*	Reference†	Original value, (Å)	Adjusted value, (Å)	Notes‡
1965		14,15	13.339		
1964		9,10	13.336		
1936	*G*	16	13.330	13.330	
1934	*G*	17	13.32	13.32	
1933	*G*	18	13.324	13.324	
1932	*G*	19	13.339	13.339	
1932	*G*	20	13.32	13.32	
1930	*G*	21	13.37	13.37	
1930	*S*	22	13.338	13.338	*d,f*
1929	*G*	23	13.35	13.32	*b,f*
1926	*S*	24	13.335	13.337	*b,h,f*
1921	*S*	25	13.3367	13.3379	*b,f*

* *G* refers to a grating spectrometer measurement; *S* refers to a single-crystal spectrometer measurement.

† The reference numbers refer to the list at the end of this article.

‡ Notation *b* indicates that the measured value was revised to include the effect of a new value for the main reference line; *d* indicates that the measured value was revised to include the effect of new values for the two or more reference lines used; *f* indicates the measurement was originally reported in terms of a relative wavelength scale but has here been converted to angstroms; *h* indicates that the measured value was revised to include the effect of a more accurate value for the grating constant of the crystal used.

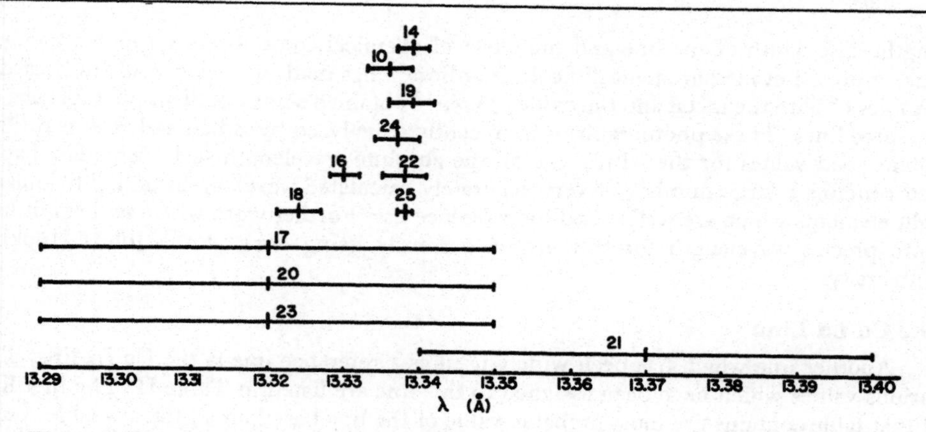

Figure 2. The values from Table II for the Cu $L\alpha$ line with approximate probable errors shown. The numbers give the references from which the value was taken and which are listed at the end of this paper.

which lines will be used as a reference as it did in a recent measurement[26] of Hg M lines when the $L\alpha$ lines of molybdenum and niobium were chosen as reference lines because they bracketed the region of interest, were strong, had substantially less uncertainty than the expected error of the measurement, and were readily available.

PROPOSED REFERENCE POINTS

A number of proposals have been made for increasing the number of accurately known reference points in the soft X-ray region. One of these[27] makes use of the first resonance absorption line of neon, which provides a much narrower line than is usually found in this wavelength region. Furthermore, there is no danger of solid state or chemical effects shifting the wavelength. The circumstance which makes this proposal particularly feasible is the fact that the broad nickel $L\beta$ line covers the region of the Ne K absorption edge so as to make a very convenient source of radiation for the absorption cell. The wavelengths of the various features of the Ne K edge are already known to better than 100 ppm, and the accuracy could be easily improved upon.

It has also been suggested by Liefeld that the high-frequency limit of the bremsstrahlung radiation would make a good and extremely convenient reference mark since it can be shifted to any desired wavelength.

The bremsstrahlung high-frequency limit represents the radiation given off when an electron which has been accelerated in an X-ray tube transfers all of its energy to a single photon in colliding with the anode; hence,

$$eV = hf \tag{1}$$

or

$$\lambda = \frac{hc}{eV} \tag{2}$$

where e is the charge on the electron; V is the voltage on the X-ray tube; h is Planck's constant; f is the photon frequency; λ is the photon wavelength; and c is the velocity of light. A typical plot of the high-frequency limit as obtainable in the soft X-ray region is given in Figure 3.

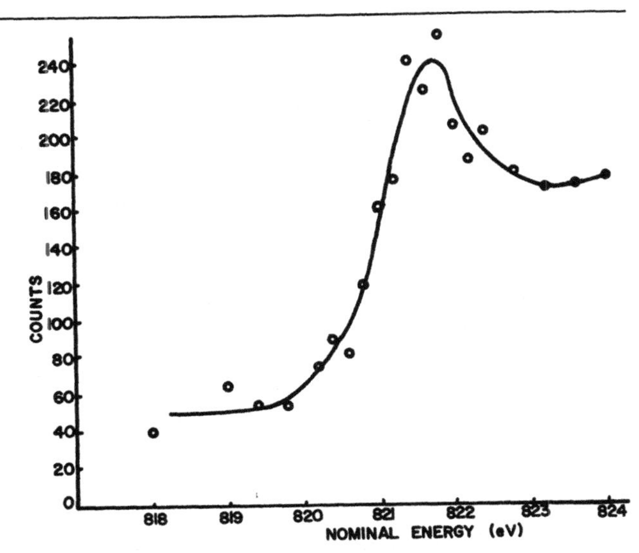

Figure 3. A typical curve showing the high-frequency limit of the bremsstrahlung radiation in the long-wavelength region. The X-ray tube anode was nickel and the tube current was 90 mA.

The data were obtained from an X-ray tube with tungsten filaments and a nickel anode run at 90 mA. The spectrometer was a double-crystal vacuum spectrometer set to a wavelength in the 15-Å region. The detector was an ordinary flow proportional counter with standard electronics. The high-voltage source for the tube was a stable, low-ripple, well-calibrated, 0 to 3000 V power supply. The output of the supply was monitored by a Leeds and Northrup K-3 potentiometer with a calibrated standard cell. In order to prevent the voltage drop across the filaments from affecting the results, data were taken only during those intervals when the filament supply was turned off. The supply was turned off for about 8.4 msec every 17 msec by means of a rectifier in the AC line.

If one assumes that the spectrometer spectral window is symmetric and that the X-ray intensity near the high-frequency limit is a constant independent of the frequency for $hf < eV$ and zero for $hf > eV$ as given by the Sommerfeld theory,[28] the voltage for equation (2) can be taken from the low-voltage inflection point of Figure 3. There are, however, two corrections which must be made to this nominal voltage so that equation (2) becomes

$$\lambda = \frac{hc}{e} \frac{1}{(V_n + V_c + V_\phi)} \tag{3}$$

where V_n is the high-voltage power-supply reading at the high-frequency limit; V_c is the correction obtained from the K-3 potentiometer to be added to the power-supply reading (in effect, calibrating the power supply); and V_ϕ is the voltage necessary to correct for the work function of the filament. In the particular case shown in Figure 3, $V_n = 821.05$ V, $V_c = 3.78$ V, and $V_\phi = 4.5$ V, which makes $\lambda = 14.950$ Å.

The accuracy obtainable from this method is suprisingly good. The value of V_n can be located to within 0.07 V, and it is easy to measure V_c to about 0.01 V so the error from these sources is less than one part in 10,000; thus, they will affect λ by less than 2 mÅ, which, in turn, is less than the error in the wavelength of the Cu $L\alpha$ line shown in Figure 2. However, two important points have so far been neglected. One, it is difficult to measure the work function with an error of the order of 0.01 V. Two, the solid-target bremsstrahlung spectrum in the region of the high-frequency limit is more complicated than the model based on the Sommerfeld theory, which requires a thin target; thus, it is possible that the true high-frequency limit may differ from the low-voltage inflection point by more than 0.05 V. Fortunately, these two difficulties can be overcome by first taking a bremsstrahlung isochromat with the spectrometer set on a well-known point such as the Al $K\alpha$ line or the first resonance peak of neon. Then another isochromat is run at the setting which it is desired to calibrate, and the new wavelength value is obtained by use of the energy difference between the two points. This method does not require any knowledge at all of the filament work function; and, as long as the same point on the isochromats is used, the exact difference between that point and the true high-frequency limit need not be known either.

APPLICATIONS

There are a number of interesting problems whose solution can be facilitated by more accurate wavelength measurements. One of these, the origin of the multiple peaks in the $M\alpha$ lines of the rare earths, has been largely solved with the aid of the data given by Fischer and Baun in the preceding paper[29] in this volume, which includes the first set of rare-earth M-series wavelength measurements in which a strong effort was made to eliminate the effects of self-absorption. Another problem concerns the only reported intrashell X-ray transitions. These $L_{\mathrm{I}}\, L_{\mathrm{II,III}}$ lines were first reported by Tomboulian and Cady[30] and later by others in the 300-Å region. Unfortunately, the energy value for the L_{I} level which one obtains by the use of this line differs considerably from the value obtained from photoelectron measurements.[31] Since the original measurements were necessarily somewhat imprecise, additional wavelength values might resolve this problem. There are some other cases where there is a disagreement between values of the L_{I} level obtained from X-ray absorption measurements and photoelectron measurements which might be resolved by better wavelength values. Of course, better wavelength values will provide better energy-level values for comparison with Hartree–Fock and other calculations and will also facilitate the comparison of work done in different laboratories.

REFERENCES

1. M. Siegbahn, "LXIII Precision-Measurements in the X-ray Spectra Part II," *Phil. Mag.* **37**: 601, 1919.
2. M. Siegbahn, "Absolute X-Ray Wavelengths," *Nature* **151**: 502, 1943.
3. M. Siegbahn, "Über das Rötgenspektrum von Wolfram," *Z. Physik* **20**: 533, 1919.
4. J. J. Merrill and J. W. M. DuMond, "Precision Measurement of the *L* X-Ray Spectra of Uranium and Plutonium," *Phys. Rev.* **110**: 79, 1958.
5. J. W. M. DuMond, "A Survey of Our Present Sources of Information on the Conversion Constant, Λ ($= \lambda g/\lambda s$), and the Absolute Wavelengths of X-Ray Emission Lines," *Proc. Natl. Acad. Sci. U.S.* **45**: 1052, 1959.
6. P. Bergvall, O. Hornfeldt, and C. Nordling, "Crystal Diffraction and Photo Electron Measurements of *K* and *L* Levels in the Elements ^{47}Ag to ^{52}Te," *Arkiv Fysik* **17**: 113, 1960.
7. J. A. Bearden, "Selection of W $K\alpha_1$ as the X-Ray Wavelength Standard," *Phys. Rev.* **137**: B455, 1965.
8. J. A. Bearden, A. Henins, J. E. Marzolf, W. C. Sauder, and J. S. Thomsen, "Precision Redetermination of Standard Reference Wavelengths for X-Ray Spectroscopy," *Phys. Rev.* **135**: A899, 1964.
9. J. A. Bearden, *X-Ray Wavelengths*, U.S. Atomic Energy Commission, NYO-10586, Springfield, Va., 1964.
10. J. A. Bearden, "X-Ray Wavelengths," *Rev. Mod. Phys.* **39**: 78, 1967.
11. B. Nordfors, "On the *K* Spectrum of Aluminium and its Oxide," *Arkiv Fysik* **10**: 279, 1956.
12. F. Tyrén, "Precision Measurements of Soft X-Rays with Concave Grating," *Nova Acta Regiae Soc. Sci. Upsaliensis* **12**(1), 1940.
13. B. Edlén and L. A. Svensson, "The Wavelengths of the Lyman Lines and a Redetermination of the X-Unit from Tyrén's Spectrograms," *Arkiv Fysik* **28**: 427, 1965.

14. J. A. Bearden and A. F. Burr, *Atomic Energy Levels*, U.S. Atomic Energy Commission, NYO-2543-1, Springfield, Va., 1965.
15. J. A. Bearden and A. F. Burr, "Reevaluation of X-Ray Atomic Energy Levels," *Rev. Mod. Phys.* **39**: 125, 1967.
16. F. Tyrén, "Wellenlängenbestimmungen der *L*-Serie an den Elementen ^{29}Cu bis ^{26}Fe," *Z. Physik* **98**: 768, 1936.
17. M. Siegbahn and T. Magnusson, "Zur Spektroskopie der Ultraweichen Röntgenstrahlung. III.," *Z. Physik* **88**: 559, 1934.
18. J. A. Prins, "Die Struktur Einigen Ultraweichen Röntgenlinien," *Z. Physik* **81**: 507, 1933.
19. R. B. Witmer and J. M. Cork, "The Measurement of X-Ray Emission Wavelengths by Means of the Ruled Grating," *Phys. Rev.* **42**: 743, 1932.
20. J. A. Prins, "Absolute Wellenlängen Messungen von Röntgenstrahlen II," *Physica* **12**: 15, 1932.
21. C. E. Howe, "The *L* Series Spectra of the Elements from Calcium to Zinc," *Phys. Rev.* **35**: 717, 1930.
22. A. Karlsson, "Die Gillerkonstante hochmolokularer, gesättigter Fettsäuren nebst den Röntgenspektrem innerhalb der *K*- und *L*-Reihen bei einigen niedrigeren Elementen," *Ark. Mat. Astr. Fys.* **22A**: (9), 1930.
23. G. Kellstrom, "Wellenlängenbestimmungen in der *L*-Reihe der Elemente ^{29}Cu bis ^{20}Ca mit Plangitterspektragraph," *Z. Physik* **58**: 511, 1929.
24. R. Thoraeus, "The X-Ray Spectra of the Lower Elements," *Phil. Mag.* **1**: 312, 1926.
25. E. Hjalmar, "Beiträge zur Kunntnis der Röntgenspektren," *Z. Physik* **7**: 341, 1921.
26. A. F. Burr, "*M* Series X-Ray Emission Lines of Mercury," *Bull. Am. Phys. Soc.* **11**: 389, 1966.
27. R. J. Liefeld, "The *K*-Shell Photon Absorption Spectrum of Gaseous Neon," *Appl. Phys. Letters* **7**: 276, 1965.
28. A. Sommerfeld, "About the Production of the Continuous X-Ray Spectrum," *Proc. Natl. Acad. Sci. U.S.* **15**: 393, 1929.
29. D. W. Fischer and W. L. Baun, "The Influence of Sample Self-Absorption on Wavelength Shifts and Shape Changes in the Soft X-Ray Region: The Rare-Earth *M* Series," in: G. R. Mallett, J. B. Newkirk, and H. G. Pfeiffer (eds.), *Advances in X-Ray Analysis, Vol. 11*, Plenum Press, New York, 1968, pp. 230–240.
30. D. H. Tomboulian and W. M. Cady, "Radiative X-Ray Transitions Within the *L* Shell," *Phys. Rev.* **59**: 422, 1941.
31. A. Fahlman, D. Hamrin, R. Nordberg, C. Nordling, and K. Seigbahn, "Revision of Electron Binding Energies in Light Elements," *Phys. Rev. Letters* **14**: 127, 1965.

DISCUSSION

Chairman H. S. Peiser (National Bureau of Standards): I wonder, have the Al $K\alpha_1$, $K\alpha_2$ lines been separated?

A. F. Burr: They have been separated, after a fashion. That is, there are a number of curves in which they are just barely resolved, if you have a good imagination. By use of the theoretical intensity relationship, one can assign a wavelength value to each component, but to say that they are really resolved is slightly stretching a point, but not much.

H. K. Herglotz (E. I. du Pont): Would you please tell me that second term in your denominator there? I missed that.

A. F. Burr: The second term is just sort of a calibration. It calibrates the power supply. The first term represents the direct reading of the power supply as it comes from the manufacturer.

H. K. Herglotz: It's probably the ripple of the DC.

A. F. Burr: No, the power supply is filtered enough so that the ripple will not make any difference. The middle term just represents the correction factors and the National Bureau of Standards volts, as determined by a K-3 potentiometer. This is just to emphasize that the method is accurate enough so that the very best voltage measurement that you can make is worthwhile.

H. K. Herglotz: My second question is that I am a little at odds with your third term, which is supposed to be the work function of the tungsten of the cathode. In the common mode of operation of X-ray tubes, the electrons are not supplied out of the tungsten filament, but from the cloud of electrons surrounding the filament. Besides, isn't the thermal energy of these electrons of about the same magnitude as the work function?

A. F. Burr: The energy spread due to the hot electrons coming off the filament is partially responsible for the slope of the curve at the high-frequency limit. The work function correction has to be made for the filament since the potential that one is measuring is between the Fermi level of the filament and the plate and some additional energy had to be supplied to the electrons in order to take them out of the tungsten into the vacuum. This is supplied by the energy that is given to the filament circuit, but the corrections have to be put in here if you want the energy that the electron has just before it stops.

A. Giamei (Pratt and Whitney Aircraft): I have a short question for Dr. Burr concerning the placement of X-ray wavelengths on a relative scale. You very briefly mentioned the possibility of using a Mössbauer source as a standard, and I was curious as to why one would choose an X-ray line as a standard rather than a nuclear source which is, of course, much sharper and therefore much easier to define.

A. F. Burr: A Mössbauer line is much sharper than any X-ray line and, hence, in principle, much easier to define. However, the intensity of any practical source is so small that precision measurement is extremely difficult. The most precise Mössbauer measurement made (on the Fe^{57} line) has a relative error of the order of 10 ppm, an order of magnitude worse than the present secondary X-ray standards.

APPLICATIONS OF A PORTABLE RADIOISOTOPE X-RAY FLUORESCENCE SPECTROMETER TO ANALYSIS OF MINERALS AND ALLOYS*

J. R. Rhodes and T. Furuta

Texas Nuclear Corporation
Austin, Texas

ABSTRACT

A portable, battery-operated X-ray fluorescence analyzer weighing 15 lb is described, consisting of a NaI(Tl) scintillation-counter probe and an electronic unit with a single-channel pulse-height analyzer and reversible scaler. Radioisotope X-ray sources are used for excitation of the sample and, where necessary, balanced filters for resolution of neighboring characteristic X-rays. Emphasis has been placed on designing and producing an instrument that is easy and convenient to operate in laboratory, factory, or field conditions and that can equally well be used to measure extended surfaces, such as rock faces, or finite samples in the form of powders, briquettes, or liquids. The feasibility of the following analyses has been studied by using for each determination the appropriate radioisotope source and filters: sulfur in coal; calcium and iron in cement raw mix; copper in copper ores; and vanadium, chromium, molybdenum, and tungsten in steels. Detection limits, based on counting statistics obtained in count times of 10 to 100 sec, range from 0.03% for copper in ores to 0.2% for sulfur in coal. Both matrix absorption and enhancement effects were encountered and were eliminated or reduced substantially by suitable choice of source energy, by the use of nomograms, or by semi-empirical correction factors based on attenuation or scattering coefficients.

INTRODUCTION

There has been a long-standing need to develop a practical and truly portable X-ray fluorescence spectrometer for rapid, nondestructive, *in situ* analysis in mining, geology, metallurgy, and engineering.

Such an instrument could have many applications in routine assay of ores, cement, and coal; in alloy analysis for identification purposes; and in "go, no-go" and other on-the-spot measurements for quality control of a wide range of processes and products. Typical problems include scrap sorting before manufacture of alloys; checking ingots, sheet, bars, tubes, and other forms of stock; and recognition of components after certain manufacturing processes, such as extrusion or forging, which obliterate ordinary markings. In geology and mining, rapid assay of pulp, core, or rock samples in the field, in a field laboratory, or on board dredgers could be used to roughly delineate ore bodies, strata, beach sands, and other economic mineral deposits. Thus, valuable information would be immediately available and decisions could be made without having to wait days or weeks for the result of laboratory analyses. Also, the large number of samples that at

* Work supported by the United States Atomic Energy Commission, Division of Isotopes Development.

present are returned to the central laboratory for assay could be reduced by orders of magnitude, with concomitant reduction of work load and operating costs.

The suitability of X-ray fluorescence spectrometry for sensitive, nondestructive, and rapid elemental analysis is well known. However, X-ray tubes require bulky high-voltage power supplies and copious cooling water. The standard dispersive method for energy selection, employing a crystal spectrometer, is also unsuitable for portable field instruments, again because of the spectrometer's size, but also because of the delicacy of the precision-engineered moving parts. On the other hand, radioisotope X-ray sources are small, compact, inexpensive, and require no external power. Their low photon output (about 10^7 sec^{-1} compared with about 10^{13} sec^{-1} from an X-ray tube) precludes the use of dispersion for energy selection because of the high geometrical losses in the narrow collimators required to obtain suitable resolution.

Thus the techniques that must be used in a truly portable X-ray analyzer involve excitation of the characteristic spectra with a radioisotope source and nondispersive energy selection. In the past 10 years these techniques have been studied in a number of laboratories and have been refined to a stage at which practical instruments have been made. In a recent review of radioisotope X-ray spectrometry,[1] it was noted that the concept of X-ray fluorescence analyzers that are simpler, cheaper, and much more compact than the conventional apparatus is going to demand a new approach to the application of X-ray analysis. Many uses not hitherto envisaged have been found and more are likely to arise as this approach becomes more universal.

In spite of the apparent wealth of potential applications, to the authors' knowledge, only one portable radioisotope X-ray fluorescence analyzer is commercially available at the time of writing.[2-4] Other earlier work in this area[5-8] does not seem to have led to such developments.

The instrument that has been developed in this laboratory (see Figure 1) is basically a scintillation-counter probe with a single-channel pulse-height analyzer for coarse energy selection and a reversible scaler for output display. The scintillation-counter probe houses balanced filters for isolation of the appropriate characteristic X-rays (when necessary), a radioisotope source, and a fail-safe shutter that operates when the sample and probe are brought together.

A number of feasibility studies have been made to test the performance of the instrument in ore, coal, cement, and alloy analysis. The optimum choice of source and filters for each analysis and the results of typical determinations are detailed in the rest of the paper.

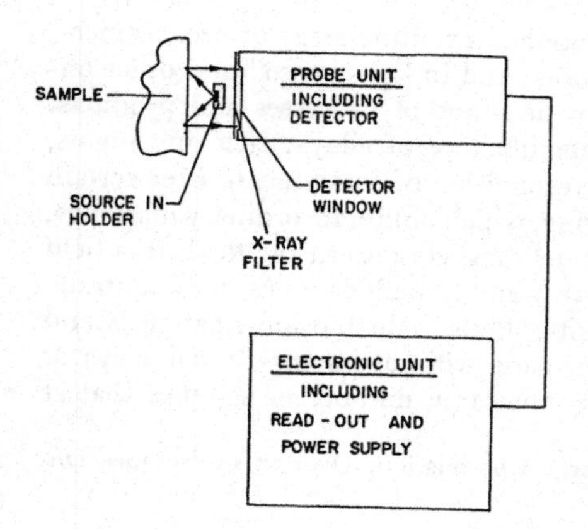

Figure 1. Basic arrangement for nondispersive radioisotope X-ray spectrometer.

SPECIFICATION AND DESCRIPTION OF INSTRUMENT

Desirable Properties of a Portable Analyzer

The instrument should be capable of use both indoors and outdoors, in laboratory, factory, or field conditions. For field use, it should be easily carried by a man and be independent of all supplies (*e.g.*, electrical power, cooling water, counter gas, liquid nitrogen). Two kinds of samples should be measurable: (1) an extended surface, such as a rock face, core sample, alloy sheet, or steel coil, and (2) a finite and small volume, such as a powdered solid, a liquid, or an alloy specimen. Thus, the probe could either be hand-held for presentation to the sample or mounted so that small samples in or out of containers can be presented to it.

In the field, the instrument is likely to experience wide temperature variations and should therefore be capable of operating in ambient temperatures from, say, −20 to +50°C. It is too much to expect the equipment to stay on calibration over this entire range, but it should be stable against diurnal temperature transients of 10 to 20°C.

The instrument should, of course, be reasonably rugged and shockproof.

Although the photon output of radioisotope X-ray sources is so low that the radiation dose at 1 ft from an unshielded source is near or below the maximum permissible level (2.5 mR/hr), some form of protection should be incorporated into the measuring head to prevent accidental exposure of, say, the fingers by coming into near contact with the emitting surface of the source. Also, the radioactivity should be sealed into the source capsule to prevent contamination in the event of breakage of the apparatus.

The analytical performance of the instrument should be such as to allow the determination of as many elements as possible down to concentrations of 0.1% or below, in measurement times of 1 min or less. Poorer sensitivity than this would seriously limit applicability for rapid analysis of major constituents. On the other hand, trace analysis demands the capability of detecting a few parts per million, and this is not the aim of the present instrument.

With the standard "central source" geometry (see Figures 1 and 2), distances between source, sample, and detector are about 1 cm, so that air absorption only becomes appreciable for energies of 2 keV (P *K* X-rays) or below. This is a fundamental limitation because evacuation of the source–sample–detector space, or flushing with hydrogen or helium, is not considered practicable in the instrument as specified.

In spite of the low emission from radioisotope X-ray sources, the high-efficiency geometry gives rise to high detected count rates. One of the features of radioisotope X-ray spectrometry is operation at relatively high count rates and low signal-to-noise ratios.[1,3] The detector and electronic equipment must, therefore, be capable of accepting count rates of up to 10^5 sec^{-1} without changes in gain, resolution, or linearity.

Energy Selection

X-ray filters are used to isolate the characteristic lines from a given element. These filters are thin metal or loaded plastic foils placed over the detector window. The detector itself, if a scintillation or proportional counter, may afford coarse energy resolution and, if so, the electronic unit would include a pulse-height analyzer.

The action of balanced filters in isolating characteristic X-rays has been described in the literature[9–11] but is reviewed here for convenience. By referring to Figure 3, it can be seen that, by thickness adjustment, the X-ray transmissions through two filters of adjacent atomic number can be made equal over a wide range of energies, except for the "pass band" between their two absorption edges. Thus, the difference between the two transmissions is proportional to the X-ray flux in the narrow energy band between the two absorption edges.

In use, balanced filters require the subtraction of two measurements. In a portable instrument, the method used is to move the two filters successively over the detector

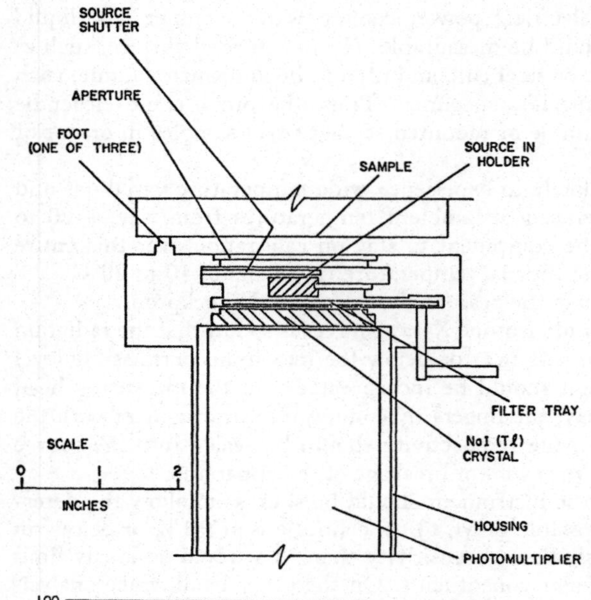

Figure 2. Layout of probe showing positions of filters, source, shutter, and sample.

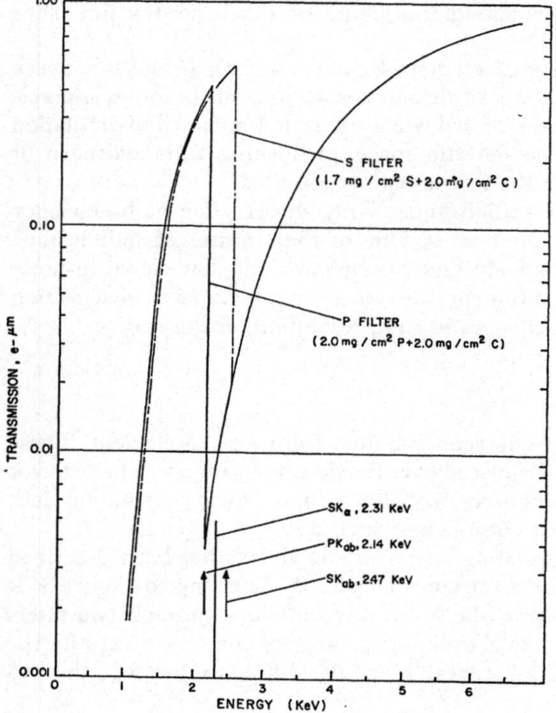

Figure 3. Calculated transmissions for sulfur and phosphorus balanced filters.

window without moving the sample. The difference reading is obtained in any convenient way (see below).

The main drawback of balanced filters is the rapid increase in statistical error in the difference count for increasingly smaller ratios of wanted to unwanted radiation incident on the filters. It has been found that, if the detector itself has a significant energy resolution, it can be used in conjunction with filters to improve the sensitivity of analysis. For example, the exciting energy from the source can often be chosen so that the back-scattered X-rays are resolved by the detector from the desired fluorescent X-rays.

Choice of Detector

The choice of detector is a fundamental step in the design of a portable X-ray analyzer because it affects the analytical performance, on the one hand, and much of the electronic specification, on the other. Many detectors of X-rays are available, but, in the light of the overall specification laid down in the previous section, only four types are worth consideration. These are Geiger–Müller tubes, proportional counters, scintillation counters, and Si(Li) or Ge(Li) solid-state detectors.

The Geiger–Müller tube is rugged, sensitive, and inexpensive, but is relatively slow (*i.e.*, has a long dead time), and its output is not proportional to X-ray energy. At small increase in cost and electronic complexity, the much higher performance proportional or scintillation counters, can be employed. The proportional counter is fast and has a relatively good energy resolution, being able to resolve characteristic X-rays from elements some four atomic numbers apart. The NaI(Tl) scintillator–photomultiplier combination is also fast enough to count 10^5 sec^{-1} without appreciable dead-time losses or non-linearity. Its resolution is not as good as that of the proportional counter but its detection efficiency for X-rays above about 20 keV greatly exceeds that of the proportional counter. The ability of the proportional counter to measure X-rays down to 1 keV and below is less relevant in view of our restriction to the use of air paths.

Although energy resolutions of 0.5 keV or better are now obtainable with the best solid-state detector systems, these results are only obtained with the use of cryogenic cooling and very low noise cooled F.E.T.* preamplifiers. This alone would not rule out these detectors completely, but they have the additional disadvantage of their small area (1 cm^2 or less for the best resolution), which precludes the use of efficient geometry. A recent paper[12] showed that the very large increase in signal-to-noise ratio obtained by using a high-resolution solid-state detector for X-ray analysis of silver ores was almost completely offset by the very low count rates obtained, which necessitated hour-long measurement times to obtain high sensitivity.

The final choice between sealed proportional counter and scintillation counter was made in favor of the scintillation counter as a result of two further studies. It has been reported that the gas multiplication and energy resolution of X-ray proportional counters are sensitive to incident X-ray flux.[13,14] We performed experiments to investigate whether or not these effects were significant in a sealed, Xe/CH_4-filled proportional counter† designed especially for radioisotope X-ray spectrometry. It was found that the gas gain was reduced progressively as the flux of Cu *K* X-rays (8.0 keV) was increased until, at a detected flux of 5×10^4 sec^{-1}, the gain had been reduced by 30%. Also, the photopeak became non-Gaussian, and the energy resolution deteriorated. These results are in accordance with the space charge effects already reported in other proportional counters.[13,14]

The second factor that influenced our choice is the recent availability of a 10-stage,

* F.E.T. denotes field effect transistor.

† Type PX 130, Twentieth Century Electronics, Ltd.

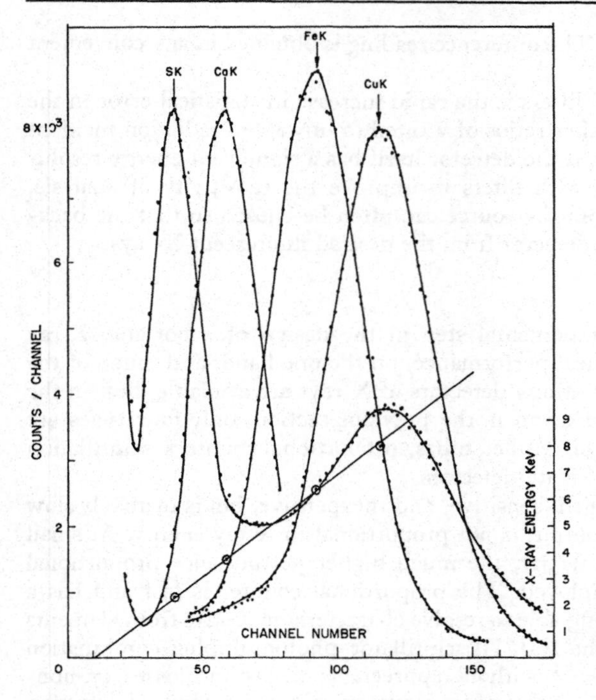

Figure 4. X-ray spectra excited by ^{3}H/Zr and detected by a Harshaw NaI(Tl) crystal coupled to an E.M.I. photomultiplier, type 9656R.

2-in.-diameter photomultiplier tube* with such high quantum efficiency and low noise that S *K* X-rays (2.3 keV) are measurable with its use. The spectra shown in Figure 4 were obtained by using this photomultiplier coupled via a 0.5-in.-thick quartz light guide to a 1.75-in.-diameter by 2-mm-thick NaI(Tl) crystal† with a 0.005-in.-thick beryllium window. Characteristic spectra were excited by a standard 4-Ci ^{3}H/Zr source‡ in samples of sulfur, calcium carbonate, iron, and copper. The S *K* X-rays were resolved from noise with a peak-to-valley ratio of 2.6 : 1.

It should be noted that the high-performance photomultiplier tube is only used when necessary, that is, for measurement of S *K* to Ca *K* X-rays. At higher energies, other, less expensive tubes can be used.§

Thus the NaI(Tl) scintillation counter is clearly the best choice of detector, combining simplicity, stability, and ruggedness with the capability of measuring characteristic X-rays over the whole energy range where air paths are feasible, namely, from the elements sulfur upward.

Choice of Electronic System

The sophistication of the electronic unit is limited by considerations of power drain, size, weight, and stability, and governed by the requirements and output of the probe unit.

The choice of the scintillation counter as detector largely fixes the specifications for pulse amplifiers and photomultiplier high-voltage supplies. In addition, inclusion, or

* Type 9656 KR, E.M.I., Ltd.
† Type Harshaw.
‡ Type TRT2, Radiochemical Centre, Amersham.
§ Such as Radio Corporation of America Types 6342A or 4518, or E.M.I. Types 9656 KS or 9656 KB.

otherwise, of a pulse-height analyzer must be decided upon and whether the output should have digital or analogue form. An earlier portable X-ray analyzer[2-4] used a scintillation-counter detector but did not include a pulse-height analyzer because the extra electronic complexity was not considered worthwhile at the time. The continuing miniaturization and improvements in reliability of electronic systems have now rendered the inclusion of a pulse-height analyzer quite straightforward. As mentioned above, this should lead to specific improvements in analytical performance.

As to the form of output display, the main choice lies between rate-meter and scaler. A rate-meter is much simpler and cheaper and is widely used in field instruments. However, a scaler does not have a drift problem, uses less power because no zero suppression circuit is necessary, and requires less time to take a reading with a given statistical error. Furthermore, experience with field rate-meters indicates that operators prefer to be presented with a displayed number rather than to read a fluctuating meter needle.

A third main point of consideration is what type or types of power supply should be available. Battery operation is essential for field use and various types of batteries are available, including rechargeable cells. In the laboratory, even if a field laboratory, and in factories and workshops, line-voltage outlets are readily accessible. For battery operation, a compromise must be struck between battery life (which must be at least one working day) and electronic performance, particularly tolerance to high count rates.

Other considerations are the number, type, and layout of the controls and switches (the number of adjustable controls should be a minimum); capacity of the scaler and whether or not it should have a "subtract" facility; choice of timing interval or intervals; range of variation of amplifier gain, photomultiplier high-voltage, and discriminator levels; and tolerance to high rates of large input pulses (non-overload characteristics). Also, the overall shape of the instrument case is an important factor when it has to be carried by hand for long distances.

Description of Instrument

Figure 1 shows the basic arrangement of the instrument, and Figure 2, a schematic layout of the probe, including the relative positions of source, filters, safety shutter, and sample.

Figures 5, 6, and 7 show three photographs of the prototype instrument. Figure 5 is a general view of the probe and electronic units.

In operation as a hand-held unit (Figure 6), depression of one of the feet by pressing the probe against the material to be measured causes the source shutter to open and so exposes the source to the sample. A second mode of operation is as a bench-mounted instrument (see Figure 7) when small samples are placed over the aperture (aperture sizes from 1.8 to 0.9-in. diameter can be used). In this case, the shutter is actuated when the sample is covered by a cap which depresses the movable foot. The shield in the source shutter is a 0.5-in.-diameter by 0.06-in.-thick tungsten-alloy disc, which attenuates the low-energy X-radiation from the sources used by a factor of about 10^5. The cap has enough space under it to accommodate the usual range of samples in the form of alloys, briquettes, powders, and liquids. Sample holders with thin Mylar windows are used for powders and liquids.

Although much work has been reported[1-4] on the characteristics of central-source geometry and the present measuring head is of standard design, it is appropriate to mention here two important properties. Firstly, because of the shadow effect of the source in the middle of the detector window, there exists an optimum sample distance where the detected flux from the sample is a maximum.[3,4] For minimum sensitivity to sample

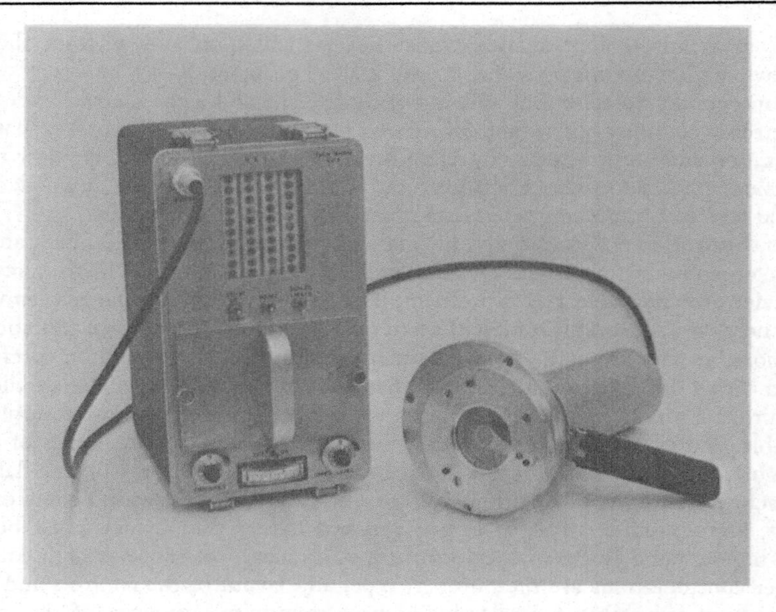

Figure 5. Photograph of portable analyzer showing probe and electronic units.

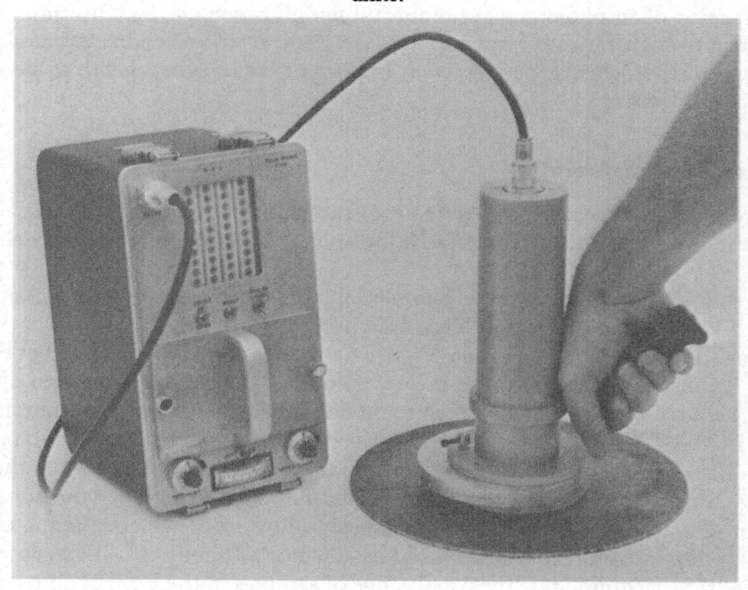

Figure 6. Photograph showing method of operation for analysis of extended surfaces.

movement, the surface of a smooth sample should be at this distance and care has been taken in the design of the measuring head to ensure this. A movement of 0.05 in. of the sample away from the surface of the measuring head affects the count rate by less than 1%.

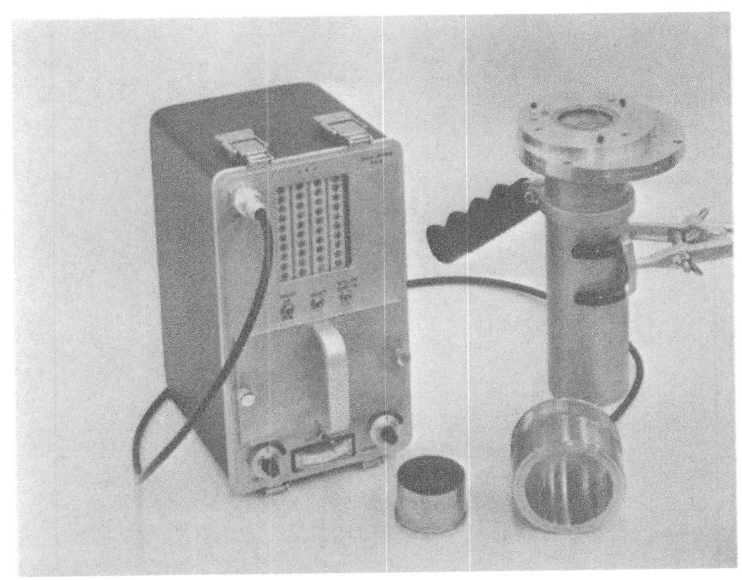

Figure 7. Tentative arrangement for analysis of small samples, including powders and liquids.

The second point concerns the aperture diameter, whose standard value in this apparatus is 1.8 in. Reduction of the aperture to measure smaller samples results in a lower sensitivity owing to reduced fluorescent intensity from the exposed sample and increased scattering from the underside of the aperture plate. It has been shown[3] that the sensitivity for a given analysis is reduced by a factor of about 2 if the aperture diameter is only 1 in. For smaller diameters, the sensitivity falls off rapidly and alternative designs of the measuring head have to be used.

Each photograph shows clearly the layout of the front panel. The scaler has four decades displayed, after two suppressed scales of eight, the assumption being that the latter are not statistically significant at the high count rates used. The total capacity is therefore 640,000 counts. The threshold and upper-level controls allow windows of variable widths to be set about given energy regions in the range of 2 to 1000 keV. The meter indicates the state of the battery charge. The battery unit holds ten 1.5-V "D-sized" dry cells which have a life of about 50 hours in continuous operation at room temperature. They can be replaced by alkaline or mercury cells of similar shape for low-temperature uses and longer life. Alternatively, a line-operated power supply is available which fits into the battery box.

The reversible scaler is designed primarily for operation with balanced filters in the probe. The procedure is, with the first filter in position, to scale *up* for one timing interval (10 sec), then change filters by flicking over the lever on the probe head, and scale *down* for one timing interval. If better counting statistics are needed, the process is repeated for more timing intervals. The basic timing interval can also be changed if required.

APPLICATIONS

Although, at present, the only investigations made have been in the laboratory with the prototype instrument and standard samples, the analyses described below serve to

Table I. Radioisotope Sources Used and Some Relevant Properties

Isotope	Half-life, years	Type and energy of radiation emitted	Approximate flux, photon/sec over 4π	X-rays excited in samples	Activity used	Maximum practicable activity
$^{3}H/Ti$	12.3	Bremsstrahlung, 3–10 keV; Ti *K* X-rays, 4–5 keV	2×10^{7}	S *K* Ca *K*	3.5 Ci	5 Ci
$^{3}H/Zr$	12.3	Bremsstrahlung, 3–10 keV	7×10^{6}	Ca *K* Cr *K* Mn *K* Fe *K*	4.5 Ci	4.5 Ci
^{55}Fe	2.7	Mn *K* X-rays, 5.9 keV	7×10^{6}	Ca *K*, V *K*	1 mCi	20 mCi
^{210}Pb	22	γ-rays, 47 keV; Bi *L* X-rays, 11 keV	4×10^{7}	Cr *K* Mn *K* Fe *K*	10 mCi	50 mCi
^{238}Pu	89.6	U *L* X-rays, 15–17 keV	3×10^{7}	Fe *K* Cr *K* Mn *K*, W *L*	20 mCi	30 mCi
^{109}Cd	1.3	Ag *K* X-rays, 22 keV	3.5×10^{7}	Fe *K* Cu *K* W *L*, Mo *K*	1 mCi	5 mCi
^{241}Am	460	γ-rays, 60 keV; Np *L* X-rays, 11–22 keV	2×10^{7}	Mo *K* Ag *K*	1 mCi	100 mCi
$^{147}Pm/Al$	2.6	Bremsstrahlung, 10–100 keV	2×10^{7}	Mo *K* Ag *K*, W *K*	0.5 Ci	5 Ci
^{153}Gd	0.65	γ-rays, 100 keV; Eu *K* X-rays, 42 keV	1.5×10^{7}	W *K*	1 mCi	5 mCi
^{57}Co	0.74	γ-rays, mainly 122 keV; Fe *K* X-rays, 6.4 keV	3.5×10^{7}	W *K*	1 mCi	10 mCi

illustrate the capabilities of the portable analyzer and to exemplify some of the modes of utilizing the two-position filter assembly and the reversible scaler.

In each feasibility study, the optimum source was chosen from those commercially available. The sources used, together with relevant data, are listed in Table I. The choice of these sources depends on the ability to excite characteristic X-rays efficiently without appreciable scattering; a half-life long enough to permit easy correction for decay and cheap source replacement; a specific activity sufficient to yield an emission of at least 10^7 photons/sec/cm^2 of source surface; and the absence of radiations of energy above 200 keV since they cannot be screened from the detector when central-source geometry is used. Although ^{153}Gd and ^{210}Pb are listed as suitable sources, they were not used in this investigation because they were temporarily unavailable.

Sulfur in Coal

Sulfur in coal has both organic and inorganic (*i.e.*, pyritic) origin. The mineral matter in coal contains significant concentrations of aluminum, silicon, calcium, iron, and sulfur so that *coal ash* consists mainly of alumino-silicates, iron oxide, and some calcium oxide. The combustible part of coal is mainly carbon, hydrogen, and oxygen. Sulfur occurs in concentrations up to 5% and is an important air pollutant.

Incidentally, it is also of practical interest to know the ash content of coal both at coal preparation plants, to control washing and blending, and at coal-fired power stations, to control combustion efficiency. Radioisotope X-ray backscattering is already in use[15] for this measurement, so it can be safely assumed that the present instruments can measure total ash content by the same method.

To measure sulfur content, ^{3}H/Ti and ^{55}Fe sources (see Table I) were compared for excitation of S *K* X-rays (2.31 keV). Artificial samples were made up by briquetting graphite powder, Al_2O_3, $CaCO_3$, and Fe_2O_3 with flowers of sulfur to give a range of sulfur contents (0 to 5%) and a range of ash contents (4 to 16%).

Although Al *K* and Si *K* X-rays are too soft to penetrate the air gap and detector window, it is necessary to use balanced sulfur and phosphorus filters to isolate S *K* X-rays from Ca *K* X-rays (3.7 keV).

Production of the sulfur and phosphorus filters posed a special problem since the required values of mass per unit area of each element are only about 2 mg/cm^2. The calculated transmission curves of a pair of balanced sulfur and phosphorus filters of optimum mass per unit area are shown in Figure 3. After a number of possible ways of making the filters had been considered, a method proposed by Dunne[16] was used. A colloidal suspension of powdered sulfur or red phosphorus was made by ball-milling the powder with a solution of polystyrene in toluene. This "paint" of known sulfur (or phosphorus) concentration was spread in a uniform, known thickness on a lubricated glass plate with the use of a film casting knife. The dry film, about 0.001-in. thick, was stripped off, and selected areas were cut out to make the filters. The two filters chosen for the experiment each contained about 50% (by weight) of the element and had values of total mass per unit area of 3.5 to 4 mg/cm^2.

Measurements on the coal samples were made by using these filters and the above mentioned sources. The best results were obtained by using ^{3}H/Ti and the single-channel pulse-height analyzer with its window set about the S *K* X-ray energy, as shown in Figure 8. A standard deviation of 0.17% sulfur was obtained in a count time of 100 sec with each filter. The calibration curve obtained is shown in Figure 9 for two values of total ash.

It is evident that the sulfur determination is insensitive to total ash content. This

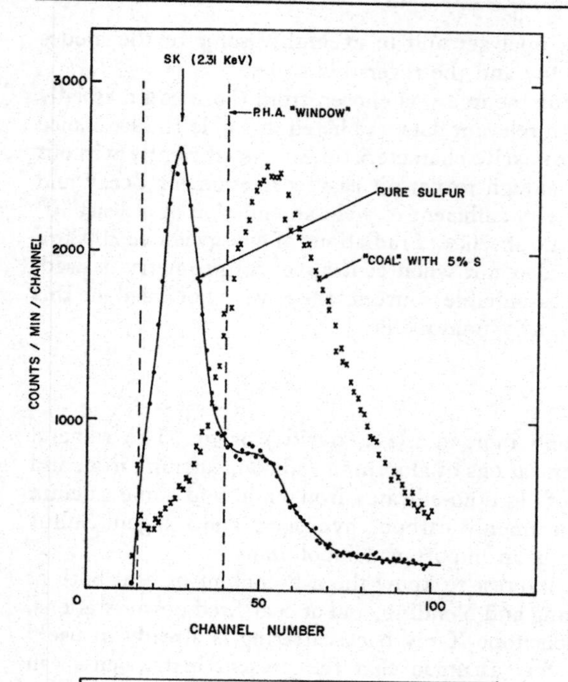

Figure 8. Spectra from "coal" and sulfur samples excited by a 3.5-Ci $^3H/Ti$ source.

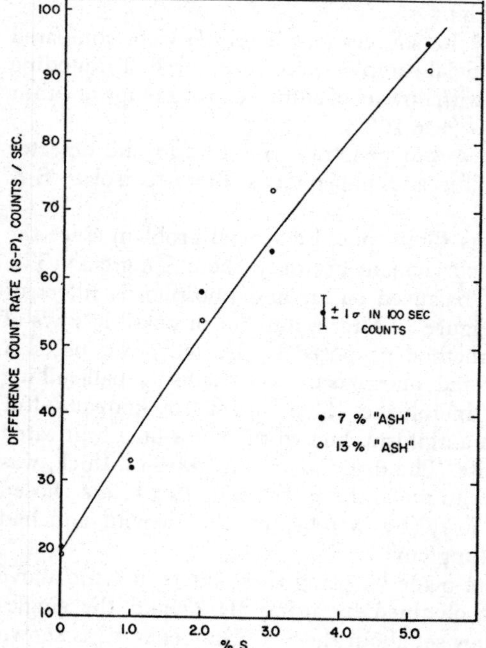

Figure 9. Calibration curve for sulfur in "coal" by using a $^3H/Ti$ source, balanced sulfur and phosphorus filters, and a pulse-height analyzer.

is not so if balanced filters are not used, when the high ash curve lies considerably above the low ash one. This is due to detection of Ca *K* X-rays, which are unresolved from S *K* X-rays and so, in part, fall in the pulse-height analyzer window.

Calcium Oxide and Iron Oxide in Cement Raw Mix

Cement raw mix contains about 44% CaO (as calcium carbonate) and 1 to 5% Fe_2O_3. It is important to have an accurate measure of CaO content, and the Fe_2O_3 content should not exceed specified values in given cements.

The best sources for excitation of Ca *K* X-rays (3.7 keV) are ^{55}Fe, $^{3}H/Ti$, and $^{3}H/Zr$. Fe *K* X-rays (6.4 keV) are best excited by $^{3}H/Zr$.

Further comparisons showed that ^{55}Fe can be used to excite Ca *K* X-rays while not exciting Fe *K* X-rays at all. On the other hand, $^{3}H/Zr$ can be used to excite Fe *K* X-rays; Ca *K* X-rays that are also excited can be cut out by a 0.0005-in.-thick iron filter. With this source, there is the possibility of monitoring both CaO and Fe_2O_3 contents by making two measurements, one with and one without the iron filter. This type of double measurement is particularly appropriate for use with the portable analyzer, as will be seen. The $^{3}H/Ti$ excites Fe *K* X-rays less efficiently and Ca *K* X-rays more efficiently than $^{3}H/Zr$ does. This source was found to be much less suitable for making the combined measurement than $^{3}H/Zr$. It is also less suitable for making an interference-free measurement of CaO than ^{55}Fe and so was not considered further.

A suite of samples with varying Fe_2O_3 and CaO contents was made up by adding Fe_2O_3, $CaCO_3$ (to increase the CaO content), or starch (to decrease the CaO content), to a cement raw-mix sample. The CaO contents of the suite ranged from 38.0 to 50.0% and the Fe_2O_3 contents from 1.45 to 4.5%.

Figure 10 shows a graph of the count rate, with the use of ^{55}Fe to excite Ca *K* X-rays, against the per cent of CaO. A number of different Fe_2O_3 concentrations are used, and it is evident that, within the limits of statistical error, there is no significant matrix absorption effect due to iron. This is to be expected, as the energies of both the exciting radiation and the fluorescent X-rays are below the Fe *K* absorption edge.

At 45% CaO, the standard deviation in a 50-sec count is equivalent to 0.1% CaO. The source activity, 1 mCi, can be readily increased by a factor of 10.

The concave shape of the calibration curve, from 38 to 44% CaO, is due to the starch addition for reducing the CaO content. The low mass absorption coefficient of starch allows more Ca *K* X-rays to be emitted from the sample than would be the case if the matrix remained constant.

Figure 11 shows the count rate against Fe_2O_3 content with the use of a 4-Ci $^{3}H/Zr$ source and 0.0005-in.-thick iron filter to cut out Ca *K* X-rays. It is seen that the correlation with iron oxide content is excellent, and interference due to variable CaO concentration is not significant. The standard deviation due to counting statistics is equivalent to 0.045% Fe_2O_3 at 3% Fe_2O_3 in a 50-sec count.

In Figure 12, $(I_0 - 2I_{Fe})$ is plotted against CaO content by using the $^{3}H/Zr$ source. Here I_0 is the count rate with no filter used; and I_{Fe}, that with the iron filter used. The iron filter absorbs 50% of the incident Fe *K* X-ray flux; thus $(I_0 - 2I_{Fe})$ represents the flux of Ca *K* X-rays. Within the limits of statistical error (one standard deviation at 45% CaO is equivalent to 0.12% CaO in a 50-sec count), the correlation with CaO content is excellent.

Since data presentation on the portable analyzer is by means of a reversible scaler timed in units of 10 sec, $I_0 - 2I_{Fe}$ can be obtained simply by counting *up* for *n* units of 10 sec with no filter, then counting *down* for 2*n* units of 10 sec with the iron filter. Thus, both iron and calcium oxide contents can be measured by using the same source and the above procedure.

Copper in Copper Ores

The range of copper content in copper ores of economic grade varies from about

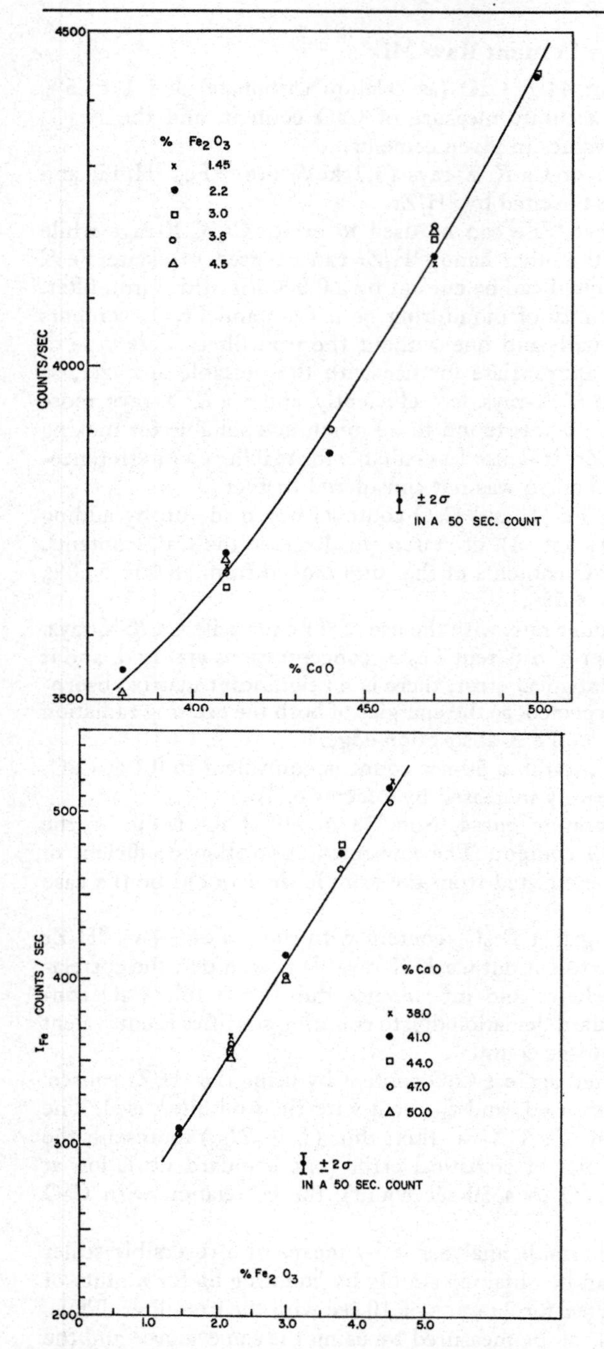

Figure 10. Count rate *versus* per cent of CaO with the use of a ^{55}Fe source and no filter.

Figure 11. Count rate *versus* per cent of Fe_2O_3 with the use of a ^{3}H/Zr source and an iron filter.

0.5% copper upward. The copper is almost always accompanied by iron (in the range of 0.5 to 7% iron, approximately), and occasionally by cobalt (up to 2% cobalt). The characteristic X-rays of these three elements are not resolved by either a scintillation or a proportional counter.

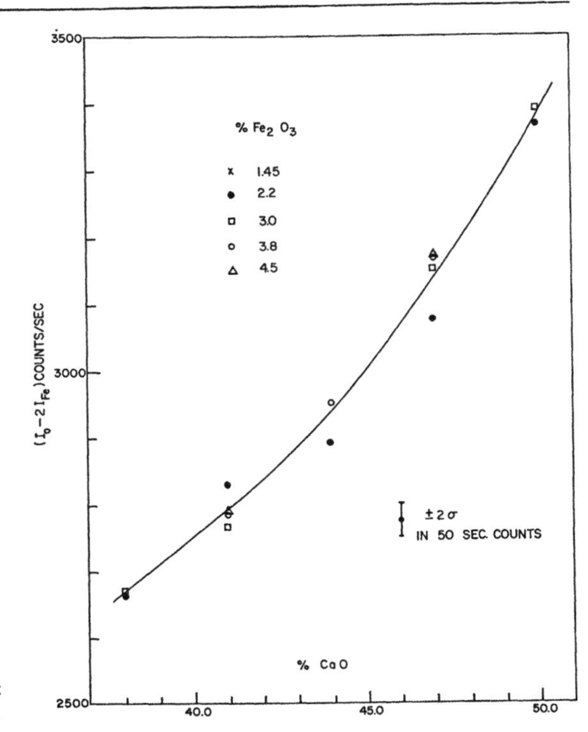

Figure 12. $(I_0 - 2I_{Fe})$ *versus* per cent of CaO with the use of a $^3H/Zr$ source.

It has been shown previously[17] that adequate sensitivity can be obtained by using ^{109}Cd as the exciting source and discriminating against the scattered radiation (20 keV) with a proportional counter. However, the only results reported to date, using a portable analyzer with a scintillation counter (without a single-channel pulse-height analyzer), have yielded the barely adequate sensitivities of 0.25 to 0.5% copper[4,5] in the rather long measurement times of 2 to 5 min.

With the present instrument, the pulse-height analyzer enables ^{109}Cd to be used.

Nickel and cobalt balanced filters (as metal foils of mass per unit area 7.7 and 8.5 mg/cm^2, respectively) were used to isolate Cu $K\alpha$ X-rays from Fe $K\alpha$ and any Co $K\alpha$ X-rays excited. Figure 13 shows the spectrum obtained from a copper ore sample containing 1.9% copper and 2.1% iron. The pulse-height analyzer is set across the fluorescent peak and the balanced filters used to select the Cu K X-rays in this peak.

A set of artificial copper ore samples was made up by "salting" a copper-ore "tails" sample with known amounts of CuO, Fe_2O_3, or CoO. Figure 14 shows the difference in count rate between the two filters plotted against copper content. The scatter of points about each calibration curve is relatively small, which indicates that, in a known matrix, the accuracy of copper determination should be good.

As to sensitivity, the standard deviation for the difference in count rate at zero copper content is 0.03% copper in count times of 20 sec per filter. Thus, the sensitivity is entirely adequate for the assay of copper ores of economic grade.

The matrix absorption effects due to variation in cobalt or iron content are considerable, however. In the case of most United States copper ores, where iron is the major interfering element, the use of a nomogram completely eliminates the effect due to iron by the graphical solution of two equations. Figure 15 shows such a nomogram. The ordinate is the difference in count rate (nickel to cobalt). The abscissa is the count rate

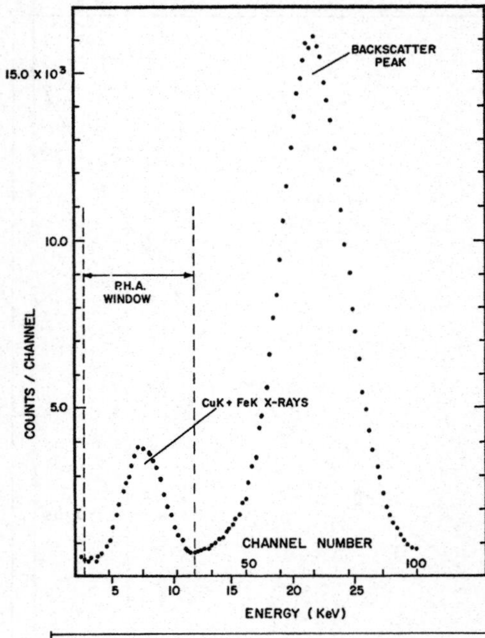

Figure 13. Spectrum from copper ore containing 1.9% copper and 2.1% iron, showing pulse-height analyzer window; source, ^{109}Cd.

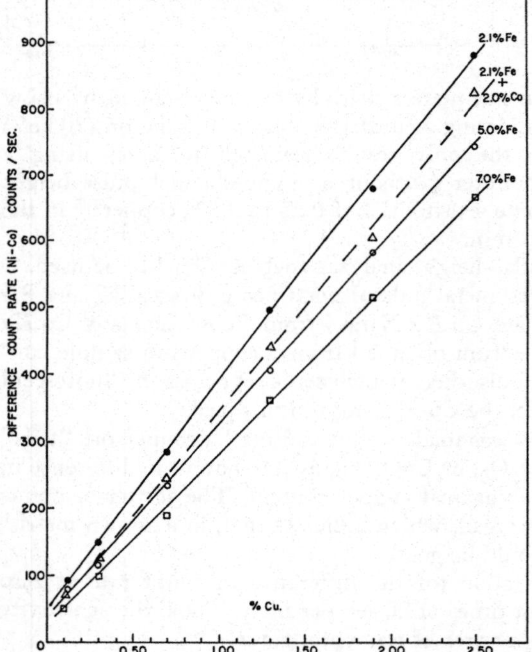

Figure 14. Calibration curves of the difference count rate against copper content of artificial ore samples.

with the cobalt filter alone. The cobalt filter transmits Fe *K* X-rays readily but heavily absorbs Cu *K* X-rays. In practice, the use of such a nomogram with the portable analyzer is straightforward, the cobalt filter and difference count rates being referred to the graph immediately they are obtained without further processing.

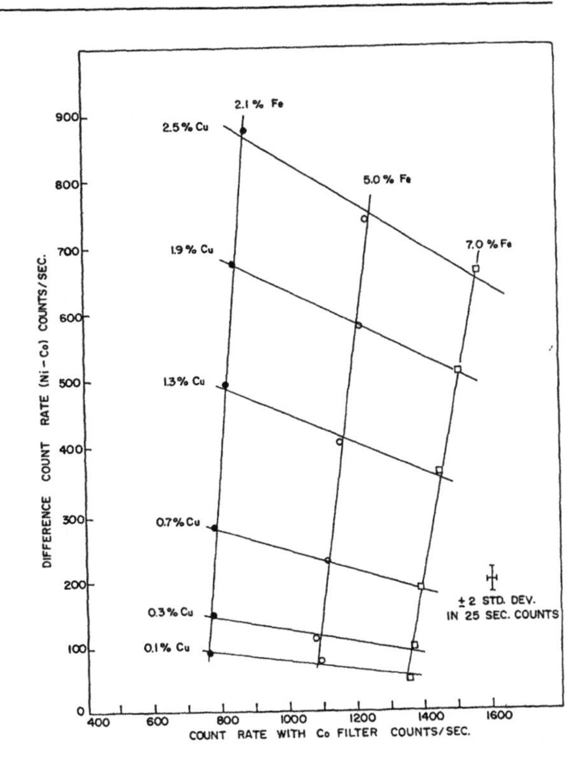

Figure 15. Nomogram for copper ores with varying iron content.

In 20-sec measurement times with each filter, the statistical error is 0.035% copper and 0.05% iron (1 standard deviation).

Vanadium, Chromium, Molybdenum, and Tungsten in Steels

Several hundred different steels exist, and even a company dealing in only one family of alloys has an increasing need for rapid quantitative analysis of stock and finished products in order to control production or to prevent costly mixups. In general, identification of a given alloy steel can be made by measuring the concentration of only one or two elements since the various groups of alloys are generally distinguishable visually. Checking specifications in production runs usually involves routine determination of one element, which is less complicated than an identification problem but usually requires greater accuracy.

In this study, a set of National Bureau of Standards steel samples was used that exhibits wide variations in the concentration of several elements that are important constituents of low- and high-alloy steels and high-speed tool steels. Table II lists the analyses of the samples. Since the concentrations of the major elements are relatively high and very variable, it is expected that matrix effects will be greater than normally encountered in analysis of a single type of steel.

The samples had a diameter of 1.25 in., which is less than the optimum diameter (1.8 in.) and necessitated reducing the aperture diameter to 1.15 in. Thus, the sensitivities obtained in the determinations described below are up to a factor of 2 lower than would be obtained by using larger samples.[3]

Table II. Important Constituents of National Bureau of Standards Steels Used in Alloy Analysis Studies*

Sample No.	Vanadium %	Chromium %	Molybdenum %	Tungsten %	Cobalt %	Manganese %
D836	0.63	6.02	2.80	9.7	...	0.21
D837	3.04	7.79	1.50	2.8	2.9	0.48
D838	1.17	4.66	8.26	1.7	4.9	0.20
D839	1.50	2.72	4.61	5.7	7.8	0.18
D840	2.11	2.12	0.07	13.0	11.8	0.15
D841	1.13	4.20	0.84	18.5	...	0.27

* The balance of the major constituents is iron.

Vanadium. V *K* X-rays (4.9 keV) are best excited by ^{55}Fe, whose 5.9-keV Mn $K\alpha$ radiation does not excite any other significant element in the steels. The main interfering radiations are backscattered X-rays at 5.8 to 5.9 keV and some Cr *K* X-rays, excited by the relatively low flux of Mn $K\beta$ X-rays from the source. A titanium filter 0.0005 in. thick was found to reduce adequately the detected flux of backscattered and Cr *K* X-rays. Since no other radiation is present, the width of the pulse-height analyzer window is unimportant as long as all the above mentioned X-ray energies are detected.

The results obtained by using the single filter and 50-sec counts are shown in Figure 16. The standard deviation due solely to counting statistics is, in per cent of vanadium, 0.034%. Since the count rate is well below the maximum tolerable (10^5 counts/sec) and the source activity is also less than the maximum practicable (20 mCi), it is clear that the statistical error could be reduced considerably. For example, with a 20-mCi source, 1 standard deviation would be 0.008% of vanadium in 50 sec.

The matrix effects on the uncorrected results are quite severe, however, and the overall standard deviation of curve *A* in Figure 16 was found to be 0.18% of vanadium by the least-squares method. By inspection of the curve and by comparison of the relevant mass attenuation coefficients (see Table III), the source of this error is seen to be the variable molybdenum and tungsten contents. Also, molybdenum and tungsten should be equally effective because the sums of their attenuation coefficient for incident and fluorescent radiation are nearly equal. When the experimental results were corrected to constant matrix mass attenuation coefficient by using the tabulated data, the overall error was reduced to a value not significantly different from counting statistics, which indicated that the matrix absorption effects and counting statistics were the only sources of appreciable error.

A suitable empirical correction was developed as follows: A linear correction factor was chosen such that

$$I_{\text{corr}} = [I_F - I_F(0)]\,(1 + AR) \tag{1}$$

where I_F is the measured scale reading; $I_F(0)$, the derived scale reading at 0% vanadium; *R*, the per cent of tungsten plus molybdenum; and *A*, an empirical factor.

By repeated least-squares regression analysis of the known data, the best value of *A* was found to be 0.033 when the minimum overall standard deviation was 0.034% vanadium. This value is not significantly different from that due to counting statistics alone.

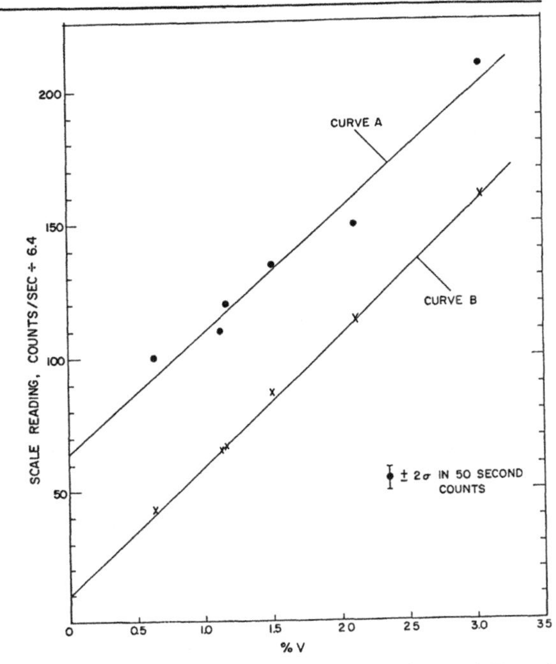

Figure 16. Calibration curves for vanadium in steel. Curve A is the uncorrected scale reading; curve B is corrected for variations in the tungsten plus molybdenum content.

In Figure 16, the corrected experimental data (by using the known values of R and A) is plotted against the per cent of vanadium, and it is seen that a good straight line is obtained with hardly any scatter.

Chromium. Of the radioisotope sources tested for efficient excitation of Cr K X-rays, ^{238}Pu gave the best results; although, in the event of ^{238}Pu's being not available ^{3}H/Zr or perhaps ^{57}Co or ^{109}Cd could be used instead, with some loss in sensitivity. Also, ^{210}Pb is a possibility but was not available for these experiments.

The spectra from the steel samples exhibited two peaks, the backscattered U L X-rays from the source (13 to 15 keV) being resolved from a group of fluorescent X-rays comprising V K, Cr K, Mn K, Fe K, Co K, and W L. The pulse-height analyzer window was set to include only the peak due to the fluorescent X-rays. Vanadium and titanium balanced filters were used to isolate the Cr $K\alpha$ X-rays in this peak. The filters took the form of metal foils, both of mass per unit area 6.14 mg/cm^2. To the titanium filter was added a 0.7-mg/cm^2 aluminum foil to obtain a better balance.

Table III. Mass Attenuation Coefficients for the Constituents of the Steel Samples at 5.0 and 5.9 keV

Element	$\mu(5.9)$	$\mu(5.0)$	$\mu(5.9) + \mu(5.0)$
Tungsten	380	530	910
Molybdenum	360	530	890
Chromium	71	113	184
Iron	91	144	235
Cobalt	98	147	245
Vanadium	530	96	626
Manganese	80	124	204

The results obtained by using the 20-mCi ^{238}Pu source, vanadium and titanium balanced filters, and 50-sec counts with each filter are shown in Figure 17. The standard deviation due to counting statistics is 0.15% chromium. The practical limit of precision is probably 0.05% chromium, obtainable by using a source of double the activity and count times of some 200 sec. At present, higher ^{238}Pu activities than 40 mCi are not practicable because of the relatively high cost of the source, but this could be reduced in the future.

As is seen from the uncorrected results in Figure 17, the matrix effects lead to very much larger errors than do the counting statistics, the overall standard deviation of curve A being 0.78% chromium. As in the case of vanadium, matrix absorption effects due to tungsten and molybdenum appear to predominate, although some enhancement by Co K X-rays might also be expected.

A similar approach to that used for vanadium was employed in an attempt to reduce the matrix effects. The correction equation was similar to equation 1, with $I_F(0)$ set at zero. The optimum value of A was found to be 0.049, and the corresponding overall standard deviation, 0.36% chromium. The corrected values of Cr $K\alpha$ flux are plotted as curve B in Figure 17.

Inspection of the scatter of these points (which is significantly greater than would be expected from counting statistics alone) and comparison with the sample analyses listed in Table II leads us to the conclusion that enhancement effects are negligible but absorption effects due to tungsten have not been completely corrected for. The attenuation coefficients for tungsten are some 25% greater than those for molybdenum, and a better correction might be achieved if the values of the tungsten content were correspondingly weighted.

Molybdenum and Tungsten Plus Molybdenum. A comparison of ^{109}Cd, $^{147}Pm/Al$, and ^{241}Am was made for excitation of Mo K X-rays. Figure 18 shows the spectrum

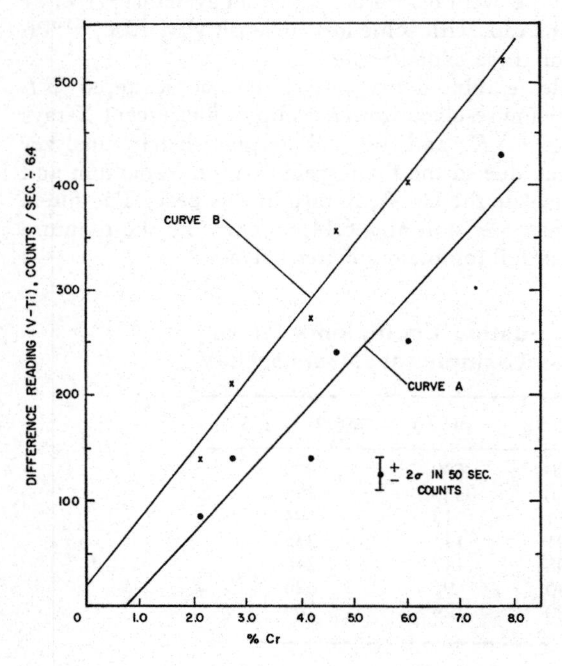

Figure 17. Calibration curves for chromium in steel. Curve A is the uncorrected scale reading; curve B is corrected for variations in the tungsten plus molybdenum content.

obtained from sample number D839 by using the ^{147}Pm/Al source. It is seen that Mo K X-rays are clearly resolved from the other characteristic X-rays excited. Thus, balanced filters are not necessary in this particular analysis.

Of the sources tested, ^{109}Cd gave the best results, these being somewhat better than those obtained with ^{147}Pm/Al. However, ^{147}Pm/Al also excites W K X-rays and can be used for a separate measurement of tungsten content or for a combined measurement of tungsten plus molybdenum content, as might be required to correct the matrix effects on vanadium and chromium determinations (see above). The ^{241}Am source gave somewhat worse results for molybdenum determination and cannot excite W K X-rays, so it was not considered further.

Figure 19 shows calibration curves for molybdenum determination by using ^{109}Cd (curve A) and ^{147}Pm/Al (curve B). The standard deviations due to counting statistics obtained in 50-sec counts were 0.012% molybdenum by using ^{109}Cd and 0.020% molybdenum by using ^{147}Pm/Al. Note that the ordinate of Figure 19 corresponds to 10-sec counts. The respective overall errors were 0.11 and 0.14% molybdenum. Thus, although the activities of both these sources could easily be increased tenfold, there is no point unless the other errors (presumably matrix effects), which are not appreciable, are reduced considerably. The use of balanced filters to isolate Mo K X-rays from unresolved scattered radiation has been found to reduce matrix effects in this analysis.[3] Also, balanced filters would be essential if niobium were present in addition and molybdenum and niobium contents were required separately.

Figure 20 (whose ordinate again corresponds to 10-sec counts) shows a curve of scale reading against molybdenum plus tungsten content with the use of ^{147}Pm/Al and the wide pulse-height analyzer window shown in Figure 18. Good correlaton is obtained, and the accuracy of the combined determination is clearly adequate for the purpose of correcting matrix absorption effects on the vanadium and chromium determinations. Experiments with pure molybdenum and tungsten samples showed that this measure-

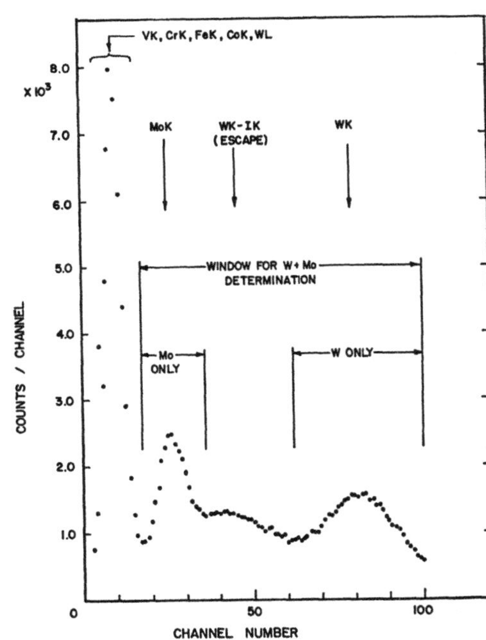

Figure 18. Spectrum from steel No. D839 with the use of a ^{147} Pm/Al source; measurement time, 15 sec.

ment was twice as sensitive to molybdenum as to tungsten. The inclusion of a single absorption type of filter to reduce the proportion of Mo *K* X-rays detected could be used to boost the relative effect of tungsten to any value desired for empirical correction purposes.

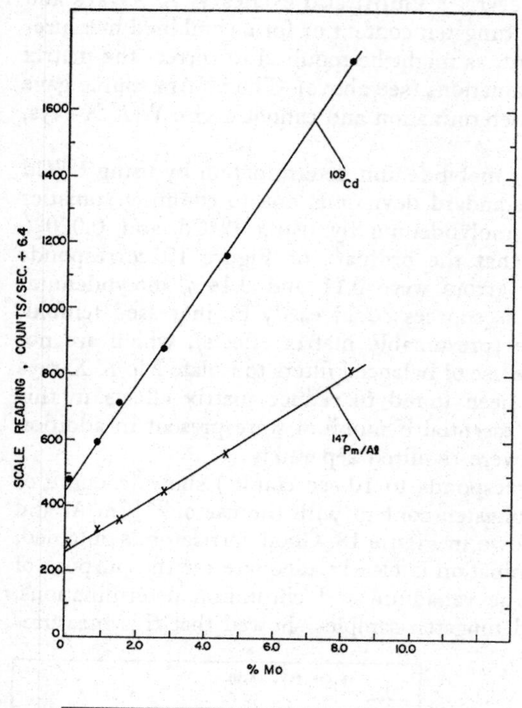

Figure 19. Calibration curves for molybdenum in steel.

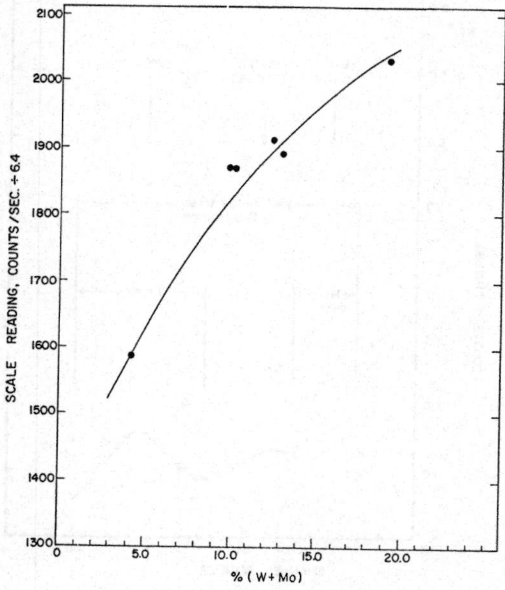

Figure 20. Calibration curve for molybdenum plus tungsten with the use of ^{147}Pm/Al and a wide channel.

Comparison of* K *and* L *X-Ray Excitation for Tungsten. The much greater penetration of the K X-rays of heavy elements would make their excitation preferable to L X-ray excitation for analysis of rough or scaled surfaces. A comparison was made of ^{147}Pm/Al and ^{57}Co for excitation of W K X-rays, no filters being used. Also, ^{153}Gd is a possibility but was not available. The best results were obtained by using ^{147}Pm/Al and a pulse-height analyzer window such as shown in Figure 18. This source is cheaper and longer lived than both ^{57}Co and ^{153}Gd.

Figure 21 shows the calibration curve obtained. As is usual with heavy elements in relatively light matrices, the calibration is convex, and so the overall standard deviation was not calculated. The standard deviation due to counting statistics in a 30-sec count was less than the size of the points on the graph. The scale reading shown in the figure corresponds to 10-sec counts. Comparison of the results with the sample analyses (Table II) indicates that there is a small matrix absorption effect due to molybdenum. For a precise tungsten determination, this would have to be corrected.

Balanced filters of copper and nickel (metal foils of mass per unit area 11.3 mg/cm^2 and 11.5 mg/cm^2, respectively) are necessary to isolate W L X-rays from the K X-rays of the transition metals in steel. For excitation of W L X-rays, ^{109}Cd, ^{147}Pm/Al, and ^{238}Pu sources were compared. All give efficient excitation, but both ^{109}Cd and ^{147}Pm/Al excite Mo K X-rays, which give rise to a marked enhancement of W L X-rays. So, for these steel samples, which contain high and variable concentrations of molybdenum, these two sources were rejected.

Figure 22 shows the calibration curve obtained by using the ^{238}Pu source, 50-sec count times, and balanced copper and nickel filters. A smooth curve is obtained without any significant matrix effects. The overall standard deviation (0.36% tungsten) was not significantly different from that due to counting statistics in a 50-sec count. Thus, if a 40-mCi source were used and 200-sec count times could be tolerated, a precision of 0.13% tungsten could be obtained.

Thus, in a given determination of tungsten in steel, the final choice between excitation of the L or the K X-rays would depend on the character of the sample surface, the accuracy required, and the amount and variability of the molybdenum content.

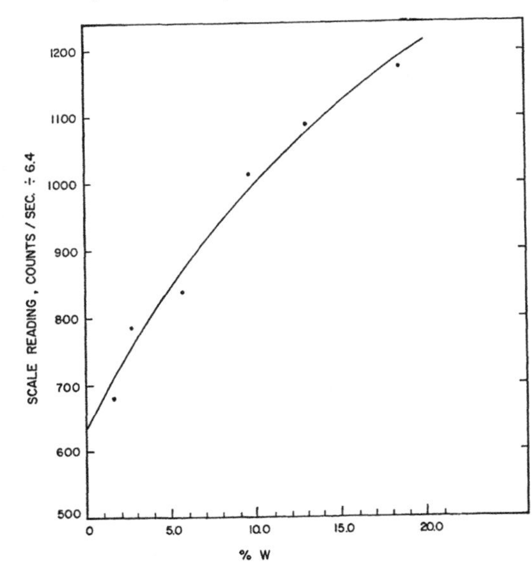

Figure 21. Calibration curve for tungsten with the use of ^{147}Pm/Al to excite W K X-rays.

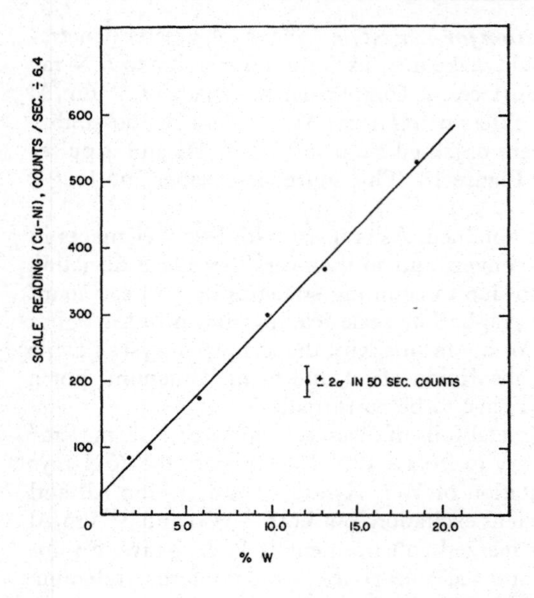

Figure 22. Calibration curve for tungsten with the use of ^{238}Pu to excite W *L* X-rays.

SUMMARY AND CONCLUSIONS

1. A high-performance, fully portable, battery-operated, radioisotope-excited X-ray analyzer of 15-lb all-up weight has been developed. Improved precision and shorter measurement times compared with those of earlier instruments have been obtained through the use of a reversible scaler for output presentation instead of a rate-meter. Higher sensitivity and versatility has been achieved through the use of a single-channel pulse-height analyzer to cut out unwanted radiation that is resolved by the detector, a NaI(Tl) scintillation counter.

2. Energy selection of neighboring characteristic X-rays is performed by X-ray filter techniques, particularly balanced filters, and the reversible scaler makes balanced-filter operation extremely convenient.

3. The range of elements measurable is sulfur upward. Determination of elements sulfur to scandium has been made possible through the recent development of low-noise photomultipliers with high quantum efficiencies.

4. Applications to ore, coal, cement, and alloy analysis have been studied. The analyses investigated were sulfur in coal; calcium and iron in cement raw mix; copper in copper ores; and vanadium, chromium, molybdenum, and tungsten in steels.

5. Sensitivities range from about 0.03% for copper in copper ores to about 0.2% for sulfur in coal. Measurement times are in the range of 10 to 100 sec. Some 10 radioisotope X-ray sources of suitable emission characteristics, half-life, and cost are available, and, for each analysis, several sources were compared to find the best one.

6. In most determinations, matrix effects were the main source of error, and a number of methods of eliminating them were investigated. The most successful for the alloy analyses was to establish a linear correction factor proportional to the concentration of the interfering element(s) and to determine the latter separately. In the case of the copper ores and cements, which were effectively two-component materials, methods were devised which determined both components in a single set of two measurements. Note that, because of the need to use balanced filters, the instrument is designed to make two successive measurements (with different filter combinations) on a given sample.

ACKNOWLEDGMENT

The authors are particularly grateful to Dr. J. D. Hall, Texas Nuclear Corporation, for designing and building the prototype electronic unit to specification.

REFERENCES

1. J. R. Rhodes, "Radioisotope X-Ray Spectrometry—A Review," *Analyst* **91**: 683, 1966.
2. S. H. U. Bowie, A. G. Darnley, and J. R. Rhodes, "Portable Radioisotope X-Ray Fluorescence Analyzer," *Trans. Inst. Mining Met.* **74**: 361, 1965.
3. J. R. Rhodes, T. W. Packer, and I. S. Boyce, "Rapid X-Ray Analysis of Copper Alloys and Steels using a Radioisotope Portable Analyzer," in: *Radioisotope Instruments in Industry and Geophysics, I*, International Atomic Energy Agency, Vienna, 1966, p. 127.
4. A. G. Darnley and C. C. Leamy, "The Analysis of Tin and Copper Ores using a Portable Radioisotope X-Ray Fluorescence Analyzer," in: *Radioisotope Instruments in Industry and Geophysics, I*, International Atomic Energy Agency, Vienna, 1966, p. 191.
5. J. Niewodniczanski, "Analysis of Copper Ores in Mine Conditions by X-Ray Fluorescence Induced by Isotope Sources," in: *Radioisotope Instruments in Industry and Geophysics, I*, International Atomic Energy Agency, Vienna, 1966, p. 173.
6. J. O. Karttunen, H. B. Evans, D. J. Henderson, P. J. Markovich, and R. L. Niemann, "A Portable Fluorescent X-Ray Instrument Utilizing Radioisotope Sources," *Anal. Chem.* **36**: 1277, 1964.
7. J. O. Karttunen and D. J. Henderson, "An Improved Portable Fluorescent X-Ray Instrument Using Radioisotope Excitation Sources," *Anal. Chem.* **37**: 307, 1965.
8. J. I. Trombka, I. Adler, R. Schmadebeck, and R. Lamothe, "Non-Dispersive X-Ray Emission Analysis for Lunar Surface Geochemical Exploration," *NASA Rept.* No. X-641-66-344, Goddard Space Flight Center, 1966.
9. J. F. Cameron and J. R. Rhodes, "Filters for Energy Selection in Radioisotope X-Ray Techniques," in: G. L. Clark (ed.), *Encyclopedia of X-Rays and Gamma Rays*, Reinhold, New York, 1963, p. 387.
10. P. Kirkpatrick, "On the Theory and Use of Ross Filters," *Rev. Sci. Instr.* **10**: 186, 1939; **15**: 223, 1944.
11. E. G. Boonstra, "Statistical Considerations in the Design of Balanced X-Ray Filters," *J. Sci. Instr.* **42**: 563, 1965.
12. J. R. Rhodes, "Optimization of Excitation and Detection Techniques for Analysis of Silver Ores by Radioisotope X-Ray Spectrometry," in: P. S. Baker (ed.) *Second Symposium on Low-Energy X- and Gamma Sources and Applications, Austin, Texas, 1967*, Vol. I, p. 442.
13. N. Spielberg, "Characteristics of Flow Proportional Counters for X-Rays," in: J. B. Newkirk and G. R. Mallett (eds.), *Advances in X-Ray Analysis, Vol. 10*, Plenum Press, New York, 1967, p. 534.
14. P. W. Sanford and J. L. Culhane, "The Stability of X-Ray Proportional Counters Under Extreme Operating Conditions," in: P. S. Baker (ed.), *Second Symposium on Low-Energy X- and Gamma Sources and Applications, Austin Texas, 1967*, Vol. I, p. 376.
15. J. R. Rhodes, J. C. Daglish, and C. G. Clayton, "A Coal Ash Monitor with Low Dependence on Ash Composition," in: *Radioisotope Instruments in Industry and Geophysics, I*, International Atomic Energy Agency, Vienna, 1966, p. 447.
16. J. A. Dunne and N. L. Nickle, "Balanced Filters for the Analysis of Al, Si, K, Ca, Fe and Ni," in: P. S. Baker (ed.), *Second Symposium on Low-Energy X- and Gamma Sources and Applications, Austin, Texas, 1967*, Vol. I, p. 336.
17. J. R. Rhodes, Thelma G. Ahier, and D. O. Poole, "Analysis of Zinc and Copper Ores by Radioisotope X-Ray Fluorescence," *U.K. At. Energy Authority Res. Group Rept. AERE* R4474, 1954.

DISCUSSION

G. Edwards (Colorado School of Mines): There was a paper given at the Spectrographic Conference earlier this week from McDonnell Aircraft. They've developed a portable power source and are using a polaroid camera for obtaining the results with this. It is quite a bit quicker. Have you tried any film techniques on your source?

J. R. Rhodes: We haven't used film directly because it is much too insensitive for use with a radioisotope source. Apparently, a lithium fluoride dispersive element was used in the camera you mention, and a 50-kV conventional, "portable" X-ray generator as the source.

L. S. Birks (Naval Research Laboratory): Could you explain why you regard a portable source as an advantage if you are talking about measuring steel, because you don't find much of that out in the field?

J. R. Rhodes: Apparently there is a need for rapid *in situ* analysis on the factory floor or in the stockyard, for checking specifications of forgings, coils, bars, and the like, before dispatch. Cutting a piece off for laboratory analysis might take too long or unacceptably damage the product. The spark or grinding-wheel test that is often done now demands a detailed knowledge of the type of spark given off by each of several hundred alloys. Even then, the test is only qualitative. Furthermore, the experienced people who can perform this test are dying out.

K. J. Cox (Kerr-McGee Corporation): Have you done any test work on uranium? What kind of excitation source would you use for something that high in the periodic chart?

J. R. Rhodes: Uranium *K* X-rays can be excited quite readily with ^{57}Co, or the *L* X-rays with ^{109}Cd. I don't think the sensitivity of analysis is as high as that using ordinary radiometric survey methods, but, of course, the analysis is specific for uranium.

K. J. Cox: What type of sensitivity are you referring to?

J. R. Rhodes: Exciting the U *L* X-rays would probably give a detection limit of 0.01% or so.

K. J. Cox: That's plenty good enough for mine exploration.

J. R. Rhodes: I'm surprised, I thought one needed better than that for exploration.

K. J. Cox: Are you talking about 0.01% of the ore, or mineral?

J. R. Rhodes: About 0.01% by weight of uranium in the rock. But, surely, in radiometric survey methods, you can get down an order of magnitude below that.

K. J. Cox: If the equilibrium is okay. It isn't always; that's why I was checking on this.

Chairman K. F. J. Heinrich: I should like to add that, in an age where we are proud of making our instruments bigger, heavier, and more complex, it is quite admirable to see the results that are obtained with very simple but very sophisticated instrumentation. Thank you very much.

THE APPLICATION OF RADIOISOTOPE NONDISPERSIVE X-RAY SPECTROMETRY TO THE ANALYSIS OF MOLYBDENUM

A. P. Langheinrich and J. W. Forster

Kennecott Research Center, Western Mining Divisions
Kennecott Copper Corporation
Salt Lake City, Utah

ABSTRACT

Radioisotope excitation, solid-state detection, and multichannel pulse-height analysis have been applied successfully to the determination of molybdenum in copper and molybdenum-process intermediates. Analytical data are presented, and comparisons are made with the conventional X-ray fluorescence approach. The equipment performed satisfactorily in molybdenum plant control and shows promise for other industrial applications with the development of suitable isotope sources.

INTRODUCTION

The use of radioisotope sources in X-ray analysis and the application of multichannel pulse-height analyzers is not new.[1-4] Only recently, however, developments in the field of solid-state detectors have made available an analytical system equal in potential to that of conventional X-ray equipment. Instrumentation of this type, utilizing a ^{125}I source, a lithium-drift silicon detector, and a 400-channel analyzer, was applied successfully to various analytical problems. Since ^{125}I was found highly efficient for molybdenum, the determination of molybdenum in copper and molybdenum-process intermediates was studied in detail.

Analytical data were obtained on powder samples and solutions. Calibration curves were established and precision values obtained for molybdenum in mine samples, in copper concentrate, in reverberatory slag, and in molybdenum plant solutions. A flow cell was built to allow for the continuous on-stream analysis of solutions and slurries. In all these tests, equivalence was demonstrated between the results obtained with radioisotope, nondispersive methods and those with conventional X-ray fluorescence methods.

Small size and ease of handling make this type of equipment a logical choice for in-plant usage. Since the activity of the radiation sources used in this work was less than 3 mCi, health hazards were minimal. Future research efforts will include the investigation of additional sources to establish maximum operating conditions in the analysis of other elements.

INSTRUMENTATION

The analytical system used and evaluated in this work was first reported during 1966.[5-7] It is sold by the Technical Measurement Corporation (TMC) and consists of a radioisotope source, a low-temperature solid-state detector, a multichannel pulse-height analyzer, and a digital printout or *x-y* recorder. A 110-V power source is needed. The whole instrument package takes up a space of less than 10 ft^3 and can readily be transported. The individual modules are described briefly below.

The source unit is a sealed artificial radioisotope. In the case of ^{125}I, it is mounted in tellurium and attached to a plastic holder ring. The sample holder, built by laboratory personnel, fits into the holder ring and is also made of plastic material. It can accept a 1.5-in.-diameter sample pellet or liquid cell. A circular sample area of 1.1875-in. diameter is exposed to the primary radiation from the source. The inside of the holder ring and the exposed side of the sample holder are both covered with tin foil. The tin foil reduces scattered radiation as compared with plastic alone and provides a more favorable peak-to-background ratio. A plastic flow cell of compatible dimensions was constructed at this laboratory to allow for on-stream analyses of liquid and slurry samples. Source unit and sample holders are shown in Figure 1.

The detector unit, or photon spectrometer, consists of a lithium-drift silicon detector, a cryogenic chamber with beryllium window, a 1-liter/sec ion pump, a liquid-nitrogen reservoir, and the associated power supplies and electronics. A picture of this unit and the analyzer and read-out equipment as well is shown in Figure 2. The detector system is

Figure 1. Flow cell (left), source unit (center), and static sample holder (right).

Figure 2. General view of instrumentation.

directly compatible with the multichannel pulse-height analyzer. Qualitative results can be obtained from the scope of the analyzer. For quantitative analyses, a scaler, a paper-tape printer, or an x-y recorder is needed.

A minimum instrument package for applied spectroscopic work would consist of the source and detector unit, a single-channel analyzer, and a scaler. For specific applications, a package of this type would cost less than $10,000. Weight and space requirements are quite small, and transportation does not present a serious problem; hence, in-plant applications and use in field exploration can be foreseen as long as a portable AC generator or a converter is made available for the field work.

OPERATION

Photon spectrometry is capable of fast nondestructive identification and quantification of elements with $Z > 20$. Its usefulness and sensitivity for a given element depends largely on the choice of radioactive isotope as the source of efficient primary radiation. Selected physical values of sources currently available through the manufacturer of this equipment are presented in Table I.

The primary radiation from the source excites characteristic fluorescent X-radiation of the target materials. The solid-state detector measures the quantity and energies of these X-rays. The lithium-drift silicon absorbs radiation with an efficiency approaching 100% in the low-energy X-ray range and delivers output pulses directly proportional to the energy of the individual photons absorbed. These pulses are separated in the multichannel pulse-height analyzer. The resolved spectrum is displayed on the scope of the analyzer and can be recorded either graphically on an x-y recorder or digitally on the printer. The resolution of the combined detector–analyzer units is better than 1 keV and allows for the identification of even neighboring elements in the periodic table if energy and concentration differences are of the right order of magnitude.

Table I. Source Data

Source	Half-life	Activity, mCi as received	Main mode of decay	Useful radiation, keV
^{241}Am	458 years	1.0	α*	59.6‡ γ-ray
				26.4 γ-ray
				14.0 Np X-ray
				17.8 Np X-ray
				20.8 Np X-ray
^{57}Co	267 days	...	E†	122‡ γ-ray
				137 γ-ray
^{159}Dy	144 days	0.6	E	44.5‡ Tb X-ray
				43.7 Tb X-ray
				50.4 Tb X-ray
^{125}I	57.4 days	2.0	E	35.5 γ-ray
				27.5‡ Te X-ray
				31.0 Te X-ray

* α, alpha decay.
† E, electron capture.
‡ Highest intensity per disintegration.

The system permits observation of the complete spectrum of elements excited by a particular source and, hence, facilitates the determination of a number of elements simultaneously.

EXPERIMENTATION

The instrument package described earlier was delivered with a ^{125}I and a ^{241}Am source. Consistent with energy relationships, initial testing showed that the iodine source would be of greater immediate value. It would, however—as shown in Figure 3—excite molybdenum much more efficiently than copper and iron, the two other elements of prime interest. Therefore, the initial study was concentrated on molybdenum.

Sample Preparation

Since sample preparation is directly related to precision, this step must be carried out with care. The general method used at this laboratory in conventional X-ray work was applied successfully. The samples, ground to −200 mesh, were pressed in 1.375-in.-diameter aluminum cups at 20,000 psi. This treatment resulted in a satisfactorily uniform sample surface, was quick, and minimized errors due to sample preparation.

For liquids, Lucite cups were used. The liquid surface was covered with 0.00025-in. Mylar film kept in place with a Lucite ring. Care must be taken that the surface is level, *i.e.*, positive or negative menisci must be avoided.

Sample Excitation

The source–sample relationship was investigated, and the sample holder was designed to give good sample-surface presentation to the source radiation while keeping a low noise

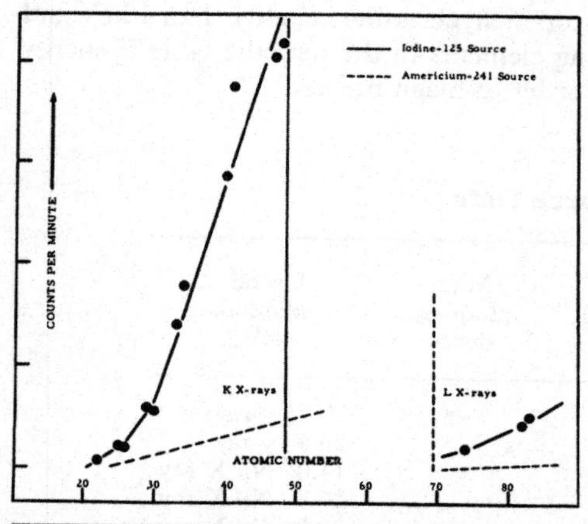

Figure 3. Excitation efficiency.

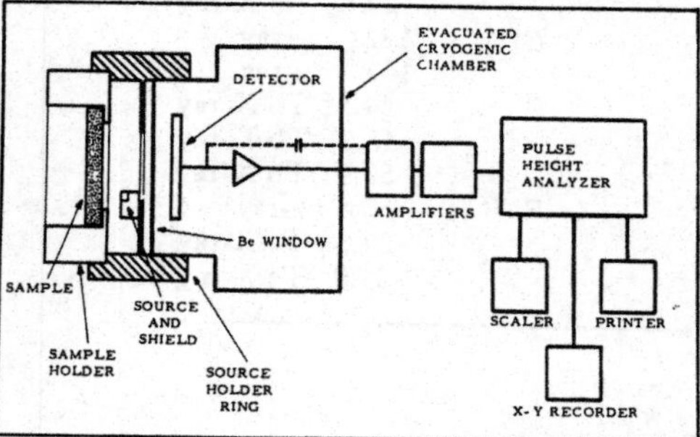

Figure 4. Source–sample–detector arrangement.

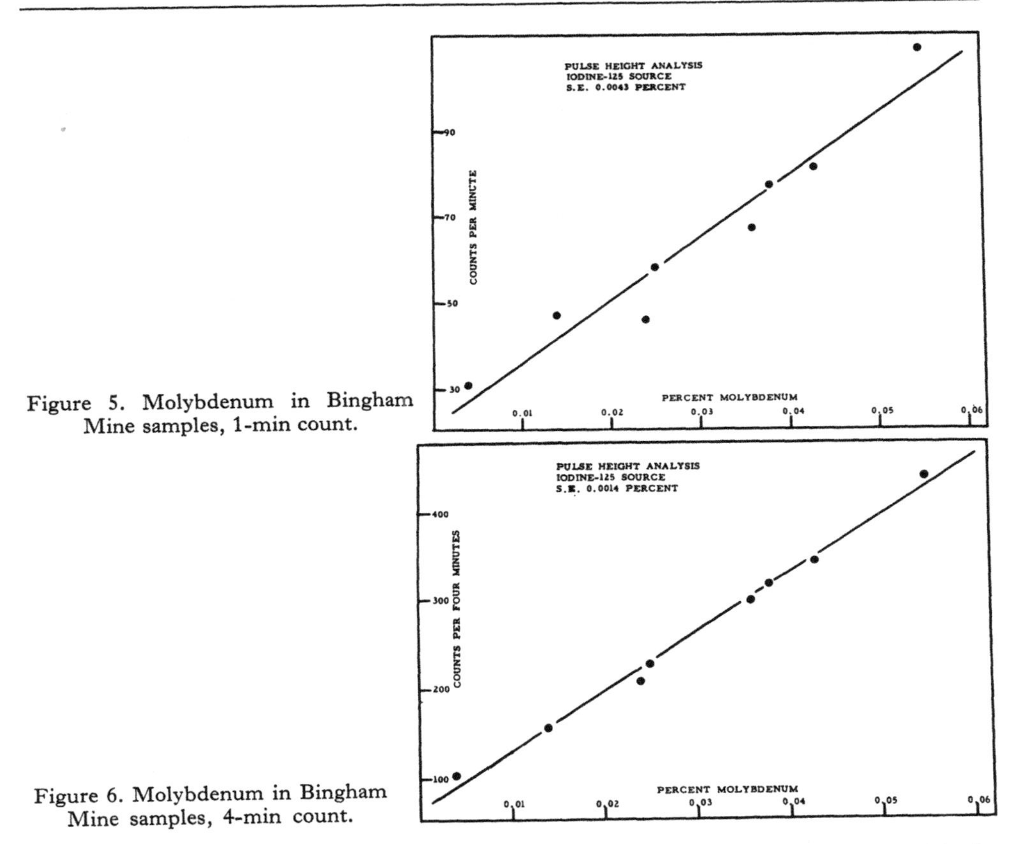

Figure 5. Molybdenum in Bingham Mine samples, 1-min count.

Figure 6. Molybdenum in Bingham Mine samples, 4-min count.

level. Photographic techniques were used in determining the best conditions with the available equipment. The source–sample–detector arrangement is shown in Figure 4. Other geometric configurations have been described in the literature.[8]

As a source, ^{125}I is readily available at moderate cost. Its half-life is 57.4 days. The radioisotope excites K-radiation of elements 20 to 49 and L-radiation of elements >50. For the test illustrated in Figure 3, pure elements were used. It was found that the intensity of the resulting fluorescent X-radiation was too low for quantitative work with elements 22 to 30 in the concentration ranges of interest. A radiation source of approximately 12 keV—a value considerably lower than the useful radiation from ^{125}I—should prove quite satisfactory for exciting these elements. Above element 30, quantitative work appeared possible.

Quantitative Studies

After the usefulness of ^{125}I had been established for molybdenum work, samples were obtained that had been analyzed previously by other means. These samples included ore, slag, copper concentrate, and molybdenum plant solutions. Mine samples were studied first. These particular samples are of considerable interest and value since they were used previously in an X-ray standards program and had been analyzed by several laboratories and methods. Molybdenum ranged from 0.004 to 0.055%. The goal was to establish operating conditions, including counting times, that were practical and would provide precisions equal to those obtained by conventional X-ray. Figures 5 and 6 show the improvement of increased count time. The relatively weak source used here was later replaced by a source of higher activity. In the latter case, satisfactory precisions were found at counting rates of less than 1 min. The figures show plots of X-ray counts *versus* chemical assay data. The curves were calculated by the least-squares method, and deviations from this curve were compared statistically to obtain standard errors of estimate.

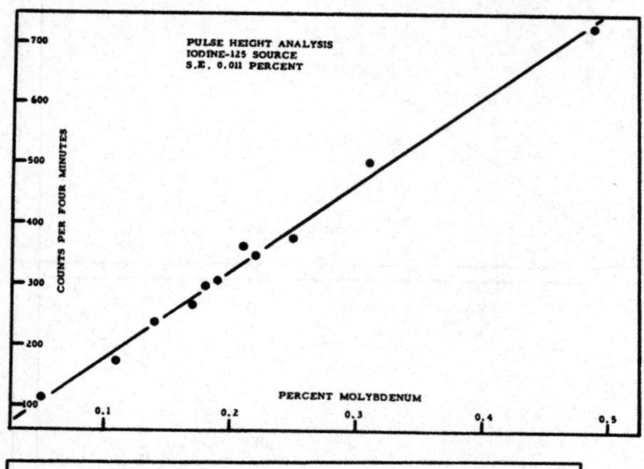

Figure 7. Molybdenum in copper concentrate.

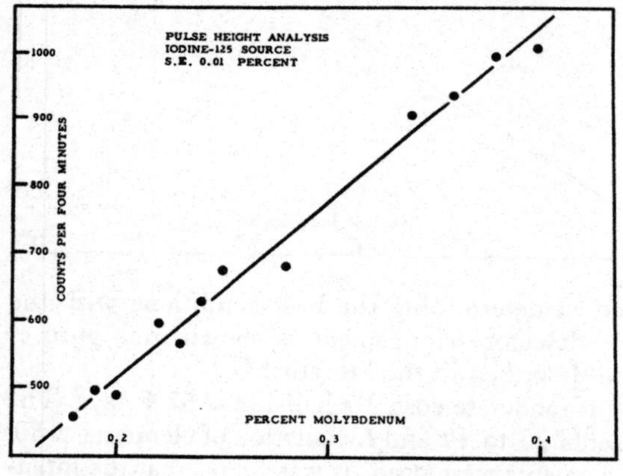

Figure 8. Molybdenum in reverberatory slag.

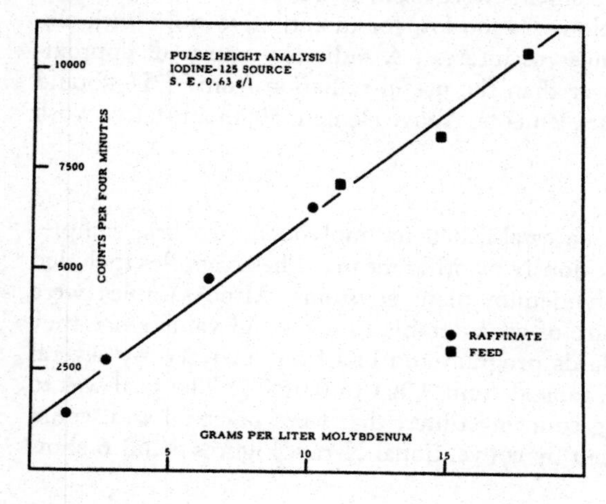

Figure 9. Molybdenum in molybdenum plant solutions.

Table II. Standard Deviations and Standard Errors of Estimate

Test	SD, % molybdenum		SE, % molybdenum	
Instrument	0.0017	Mine samples, 4 min		
	0.0020			
Instrument + sample preparation	0.0022			
Instrument + sample positioning	0.0023			
Count time				
Mine samples			0.0043	1 min
			0.0016	4 min
			0.0014	
			0.0013	
Slag samples			0.021	1 min
			0.010	4 min

Table III. Comparison of Conventional X-Ray with Photon Spectrometry, Per Cent Molybdenum

Sample type	Range	SE, X-ray, 10 sec	SE, photon spec., 4 min
Mine samples	0.004–0.055	0.0020	0.0014
Copper concentrate	0.05–0.49	0.011	0.011
Reverb. slag	0.18–0.40	0.011	0.010
Solutions	1.35–18 g/l	0.70 g/l	0.63 g/l

Similar tests were repeated for molybdenum in copper concentrate, in reverberatory slag, and in solutions. Examples of this work are shown in Figures 7, 8, and 9. The instrument standard deviation, the standard deviations of sample preparation and sample positioning, and the standard error of estimate of counting time were determined. Results are listed in Table II. Table III presents a comparison of standard error data for conventional X-ray and photon spectrometry and shows the equivalence of both methods.

After this preparatory work, a number of "unknown" samples were selected and molybdenum assays were made. Results were compared with chemical, conventional X-ray, and atomic absorption methods, as shown in Table IV. The data indicate that, on the whole, the new approach compares favorably with the other methods.

Table IV. Per Cent of Molybdenum in Unknown Samples

Sample	Gravimetric	Colorimetric	Atomic absorption	Conventional X-ray	Photon spectrometry
1	0.035	0.039	0.049*	0.033	0.038
2	0.046	0.054	0.053*	0.046	0.046
3	0.025	0.029	0.034*	0.031	0.028
4	0.011	0.022*	0.024*	0.024	0.026
5			0.033	0.038	0.034
6			0.16*	0.15	0.16
7			0.76	0.76	0.76
8	0.005	0.006	0.007*	0.012	0.005
9			0.008*	0.012	0.012
10			0.009*	0.013	0.012

* Average of several determinations.

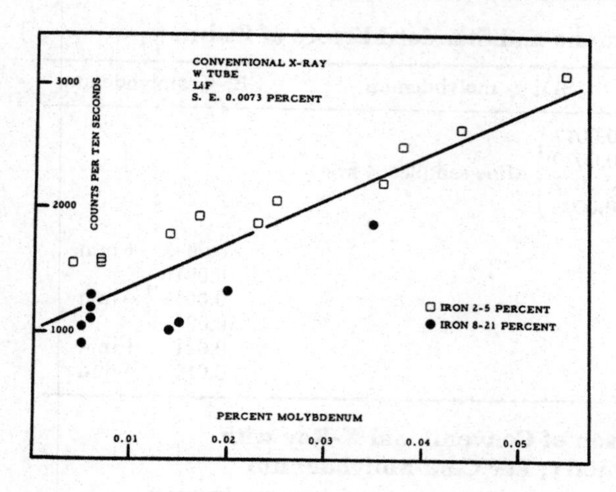

Figure 10. The effect of iron on molybdenum analyses by X-ray counts.

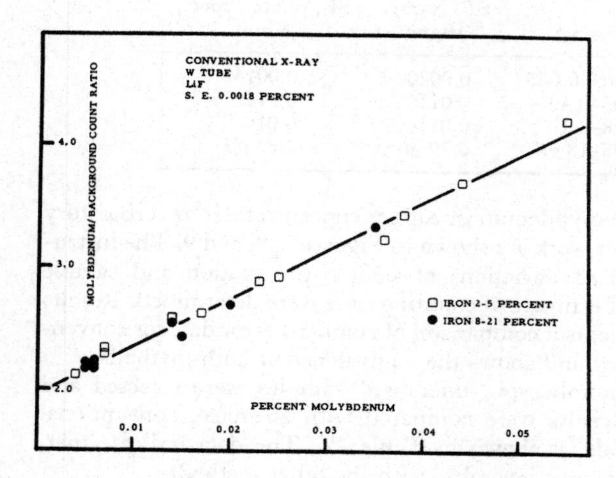

Figure 11. The effect of iron on molybdenum analyses by X-ray count ratios.

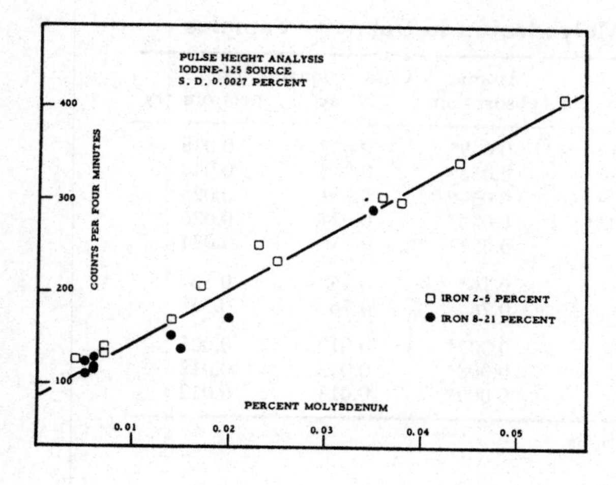

Figure 12. The effect of iron on molybdenum analyses by photon spectrometry.

Earlier work at this laboratory established the detrimental effect of varying iron concentrations on molybdenum assays when conventional X-ray methods were used without compensation steps. The interelement problem is illustrated in Figure 10. Various experimental and mathematical means have been reported to overcome difficulties of this type.[9–12] A simple compensating method successfully used in a variety of applications is illustrated in Figure 11. Here chemical element percentages are related to molybdenum-to-background ratios. The use of this approach reduces the standard error from 0.0073 to 0.0018%. In conventional X-ray work, the primary radiation from the tube consists of a continuum plus target X-rays. Since, however, the primary radiation in radioisotope work approaches monochromaticity, the interelement effect under such conditions was thought to be worth checking. As is shown in Figure 12, the iron effect is minimized, and the standard error of estimate is reduced to 0.0027%. The apparent discrepancy of this value with the value of 0.0014% in Figure 6 stems from the fact that the latter was obtained with samples of a normal iron range of 2 to 5%.

Source Decay

Iodine-125 has a half-life of less than 60 days. For this reason the effect of source decay on calibration was checked experimentally. Since the observed intensity loss showed only fair agreement with theoretical data, it is recommended that at least two standards should be utilized every working day for slope correction purposes.

Exploration Work

The instrument package has specific value in exploration work where two techniques of sample identification are of particular interest.[13] In the first case, the sample is exposed long enough to develop peaks of all trace elements displayed by the *x*-*y* recorder. At this point, two known elements are counted and plotted on the same chart. These two peaks with their known energy values in keV can be used to determine graphically the identity of all unknown elements. The procedure is illustrated in Figure 13.

The second approach consists of storing the spectrum of the unknown sample in the first one-half number of channels available in the pulse-height analyzer. A suspected element is then recorded in the second set of channels. By superimposing the two sets, the investigator is able to determine the presence or absence of a particular element. This procedure is shown in Figure 14. In the example, an unknown material gave peaks

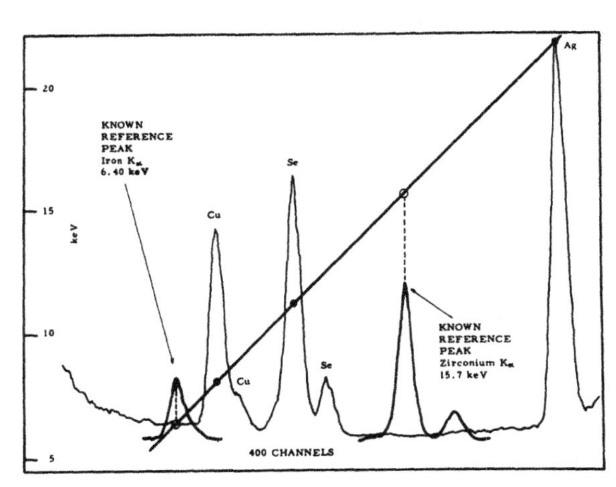

Figure 13. Identification of elements.

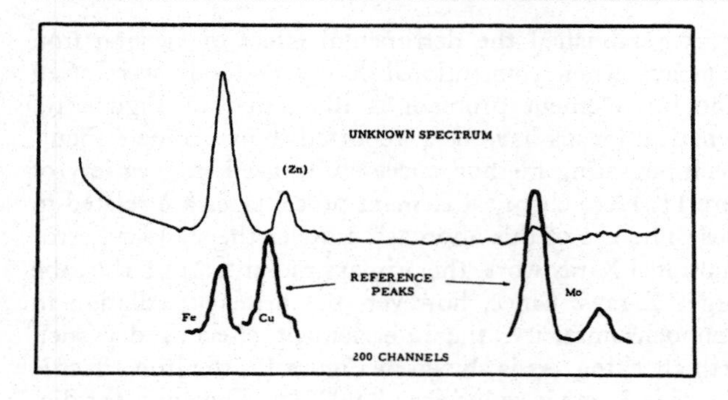

Figure 14. Element identification through superimposition.

believed to be due to iron, copper, and molybdenum. By superimposition, only the presence of iron and molybdenum could be verified. The remaining peak in the unknown spectrum was later identified as zinc. After identification, the usual quantitative work can begin.

On-Stream Analysis

Process control through on-stream measurements is another projected area of application for this equipment. A flow cell was constructed and preliminary feasibility tests were carried out. The reliability of the cell and the reproducibility of results were satisfactory. The experimental arrangement is shown in Figure 15; data are presented in Table V.

In the reproducibility tests, solid mine samples of known molybdenum concentration were mixed with water to result in a slurry sample containing 30% solids. Counts were taken during repeated periods of 10 min, and standard deviations were calculated.

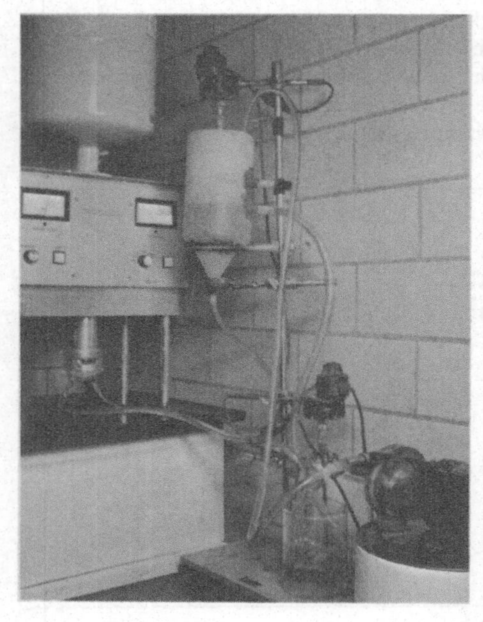

Figure 15. On-stream arrangement.

Table V. On-Stream Data

1. Example of reproducibility test
   Pulp, 30% solids
   Weak source
   Molybdenum, 0.038%
   Ten 10-min counts
   Mean 164.1 counts
   SD 10.6 counts = 6.5% (solids 4.4%)
2. Examples of solution test
   Molybdenum plant solutions
   New source
   Molybdenum range 0.23–22.8 g/liter
   SE, 0.23%, 0.4-min count
   SE, 0.10%, 1-min count

Future on-stream analysis of slurries will depend on practical means of density correction still to be developed for such a system. The solution tests covered wide concentration ranges in actual molybdenum plant solutions and provided standard errors of estimate acceptable to operating personnel.

CONCLUSION

The TMC instrument package consisting of a radioisotope source, a solid-state detector, a multichannel pulse-height analyzer, a tape printer, an *x-y* recorder, and a scaler has been used successfully in the qualitative and quantitative determination of molybdenum in a number of materials. The instrument quality and stability were excellent. No repairs were necessary during a 6-month evaluation period. The instrumentation is compact and simple to handle and has low power requirements; hence, it is adaptable to on-site usage in production processes and field exploration work. Personnel of average intelligence can be instructed within a few hours to do routine work. Radiation danger was checked by means of regular badge tests and, based on safety levels set by health authorities, was found practically nonexistent. The resolution of the particular detector–analyzer combination utilized in this work was 0.64 keV.

Shortcomings of the system center around the half-life of the iodine source and the fact that needed specific sources are not readily available. For maximum element excitation, the source should emit radiation that is slightly more energetic than the photoelectric absorption-edge energy of the element being analyzed. Available sources leave serious gaps, especially with regard to quantitative work in the iron–copper region of the periodic table. As soon as manufacturers or source vendors overcome this deficiency, wide-range applications can be foreseen. The ideal source would be specific, would have low activity for safety reasons, and would possess a long half-life so that corrections can be avoided. In certain applications, especially in the case of industrial plant applications where appropriate safety measures are readily taken, stronger sources may be used. The use of liquid nitrogen in the detector unit has not caused problems. A refill is recommended every third day.

It should be noted that the results of the quantitative studies, expressed as standard errors of estimate, reflect analytical problems associated with both X-ray and chemical procedures. Table IV provides a measure of the differences that can be found where several laboratories and methods are involved.

Currently, the mathematical interpretation and elimination of matrix interferences is being studied. Figure 12, a plot of concentration *versus* counts by photon spectrometry, shows a significant decrease of the standard error over conventional X-ray data in Figure 10, namely, 0.0027 *versus* 0.0073%. This is believed to be due to the short wavelengths of the radiations involved, the simplicity of the primary spectrum, the absence of an analyzing crystal and its scatter, and the use of pulse-height discrimination. Residual absorption effects still recognizable cannot be eliminated by simple ratio techniques because of the low activity of the source and its resulting low count rates. If further improvement is desired, other compensating techniques—perhaps based upon iron counts or integrated background values—need to be applied.

Additional work is being conducted in the analysis of lead and zinc, the analysis of tellurium by use of a dysprosium source, and the utilization of ^{241}Am. Also, extensive on-stream tests have been scheduled.

ACKNOWLEDGMENTS

The authors wish to thank the Kennecott Copper Corporation for permission to publish this paper and acknowledge the valuable suggestions and encouragement of W. M. Tuddenham.

REFERENCES

1. W. J. Campbell, T. D. Brown, and J. W. Thatcher, "X-Ray Absorption and Emission," *Anal. Chem.* **38**: 418R–420R, April, 1966.
2. J. O. Karttunen and D. J. Henderson, "An Improved Portable Fluorescent X-Ray Instrument Using Radioisotope Excitation Sources," *Anal. Chem.* **37**: 307–309, Feb., 1965.
3. A. G. Darnley and M. J. Gallagher, "Progress Report on Use of Portable Radioisotope X-Ray Fluorescence Analyser," *Trans. Inst. Mining Met.* **711**: B105–106, Feb., 1966.
4. J. R. Rhodes, "Radioisotope X-Ray Spectrometry, A Review," *Analyst* **91**: 683–699, Nov., 1966.
5. "X-Ray Unit Uses Solid-State Detector," *Chemical and Engineering News* **42**: 115A, Jan. 3, 1966.
6. H. R. Bowman, E. K. Hyde, S. G. Thompson, and R. C. Jared, "Application of High-Resolution Semiconductor Detectors in X-Ray Emission Spectrography," *Science* **151**: 562–568, Feb. 4, 1966.
7. R. H. Muller, "Characteristic X-Rays Excited by Radioisotopes," *Anal. Chem.* **38**: A115–117, Oct., 1966.
8. J. R. Rhodes, "Radioisotope X-Ray Spectrometry, A Review," *Analyst* **91**: 684–686, 690, Nov. 1966.
9. G. Andermann, "Semitheoretical Approach to Interelement Correction Factors in Secondary X-Ray Emission Analysis," *Anal. Chem.* **38**: 82–86, Jan., 1966.
10. B. J. Alley and R. H. Myers, "Corrections for Matrix Effects in X-Ray Fluorescence Analysis, Using Multiple Regression Methods," *Anal. Chem.* **37**: 1685–1690, Dec. 1965.
11. R. O. French, R. W. Vaughn, A. P. Langheinrich, and J. F. Baum, "Application of Continuous XRF Analysis to Mill Control," paper presented at the *Symposium on Continuous Process Control*, AIME, Philadelphia, Pa., Dec. 5, 1966.
12. H. A. Liebhafsky, H. G. Pfeiffer, E. H. Winslow, and P. D. Zemany, *X-Ray Absorption and Emission in Analytical Chemistry*, John Wiley and Sons, Inc., New York, 1960, pp. 162–191.
13. Technical Measurement Corporation, private communication, Nov., 1966.

QUANTITATIVE MICROPROBE ANALYSIS OF THIN INSULATING FILMS

J. W. Colby

Bell Telephone Laboratories, Incorporated
Allentown, Pennsylvania

ABSTRACT

The analysis of thin insulating films occurring in various stages of the manufacture of integrated circuits has in the past been difficult, if not impossible. Their dimensions usually preclude analysis by conventional chemical techniques, hence very little is known concerning their relative stoichiometries. However, by microprobe analysis, films as thin as 540 Å ($\sim$18 $\mu g/cm^2$) have been successfully analyzed. In general, two different approaches are taken, depending on the film thickness. Usually, films thicker than 2500 Å ($\sim$50 $\mu g/cm^2$) are analyzed by conventional microprobe techniques. A computer program, called MAGIC, has been written, which converts the raw X-ray intensities to chemical composition. All data are corrected for dead time, background, absorption, atomic number effects, and fluorescence by characteristic radiation, if required. A minimum of input is required, only the chemical symbols, X-ray lines employed, and the accelerating voltage being necessary in addition to the raw X-ray intensities. All constants such as atomic weights, critical excitation potentials, X-ray wavelengths, and absorption coefficients are stored or calculated internally, which reduces the errors and time associated with looking up and key-punching these values. New fluorescent yields are used for *K* and *L* radiation. No fluorescence correction is made for *M* radiation. The use of this correction program generally gives results, accurate to about 2 to 4%, relative to the amount present. For films thinner than 2500 Å, a new model is proposed which allows the X-ray spectra from the film and substrate to be unfolded to give the composition of the film only. To accomplish this, the atomic number correction of Duncumb and deCasa employed in MAGIC has been used in conjunction with a mean electron energy concept. Results obtained to date through the use of this model have been surprisingly good, being of the order of 5% relative to the amount present. Potential errors and uncertainties are discussed, and results given which illustrate the accuracy of the two methods.

INTRODUCTION

There are many different processes for the preparation of thin insulating films used in semiconductor devices, and, in each process, there are usually several variables. It may be expected, therefore, that some differences in stoichiometry may occur in films prepared by different processes and perhaps even in films prepared by the same process. Variations in physical and electronic properties have been shown to exist between such films, and it is logical to assume that some of these variations may be due to variations in stoichiometry or to the presence or absence of trace impurities.

Because of their extreme thinness, usually 1000 to 5000 Å, it has been quite difficult to provide any quantitative chemical data about these films. Compositions of such films are usually specified on the basis of thermodynamic principles, gas-phase composition of the reactants, *etc.* While these may be sufficient to describe the gross product, minor fluctuations may alter slightly the stoichiometry of the film and, consequently, some of its properties. In some instances, glancing-angle electron diffraction may be employed to determine differences between films which are crystalline, and, although this may be extremely valuable, it does not necessarily convey information concerning small variations in stoichiometry.

Although microprobe analysis should lend itself very well to such determinations, it has only recently been applied to thin films.[1] Most microprobe analysts operate at accelerating potentials of 10 to 30 keV, where the depth of penetration is on the order of several microns. It was found, however, that, by lowering the accelerating potential to 4 to 6 keV, the depth of penetration could be limited to approximately 2500 Å in the case of silica,[2] which would make possible the analysis of fairly thin films with conventional microprobe techniques. To analyze thinner films required the preparation of calibration curves.[1] To avoid such calibration curves, a new approach was sought which utilized basic principles and previously obtained experimental data. The resulting model, to be described herein, is completely general and should allow almost any film of any thickness on any substrate to be quantitatively analyzed. The precision should be comparable to that of conventional techniques of microprobe analysis, and the results obtained to date indicate the accuracy to be nearly as good.

BASIC PRINCIPLES OF X-RAY GENERATION

When a beam of electrons impinges on the surface of the sample to be analyzed, a portion of the electrons penetrate the specimen and are brought to rest within the specimen, while a smaller portion of the electrons are backscattered from the surface. This is illustrated in Figure 1 for electrons of initial energy E_0 incident normally on a specimen. If the energy of an electron in the specimen is E as it traverses the specimen along the path shown, then the number of ionizations, dn, per unit of path length, dx, may be expressed as

$$\frac{dn}{dx} = \Psi\left(\frac{C_A N_0 \rho}{A}\right) \tag{1}$$

where Ψ is the classical ionization cross section for X-rays in the element A; ρ is the density of the alloy; N_0 is Avogadro's number; and A is the atomic weight of element A. Since ionizations can occur along the path length of the electron to a point at which the energy has dropped below E_c, the critical excitation potential, the total ionizations produced by each electron remaining within the specimen is

$$n = \frac{C_A N_0}{A} \int_{E_0}^{E_c} \frac{\rho \Psi}{dE/dx}\, dE \tag{2}$$

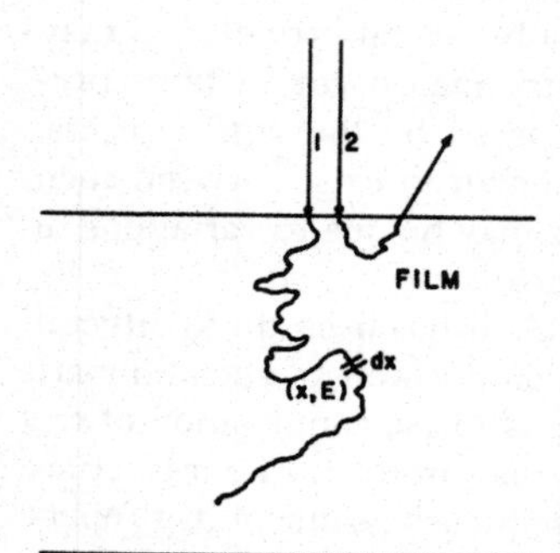

Figure 1. Electron paths in thick films or bulk targets.

where dE/dx is the loss of energy per unit of path length. If the stopping power S is defined as[3]

$$S = -\frac{1}{\rho}\frac{dE}{dx} \tag{3}$$

then equation (2) becomes

$$n = \frac{C_A N_0}{A}\int_{E_c}^{E_0} \Psi/S \cdot dE \tag{4}$$

Since some of the electrons backscattered from the specimen have energies greater than E, Duncumb and Shields[4] introduced a factor R, following Green and Cosslett,[5] which is the fraction of ionizations remaining when the loss due to backscattering is removed. Thus the total ionizations per incident electron is

$$n = \frac{C_A N_0}{A} R \int_{E_c}^{E_0} \Psi/S \cdot dE \tag{5}$$

A similar expression may be derived for pure A, so that, in the absence of other effects,

$$\frac{n(\text{alloy})}{n(\text{pure})} = k_A = \frac{C_A R_{AB}\int_{E_c}^{E_0} \Psi/S_{AB} \cdot dE}{R_A \int_{E_c}^{E_0} \Psi/S_A \cdot dE} \tag{6}$$

The ionization cross section is calculated from[5]

$$\Psi E_c^2 = \text{const} \cdot \frac{1}{U} \ln U \tag{7}$$

where $U = E/E_c$. The stopping power is calculated from the Bethe[6] law as given by Nelms[7]

$$S = \text{const} \cdot \frac{Z}{A}\frac{1}{E} \ln\left(\frac{\sqrt{e/2} \cdot E}{J}\right) \tag{8}$$

where Z is the atomic number; e, the base of the natural logarithms; and J is the mean ionization potential as given by Duncumb and deCasa,[8]

$$J/Z = 14.0(1.0 - e^{0.1Z}) + \frac{75.5}{Z^{Z/7.5}} - \frac{Z}{100 + Z} \tag{9}$$

The backscatter loss factors were calculated by Duncumb[9] from[4]

$$R = 1 - \frac{\int_{W_c}^{1} d\eta/dW \int_{E_c}^{WE_0} \Psi/S \cdot dE \cdot dW}{\int_{E_c}^{E_0} \Psi/S \cdot dE} \tag{10}$$

by use of the backscattered electron-energy distribution data and backscattered-electron yields of Bishop.[10] Here, $W = E/E_0$ and $d\eta/dW$ is the number of electrons backscattered per incident electron per unit of energy interval. In the above expression, all energies and potentials are in electron volts.

The two factors are lumped together and called the atomic number effect. The atomic number effect is thus calculated from equation (6) by performing the numerical integration indicated. Whereas the numerical integration is not particularly tedious or difficult (on a digital computer), a simplification may be introduced with little or no loss in accuracy.

If, after substituting equations (7) and (8) in equation (6), it is assumed that the energy E may be replaced by a mean energy value $\bar{E}$ which is equal to $(E_0 + E_c)/2$, then the integral signs may be removed and, equation (6) reduces to

$$k_A = C_A \frac{R_{AB}}{R_A} \frac{\dfrac{1}{(Z/A)_{AB}} \dfrac{1}{\ln[(\sqrt{e/2} \cdot E)/J_{AB}]}}{\dfrac{1}{(Z/A)_A} \dfrac{1}{\ln[(\sqrt{e/2} \cdot E)/J_A]}} \tag{11}$$

which may be further simplified to

$$k_A = C_A \frac{R_{AB}}{R_A} \frac{(Z/A)_A}{(Z/A)_{AB}} \frac{(7.061 + \ln \bar{E} - \ln J_A)}{(7.061 + \ln \bar{E} - \ln J_{AB})} \tag{12}$$

In equation (12), the mean energy is in kiloelectron volts, and the constant is equal to $\ln \sqrt{e/2} + \ln 1000$. The factor 1000 is introduced to permit expressing the mean energy $\bar{E}$ in kiloelectron volts.

In Reed and Long's expression as given by Smith,[11] the constant was 5.16, and J was replaced by Z. Smith believed this factor to be semiempirical, being obtained by experiment, and consequently modified the expression to fit his data. However, Reed and Long were using the original form of the Bethe law, in which the logarithmic term was given by $\ln(2E/J)$, where $J = 11.5Z$ eV and E is also in electron volts. On substitution of $\bar{E}$ for E and by expressing $\bar{E}$ in kiloelectron volts instead of electron volts, this becomes $\ln 2 + \ln 1000 + \ln \bar{E} - \ln 11.5 - \ln Z$, which reduces to $5.159 + \ln \bar{E} - \ln Z$. Thus, the constant as given by Reed and Long was not experimental, and Smith's modification on the basis given was not justifiable.

The use of equation (12) or equation (6) gives results which differ by less than 1%. In general, the atomic number effect acts in such a manner as to reduce the apparent composition of heavier elements and to increase the apparent composition of lighter elements. Although backscatter and penetration effects act in opposition to each other, complete compensation is rarely achieved, so that an atomic-number correction should always be applied to microprobe X-ray data.

Absorption effects are also always present, and an absorption correction should similarly be applied to all microprobe X-ray data. In many cases, data must also be corrected for fluorescence by characteristic radiation. Since these corrections have been adequately discussed in the literature previously, it will not be necessary to discuss them at this time.

MAGIC

A computer program called MAGIC, an acronym for Microprobe Analysis General Intensity Corrections, has been written in FORTRAN II for use on an IBM 7094 computer.*

* A FORTRAN IV version for use on an IBM 360 computer is also available.

The program is completely general in that it applies all corrections required for any system containing up to eight elements. A ninth element may also be determined by difference. Data are corrected for absorption by using Duncumb and Shields'[12] modification of Philibert's[13] expression, *i.e.*, the fraction of radiation emerging from the specimen is

$$f(\chi) = \frac{1 + h}{(1 + \chi/\sigma_c)[1 + h(1 + \chi/\sigma_c)]} \tag{13}$$

where $\chi = (\mu/\rho)$ cosec θ; μ/ρ is the mass absorption coefficient; θ is the X-ray emergence angle; $h = 1.2A/Z^2$; and σ_c is the modified Lenard coefficient as given by Heinrich[14] as

$$\sigma_c = \frac{4.5 \times 10^5}{E_0^{1.67} - E_c^{1.67}}$$

The program checks for fluorescence by characteristic radiation by comparing the absorption edge of each element analyzed against the wavelengths of the lines analyzed. If fluorescence effects are present, the program corrects the data, using the method of Reed[15]

$$\frac{I_f}{I_d} = 0.5P_{ij}C_B\frac{r_A - 1}{r_A}\omega_B\frac{A}{B}\left(\frac{U_B - 1}{U_A - 1}\right)^{1.67}\frac{(\mu/\rho)_B{}^A}{(\mu/\rho)_B}\left[\frac{\ln(1 + y)}{y} + \frac{\ln(1 + v)}{v}\right] \tag{14}$$

where the symbols have the same meaning as given by Reed.[15]

All data are corrected for atomic number effects by using the method outlined above, so that the complete correction procedure is

$$C_A = k_A\left[\frac{R_A(Z/A)_{AB}(7.061 + \ln\bar{E} - \ln J_{AB})}{R_{AB}(Z/A)_A(7.061 + \ln\bar{E} - \ln J_A)}\right]\left[\frac{f(\chi)_A}{f(\chi)_{AB}}\right]\left[\frac{1}{1 + I_f/I_d}\right] \tag{15}$$

The first term in brackets is the complete atomic-number correction, the second term is the absorption correction, and the third term is the fluorescence correction (when applicable) and k is the intensity ratio corrected for dead time and background.

The program is different than other programs in that it does not allow one to choose a method which will make the answer come out right, nor does it allow one to enter various values of parameters such as absorption coefficients and fluorescence yields to make the data come out right. The program is completely self-contained, all parameters being stored or calculated within the program. Thus, the input to the program is extremely simple; only the chemical symbols, lines analyzed (K, L, or M), and the accelerating potential are required input. An example of the input is shown in Figure 2. In addition to the composition (mass and atomic), the output also lists the mean k ratios and two sigma limits, the peak-to-back-ground ratios, the minimum detectability limits (based on a signal 20% above background), the individual k ratios, the number of iterations required to converge, all the basic parameters such as absorption coefficients and fluorescence yields that were used in correcting the data, the raw X-ray intensities, the backgrounds, and the dead times used. If an element was determined by difference or if other than pure elemental standards were used, there is a note to that effect. If, in addition to the regular input, the density is included, the depth of penetration is also calculated and printed out.

A complete listing of the input and output for a sample problem is given in the Appendix. The absorption coefficients and jump ratios are calculated by using Heinrich's[16] data. The backscatter-loss factors of Duncumb[9] have been curve fitted to a fifth-degree polynomial, and the coefficients for the polynomials are also stored in the computer. The

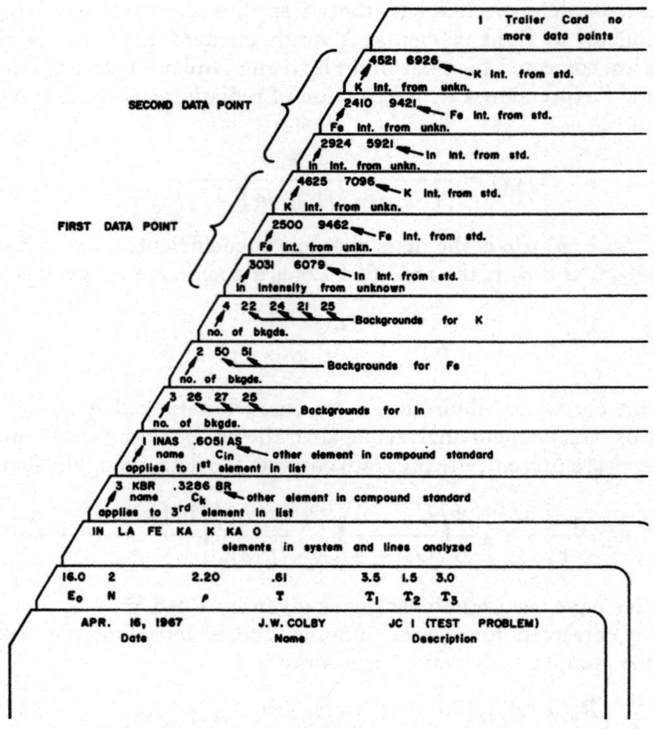

Figure 2. Input data for sample problem.

fluorescent yields are calculated from a semiempirical expression due to Burhop,[17]

$$\left(\frac{\omega}{1-\omega}\right)^{\frac{1}{4}} = A + BZ + CZ^3 \tag{16}$$

The experimental yields given in the recent, very excellent compilation of Fink *et al.*[18] were used to calculate the coefficients A, B, and C in equation (16) for K, L, and M yields. The theoretical yields given in Fink *et al.*[18] were not used, primarily because they are, in general, higher than the experimental values, which is probably caused by the theoretically calculated values not having included relativity effects. The constants for the K, L, and M yields are given in Table I. The experimental points and the fitted curves are shown in Figure 3, which was obtained on a Calcomp plotter and 360 computer. The fluorescent yields were calculated from equation (16) and the constants in Table I, and they are tabulated in Table II. The standard error of estimate on the tabulated values is less than 0.02 in all cases.

Table I. Constants for Calculating the Fluorescent Yields

	K	L	M
A	−0.03795	−0.11107	−0.00036
B	0.03426	0.01368	0.00386
C	-0.11634×10^{-5}	-0.21772×10^{-6}	0.20101×10^{-6}

Table II. Fluorescent Yields

Z	K	L	M	Z	K	L	M
1				51	0.854	0.088	0.002
2				52	0.862	0.095	0.003
3				53	0.869	0.102	0.003
4				54	0.876	0.110	0.003
5				55	0.882	0.118	0.004
6	0.001			56	0.888	0.126	0.004
7	0.002			57	0.893	0.135	0.004
8	0.003			58	0.898	0.143	0.005
9	0.005			59	0.902	0.152	0.005
10	0.008			60	0.907	0.161	0.006
11	0.013			61	0.911	0.171	0.006
12	0.019			62	0.915	0.180	0.007
13	0.026			63	0.918	0.190	0.007
14	0.036			64	0.921	0.200	0.008
15	0.047			65	0.924	0.210	0.009
16	0.061			66	0.927	0.220	0.009
17	0.078			67	0.930	0.231	0.010
18	0.097			68	0.932	0.240	0.011
19	0.118			69	0.934	0.251	0.012
20	0.142	0.001		70	0.937	0.262	0.013
21	0.168	0.001		71	0.939	0.272	0.014
22	0.197	0.001		72	0.941	0.283	0.015
23	0.227	0.002		73	0.942	0.293	0.016
24	0.258	0.002		74	0.944	0.304	0.018
25	0.291	0.003		75	0.945	0.314	0.019
26	0.324	0.003		76	0.947	0.325	0.020
27	0.358	0.004		77	0.948	0.335	0.022
28	0.392	0.005		78	0.949	0.345	0.024
29	0.425	0.006		79	0.951	0.356	0.026
30	0.458	0.007		80	0.952	0.366	0.028
31	0.489	0.009		81	0.953	0.376	0.030
32	0.520	0.010		82	0.954	0.386	0.032
33	0.549	0.012		83	0.954	0.396	0.034
34	0.577	0.014		84	0.955	0.405	0.037
35	0.604	0.016		85	0.956	0.415	0.040
36	0.629	0.019		86	0.957	0.425	0.043
37	0.653	0.021	0.001	87	0.957	0.434	0.046
38	0.675	0.024	0.001	88	0.958	0.443	0.049
39	0.695	0.027	0.001	89	0.958	0.452	0.052
40	0.715	0.031	0.001	90	0.959	0.461	0.056
41	0.732	0.035	0.001	91	0.959	0.469	0.060
42	0.749	0.039	0.001	92	0.960	0.478	0.064
43	0.765	0.043	0.001	93	0.960	0.486	0.068
44	0.779	0.047	0.001	94	0.960	0.494	0.073
45	0.792	0.052	0.001	95	0.960	0.502	0.077
46	0.805	0.058	0.001	96	0.961	0.510	0.083
47	0.816	0.063	0.002	97	0.961	0.517	0.088
48	0.827	0.069	0.002	98	0.961	0.524	0.093
49	0.836	0.075	0.002	99	0.961	0.531	0.099
50	0.845	0.081	0.002	100	0.961	0.538	0.106

In use, the program counts the chemical symbols and lines to determine the number of elements in the system and the number analyzed. It then compares each symbol and line against those in memory, until a match is obtained. All parameters are then calculated. Raw X-ray data are then read in, corrected, and averaged and all data printed out.

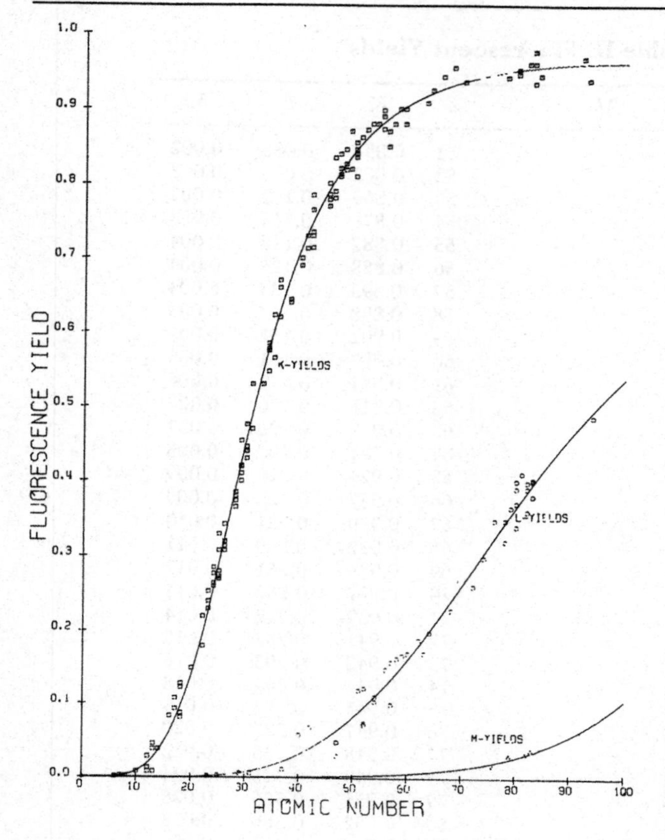

Figure 3. Fluorescent yields for K, L, and M lines.

As an example of the program's speed and utility, for a nine-element system recently analyzed, it required 45 min to look up and key-punch the 135 parameters required. The program described herein "looked up" the constants for and completely solved eight such problems in less than 20 sec.

ANALYSIS OF INSULATING FILMS

Although MAGIC may be used with any system, bulk or film, it was primarily intended for use in the analysis of films such as are used in the semiconductor industry. By employing accelerating potentials of ~5 keV, film as thin as 2500 Å (~50 μg/cm²) may be routinely analyzed. The depth of penetration, X, is calculated from

$$X(\text{Å}) = \frac{330\ \text{Å}}{\rho Z} E_0^{1.5} \qquad (17)$$

where ρ is the density of the film. Castaing[19] used a 1.67-power-voltage dependence; however, we find a 1.5-power-voltage dependence more satisfactory. Also, Cosslett and Thomas[20,21] have shown that the mean range and extrapolated range both follow a 1.5-power law. The constant 330 was experimentally determined for our instrument by measuring the penetration as a function of accelerating potential, for aluminum films evaporated on silicon substrates.

Table III. Microprobe Analysis of Silicon in Steam-Grown Silica Films

Accelerating potential, keV	Intensity ratio	Mass concentration	Atomic concentration	Theoretical mass concentration
4.0	0.427 ± 0.006	45.9 ± 0.6	32.6 ± 0.5	46.7
5.0	0.426 ± 0.002	45.8 ± 0.2	32.5 ± 0.2	46.7
5.5	0.428 ± 0.003	46.0 ± 0.3	32.7 ± 0.2	46.7
6.0	0.429 ± 0.003	46.1 ± 0.3	32.8 ± 0.3	46.7

Steam-grown silica (~5000 Å) on a silicon substrate was analyzed at an accelerating potential of 5 keV, with carbon-coated, bulk, fused silica as a standard, to be certain that films of steam-grown silica were stoichiometric. The intensity ratios (*k* ratios) obtained for silicon and oxygen were 0.9891 ± 0.0106 and 1.0064 ± 0.0158, respectively. Films of steam-grown silica (2900 to 6000 Å) on silicon substrates were analyzed at 4.0, 5.0, 5.5, and 6.0 keV, with pure silicon as a standard. A carbon coating was not required. The results obtained after correction by MAGIC are shown in Table III.

It may be noted that, although the *k* ratios are approximately 8 to 9% low owing primarily to the atomic-number effect, the chemical compositions obtained after correction agree quite well with the theoretical stoichiometry, being within 2% in all cases. These results are based on 300 analyses of 17 samples and, consequently, are well representative.

Silica films (3600 to 4500 Å) on silicon substrates, produced by a DC plasma-discharge process (oxygen plasma with $SiBr_4$), were analyzed by microprobe analysis and compared with steam-grown silica. All three samples were found to contain approximately 2% (by weight) more silicon than steam silica. These results are shown in Table IV.

Silicon nitride films (~5000 Å) on silicon substrates, prepared pyrolytically by the reaction of silane (SiH_4) and ammonia, were analyzed for silicon content and were found to contain 60.4 wt.% silicon (43.2 at.%). Subsequent emission spectrographic analyses of similar films confirmed the microprobe analyses, differing by approximately 1% (relative).

Silicon nitride films prepared by DC plasma discharge from $SiBr_4$ are in general silicon poor and contain bromine. A special series of films varying in thickness from 1900 to 4000 Å was prepared by this method on silicon substrates. The substrate temperature was the only deliberately introduced variable. The results of microprobe analyses of these

Table IV. Comparison of Plasma and Steam Silica

Mass Concentration

Plasma	Steam	Difference
47.9	46.0	1.9
47.5	45.6	1.9
47.9	45.7	2.2

Table V. Microprobe Analyses of Silicon Nitride Films Produced by DC Plasma Discharge

Sample	Substrate temperature, °C	Silicon, wt. %	Bromine, wt. %
I	700	58.6	0.4
II	650	60.8	0.3
III	600	58.3	0.3
IV	550	55.8	1.0
V	500	62.0	0.5
VI	450	61.2	1.2
VII	400	56.9	1.2

Table VI. Microprobe Analysis of Aluminum in Alumina Films

Intensity ratio	Mass concentration
0.519	56.1
0.524	56.6

films are shown in Table V, where it may be noted that considerable variations in stoichiometry occur.

Alumina films (3200 to 4000 Å) on silicon substrates, prepared pyrolytically by the reduction of $AlCl_3$, were analyzed in the microprobe, and the results are shown in Table VI.

Although the films were identified by glancing-angle electron diffraction as κ-alumina, they are definitely aluminum rich. This may be explained in part, since the depth of penetration in the case of electron diffraction is of the order of 20 to 100 Å at 50 keV, whereas the depth of penetration of the microprobe analysis was approximately 3000 Å. Thus the surface may have been κ-alumina, but the bulk of the film was aluminum-rich.

In all of these experiments, it is absolutely necessary to peak the spectrometer on both the unknown and the standard to avoid the effects of wavelength shift. As examples of the magnitudes of the errors involved, steam-grown silica was analyzed first by peaking the spectrometer on the pure silicon standard only, then recording the intensities from the unknown (silica) and standard (silicon). Next, the spectrometer was peaked for the silicon wavelength on silica, and the intensities were recorded from the silica and silicon. Finally, the analysis was made in the normal way by peaking the spectrometer on both the silicon standard and the silica film and recording the intensities from each. The results after being corrected by MAGIC are shown in Table VII. A similar comparison was made for the analysis of silicon in silicon nitride (Si_3N_4), by using silicon as a standard. The results are shown in Table VIII.

Table VII. Effects of Wavelength Shift on the Analysis of Silica with the Use of a Silicon Standard

	Intensity ratio	Silicon content	
		wt. %	at. %
Peaked on silicon only	0.347	37.6	25.6
Peaked on silica only	0.646	67.8	54.6
Normal, peaked on both	0.425	45.8	32.5

Table VIII. Effects of Wavelength Shift on the Analysis of Silicon Nitride

	Intensity ratio	Silicon content	
		wt. %	at. %
Peaked on silicon only	0.491	52.0	35.1
Peaked on silicon nitride only	0.705	73.2	57.6
Normal, peaked on both	0.565	59.5	42.3

From these two examples, it may be seen that drastic errors can occur if the analysis is performed improperly. In general, these differences are much larger than those reported by Hart;[22] however, he was using a poor-resolution spectrometer. The silicon measurements were made in our case with a PET crystal and a flow proportional counter, with peak-to-background ratios at 5 keV of approximately 700 : 1. The magnitudes of the wavelength shifts were 2.2 X units ($\sim$0.002 Å) toward the short-wavelength side in the case of silica and 0.9 X units ($\sim$0.0009 Å) also toward the short-wavelength side for silicon nitride.

In all these measurements, the film was on a conducting substrate, so carbon-coating was unnecessary. However, in the analysis of some films, another effect can occur which can lead to erroneous results. At low accelerating potentials, the secondary electron yield is quite high,[23] especially for some of these insulators. Consequently, the space-charge effect is offset by the secondary electron emission of the sample, and a self-regulating system is obtained. However, in a recent analysis of an aluminum nitride film (2800 Å) on a silicon substrate, it was noted that consecutive 10-sec aluminum-intensity readings taken on the same spot increased markedly. This is illustrated in Figure 4, where the sample current and aluminum X-ray intensity are plotted as a function of time. Initially, a space-charge is set up at the surface, but, as secondary electron emission increases, the sample current decreases and the X-ray intensity increases until a self-regulating system is obtained. Results obtained from the first 10-sec readings and results obtained from the intensity readings after stability had been reached are compared in Table IX. These results were obtained at 4 keV.

In the preceding analyses, it is noted that the correction procedures employed in MAGIC generally give results which are biased slightly low for the heavier elements and slightly high for the lighter elements. However, in all cases, the mass concentrations obtained differ from reported compositions (theoretical and chemical) by less than 2%, relative. Results obtained in analyzing much heavier elements such as tantalum in tantalum

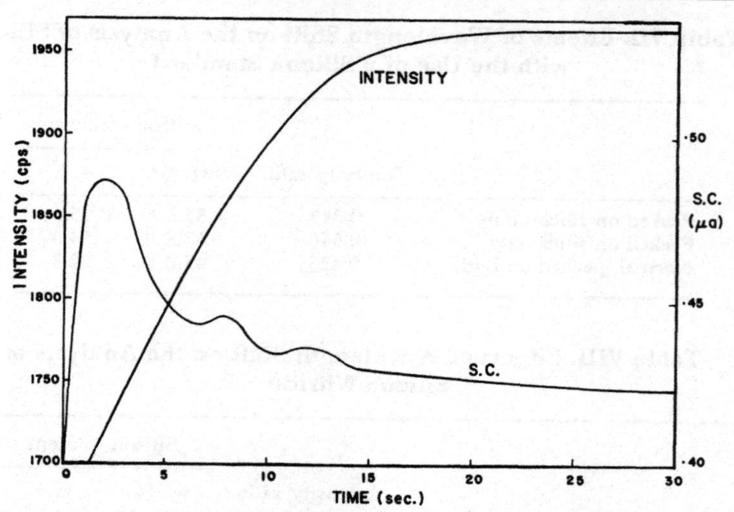

Figure 4. Effects of secondary electron emission on X-ray intensity from aluminum nitride film on silicon.

Table IX. Effects Due to Secondary Electron Emission on the Analysis of Aluminum Nitride (AlN)

		Aluminum	
	Intensity ratio	wt. %	at. %
Initial intensity	0.562	59.6	43.4
Final intensity	0.612	64.6	48.6

nitride films (TaN and Ta_2N) are also biased by about the same amount in the same direction. This is shown in Table X. Here, again, it may be noted that the relative error is less than 2%. These latter films were analyzed at 10 keV, whereas all the previous films were analyzed at 4 to 6 keV.

Bulk intermediate alloys have been analyzed at higher accelerating potentials with about the same accuracy, except, gold–copper alloys. Some typical results are shown in Table XI for systems analyzed at 20 keV. With the exception of the gold analysis, as

Table X. Microprobe Analyses of Tantalum in Tantalum Nitrides

Film	Tantalum, wt. %	Theoretical, wt. %
TaN	91.5	92.8
TaN	92.4	92.8
TaN	92.2	92.8
Ta_2N	95.2	96.3

Table XI. Microprobe Analyses of Bulk Systems

System	Element and line analyzed	Intensity ratio	Mass concentration	Reported composition
GaAs	As *K*	0.500	52.4	51.8
GaAs	As *K*	0.503	52.7	51.8
Fe–Cr	Cr *K*	0.127	10.2	10.1
Fe–Cr	Fe *K*	0.894	90.9	89.8
Au–Cu	Au *L*	0.411	48.5	52.4

noted above, the relative error is less than 2%. In the case of the gold, however, the relative error is approximately 7.5%. This is in good agreement with Brown and Wittry,[24] who found that the fluorescence by the continuum caused a relative error of the order of 10% in the case of a similar alloy. Whereas, in the past, it has been claimed that the fluorescence by the continuum was negligible (Kirianenko *et al.*),[25] it is apparently only true for microprobes having low X-ray emergence angles (15 to 18°). The instrument used by Brown and Wittry[24] had an X-ray emergence angle of 52.5°, while ours is 41° (effective). Consequently, it is planned to add Henoc's correction[26] for fluorescence by the continuum to MAGIC, as it is apparent that it must be taken into account when analyzing systems containing heavy elements ($Z > 30$) in instruments having high X-ray emergence angles ($\theta > 20°$).

While the above analyses were done on bulk specimens or films of sufficient thickness that the analyzed region could be maintained within the thickness of the film, it is often desirable to analyze even thinner films. In such cases, it may not be possible to lower the accelerating potential sufficiently to preclude complete penetration of the film and still excite the X-rays of interest. Consequently, a new model is proposed which, although a gross oversimplification, has given reasonably accurate results to date.

THIN-FILM MODEL

In Figure 1, two electron paths were considered. In the proposed model, it is necessary to consider a third path, which is shown in Figure 5. Assuming initially that the substrate upon which the film of interest is deposited does not contain any of the elements to be analyzed for in the film, we can proceed as before. A portion of the electrons impinging on the surface are backscattered (electron 3 in Figure 5), while those that penetrate the film may take either of two courses. The number of ionizations produced by electrons

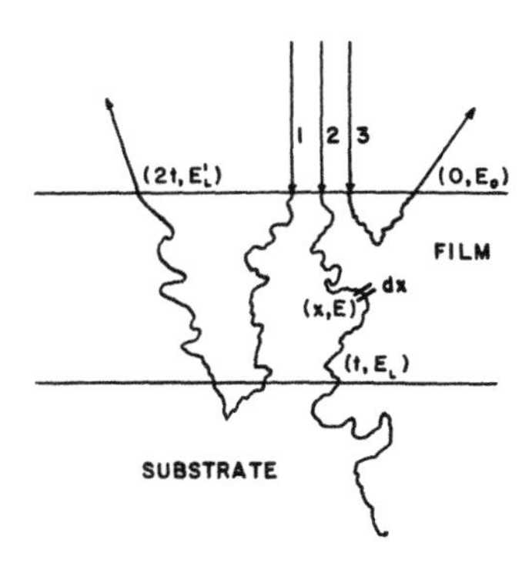

Figure 5. Electron paths in thin film on substrate.

of type 2, from equation (4) and Figure 5, is

$$n = \frac{C_A N_0}{A} \int_{E_L}^{E_0} \Psi/S \cdot dE \tag{18}$$

where E_L is the mean energy of the electrons at the interface as they are about to leave the film. The number of ionizations produced by electrons of type 1 is

$$n = \frac{C_A N_0}{A} \int_{E_L}^{E_0} \Psi/S \cdot dE + \frac{C_A N_0}{A} \int_{E_L'}^{E_L} \Psi/S \cdot dE \tag{19}$$

where E_L is the mean energy of the electrons at the interface and E_L' is the energy of the electrons as they leave the film at the upper surface. The number of electrons backscattered from the substrate is given by the backscattered electron yield of the substrate, and the number of ionizations lost owing to electrons backscattered may be taken into account by introducing the factor R, as before. Consequently, the total ionizations produced in the film are

$$n = \frac{N_0 C_A R_{AB}}{A} \left(\int_{E_L}^{E_0} \Psi/S_{AB} \cdot dE + \eta_S \int_{E_L'}^{E_L} \Psi/S_{AB} \cdot dE \right) \tag{20}$$

If the intensity from the film is compared with the intensity from a pure bulk standard, then, in the absence of other effects,

$$k_A = C_A \frac{R_{AB} \left(\int_{E_L}^{E_0} \Psi/S_{AB} \cdot dE + \eta_S \int_{E_L'}^{E_L} \Psi/S_{AB} \cdot dE \right)}{R_A \int_{E_c}^{E_0} \Psi/S_A \cdot dE} \tag{21}$$

In equations (20) and (21), η_S is the backscattered electron coefficient of the substrate. Electron range equations may be derived of the type given by Castaing[19]

$$\rho x = \text{const} \cdot \frac{A}{Z} (E_0{}^n - E_c{}^n) \tag{22}$$

which give the depth of the electrons when their mean energy has been reduced from E_0 to E_c. It may be assumed therefore that the depth to which the electrons have penetrated when their mean energy has been reduced from E_0 to E_L is

$$\rho t = \text{const} \cdot \frac{A}{Z} (E_0{}^n - E_L{}^n) \tag{23}$$

where t is the film thickness. The mean electron range follows a $\frac{3}{2}$ power-voltage dependence, as shown by Cosslett and Thomas,[20,21] and, in our microprobe, the constant was found to be 330. Consequently, we may write

$$E_L = \left(E_0{}^{\frac{3}{2}} - \frac{\rho t Z}{330\ \text{Å}} \right)^{\frac{2}{3}} \tag{24}$$

$$E_L' = \left(E_0{}^{\frac{3}{2}} - \frac{\rho t Z}{165\ \text{Å}} \right)^{\frac{2}{3}} \tag{25}$$

Absorption may be taken into account by employing the subtlety introduced by Duncumb and Shields.[12] A modified Lenard coefficient is employed, defined by

$$\sigma_c = \frac{4.5 \times 10^5}{E_0^{1.67} - E_L^{1.67}} \tag{26}$$

to take into account the fact that the X-rays are not generated so deeply in the film and therefore will be less absorbed. For instance, by using this model for a 1000-Å aluminum film on a silicon substrate at 20 keV, $f(\chi)$ for the film is 0.991, while, for bulk aluminum under the same conditions, $f(\chi)$ is 0.836.

The proposed model was first tested for pure aluminum films evaporated onto silicon substrates, and the agreement between theory and experiment was found to be satisfactory. In general, the model predicts smaller k ratios than are obtained experimentally, probably owing to electrons straggling and perhaps fluorescence of the aluminum by the silicon substrate.

The model was then tested for several thin films on silicon substrates at an accelerating potential of 20 keV. The results are shown in Table XII. As can be seen, the results agree fairly well with the reported stoichiometries, especially considering the nature of the assumptions involved. Whereas the results from the alumina films appear to be higher than the results obtained for the other films, it was found by analyzing thicker films that these films generally are aluminum-rich (see Table VI). The one exception in the above alumina films was produced under different conditions and has a lower index of refraction, which indicates that it is richer in oxygen (aluminum-poor), than the other films.

The case considered so far was simple in that fluorescence effects from the film and the substrate were not considered and neither was the situation in which the substrate contained some of the elements analyzed. In practice, this is seldom the case. The latter situation can be treated simply by adding a third term in the numerator of equation (21), *i.e.*,

$$(1 - \eta_S)R_S \int_{E_c}^{E_L} \Psi/S_s \cdot dE \tag{27}$$

where R_S is the backscatter loss factor of the substrate; and S_s, the stopping power of the substrate. Fluorescence effects within the film would be treated in the conventional manner, whereas fluorescence effects due to the substrate may possibly be taken into account by modifying the voltage-dependent part of Reed's correction, *i.e.*, by defining the overvoltage U as E_L/E_c instead of E_0/E_c.

Table XII. Results of Microprobe Analysis with a Thin Film Model

Film	Thickness, Å	Element analyzed	Intensity ratio	Mass concentration	Theoretical mass concentration
Andalusite	2400	Al	0.052	34.6	31.9–32.4
Andalusite	1800	Al	0.039	34.5	31.9–32.4
AlN	2800	Al	0.124	66.9	65.8
Al_2O_3	937	Al	0.038	55.0	52.9
Al_2O_3	540	Al	0.021	52.9	52.9
Al_2O_3	585	Al	0.024	56.2	52.9

The model proposed is not meant to provide absolutely quantitative data at the 2% level of accuracy but is meant to be used to gain at least semiquantitative results that are reasonably accurate and permit comparisons between extremely thin films. The proposed model is able to satisfy this requirement, as is evident in comparing the results of the alumina films, and so is at least partially successful.

CONCLUSIONS

In summary, a computer program has been written which is completely general, is applicable to any system, is extremely simple to use, and usually provides results accurate to 2 to 4% relative to the amount present, expect for cases noted, where fluorescence by the continuum is present. It has been shown that films as thin as 2500 Å may be routinely analyzed by using conventional microprobe techniques with sufficient accuracy to permit process parameters to be evaluated. New fluorescent yields have been calculated and are tabulated for K, L, and M lines. Finally, a new model has been proposed to permit analyzing films as thin as 500 Å ($\sim$17 $\mu g/cm^2$), again with sufficient precision to permit process evaluation. It is expected that the model could be improved by the use of more exact laws for the electron energy losses in complex targets.

ACKNOWLEDGMENTS

I am indebted to D. R. Wonsidler, who made the microprobe measurements on all the analyses reported herein; to M. J. Rand, V. E. Hauser, S. K. Tung, and W. C. Erdman for supplying the deposited films; to J. D. Ashner for supplying the evaporated aluminum films and their thicknesses; and to J. Drobek for electron-diffraction evaluation of some of the films studied.

REFERENCES

1. J. W. Colby and D. R. Wonsidler, "Stoichiometry of Thin Insulating Films," paper presented at Dallas Electrochemical Society Meeting, May, 1967.
2. C. A. Anderson, "Electron Probe Microanalysis of Thin Layers and Small Particles with Emphasis on Light-Element Determinations," in: T. D. McKinley, K. F. J. Heinrich, and D. B. Wittry (eds.), *The Electron Microprobe*, John Wiley & Sons, Inc., New York, 1966, p. 58.
3. D. M. Poole and P. M. Thomas, "Quantitative Electron-Probe Microanalysis," *J. Inst. Metals* **90**: 288, 1962.
4. P. Duncumb and P. K. Shields, "The Present State of Quantitative X-Ray Microanalysis. Part I: Physical Basis," *Brit. J. Appl. Phys.* **14**: 617, 1963.
5. M. Green and V. E. Cosslett, "The Efficiency of Production of Characteristic X-Radiation in Thick Targets of a Pure Element," *Proc. Phys. Soc. (London)* **78**: 1206, 1961.
6. H. A. Bethe, "Zur Theorie des Durchgangs Schneller Korpuskalarstrahlen durch Materie," *Ann. Physik* **5**: 325, 1930.
7. A. T. Nelms, "Energy Loss and Range of Electrons and Positron," *Natl. Bur. Std. (U.S.), Circ.* 577, 1956, and *Suppl. Circ.* 577, 1958.
8. P. Duncumb and C. deCasa, "Atomic Number and Absorption Corrections: Accuracy Obtained in Practice," paper presented at the 2nd Natl. Conf. on Electron Microprobe Analysis, Boston, Mass., 1967.
9. P. Duncumb, Private communication, 1966.
10. H. E. Bishop, "Some Electron Backscattering Measurements for Solid Targets," in: *X-Ray Optics and Microanalysis*, Hermann, Paris, 1967.
11. J. V. Smith, "Production of X-Rays," notes from course, *Microprobe Analysis*, University of Chicago, 1965.
12. P. Duncumb and P. K. Shields, "Effect of Critical Excitation Potential on the Absorption Correction," in: T. D. McKinley, K. F. J. Heinrich, and D. B. Wittry (eds.), *The Electron Microprobe*, John Wiley & Sons, Inc., New York, 1966.
13. J. Philibert, "A Method for Calculating the Absorption Correction in Electron-Probe Microanalysis," in: H. H. Pattee, V. E. Cosslett, and A. Engstrom (eds.), *X-Ray Optics and X-Ray Microanalysis*, Academic Press, New York, 1963, p. 379.

14. K. F. J. Heinrich, "The Absorption Correction Model for Microprobe Analysis," paper presented at 2nd Natl. Conf. on Electron Microprobe Analysis, Boston, Mass., 1967.
15. S. J. B. Reed, "Characteristic Fluorescence Corrections in Electronprobe Microanalysis," *Brit. J. Appl. Phys.* **16**: 913, 1965.
16. K. F. J. Heinrich, "X-Ray Absorption Uncertainty," in: T. D. McKinley, K. F. J. Heinrich, and D. B. Wittry (eds.), *The Electron Microprobe*, John Wiley & Sons, Inc., New York, 1966, p. 296.
17. E. H. S. Burhop, "Le Rendement de Fluorescence," *J. Phys. Radium* **16**: 625, 1955.
18. R. W. Fink, R. C. Jopson, H. Mark, and C. D. Swift, "Atomic Fluorescence Yields," *Rev. Mod. Phys.* **39**: 513, 1966.
19. R. Castaing, "Electron Probe Microanalysis," in: *Advan. Electron. Electron Phys.* **13**: 317, 1960.
20. V. E. Cosslett and R. N. Thomas, "Penetration and Energy Loss of Electrons in Solid Targets," in: T. D. McKinley, K. F. J. Heinrich, and D. B. Wittry (eds.), *The Electron Microprobe*, John Wiley & Sons, Inc., New York, 1966, p. 248.
21. V. E. Cosslett and R. N. Thomas, "Multiple Scattering of 5–30 keV Electrons in Evaporated Metal Films II: Range-Energy Relations," *Brit. J. Appl. Phys.* **15**: 1283, 1964.
22. R. K. Hart and D. G. Pilney, "Effect of Spectral Line Shift on Microprobe Data," paper presented at 2nd Natl. Conf. on Electron Microprobe Analysis, Boston, Mass., 1967.
23. J. W. Colby, W. N. Wise, and D. K. Conley," Backscatter and Secondary Electron Measurements in the Microprobe Analyzer," in: J. B. Newkirk and G. R. Mallett (eds.), *Advances in X-Ray Analysis, Vol. 10*, Plenum Press, New York, 1967, p. 447.
24. D. B. Brown and D. W. Wittry, "A Transport Equation Program and its Application to Electron Microprobe Analysis," paper presented at 2nd Natl. Cong. on Electron Microprobe Analysis, Boston, Mass., 1967.
25. A. Kirianenko, F. Maurice, D. Calais, and Y. Adda, "Analysis of Heavy Elements (Z 80) with the Castaing Microprobe: Application to the Analysis of Binary Systems Containing Uranium," in: H. H. Pattee, V. E. Cosslett, and A. Engstrom (eds.), *X-Ray Optics and X-Ray Microanalysis*, Academic Press, New York, 1963, p. 559.
26. J. Henoc, "Contribution to Electron Probe Microanalysis," Thesis, Paris, 1962.

APPENDIX

Sample Problem with Listing of Input Data and Corresponding Output Format

```
A 234  JUNE 20, 1967   J.W.COLBY                JC 1 (TEST PROBLEM)
16.0   2         2.20    .61         3.51.53.0
1N LA  FE KA   K KA   0
 3KBR   .3286BR
 11NAS  .6051AS
 3 26 27 25
 2 50 51
 4 22 24 21 25
 3031 6079
 2500 9462
 4625 7096
 2924 5921
 2410 9421
 4521 6926
 3141 6121
 262510320
 4322 6722
 3120 6234
 2519 9961
 4202 6621
 3202 6301
 2496 9862
 4178 6520
 3147 6229
 2334 9761
 4209 6602
 2900 6024
 2224 9641
 4064 6500
             2
```

PROBLEM NUMBER A 234

JUNE 20, 1967

SUBMITTED BY J.W.COLBY

DESCRIPTION – JC 1 (TEST PROBLEM)

MEAN CHEMICAL COMPOSITION AND TWO SIGMA LIMITS BASED ON 7 ANALYSES

ELEMENT	WEIGHT PERCENT	ATOMIC PERCENT
IN	32.12 – 0.45	10.74 – 0.34
FE	25.79 – 0.85	17.75 – 0.92
K	20.75 – 0.21	20.38 – 0.58
O *	21.35 – 1.24	51.13 – 1.80

* DETERMINED BY DIFFERENCE

MEAN INTENSITY RATIOS AND TWO SIGMA LIMITS

ELEMENT	K
IN	0.4929 – 0.0079
FE	0.2434 – 0.0085
K	0.6349 – 0.0073

ACCELERATING VOLTAGE	16.0 KEV
X-RAY EMERGENCE ANGLE	41.0 DEGREES
FILM THICKNESS	0.61 MICRONS
DENSITY	2.20
DEPTH OF ANALYZED REGION	2.07 MICRONS

PEAK-TO-BACKGROUND RATIOS (P/B), SENSITIVITIES (SENS), AND MINIMUM DETECTABILITY LIMITS (MDL)

ELEMENT	P/B	SENS	MDL
IN	401/1	1.12 %	688 PPM
FE	195/1	1.03 %	966 PPM
K	833/1	0.63 %	278 PPM

PROBLEM NUMBER A 234

JUNE 20, 1967

ELEMENT	ATOMIC NUMBER	ATOMIC WEIGHT	BACKSCATTER FACTOR	EXCITATION POTENTIAL	ABSORPTION JUMP RATIO	FLUORESCENT YIELD
IN	49	114.820	0.741	3.730	4.15	0.075
FE	26	55.847	0.897	7.112	8.11	0.324
K	19	39.102	0.895	3.607	10.11	0.118
O	8	15.999	0.952	0.532	29.68	0.003

MASS ABSORPTION COEFFICIENTS

RADIATION / ABSORBER	IN LA	FE KA	K KA
IN	443.	452.	451.
FE	441.	71.	431.
K	162.	261.	159.
O	161.	24.	157.

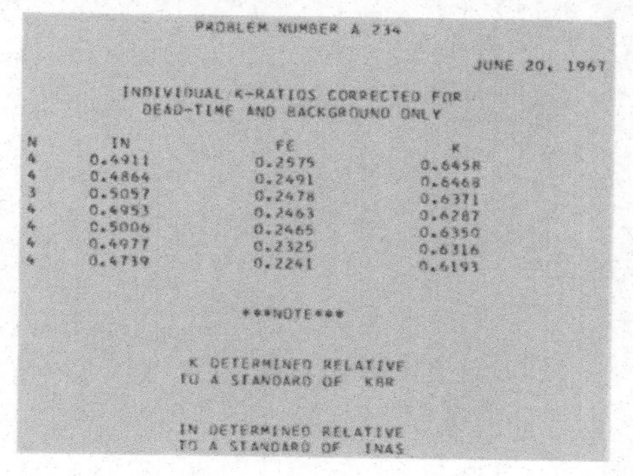

PROBLEM NUMBER A 234

JUNE 20, 1967

INDIVIDUAL K-RATIOS CORRECTED FOR DEAD-TIME AND BACKGROUND ONLY

N	IN	FE	K
4	0.4911	0.2575	0.6458
4	0.4864	0.2491	0.6468
3	0.5057	0.2478	0.6371
4	0.4953	0.2463	0.6287
4	0.5006	0.2465	0.6350
4	0.4977	0.2325	0.6316
4	0.4739	0.2241	0.6193

NOTE

K DETERMINED RELATIVE TO A STANDARD OF KBR

IN DETERMINED RELATIVE TO A STANDARD OF INAS

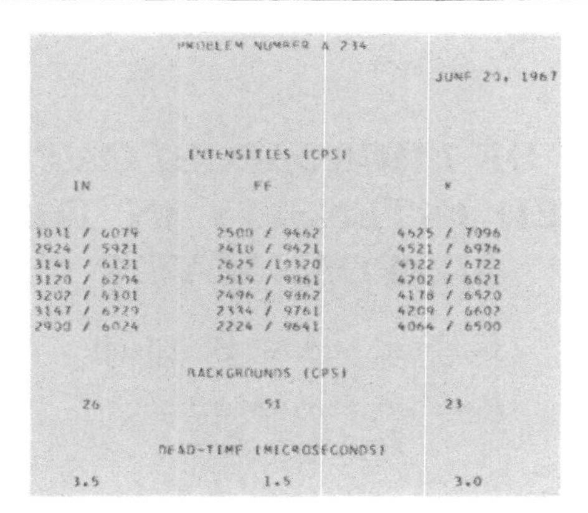

PROBLEM NUMBER A 234

JUNF 23, 1967

INTENSITIES (CPS)

IN	FE	K
3031 / 6079	2500 / 9462	4625 / 7096
2924 / 5921	2410 / 9421	4521 / 6926
3141 / 6121	2625 /10320	4322 / 6722
3120 / 6294	2519 / 9961	4202 / 6621
3202 / 6301	2496 / 9462	4178 / 6520
3147 / 6229	2334 / 9761	4209 / 6602
2930 / 6024	2224 / 9641	4064 / 6500

BACKGROUNDS (CPS)

IN	FE	K
26	51	23

DEAD-TIME (MICROSECONDS)

IN	FE	K
3.5	1.5	3.0

DISCUSSION

K. F. J. Heinrich (National Bureau of Standards): Perhaps I have missed a detail, but can you combine this with a simultaneous measurement of the thickness of the film?

J. W. Colby: As it turns out in the assumptions that are made, the concentration and the thickness both enter into the calculation. If you make some more simplified assumptions, they enter into it closely as a product so that you can't really separate the two. By making two analyses at two different accelerating voltages which are far enough apart, one could perhaps develop some sort of empirical relationship to take account of and measure the thickness with quite a bit of sensitivity. However, what we're doing now is measuring the intensity from the substrate, and we find that this is also calculatable and it gives us quite a good relationship. What we hope to do eventually is get away from film-thickness measurement entirely and do everything in the probe where, at first, nothing may be absolute but everything is relative and will allow a much better comparison between films. Right now we are dependent very much on the film thickness, and the index of refraction of these films varies quite a bit, which makes it difficult to get a good film thickness measurement. We hope eventually to be able to get completely independent of any other system.

H. K. Herglotz (E. I. du Pont): Owing to the nature of your work, you are not interested in very fine spatial resolution, but I ask this question because of pure scientific curiosity. At these very low voltages, does your beam spread?

J. W. Colby: I don't really know what the spot size is, but it is probably on the order of a couple of microns, something like that. Normally we would expand it anyway a little bit and drive the specimens while we're taking counts to avoid contamination and to avoid any inhomogeneities.

H. K. Herglotz: Is this enlarged size due to space-charge repulsion of the electrons in the beam?

J. W. Colby: We don't seem to be bothered by space charge. If we are, then our program is compensating for it and we don't know how.

H. K. Herglotz: Well, of course, you are interested only in the chemical analysis of the layer, not in resolution.

J. W. Colby: Not at all.

THE EFFECT OF MICROSEGREGATION ON THE OBSERVED INTENSITY IN THIN-FILM MICROANALYSIS

G. Judd and G. S. Ansell

Rensselaer Polytechnic Institute
Troy, New York

ABSTRACT

In the electron-probe microanalysis of thin alloy samples, particularly for areas which contain a high degree of solute segregation as are often obtained in extraction replicas or in electrochemically thinned foils, a correlation between the measured intensity and the sample microstructure has been difficult to perform. The purpose of this study was to isolate the effect of microsegregation, *i.e.*, segregation with respect to the beam size ($\sim 2\ \mu$) itself, on the characteristic intensity generated by a microvolume. To accomplish this, a model system consisting of silver-powder particles ($\sim 1\ \mu$ diameter) supported on an evaporated carbon film was investigated by using a thin-film microprobe accessory of an electron microscope (Hitachi HXA-1). The silver particles served as analogs to a highly concentrated segregation area, and the carbon film was analogous to the corresponding matrix. The simplicity of this system readily allowed for an analysis of the geometric positioning factor dependence of the generated intensity.

It was observed that the intensity measured for a given particle could vary by as much as 1000% depending only on the relative position of that particle within the electron beam. This variation was derived to be due to both the distribution of electrons within the beam and the amount of mass occupied by the particle at each distance from the beam center. A computer program was developed to perform the necessary calculation to predict the relative intensity as a function of position within the beam. These calculated intensities were found to compare favorably with the experimental measurements.

INTRODUCTION

Although both electron microscopy and electron-probe microanalysis have basically the same underlying instrumentation principles, it has been only recently that they have been combined in a single instrument of high resolution[1-3]. Concurrent examination by these two analysis techniques can now be applied to suitably prepared thin specimens.

In using a combination electron microscope–microprobe, the accelerating voltages employed are usually high (50 to 100 kV) compared with the voltages used in bulk-specimen microprobe work (5 to 35 kV). The specimen thickness is considerably reduced ($\sim$1000 Å). While the combination of accelerating voltages and restricted specimen thickness allows sufficient penetration for imaging, the efficiency of characteristic X-ray production will be substantially decreased. A benefit from these experimental conditions is the relaxation of the necessity for making absorption and fluorescence corrections.

Using such an experimental tool, one can not only characterize the morphology and structure of the thin foil in microscopic observation and selected area diffraction but can also determine its chemical composition by X-ray fluorescent microanalysis. Precipitates, chemical segregation, diffusion, and other solid-state phenomena can be studied simultaneously on the same spot by both methods. In a combined instrument with both electron microscope and microprobe capacities, structural features which can only be resolved with high-resolution electron microscopy can be precisely positioned within the beam for microanalysis. The importance of this feature cannot be overstressed since a correlation between microanalysis data from a microprobe and structural features observed separately by an electron microscope has been difficult to perform, the main problem being the lack of imaging resolution in the microprobe compared with the resolution of the electron microscope and the manipulation and consequent effects on the sample of being moved from one instrument to another. To ensure a complete knowledge of the physical system analyzed, it is essential that the unique features of both the thin-film and small-particle electron probe microanalysis be understood.

The enhancement of structural resolution is not without its drawbacks. For bulk-specimen microanalysis, the chemical analysis is performed on a finer scale than is the structural imaging. However, in the combined instrumentation, the structural resolution is of a higher order than is the chemical analysis. In the bulk sample, the microvolume analyzed is assumed to be relatively homogeneous with respect to the beam as a result of restricted imaging resolution. However, in the thin-film or small-particle analysis, localized regions displaying high degrees of segregation are readily resolved on a scale finer than is the volume sampled for microanalysis, and they are usually the areas of greatest interest; *e.g.*, the distribution of a phase is generally nonuniform on this microscale. Thus, by the inherent nature of the sampling technique used in microanalysis of a thin section, results are obtained displaying apparently exaggerated or nonrepresentative concentrations. As a result, it is extremely important to be able to characterize the interrelationship between the positioning of the structural features of the sampled area in the beam and the resultant X-ray intensity measured. In the absence of correction methods for this apparent concentration effect, any form of quantitative work on these samples could not be performed.

For these reasons, an approach has been developed to increase the resolution and improve the reproducibility of intensity measurements in the electron-probe microanalysis of samples which display a high degree of segregation relative to the size of the electron beam. The term *segregation* should be interpreted as either individual submicron-sized particles occupying a substantial portion of the beam or as a sharp-sloping concentration gradient within the area of the beam. This approach involved the analyses of the X-ray intensity generated by a model system consisting of silver-powder particles supported on a carbon film in terms of the theoretical expected intensity function. The results of this type of experimental work can then be used in conjunction with quantitative metallographic techniques of structure to yield quantitative analysis data for a variety of specimens. A thin-film microanalyzer attachment for a Hitachi electron microscope was used.

THEORY

It is convenient to refer to the intensity generated from a sample of thickness t measured by a counter at an angle θ (after the notation of Theisen[4]) as

$$I = K\frac{m_a N_0}{A} I_0 \int_0^t e^{-\sigma\rho z} e^{-(\mu/\rho)\rho z \operatorname{cosec}\theta} \, d(\rho z) \tag{1}$$

where K is a constant dependent on V/V_K and V and V_K are the accelerating and critical voltages, respectively; $(m_a N_0)/A$ is the number of atoms of a per unit of volume; I_0 is the incident electron beam intensity; σ is the effective Lenard coefficient;[5] μ/ρ is the mass absorption coefficient for the X-rays generated; ρ is the density; and z is the depth.

For a sufficiently small fluorescent volume, the beam no longer interacts in its entirety with the specimen, and, therefore, the term I_0 in equation (1) must be modified. For an area at a distance r from the beam center, I_0 should be replaced by I_{0r}, the beam intensity at r. Similarly, the amount of material present at a distance r should be incorporated into equation (1) by the introduction of a dimensionless factor V_r/V_{total}. This represents the ratio of the mass at r to the total mass of the particle. Thus the intensity contribution of the microvolume at r can be expressed as

$$I_r = K\frac{m_a N_0}{A} I_{0r} \frac{V_r}{V_{\text{total}}} \int_0^t e^{-\sigma\rho z} e^{-(\mu/\rho)\rho z \operatorname{cosec}\theta}\, d(\rho z) \tag{2}$$

The total intensity can then be found by summing over all values of r. Since all the terms, with the exception of V_0 and I_{0r}, are not a function of the radial distance from the beam center, they can be placed outside the summation. The assumption here is that the microvolume adjacent to the volume in consideration does not substantially contribute to the intensity measured from its neighbor. This is a particularly useful approximation when dealing with very small particles, thin films, and high accelerating voltages since absorption and fluorescence effects are negligible. Thus,

$$I_{\text{total}} = \sum_{r=i}^{r=j} I_r = \left[K\frac{m_a N_0}{A V_{\text{total}}} \int_0^t e^{-\sigma\rho z} e^{-(\mu/\rho)\rho z \operatorname{cosec}\theta}\, d(\rho z)\right] \sum_{r=i}^{r=j} I_{0r} V_r \tag{3}$$

The ratio of two identical particles at two different positions within the beam is obtained from equation (3). The presummation terms will cancel and yield

$$\frac{I_1}{I_2} = \frac{\sum_{r=i}^{r=j} I_r}{\sum_{r=k}^{r=l} I_r} = \frac{\sum_{r=i}^{r=j} I_{0r} V_r}{\sum_{r=k}^{r=l} I_{0r} V_r} \tag{4}$$

where i through j and k through l are the defining limits of r for the particle positions.

For thin films or particles of approximately uniform thickness, V_r can be replaced by the projected area A_r. The projected area is easily measured by using an electron microscope–microprobe combination. Ideally, a cylinder with its axis of rotation perpendicular to the beam should be used to test the geometrical relationship described herein. However, since spherical particles are far more easily obtained, they were used in this work. The important consideration for the particle dimensions is that the thickness of the particle be relatively uniform.

EXPERIMENTAL PROCEDURE

Extremely fine silver powder ($\sim 1\ \mu$) was supported on a carbon film. The powder was analogous to a solute phase displaying a high degree of segregation; and the carbon film, to the matrix phase. The interesting aspect of this system is that the matrix does not contribute to any of the characteristic intensity observed. The carbon, however, does act as both a support and conduction layer for the particles, without which factors the particles would be unstable and therefore unacceptable for this experiment.

Specimens of this type were prepared by first evaporating carbon onto a glass slide. The film was scored, stripped in distilled water, and carefully lifted onto a 100-mesh

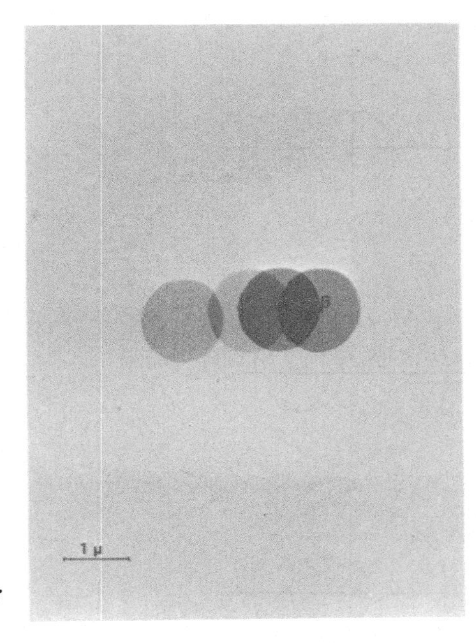

Figure 1. Multiple exposure of silver-powder particles; beam center at *B*.

gold grid. The grid was then dipped into 30-μ silver powder (99.99%) obtained from the Federal Mogul Corporation. The powder is only nominally 30 μ; there are actually many finer particles in the powder mixture. By careful inspection in the electron microscope, isolated particles of approximately 1-μ diameter were located for the intensity measurement. A Hitachi HU-11 electron microscope with a HXA-1 thin-film microanalyzer attachment was employed. Readings were taken by moving the particle within the beam, approximately 2 μ in diameter, and recording through peak intensity and specimen current. The Ag $L\alpha_1$[2] characteristic peak was measured. A photograph was used to record each position, and a multiple-exposure composite picture was made for every data set (Figure 1). This method gave a position-within-beam reference for corresponding intensity readings. This experiment was performed for several particles at both 50 and 100 kV.

METHODS OF ANALYSES

Two possible approaches existed to determine the method of analysis. The first is an interpretation of the measured characteristic-intensity results in terms of an assumed intensity-distribution function; the second, a derivation of the I_{0r} function from the experimental intensity data.

For both approaches, an accurate method of describing particle position within the beam, *i.e.*, the fraction of the projected area of the particle within each area interval in the beam, had to be developed.

Method for Intensity Comparisons

At condenser crossover, the beam is at its minimum size and is circular. Thus, for the spheres used, three placements of the center of the powder particle relative to the center of the beam were possible. The two centers could coincide [Figure 2(a)]; the centers could be sufficiently offset so that the center of the beam was outside the projected

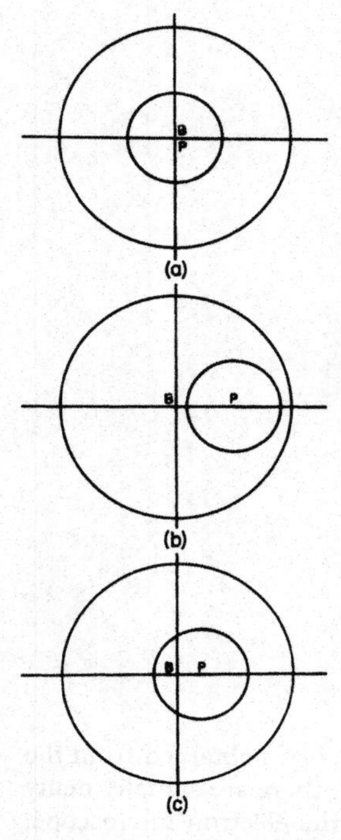

Figure 2. Relative placements of beam center B and particle center P.

area of the particle [Figure 2(b)]; or the centers could be slightly offset so the center of the beam was inside the projected area of the particle [Figure 2(c)]. At all times, the powder particle was entirely within the beam. This ensured that the same total mass of material was probed.

In terms of equation (4), these three cases can be represented as

Case I

$$I_{c=0} = \sum_{r=0}^{r=R} I_{0r}\, \Delta A_r \tag{5a}$$

where the subscript for the intensity denotes the position of the particle center ($c = 0$, the center of the particle is at the beam center) and R is the particle radius. The coordinate system chosen has its origin at the beam center for all three cases, with the units being 0 at the center and 50 at the edge.

Case II

$$I_{c=a} = \sum_{r=a}^{r=R+a} I_{0r}\, \Delta A_r \tag{5b}$$

where a is the displacement between the beam center and the particle center.

Case III

$$I_{c=a} = \sum_{r=0}^{r=R+a} I_{0r}\, \Delta A_r \tag{5c}$$

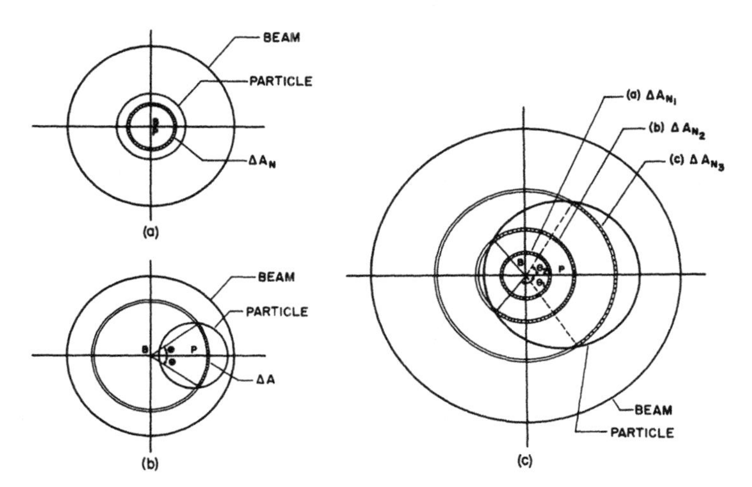

Figure 3. Area interval ΔA_n for (a) case I, (b) case II, and (c) case III.

According to Shimizu and Shinoda,[6] the electron beam was assumed to have a Gaussian electron-density distribution. This assumption also conformed to the theoretical electron beam calculations done by Ogilvie.[7] The exponential in the Gaussian function was chosen so that the intensity was unity at the center and approximately zero at the edge. The precise edge-intensity value used for the calculations was varied to determine the best value for this constant. The diameter of the beam was set at 100 units, and the cross-sectional area of the beam was divided into 500 concentric circles, each having a radius 0.1 unit greater than the preceding circle. All measurements of beam-spot size and particle-to-center distances were adjusted to these dimensions.

The method for determining the area intercepted in any one of these intervals varied for each of the three cases listed above. For case I, the area within the interval n was obtained by

$$\Delta A_n = \pi({r_n}^2 - r_{n-1}^2) \tag{6a}$$

where r was the radius of the nth circular division. This area is shaded in Figure 3(a).

For case II, the area inside an interval n was obtained by using

$$\Delta A_n = \theta({r_n}^2 - r_{n-1}^2) \tag{6b}$$

where θ was one-half the annular angle and was measured in radians. Formula (6b) calculates the area of a sector of an annular ring. This area is shaded in Figure 3(b). Note that, for this case, 2θ is always less than 180°, *i.e.*, θ is acute.

For case III, several conditions had to be considered. Here, the area interval n was calculated by using equation (6a) during the range $R - a \geq r_n \geq 0$ and by using equation (6b) for $a + R \geq r_n \geq R - a$. This is depicted in Figure 3(c). Note that the shaded areas are defined by an annular cylinder (*a*), then by an obtuse wedge cylinder (*b*) with $\theta_1 > 90°$, and finally by an acute wedge of an annular cylinder (*c*) with $\theta_2 > 90°$.

A computer program was written in FORTRAN IV language to carry out these area calculations, multiply the area with the proper electron density, and sum this product for the entire particle. This program is reproduced in Figure 4.

```
BPS FORTRAN IVD  COMPILER  VERSION 2 LEVEL 3.3  OCT   1966

/JOB             1700          JUDD.
BEGIN COMPILATION
 S.0001           2 READ(1,10) A,B
 S.0002          10 FORMAT(2F20.8)
 S.0003             IF(A) 200,100,200
 S.0004         200 IF(A-B) 101,102,102
 S.0005      102    WGTOT=0.0
 S.0006             R2 =A-B
 S.0007           1 R1=R2&0.1
 S.0008             RBAR =(R1&R2)/2.0
 S.0009             GAUSS=EXP(-RBAR**2.0/455.56)
 S.0010             X=(A**2.0&RBAR**2.0-B**2.0)/(2.0*A)
 S.0011             X=ABS(X)
 S.0012             YSQR=ABS(RBAR**2.0-X**2.0)
 S.0013             Y=SQRT(YSQR)
 S.0014             IF(X) 300,301,300
 S.0015         300 THETA=ATAN(Y/X)
 S.0016             GO TO 303
 S.0017         301 THETA=3.14159/2.0
 S.0018         303 AREA=THETA*(R1+R2)*(R1-R2)
 S.0019             GAUSS=EXP(-RBAR**2.0/455.56)
 S.0020             WGT=AREA*GAUSS
 S.0021             WGTOT=WGTOT&WGT
 S.0022             IF(R1-(A&B))103,104,104
 S.0023         103 R2=R2&0.1
 S.0024             GO TO 1
 S.0025         104 WRITE(3,11) A,B,WGTOT
 S.0026          11 FORMAT(3F20.8)
 S.0027             GO TO 2
 S.0028         101 R2=0.0
 S.0029             WGTOT=0.0
 S.0030           3 R1=R2&0.1
 S.0031             RBAR=(R1&R2)/2.0
 S.0032             GAUSS=EXP(-RBAR**2.0/455.56)
 S.0033             IF(R2-(B-A)) 105,106,106
 S.0034         105 AREA=3.14159*(R1**2.0-R2**2.0)
 S.0035             WGT=AREA*GAUSS
 S.0036             WGTOT=WGTOT&WGT
 S.0037             R2=R2&0.1
 S.0038             GO TO 3
 S.0039         106 X=(A**2.0&RBAR**2.0-B**2.0)/(2.0*A)
 S.0040             POSX=ABS(X)
 S.0041             YSQR=ABS(RBAR**2.0-POSX**2.0)
 S.0042             Y=SQRT(YSQR)
 S.0043             IF(X)400,401,402
 S.0044         400 THETA=ATAN(Y/POSX)
 S.0045             GO TO 107
 S.0046         401 THETA=3.14159/2.0
 S.0047         107 CHI=3.14159-THETA
 S.0048             AREA=CHI*(R2&R1)*(R1-R2)
 S.0049             WGT=AREA*GAUSS
 S.0050             WGTOT=WGTOT&WGT
 S.0051             R2=R2&0.1
 S.0052             GO TO 3
 S.0053         402 THETA=ATAN(Y/X)


BPS FORTRAN IVD  COMPILER  VERSION 2 LEVEL 3.3  OCT   1966

 S.0054         108 AREA=THETA*(R2&R1)*(R1-R2)
 S.0055             WGT=AREA*GAUSS
 S.0056             WGTOT=WGTOT&WGT
 S.0057             IF(R1-(A&B))109,110,110
 S.0058         109 R2=R2&0.1
 S.0059             GO TO 3
 S.0060         110 WRITE(3,12)A,B,WGTOT
 S.0061          12 FORMAT(3F20.8)
 S.0062             GO TO 2
 S.0063         100 WGTOT=0.0
 S.0064             R2=0.0
 S.0065          99 R1=R2&0.1
 S.0066             IF(R1-B) 888,202,202
 S.0067         888 RBAR=(R1&R2)/2.0
 S.0068             GAUSS=EXP(-RBAR**2.0/455.56)
 S.0069             AREA=3.14159*(R1**2.0-R2**2.0)
 S.0070             WGT=AREA*GAUSS
 S.0071             WGTOT=WGTOT+WGT
 S.0072             R2=R2&0.1
 S.0073             GO TO 99
 S.0074         202 WRITE(3,15) A,B,WGTOT
 S.0075          15 FORMAT(3F20.8)
 S.0076             GO TO 2
 S.0077             END
```

Figure 4. FORTRAN IV program used in method for intensity comparisons.

Method for Deriving Electron-Density Function

An alternative approach using the above reasoning was also attempted. In this method, the electron-density distribution was derived by using the experimental intensity measurements as the starting point. A curve was fitted to the experimental intensity data of a given powder particle. An adaptation of the above program was written to calculate the area that this particle would occupy in the beam at each area interval for 50 particle positions. Since a continuous curve was fitted to the data, the intensity for each position was read from the curve. Thus a matrix (50 × 50) was established of the form

$$I_1 = g_1 \Delta A_1 + g_2 \Delta A_2 + \cdots + g_n \Delta A_n + \cdots + g_{50} \Delta A_{50}$$

$$\vdots$$

$$I_n = g_1 \Delta A_1 + g_2 \Delta A_2 + \cdots + g_n \Delta A_n + \cdots + g_{50} \Delta A_{50}$$

$$\vdots$$

$$I_{50} = g_1 \Delta A_1 + g_2 \Delta A_2 + \cdots + g_n \Delta A_n + \cdots + g_{50} \Delta A_{50}$$

where I_n is the intensity from the fitted curve for a particle whose center is at n; g_n is the unknown distribution function for the electron density at a distance n from the beam center (corresponds to I_{0r} in the previous calculation); and A_n is the area occupied by the particle n in the nth interval. Note that, since the particle was smaller than the beam, many of these area values were zero.

A subprogram for the solution of simultaneous equations (code name SIMQ) was used to solve 50 equations for the values g_1 through g_{50}. The results were checked by solving the same matrix with a matrix inverse subprogram (code name MINV), and no appreciable round off errors were discovered.

RESULTS AND DISCUSSION

Experimentally measured characteristic X-ray intensities corrected for background and counter efficiency were observed to vary markedly as a function of position. Changes of as much as an order of magnitude in measured intensity were recorded. The necessity, therefore, of a normalization procedure was demonstrated. The corrected intensity, background, peak height, and specimen current were all observed to go through a Gaussian type of distribution. The corrected intensities were analyzed by using the computer program described earlier (Figure 4). The normalized intensity predicted by the computer calculation was compared with the normalized intensity of the experimental readings. Using this procedure, one could readily show that the same mass of silver was always present within the beam and that only a positioning effect had to be considered. Table I lists a few of the experimental results (raw data and normalized) and the corresponding predicted intensity values for the computer work (method for intensity comparisons). For all readings taken, the absolute deviation of the predicted value from the experimentally measured value was within 2% absolute in 36% of the calculations and within 5% absolute in 70% of the calculations. It should also be noted that exact end value of the Gaussian was not too critical as approximately the same results were obtained for values of I_{0r} at 50 of 0.004 to 0.020. This can be understood from the fact that the major interaction occurs near the center of the beam and a minor "flattening" of the Gaussian does not appreciably alter these findings.

By using the matrix-solution approach discussed (method for deriving electron-density function), confirmation for the Gaussian approximation was obtained. This is shown in Figure 5. A smooth curve was drawn through the calculated points. It should be stated that, though the results shown agree most favorably with the theory used, there were other instances where uninterpretable results (*i.e.*, high negative values for g) were obtained. A similar analysis has been recently performed by Rapperport[8] for specimen current measurements. He found that, when noise was superimposed upon his input distribution, what appeared to be meaningless solutions were obtained. These could then

Table I. Sample Data of Silver-Powder Experimental and Calculation Sets

Radius, arbitrary units	Voltage, kV	Characteristic radiation measured	Distance of particle center from beam center	I_e*	I_{eN}†	I_{pn}‡
11.6	100	$L\alpha_1{}^2$	0.0	95	1.00	1.00
			11.0	85	0.89	0.80
			19.4	50	0.52	0.49
			37.9	15	0.16	0.07
11.7	50	$L\alpha_1{}^2$	0.0	84	1.00	1.00
			9.5	72	0.85	0.84
			19.0	41	0.48	0.51
			24.0	27	0.32	0.33
			38.2	7	0.08	0.06

* I_e, experimental intensity.
† I_{eN}, normalized experimental intensity.
‡ I_{pn}, normalized predicted intensity.

be improved by employing *ridge analysis*. The authors plan to adapt this method for the particular application involving characteristic X-ray segregation analysis to further correct the matrix program solution.

Both the relatively high absolute error in 30% of the readings and the deviation observed in the matrix solution can readily be attributed to the low counting statistics and high background obtained for those small particles at the high voltages at which they were observed. On the other hand, one should be aware that the primary use of this type of instrumentation is for transmission and replica work which are normally performed at high accelerating voltages. Thus, the compromise of high background must be made

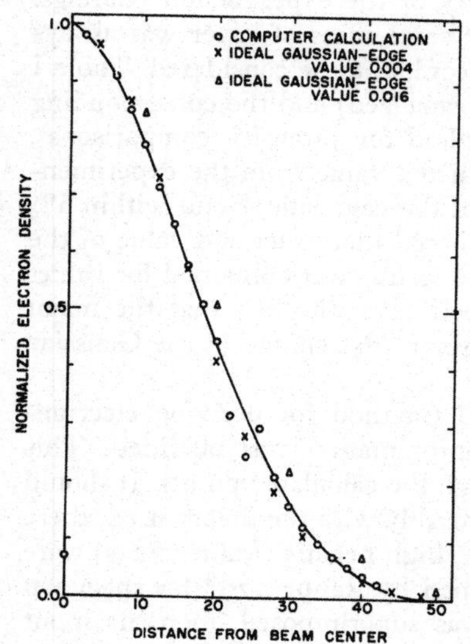

Figure 5. Computer solution for the electron-beam intensity distribution (solid line) from the experimental data set (method for deriving electron-density function). The symbols △ and × denote theoretical values for two different Gaussian functions and are used for comparison purposes.

to obtain the resolution of the exact beam position and the relative placement of the area of interest for microanalysis with respect to the beam. This is particularly true for the concurrent performance of microanalysis, transmission microscopy, and selective area diffraction.

CONCLUSIONS

The X-ray intensity generated by an area of high segregation, for example, a single particle smaller than the beam yet totally contained within the beam, was found to be a strong function of position in the beam. This positioning effect was assumed to be a result of the combined product of the electron-intensity distribution and the amount of material at each point in the beam. This hypothesis could then be used to explain adequately the observed readings from any relatively evenly thick particles. A computer program was developed that could perform this calculation. This program, together with quantitative metallography, could further be used to perform quantitative microanalysis on areas of high segregation relative to the beam size.

The Gaussian nature of the electron-density distribution within the beam was verified both by the success of correlating the intensities measured for isolated particles as a function of position within the beam and by the close similarity of the derived distribution function to a Gaussian distribution. The exact Gaussian function chosen was not found to affect the results appreciably for functions within a narrow, 0.004 to 0.020, range of end values. It is anticipated that further precision can be introduced into this analysis by applying advanced statistical methods.

ACKNOWLEDGMENT

The authors would like to thank Mr. Marc Costantino for his assistance with the computer programming. This work was supported by both the Army Office of Research, Durham, and the National Institute of Dental Research through Program 5 PO1 DEO2336-01. The experimentation was performed in the Interdisciplinary Materials Research Center at Rensselaer Polytechnic Institute, which is funded by the National Aeronautics and Space Administration. Part of the student support was obtained from the National Science Foundation.

REFERENCES

1. P. Duncumb, "An Electron Optical Bench for Microscopy, Diffraction and X-Ray Microanalysis," in: S.S. Breese (ed.), *Fifth International Congress for Electron Microscopy, Philadelphia, 1962*, Academic Press, New York, 1962, p. KK-4.
2. W. C. Nixon and R. Buchanan, "An Experimental Electron Optical Bench for Electron Microscopy and X-Ray Microanalysis," in: H. H. Pattee, V. Cosslett, and A. Engstrom (eds.), *Third International Symposium of X-Ray Optics and X-Ray Microanalysis, Stanford, California, 1962*, Academic Press, New York, 1963, pp. 255–262.
3. H. Akahori, S. Katagiri, S. Ozasa, and I. Fuhiyasu, "Improvement of X-Ray Microanalysis Attachment for Electron Microscope," in: R. Uyeda (ed.), *Sixth International Congress for Electron Microscopy, Kyoto, Japan (1966)*, Mazuron Company, Ltd., Tokyo, 1966, pp. 187–188.
4. R. Theisen, *Quantitative Electron Probe Microanalysis*, Springer-Verlag, New York, 1965, pp. 20–22.
5. P. Duncumb and P. K. Shields, "Effect of Critical Excitation Potential on the Absorption Correction," in: T. McKinley (ed.), *The Electron Microprobe*, John Wiley & Sons, Inc., New York, 1966, pp. 284–295.
6. R. Shimizu and G. Shinoda, "A Study of Electron Diffusion in Microanalysis Specimens," in: T. McKinley (ed.), *The Electron Microprobe*, John Wiley & Sons, Inc., New York, 1966, pp. 480–489.
7. R. E. Ogilvie, "Electron Beam in Microanalysis," in: R. Bakish (ed.), *Introduction to Electron Beam Technology*, John Wiley & Sons, Inc., New York, 1962, pp. 414–417.
8. E. J. Rapperport, "Deconvolution: A New Technique to Increase Electron Microprobe Resolution," Second National Conference on Electron Probe Microanalysis, Boston, Mass., 1967.

MULTISTEP INTENSITY INDICATION IN SCANNING MICROANALYSIS

Teruichi Tomura and Hiroshi Okano

Hitachi Central Research Laboratory
Kokubunji, Tokyo, Japan

and Koichi Hara and Tadao Watanabe

Hitachi, Limited, Naka Works
Katsuta-shi, Ibaragi, Japan

ABSTRACT

Improvements on the semiquantitative analysis in a scanning electron-probe microanalysis will be discussed. Improvements have been attained by the development of a new *content mapping* attachment. By using this attachment, the X-ray intensity is indicated on a cathode-ray tube in nine steps of brightness. A discrimination operation on the intensity information can be done to help the observation of fine concentration variation in the specimen. The scanning image about an element in the specimen is displayed on the screen of the cathode-ray tube or on a paper tape by a typeout recording. The image is composed of 2500 picture elements. The brightness of each picture element corresponds to an exponentially modulated X-ray intensity emitted from the corresponding point of the specimen surface. The correct information about the X-ray intensity to the observer is given by using this kind of modulation. In the image display on typewritten paper, the X-ray intensity is indicated with numbers from 0 to 9 according to the intensity grade. Each number is printed with an individual color to present a contour map of the element distribution. The time necessary to obtain one complete contour map is about 8.5 min.

INTRODUCTION

The scanning analysis which has proved to be an essential function of the electron-probe microanalyzer is, of course, an effective means of carrying out two-dimensional analysis of a specimen. However, a conventional X-ray image gives, as was pointed out by Heinrich[1] and others, unsatisfactory representation from the standpoint not only of quantitative but also qualitative information, chiefly because the optical-sensitive characteristics of the human eye are not adaptable to the contrast of such an image.

Accordingly, with such an image, the distribution of elements cannot be properly observed, and it is much more difficult to obtain accurate quantitative information.

For the purpose of overcoming these difficulties, Melford[2] developed the *enhanced contrast method*; Heinrich,[1] the *concentration mapping procedure* or the *colored display*; and Birks and Batt,[3] the *typeout method*. These, however, have not been considered satisfactory methods for obtaining accurate quantitative information because most of these methods have only two or three intensity steps. Furthermore, these methods and instruments are complicated in design and construction.

A *content-mapping system* recently developed by the authors solved the aforementioned difficulties. By the development of this system, a multistep indication of X-ray intensity, a discrimination facility concerned with intensity, an image with exponentially enhanced contrast, and a colored typeout recording are materialized with a very compact design. Consequently, a rapid semiquantitative analysis and an accurate qualitative observation have been obtained in practical use. These improvements were obtained mainly by the adaptation of a newly developed pulser which has a variable range in pulse width of 1 to 1000 instead of 1 to 10 for the conventional type. This paper covers the details of the instrumentation of the new content-mapping system and describes some examples of applications of this method for electron-probe microanalysis.

FUNDAMENTAL CONSIDERATIONS

Reliability of the Indicated Intensity

To determine the integration time for each picture element, the statistical fluctuation in the X-ray intensity must be taken into account. The intensity distribution R, which contributes to make a brightness corresponding to the intensity step position of A, is given by the following expression:

$$R(Nt) = \frac{1}{\sqrt{2\pi}} \int_{\alpha_1}^{\alpha_2} \exp(-\alpha^2/2)\, d\alpha \tag{1}$$

$$\alpha_1 = (b + nA - N)\sqrt{\frac{t}{N}} \tag{2}$$

$$\alpha_2 = [b + n(A + 1) - N]\sqrt{\frac{t}{N}} \tag{3}$$

where b is the corresponding X-ray intensity (intensity per unit of time) of the discrimination level; n is the set up X-ray intensity for an intensity step; A is the position of the intensity step under consideration; N is the X-ray intensity per unit of time from the specimen; and t is the integration time for a picture element. The indicated intensity N_i gives the observer information that the true intensity N will be between $(b + nA)$ and $[b + n(A + 1)]$. However, the actual X-ray intensity is distributed to an indicated intensity, as shown in Figure 1. Curves in Figure 1 were calculated for the indicated intensity of 1000 ± 100 counts/sec (step width, 200 counts/sec) according to equation (1).

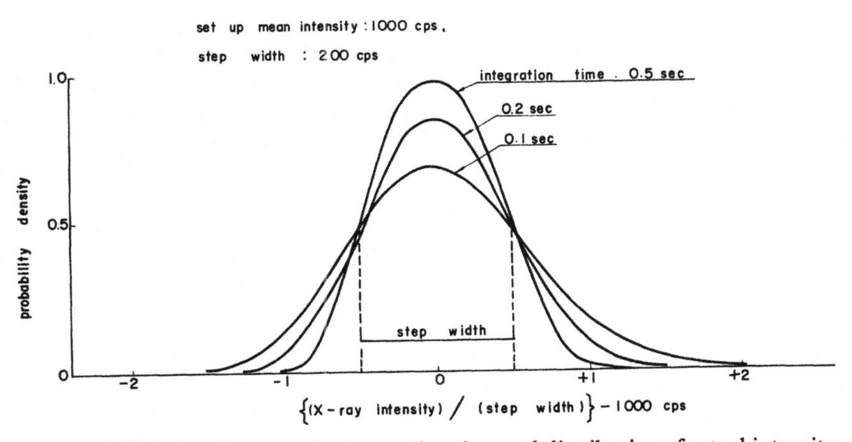

Figure 1. Relations between the integration time and distribution of actual intensity.

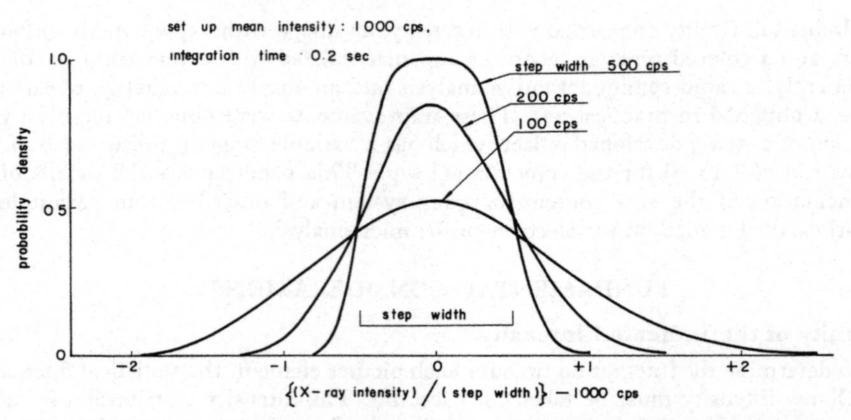

Figure 2. Relations between the step width and distribution of actual intensity.

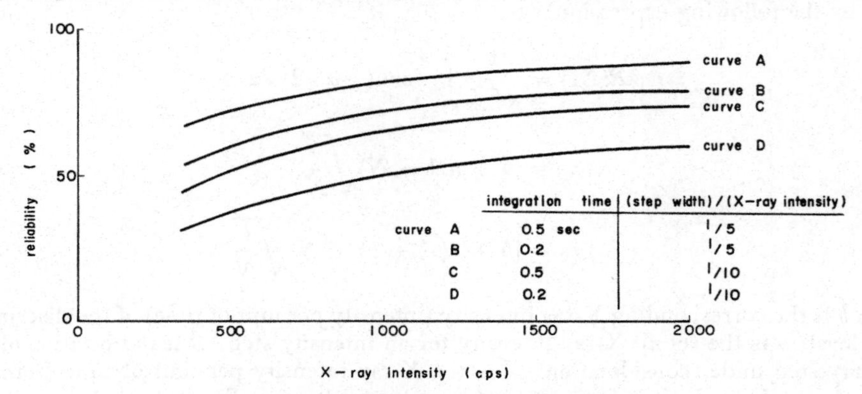

Figure 3. The reliability of indicated intensity.

The area surrounded by the upper and lower level of the step and the probability curve corresponds to a reliability of the intensity indication. From these curves the reliabilities of the indication for the integration times of 0.5, 0.2, or 0.1 sec were obtained as 91.7, 73.5, or 59.1%, respectively.

The statistical fluctuation in the X-ray intensity must also be taken into consideration in the determination of the width of the intensity step. The relation between the reliability of the indication and the step width can be obtained from equation (1). Figure 2 shows how the probability density of the actual intensity is distributed to the indicated intensity when an integration time of 0.2 sec and a mean intensity of 1000 counts/sec are assumed. In the figure, the X-ray intensity is normalized by the step width. From these curves, the reliability of the indications for the intensity-step widths of 500, 200, or 100 counts/sec were obtained as 95.2, 73.5, or 48.9%, respectively. Figure 3 shows the reliability curves *versus* X-ray intensity under various conditions of the integration time and the step width.

Picture Element

In most scanning microprobe analysis, the analyzing area at the specimen surface is an order of magnitude of 50 μ^2. If this area is analyzed with an electron probe whose diameter is about 1 μ, several thousands of picture elements are required for a frame.

Each picture element must provide the correct brightness on the screen of the cathode-ray tube without any halation. Therefore, all the area assigned to a picture element should be utilized. That is, it is obvious that a square shape is the most desirable as the figure of the picture element.

Brightness Modulation

Three methods of modulating the brightness of a cathode-ray tube are as follows:

1. Spot density controlling.
2. Beam intensity controlling.
3. Time brightening controlling.

Method 1 requires a complicated A–D converter, and method 2 requires the control of a nonlinear modulation characteristic of the tube. In our point of view, method 3 is the best modulation if a sufficiently short modulation time compared with the integration time can be adopted. The human eye has a logarithmic sensitivity to a linear variation in brightness. Therefore, the *density modulation* which is used in the usual X-ray image gives the observer an obscure image of concentration distribution. To solve this problem with method 3, a special pulser which has an output pulse width exponentially modulated to a linear variation in input-signal voltage is required. In this case, a variation range in pulse width of 1 to 1000 is desirable, because the linear range of the H. D. curve is up to 3 in the conventional photographic plate or film.

The distinguishable minimum difference of the optical density and the linear range of that H. D. curve determine the maximum number of brightness steps which can be utilized in practical analysis. An optical density difference of 0.3 between the picture elements is easy to distinguish; so, 10 brightness steps in the maximum case would be an optimum number.

DESIGN OF NEW CONTENT-MAPPING ATTACHMENT

General Construction

Figure 4 illustrates a block diagram of an electron-probe microanalyzer with the new content-mapping system attached. In Figure 5, a general view of an electron-probe microanalyzer including the content-mapping system attached to the scanning unit is shown. Figure 6 shows a schematic diagram of the content-mapping system which has been developed for the reasons mentioned in the previous section. The principle of performance is described as following.

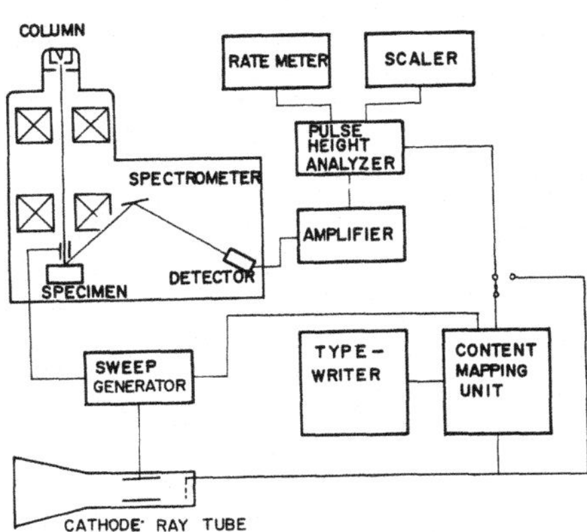

Figure 4. Schematic diagram of an electron-probe microanalyzer with a content-mapping system attached.

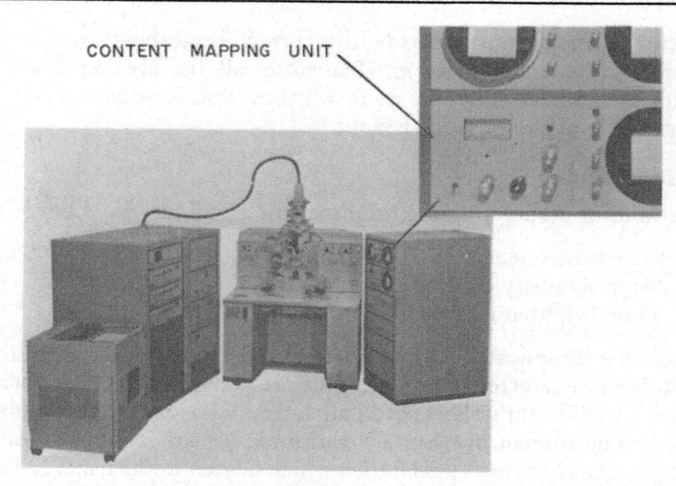

Figure 5. General view of the Hitachi-XMA 5 electron-probe microanalyzer with the content-mapping system.

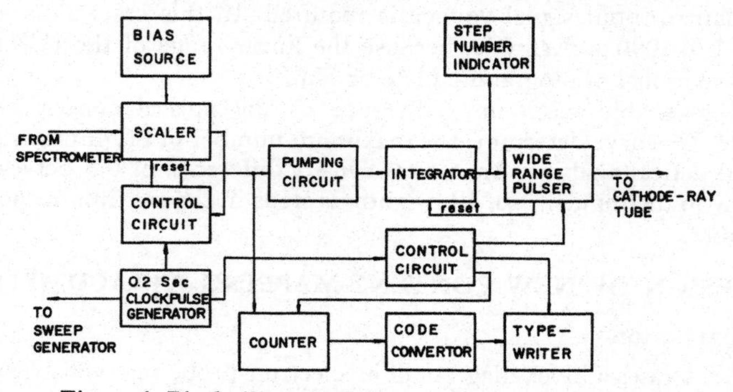

Figure 6. Block diagram of the content-mapping system.

The output pulses from a spectrometer are transmitted through an amplifier and a pulse-height analyzer to the scaler of the content-mapping unit. This scaler consists of a pumping circuit and an integrator. The width of the intensity step is set up by selecting the pumping-step height; and the intensity-bias level, by adjusting the level of the integrator. The pumping circuit gives the pulser a voltage which is proportional to the counting rate of the output pulse of the scaler, in other words, it works as a type of D–A converter. The number of steps desired to be programmed are determined by selecting a pumping constant of the pumping circuit. The pulser provides a means of converting an input voltage signal into a pulse-width signal and controls the brightness modulation time of the cathode-ray tube. On the other hand, a clock pulse generated at an interval of 0.2 sec is used for controlling the sweep generator to determine the integration time per picture element, and for, at the same time, controlling the scaler and the pumping circuit. The brightness graduation can be varied within a range of the first to ninth step. The counting rate per intensity step can be set up within a range of 50 to 500 counts/sec. The range within which the bias intensity can be set up is between 0 and 10 times the counting rate per step. The time required for obtaining one frame is about 8.5 min. The image produced by this system consists of 2500 picture elements, each of them having a square shape. The recording of the content map is carried out by means of a camera or typewriter. The image obtained by means of the typewriter is represented by figures 0 through 9, according to the step to which the intensity of the picture element belongs. The typewriter is equipped with a specially designed ink tank, which gives a color specific to each figure to be printed out.

Pulser

Figure 7 shows a block diagram of the newly developed pulser. The output-signal pulse is produced by the bistable circuit, which is controlled by a start signal and a stop signal sent from the comparator at an interval of 0.2 sec. In the comparator, an input-voltage signal, which is delivered from the pumping circuit and is in proportion to the X-ray intensity, is compared with an output voltage of the function generator. This output voltage of the function generator varies approximately logarithmically with time, so that the width of the output pulse of the bistable circuit, or, in other words, the output pulse of the pulser, varies exponentially with the input-voltage signal to the pulser. The generation of the logarithmic function was designed to use a combination of condenser–resistor circuits.

Figure 8 shows the characteristic curve of the pulser. In the conventional pulsers, the variable range of pulse width is 1 to 10, or so. This pulser, however, can produce a range of 1 to 1000. Therefore, this will make it possible to realize more than 50 steps of brightness. Furthermore, it is obvious from Figure 8 that a satisfactory exponentially linear variation of output-pulse width is obtained.

Type-Out Recording

Figure 9 shows the block diagram of the typing-out system. A signal corresponding to the X-ray intensity for a picture element from the scanning equipment is converted into a binary code by the signal converter. By the memory circuit, this binary-code signal is memorized for the time that is necessary for the typing operation of a figure.

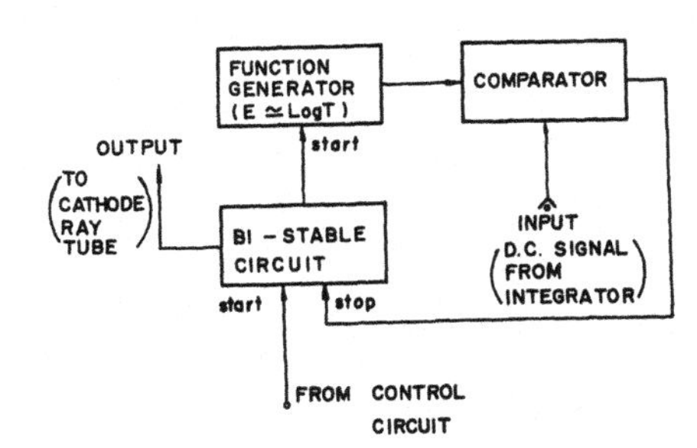

Figure 7. Block diagram of the wide-range pulser.

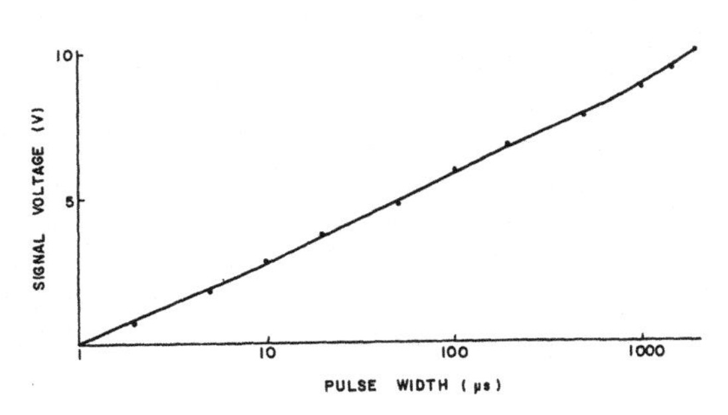

Figure 8. Characteristic curve of the wide-range pulser.

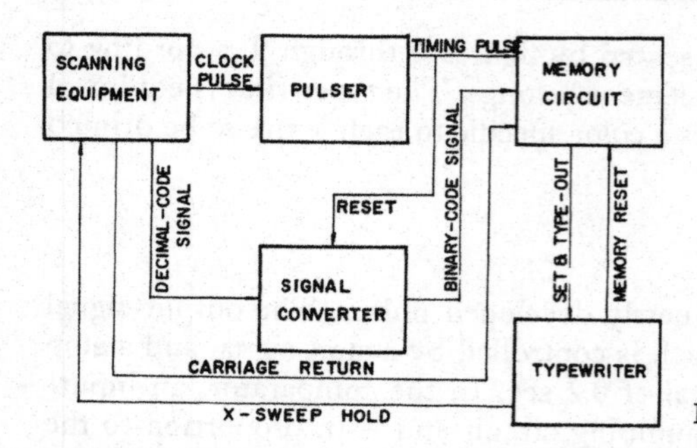

Figure 9. Block diagram of the typeout system.

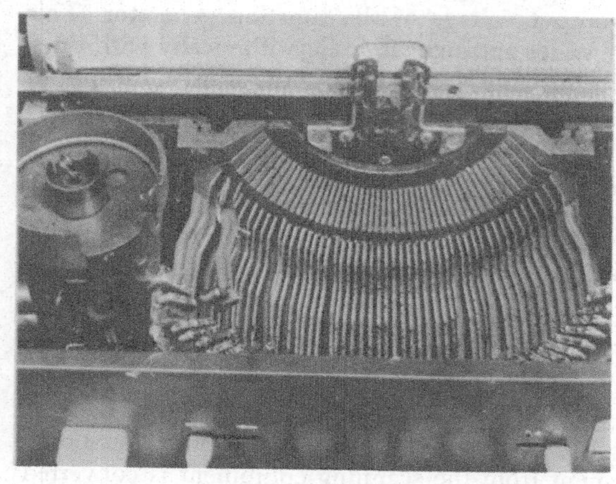

Figure 10. Colored-type equipment.

Figure 10 shows the upper part of the typewriter equipped with a special ink tank. This ink tank makes it possible to give a specific color to each figure printed out.

APPLICATIONS

Figures 11 through Figure 14 show some examples of the content map which have been obtained with the new system.

Figure 11 illustrates examples of the content map obtained by photographic recording. The specimen used is the mutual diffusion zone of copper and zinc. The image shown is the distribution analysis data of zinc. As noted in these data, the fine variation of concentration is clearly indicated quantitatively. Furthermore, if the brightness is recorded even in eight steps, each step can easily be identified. The newly developed system is greatly advanced beyond the two- or three-step indicating system. The square-shaped picture element provides the right brightness without any halation, which makes it very easy to compare the brightness of a particular step with that of another one.

Figure 12 shows the data obtained by the analysis of a copper–gold alloy which has primary crystals. The content map has distinctly shown the existence of the primary crystal that could not be clearly observed in the X-ray image. The line analysis, which is shown with the X-ray image, has hitherto been necessary to observe the fine variation of concentration. With this system, it is possible to obtain instantly the quantitative information in two dimensions. In the case of copper, the bias level was set up at 1500 counts/sec in order to expand the variation of concentration of approximately 29%.

Figure 13 shows an example of applications to the light-element analysis. The distributions of iron, chromium, and carbon in a steel specimen are displayed. In the case

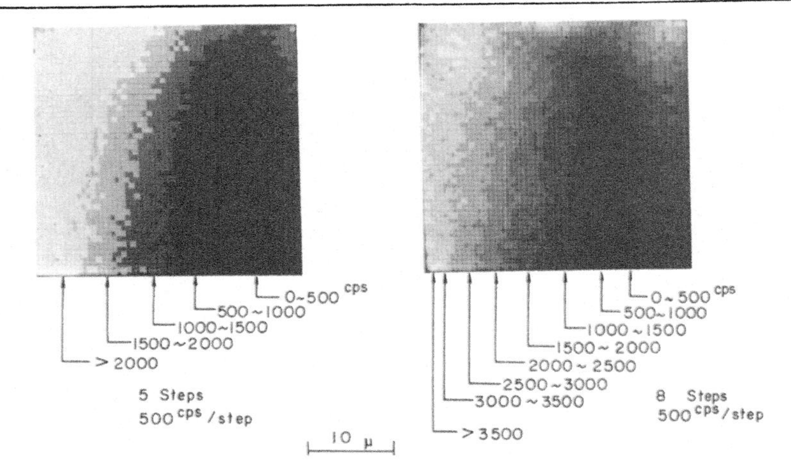

Figure 11. Example of content map.

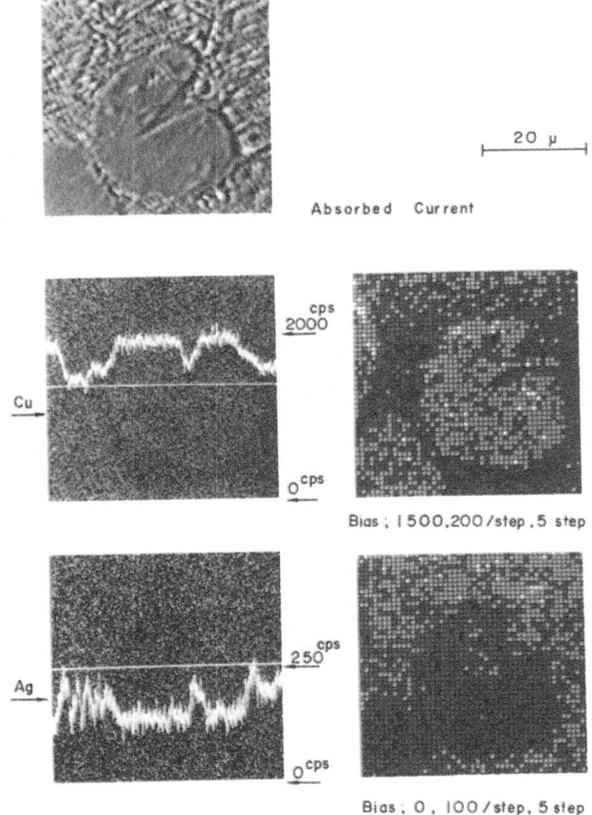

Figure 12. Analysis of copper-gold alloy.

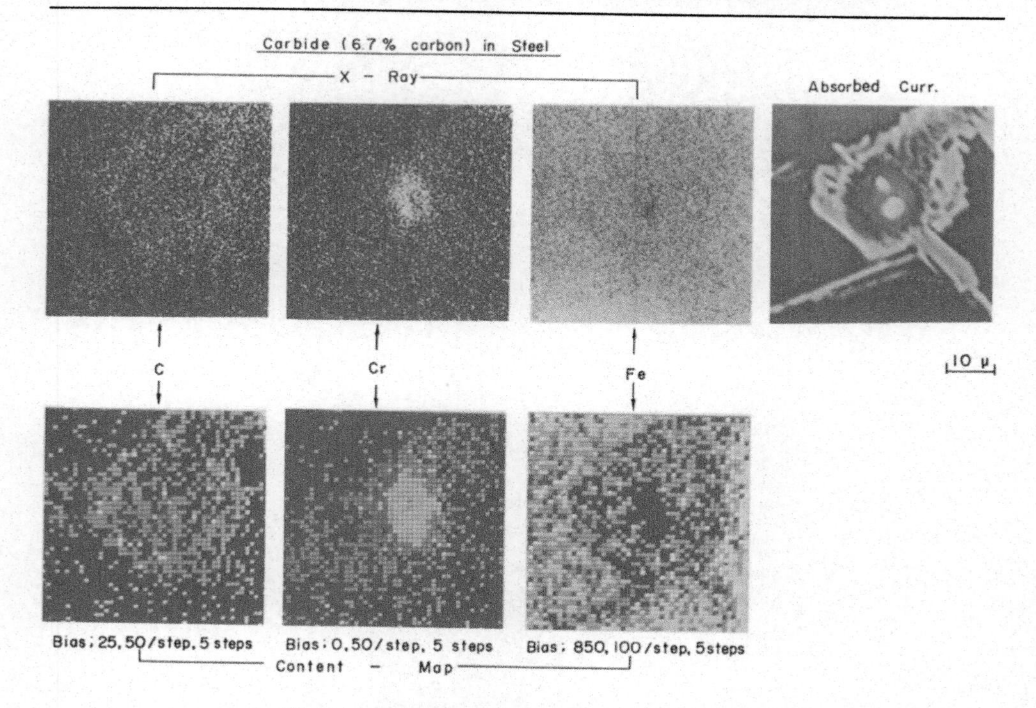

Figure 13. Analysis of the cementite phase of carbon.

of carbon, the form of the cementite phase, whose concentration of carbon is 6.7 wt.%, has been visualized clearly by the help of the content-mapping method. Moreover, from the concentration variations of iron and chromium, a state of substitution with each other is distinctly shown.

In Figure 14, an example of colored typeout recording is shown. The result shows an image with a concentration distribution of iron and aluminum in a cement clinker. In this image, the X-ray intensity per image element is represented by numerals from 0 to 7 and each figure has an individual color; for example, the first step has a red color; the second, green; and the third, violet. It is obvious that, by this colored typeout, the element distribution can easily be observed and that the intensity comparison between the picture elements can be done without comparing the brightness of others differing from the photographic film recording.

CONCLUSION

The adoption of the content-mapping system described in this paper has removed the difficulties in the conventional method of scanning microprobe analysis and the distinct visualizations of the fine variation of concentration in the scanning image were materialized without loss of rapidity in analysis.

REFERENCES

1. K. F. J. Heinrich, "Oscilloscope Readout of Electron Microprobe Data," in: W. M. Mueller and M. J. Fay (eds.), *Advances in X-Ray Analysis, Vol. 6,* Plenum Press, New York, 1963, p. 291.

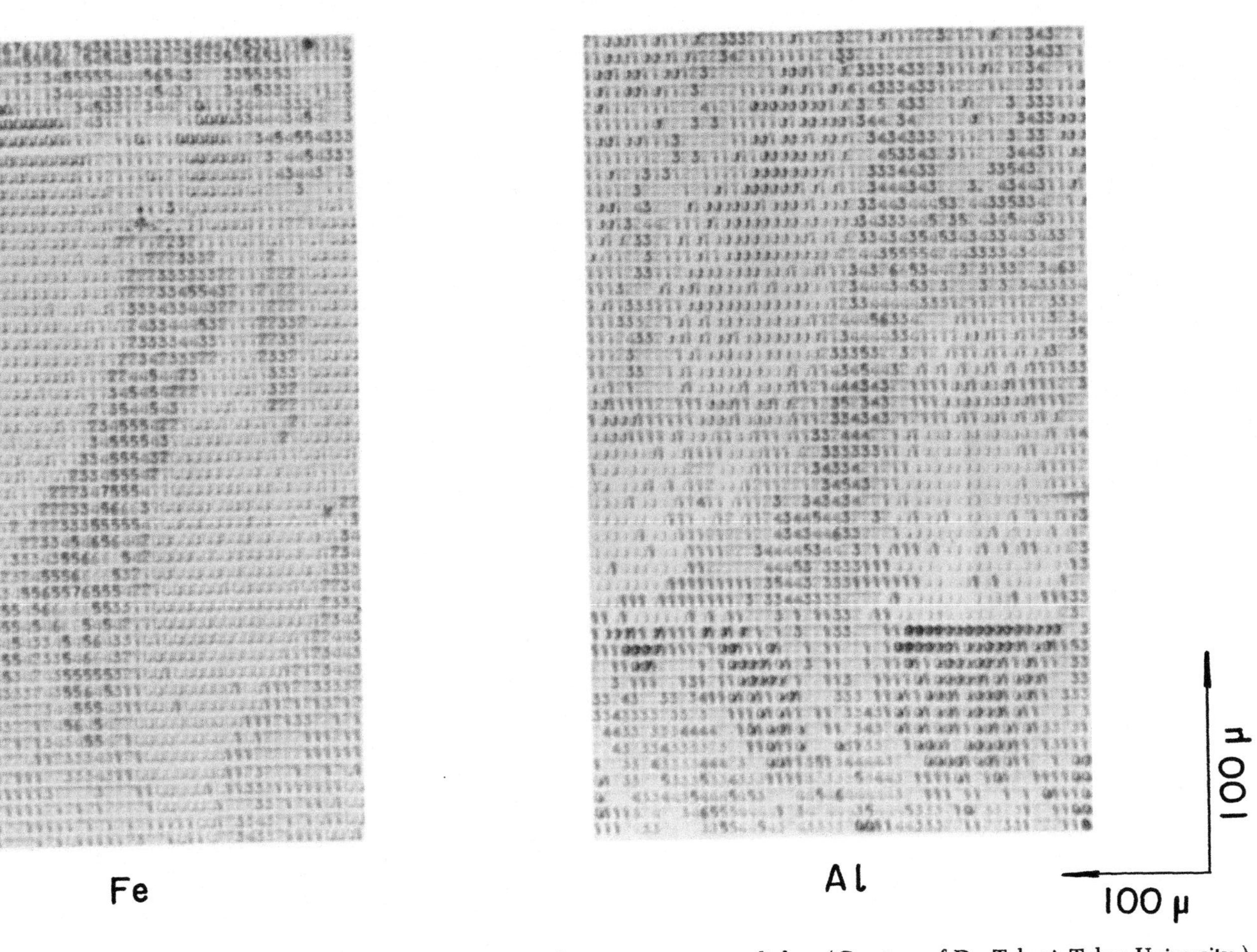

Figure 14. Example of colored typeout recording. Analysis of iron and aluminum in cement clinker. (Courtesy of Dr. Takagi, Tokyo University.)

2. D. A. Melford, "The Use of Electron Probe Microanalysis in Physical Metallurgy," *J. Inst. Metals* **90**: 217, 1962.
3. L. S. Birks and A. P. Batt, "Use of a Multichannel Analyzer for Electron Probe Microanalysis," *Anal. Chem.* **35**: 778, 1963.

DISCUSSION

K. F. J. Heinrich (National Bureau of Standards): I think that these are very beautiful maps, but, since this technique has been introduced under the name of concentration mapping and, in fact, has been reported a few years ago right here in this room, it seems lamentable that we cause more confusion by showing practically the same technique under a new name.

R. Seibert:* I realize that you might be confused. I didn't repeat all the names that were given in the bibliography of the paper, but you were mentioned several times. I condensed the paper to cut it short, but this is true, a form of the technique has been described before. The advantage that the technique described in this paper has is the wider range in concentration. In other words, more and wider steps of integrated intensity.

L. S. Birks (Naval Research Laboratory): I regret having to make comments on a paper without the authors' presence, but I think that, while the spectacular nature of the color printing certainly lends a certain quality to presenting the data, I don't believe that the quantitative nature of it can approach that of the printout from the multiscaler mode of a multichannel analyzer, where the actual number of photons at each position is printed out by an electric typewriter. As far as the equipment simplicity goes, I don't believe it can approach the simplicity of a standard multichannel analyzer with a standard electric typewriter printout that prints on ordinary paper in ordinary numbers.

R. Seibert: This may be true also, and I wish the authors were here to talk to you about it.

* This paper was presented by R. Seibert, Southern Regional Manager of Perkin-Elmer.

INVESTIGATION AND DEMONSTRATION OF SINGLE-CRYSTAL AND POWDER DIFFRACTION BY USING ZERO-POWER BETA-EXCITED X-RAY AND ^{55}Fe ISOTOPIC SOURCES*

Luther E. Preuss, William S. Toothacker, and Claudius K. Bugenis

Division of Nuclear Spectroscopy and Radiation Physics
Department of Physics
Edsel B. Ford Institute for Medical Research
Detroit, Michigan

ABSTRACT

Demonstrable diffraction (powder or single-crystal) with the use of zero-power isotopic sources would provide important data on such systems and may have implications in analysis (in this work, we fabricated 100-mCi ^{55}Fe, accelerator-produced, and 4-mCi ^{147}Pm–Cu beta-excited sources). Powder-diffraction efficiency with the use of a special diffractometer adapted to these sources is about 10^{-5} (calculated) for the Mn $K\alpha$ X-rays (^{55}Fe) on the NaCl 200 plane. This beam flux at three times natural background intensity was experimentally achieved by using photon spectrometry with proportional counters and multichannel analysis. The LiF diffraction pattern was demonstrated with intensities roughly in agreement with prediction. Lithium fluoride single-crystal diffraction intensity (200 plane) exceeded by 250-fold that of powder. Powder main-peak intensity ($\theta = 31.5°$) produced 9.2 photons per minute at the receiving slit *versus* 2.3×10^3 events for the single crystal. This single-crystal intensity with 60-sec runs provided a peak resolution of 0.29° full width at half maximum. The first-order $K\alpha$, $K\beta_1$ first order, and $K\beta_1$ second-order beams ($\theta = 31.5°$, 28.3°, and 71.6°, respectively) were demonstrated, with 1900 events per minute in the main peak ($K\alpha$). X-ray no-screen emulsion was used to study photographic recording on the LiF single crystal with exposures as short as 2.5 hr. A 2-in.-diameter camera, newly designed for the ^{55}Fe source, is collecting further emulsion data.

The isotopic source is stable, highly predictable in intensity, reliable, and compact. Requiring zero power, it thus has a potential for remote applications. Where its low output can be tolerated, it may hold promise for analysis, especially where its size, weight, and stability are important criteria. Source work here points to possible ^{55}Fe improvement in photon output of 10^2 over the source described here. Such an increase may make certain special analytical applications feasible.

INTRODUCTION

Our laboratory proposed to demonstrate and study the intensity levels of Bragg diffraction of X-rays from certain of the electron-capture isotopes and also by using the characteristic $K\alpha$ X-rays from beta-excited X-ray sources. A ^{147}Pm–Cu beta-excited source of 4.2 mCi was successfully used in the diffraction of the Cu $K\alpha$ X-rays on LiF. Main-beam intensity was only 16 $K\alpha$ X-rays per minute (approximating prediction)

* Supported in part by the U.S. Atomic Energy Commission, Division of Isotopes Development, Project No. AT(11-1)-1239.

at the receiving slit. This represented the first diffraction demonstrated by using a beta-excited X-ray source. Promethium-147 beta activities up to 300 mCi are feasible and currently are being tested.

This report shall be primarily concerned with the use of ^{55}Fe as a source of X-rays for diffraction.

One logically may question the use of such isotopic sources in diffraction when superlative X-ray sources in the form of highly refined X-ray tubes are commercially available. The first answer to this is that, although clearly predicted, diffraction for analytical purposes had not been experimentally demonstrated or evaluated with the use of such sources. For the academic, this may be a sufficiently good reason. Secondly, however, is the fact that such isotopic sources do carry unique and positive attributes. The electron-capture isotopes (and specifically ^{55}Fe) exhibit the ultimate in stability and thus are quite predictable in X-ray output. They are very small and exceedingly light, being of a few millimeters in dimension and 1 or 2 mg in weight. Our experimental X-ray sources with their supporting substrates are smaller than a dime. The flux of such electron-capture sources is monochromatic (exhibiting only the daughter-product X-rays). Perhaps the most important attribute is that they require no source of power for their operation. These qualities imply certain potential applications in analysis, such as remote-area use where electrical power is limited or nonexistent.

The primary debit of such an electron-capture source type is that the X-ray intensity is relatively low. It is to this point that this report is directed.

Iron-55 decays entirely through the electron-capture process to a nonradioactive isotope of manganese (^{55}Mn) with a 2.7-year half-life. Its decay is followed by the emission of manganese-characteristic X-rays. It exhibits a K fluorescence yield of about 31% (the Mn $K\alpha_1$ energy is 5.898 keV). Iron-55 may be pile-produced through a $^{54}Fe\ (n,\gamma)\ ^{55}Fe$ reaction with the use of enriched ^{54}Fe. The reactor production of this nuclide involves the long-term irradiation of isotopically enriched ^{54}Fe, as $^{54}Fe_2O_3$. Enrichment in the range of 50% ^{54}Fe is common (a tenfold enrichment). Higher enrichments are available. Reactor production is accomplished by the (n,p) reaction which, owing to the presence of ^{58}Fe in the target, produces ^{59}Fe. This isotope of iron is a most undesirable contaminent because of its half-life and radiation. Specific activities available by this technique are relatively low (Oak Ridge irradiations appear to have ranged from 10 to 50 Ci/g).

Cyclotron production of ^{55}Fe [with the use of the $^{55}Mn\ (p,n)\ ^{55}Fe$ reaction] has certain distinct advantages. The reaction is free of radiocontaminants and theoretically provides carrier-free material. Its uniqueness and simplicity is attributable to the single, stable isotope of manganese. A practical cyclotron experiment will provide about 750 Ci/g, which is, of course, considerably less than the carrier-free concentration of 2400 Ci/g.

Self-absorption of the Mn K X-rays in the ^{55}Fe deposit is the primary design hurdle in the fabrication of such an X-ray emitter. Platings of 2 Ci with 75% transmission have been made on an area of less than 1 cm^2.*

If one assumes a 100-mCi ^{55}Fe X-ray source with a 10- by 2-mm geometry (simulating an X-ray diffraction-tube source) applies the conventional expression[1] to determine Mn $K\alpha$ X-ray diffracted intensity by an NaCl powder specimen in our diffractometer geometry, and applies the various factors (scattering, diffractometer, unit cell, structure, multiplicity, temperature, absorption, and Lorentz polarization factor), one arrives at a prediction of 10 to 100 photons per minute at the detector.

* Fabricated by I. Gruverman of New England Nuclear Corp. for use at the University of California at San Diego.

On the basis of these predictions we designed and had fabricated a cyclotron-produced 100-mCi ^{55}Fe source in collaboration with Oak Ridge National Laboratory.* After solution of the cyclotron target, the $^{55}FeNO_3$ in water solution was evaporated to a few lambdas of volume. The concentrated solution was deposited through low-temperature evaporation of the aliquot in a milled-out recess in a platinum disc. Drying was accomplished at 150 to 200°C. The nitrate was then transformed to $^{55}Fe_2O_3$. A hydrogen flush at 800°C reduced this to metallic ^{55}Fe. An aluminum vacuum evaporated film approximately 5000 Å in thickness over the ^{55}Fe deposit protected against oxidation. A 51-μ-thick beryllium foil serves as a seal to the source and an exit window with high transmission to the Mn K X-rays.

We first tested our predictions in a laboratory-design diffractometer, using gross counting techniques and simple amplifying and integrating electronic gear. We were unsucessful in this approach for demonstrating diffraction on NaCl powder samples. This was attributed to a high area background and an unanticipated low transmission in the ^{55}Fe source (due to nonuniformity of deposit and to a high concentration of stable iron).

The experiment was redesigned to study the diffracted intensity by using fixed-position, timed runs and by doing spectroscopy on the Mn $K\alpha$ X-rays with the use of the setup shown in Figure 1. Spectral analysis was carried out with a thin-window beryllium counter (Amperex PC–303), a Fluke high-voltage supply, Hamner linear amplifiers (N–361 and N–308), and a 256–channel Nuclear Data analyzer. Studies with this system and the use of specimens of NaCl powder clearly demonstrated the Mn $K\alpha$ diffracted beam from the 200 plane at roughly the intensity levels predicted.

Lithium fluoride powder was substituted for the NaCl samples, and fixed-position runs were made. Integration of the area under the analyzer Gaussian peaks, representative of the Mn $K\alpha$ doublet, provided total intensity values which were transposed to a separate plot shown in Figure 2. This is the first such LiF diffraction pattern thus obtained.[2] The θ values are 26.9, 31.5, and 47.6° corresponding to diffraction of the $K\alpha$ doublet on the 111, 200, and 220 planes, respectively. Sample problems prohibited tests at higher θ values in this particular study.

To facilitate this evaluation, single-crystal LiF was substituted for powder, and diffraction from the 200 plane was used as a standard of comparison. Figure 3 illustrates

* The ^{55}Fe source was fabricated at Oak Ridge National Laboratory, Oak Ridge, Tenn. Acknowledgement is made to F. Case, E. Beauchampe, and J. Ratledge for aid in source design and construction.

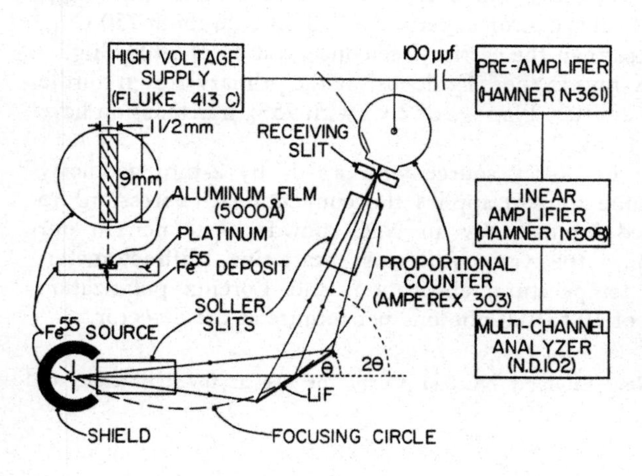

Figure 1. The experimental system used for the fixed-position diffracted-beam-intensity studies. Specimens were both LiF and NaCl. Spectroscopy was carried out by proportional counting and multichannel analysis. Reproduced by permission of the American Institute of Physics (*Journal of Applied Physics*, "Letters").

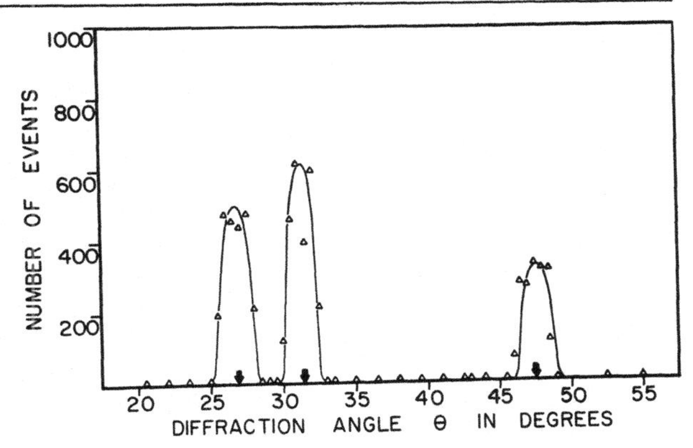

Figure 2. Diffraction pattern made up of the three most intense lines from LiF for the Mn $K\alpha$ doublet from the electron-capture decay of 100-mCi ^{55}Fe. Relative and absolute intensities agree with prediction. Reproduced by permission of the American Institute of Physics (*Journal of Applied Physics*, "Letters").

the first pattern obtained by using 1-min fixed-position runs. Diffraction was demonstrated for the 200 plane for the $K\alpha$ doublet as well as for $K\beta_1$, first and second order. Our relatively crude diffractometer exhibited a resolution of about 0.3° expressed as full width at half maximum, for the $K\alpha$ beam on the 200 plane.[3,4] This, doubtless, might be improved by using a more precise diffractometer.

At this point in the experiment, we became interested in the possibility of recording the diffracted beams by means of photographic emulsion. As a preliminary step, we exposed X-ray no-screen emulsion for about 1 hr to the focused Mn $K\alpha$ beam for the LiF 200 plane and obtained a well-defined line of good photographic density.[4] This encouraged us to design a much simplified and rudimentary Debye–Scherrer camera, which was fabricated to take the 100-mCi ^{55}Fe source. Figure 4 is a view of this instrument. The camera has a 50.8-mm internal diameter. The 100-mCi source holder is integral with a slit-design aperture (the 1- by 10-mm aperture is varied in dimension through the use of shims). The cylindrical body which is the film holder is made rotatable on the base, and this allows for ancillary studies with a single crystal (since the vertical peri-

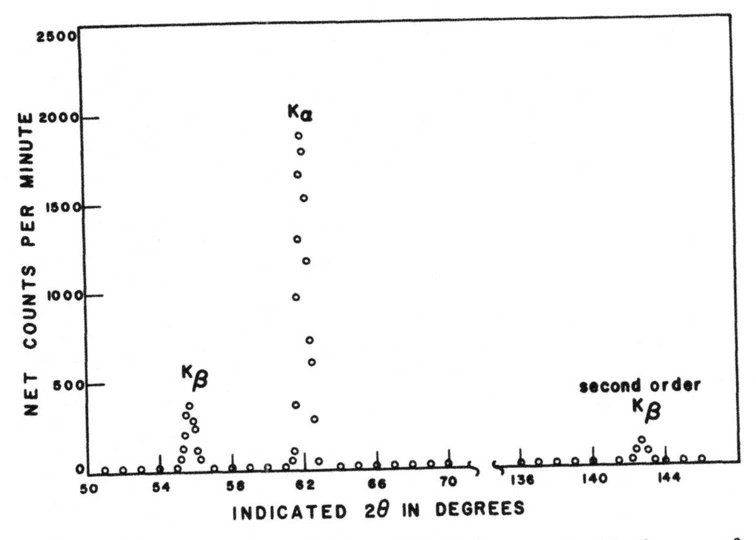

Figure 3. Diffraction pattern from LiF single crystal with the use of the 100-mCi ^{55}Fe source. Best peak-intensity values are about 2.5×10^2 times that noted with LiF powder.

pheral positioning of the source provides for focusing at the emulsion when using a small crystal).

One-hour exposures on X-ray no-screen emulsion with the use of a LiF single crystal produced a dense photographic exposure for the line corresponding to the 200 diffraction plane. Lithium fluoride-powder specimens of 200 mesh in a 1.0-mm Lindemann glass sample capillary produced diffraction patterns but required 100-hr exposure periods. Seven diffraction lines were demonstrable visually with these exposures and were clearly shown by densitometry. The typical line "feathering" due to slit-style apertures was exhibited.

Graphite-powder camera exposures showing eight identifiable lines were made in geometry equivalent to the above LiF studies (exposures were also of the order of 10^2 hr). Figure 5 shows graphite with lines from the 112, 110, 004, 100, 101, and 002 planes for the $K\alpha$ radiation (self-absorption and radioactive decay had reduced this ^{55}Fe source to the equivalent of about a useful 10 mCi at the time of these exposures).

Despite the technological struggle to elicit diffraction patterns by using the weak ^{55}Fe source, the worker has reason to be encouraged by the diffraction which has been demonstrated both by proportional-counter detection and by photographic emulsions.

If one considers the direct competition between the X-ray tube designed for remote-area low-wattage use and the ^{55}Fe source, one has further reason to be cautiously optimistic.

J. Dunne[5] has published a feasibility study discussing a low-power-consumption X-ray tube in which the device operates at a flux of 10^{12} photons per second per 4π steradians.

Now, consider a 2-Ci cyclotron-produced ^{55}Fe source with 75% transmission (such a source is an accomplished fact). Simple calculations show that about 2.3×10^{10} photons are produced in such a source, and roughly 1.7×10^{10} of these per second per 4π steradians are available. The experimenter using accelerator-produced ^{55}Fe is thus within a factor of about 50 of the output of the tube designed for low power consumption and remote-area use. Should the worker be willing to sacrifice a degree of resolution or

Figure 4. The laboratory Debye–Scherrer camera designed for use with the ^{55}Fe source. The cylindrical source housing (center) positions the source at the emulsion circle.

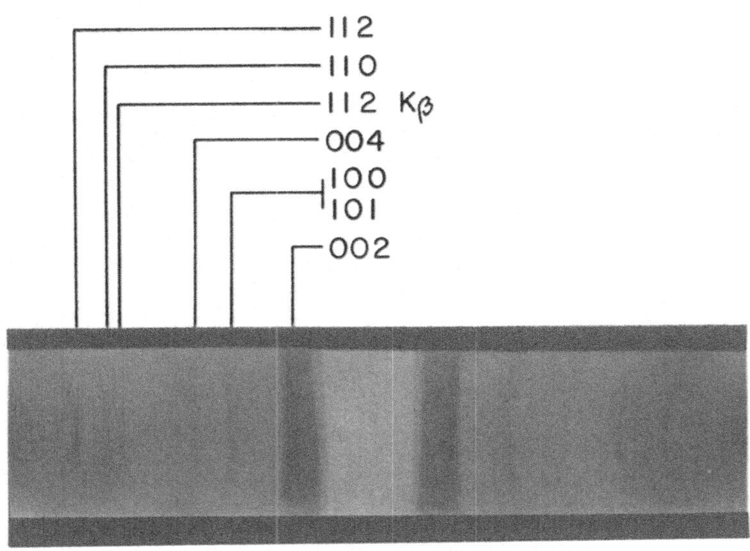

Figure 5. This is an exposure in the laboratory-fabricated Debye–Scherrer camera with graphite as the specimen and exposure for a 100-hr period. The ^{55}Fe source at this point was equivalent to only about 10 mCi (owing to half-life and self-absorption).

to use multiple detectors (with the diffractometer) or special film techniques, he then might be able to attain a better ratio of equity in comparing this electron-capture source with the conventional X-ray device.

Our laboratory has in the design stage a multicurie cyclotron-produced ^{55}Fe source which would provide for conventional diffractometer motorized runs on crystalline samples with run times competitive with existing systems and with resolutions only slightly poorer than those currently acceptable.

In our camera studies, the use of this 2-Ci source would have meant a 0.5-min exposure with the use of the single-crystal LiF sample in our rudimentary 2-in. camera and about 0.5-hr exposure for the LiF powder sample. The intensity factor of 50 difference which exists is a far cry from the death knell of the X-ray tube. Its intensities cannot be completely matched. The isotopic source, however, does hold some promise for remote-area, zero-power application. It conceivably could be used to produce meaningful analytical data in on-the-site lunar and planetary exploration where wattage is in short supply or in geological studies in remote areas.

REFERENCES

1. B. D. Cullity, *Elements of X-ray Diffraction*, Addison-Wesley, Reading, Mass., 1956, p. 389.
2. L. E. Preuss, W. S. Toothacker, and C. K. Bugenis, "Demonstration of X-Ray Diffraction by LiF using the Mn $K\alpha$ X-Rays Resulting from Fe-55 Decay," *Appl. Phys. Letters* **9**: 159, 1966.
3. L. E. Preuss, W. S. Toothacker, and C. K. Bugenis, "Single-Crystal X-Ray Diffraction Using Photons from Two Radioactive Isotopic Sources," *J. Appl. Phys.* **38**: 3404, 1967.
4. W. S. Toothacker, L. E. Preuss, and C. K. Bugenis, in: P. S. Baker (ed.), *Proceedings of the Symposium on Low Energy X- and Gamma Sources and Applications*, published by ORNL, Austin, Texas, Mar., 1967.
5. J. Dunne, Jet Propulsion Laboratory Techn. Memo. No. 33-218.

AN X-RAY SMALL-ANGLE SCATTERING INSTRUMENT

Donald M. Koffman

Advanced Metals Research Corporation
Burlington, Massachusetts

ABSTRACT

An X-ray small-angle scattering instrument is described which is used for recording X-ray diffraction patterns or small-angle X-ray scattering curves in an angular region very close to the direct beam. The measurement of X-ray intensity is accomplished with standard geiger or scintillation counter techniques. The instrument is designed for use with a spot-focus or vertical-line X-ray source. In essence, it is a multiple-reflection double-crystal diffractometer, based on a concept developed by Bonse and Hart, employing two grooved perfect germanium crystals arranged in the parallel position. Multiple diffraction from these crystals produces a monochromated X-ray beam which can be several millimeters wide while still exhibiting extremely high angular resolution. As a result, effective sample volumes can be employed with maximum volume-to-thickness ratios. The principal features of the instrument are discussed with emphasis on the advantages of this device over those employing complex slit systems and film-recording techniques. Data are presented to illustrate the operation, intensity, and resolution of the unit.

INTRODUCTION

Small-angle X-ray scattering has been successfully employed for the study of the size of well-defined submicroscopic particles, *e.g.*, the fine-grained catalysts, macromolecules, colloidal micelles in solution, and clustering of atoms in solid solution. These cases generally afford large scattered intensities because the variations of the electronic densities in the samples are considerable.

The detection of large lattice spacings found in some minerals and in certain complex molecules, such as the high polymers or proteins, has also led to the development of many experimental arrangements for small-angle X-ray-scattering measurements. This equipment has been more than adequately reviewed by Guinier and Fournet[1] and Beeman *et al.*[2]

The fundamental requirement of these devices is that they should provide monochromatic X-ray beams with high angular resolution. In most cases, the degree of collimation and monochromatization are dictated by intensity considerations. These considerations often lead to the design of slit systems which provide angular collimation only in the direction normal to the slit.

Three major conditions govern the design of a small-angle scattering instrument. These can be summarized as follows:

1. The effective cross section and divergence of the primary beam should be small.

2. The parasitic scattering (radiation received at the point of observation with no sample in the beam) should be minimal.
3. The spectral purity of the primary beam should be maximized.

The first condition is dictated by the fact that, if the angular divergence and beam cross section (for the slit-system device) are not limited, a good definition of the scattering angle θ is not possible since X-rays arriving at the point of observation will originate from an area rather than a point on the sample.

The second condition is important because the measurement of the intensity received at the point of observation is a correct measurement of the intensity scattered by the sample only if no parasitic scattering exists. If detector methods are employed, the parasitic scattering can be subtracted from the observed scattering only if this correction is small.

The condition of spectral purity is very important in the measurement of continuous scattering since, if the X-ray beam is not monochromatic, the influence of the continuous spectrum may be considerable and the observed X-ray scattering may have an appreciable contribution from the X-rays in the continuous spectrum. Crystal monochromators or balanced (Ross) filters in conjunction with pulse-height analysis should be employed. For small-angle crystalline diffraction which is analogous to the usual high-angle phenomena, the filtering of the $K\beta$ is sufficient. The diffraction effects due to the more intense characteristic radiation emerge from the general background of diffracted and scattered radiation produced by the continuous spectrum.

In attempting to meet the above conditions, certain disadvantages are inherent in a slit-system device. Since the angular resolution of such a device is determined by the degree of collimation, high resolution is always obtained at the expense of X-ray intensity. In addition, the elimination of parasitic scattering from the slits defining the X-ray beam requires the employment of complex slit systems, which become more and more difficult to align as resolution requirements become more rigorous.

To circumvent these disadvantages, small-angle scattering experiments have been performed with the use of a double-crystal diffractometer.[3] The two crystals are set in the parallel position with the sample placed between them. The X-ray beam diffracted from the first crystal is used to irradiate the sample. The scattered beam is then analyzed by diffracting it from the second crystal into a suitable detector. This system has the advantage of exhibiting the highest angular resolution [3–5] when nearly perfect crystals are employed. Moreover, the crystals not only produce monochromatic radiation but also completely utilize the entire X-ray source without sacrificing resolution. As a result, high X-ray intensities are obtained by employing an X-ray beam of large dimensions.

In contrast to slit systems, this large X-ray beam permits the examination of larger volumes of scattering material without increasing their thickness.

Although the simple double crystal diffractometer eliminates the need for parasitic scattering slits, the system is not well suited for small-angle scattering measurements since the reflection curves of perfect crystals have tails extending to high angles. In effect, the tails transmit too much of the primary beam and completely obscure the scattered intensity.

A system which has all the advantages of the simple double-crystal diffractometer without the troublesome background of the perfect-crystal reflection curve has been demonstrated by Bonse and Hart.[6,7] By employing grooved, perfect, germanium crystals, multiple Bragg reflections are obtained between the walls of the groove which results in the effective elimination of the tails of the reflection curve. These multiple reflections are shown schematically in Figure 1.

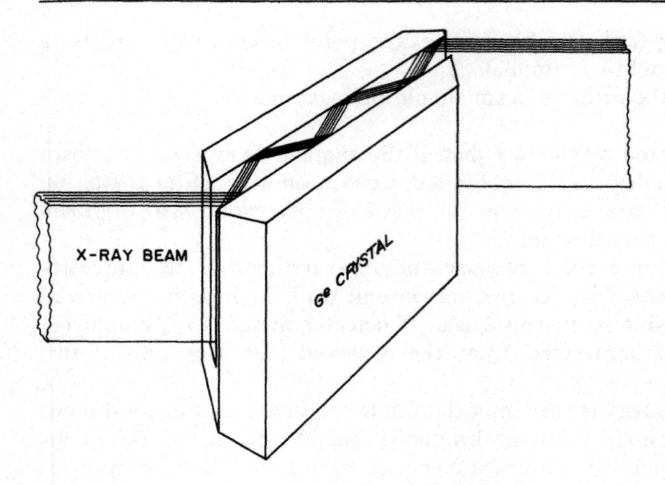

Figure 1. Multiple reflections within a groove cut into a germanium crystal.

THEORY

The shape of the Bragg reflection curve R of a perfect crystal is well known from the dynamical theory of X-ray diffraction as well as from recent experiments.[8–10] R is defined as the ratio of the reflected intensity I_h to the incident intensity I_0 for a particular direction of incidence. Bonse and Hart[6] have calculated the shape of the reflection curve for one reflection and five reflections from the (220) planes of silicon as shown in Figure 2. The points $y = \pm 1$ correspond to the limit of total reflection in the absence of absorption.

The significant features of these curves are that multiple reflection does not significantly alter the rocking breadth while the intensity of the tails, I_t, is found to be proportional to y^{-m}, where m is the total number of reflections.[6] As a result, for a total of 12 reflections, the intensity in the tails should decrease with the twelfth power of the angular deviation from the parallel position. The advantage of employing 12 reflections (six reflections per grooved crystal) is discussed later.

On the other hand, the peak intensity after m reflections can be expressed as

$$I_{\text{peak}} = \alpha^m I_0 \tag{1}$$

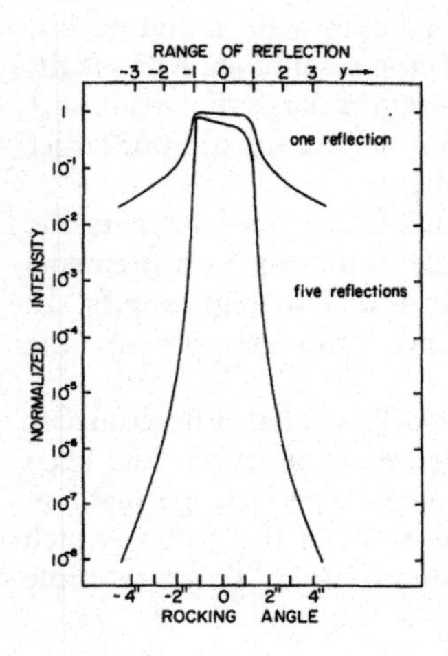

Figure 2. Calculated rocking curve for (220) reflection in a silicon perfect crystal.

where α is the fraction of the incident beam diffracted per reflection. By employing a value of α equal to 0.9, which is not uncommon for nearly perfect crystals, the peak intensity has only decreased by 70% after 12 reflections, which results in an X-ray beam of high intensity and low background.

EXPERIMENTAL APPARATUS AND PROCEDURE

Based on the above principles, a multiple reflection instrument was developed for small-angle scattering and diffraction studies. Shown in Figure 3, the unit is equipped with a beam-defining slit which consists of two sets of orthogonal micrometers. With this arrangement, the slit can be adjusted both horizontally and vertically to within 0.001 in. and can also be precisely adjusted around the X-ray beam. After passing through the defining slit, the X-rays emitted by the source are reflected six times in the groove of the first crystal and emerge parallel to the entrance beam (see Figure 1).

This exit beam, which is employed as the primary beam, passes through the sample and enters a second grooved germanium crystal where it is again reflected six times before passing into the detector. The second crystal acts as an analyzing slit and is capable of being rotated around the center of the sample by means of a large barrel micrometer calibrated in steps of 1 sec of arc (θ).

Each crystal is provided with independent rocking and tilting adjustments as well as crystal and detector translations for initial alignment purposes. The unit is supported on a solid base with three adjustable leveling feet. Since the beam entering each crystal is parallel to the exit beam, only simple translation of the crystals and detector are necessary for alignment. The translation is accomplished by consecutively measuring the intensity from the X-ray source, the first crystal, and the second crystal. The crystals can also be rocked and tilted without the need for repositioning them in the X-ray beam since the axis of rotation of these adjustments coincides with the first "bounce" position in each crystal groove.

The crystals are designed to diffract Cu $K\alpha$ radiation emitted by a spot or vertical-line X-ray source up to 3 mm in width. The walls of the crystal grooves are parallel to the (220) planes of germanium. An angular range of 6° (2θ) is provided with the capability of examining the X-ray intensity at 0° (2θ). As a result, absolute-intensity measurements can readily be made for sample-absorption determinations and the direct-beam position can be precisely measured.

Figure 3. The Advanced Metals Research Corporation's (AMR) X-ray small-angle scattering instrument.

The instrument is provided with a cover for vacuum and inert-gas operation and can also be operated in air.

To determine the scattered intensity and resolution available from this instrument, samples of Dow latex spheres were dried and the resulting powders were supported on 0.00025-in. Mylar film. Particle sizes of 0.796 and 0.109 μ were examined with the use of the spot source of a Philips standard copper X-ray tube operated at 40 kV and 20 mA. A Philips solid-state scintillation detector, pulse-height analyzer, and scaler were used for data accumulation. In order to remain within the linearity of the electronics, X-ray counting rates were maintained below 10^4 counts/sec with carefully calibrated nickel foils. All data were accumulated by using an air path.

EXPERIMENTAL RESULTS AND DISCUSSION

The results obtained from the 0.109- and 0.796-μ latex spheres are shown in Figures 4 and 5, respectively. The results obtained with no sample in place are also shown. This intensity has been corrected to account for sample absorption.

Since the latex particles are spherical and very uniform in size, the scattering curve of each exhibits a series of maxima. Bonse and Hart[7] have calculated the position and intensity of these maxima as a function of ha, employing the Rayleigh–Gans[11, 12] scattered intensity from a single sphere of radius a

$$i(ha) = \left(3\,\frac{\sin ha - ha\cos ha}{h^3a^3}\right)^2 = \frac{9\pi}{2}\left[\frac{J_{\frac{3}{2}}(ha)}{h^{\frac{3}{2}}a^{\frac{3}{2}}}\right]^2 \tag{2}$$

where $h = 4\pi \sin\theta/\lambda$ and θ is one-half the scattering angle. The intensity measured with an infinite slit height is given by

$$I(ha) = \int_0^\infty i_0(y)\, i(h^2a^2 + y^2)^{\frac{1}{2}}\, dy \tag{3}$$

where $i_0(y)$ is a constant.

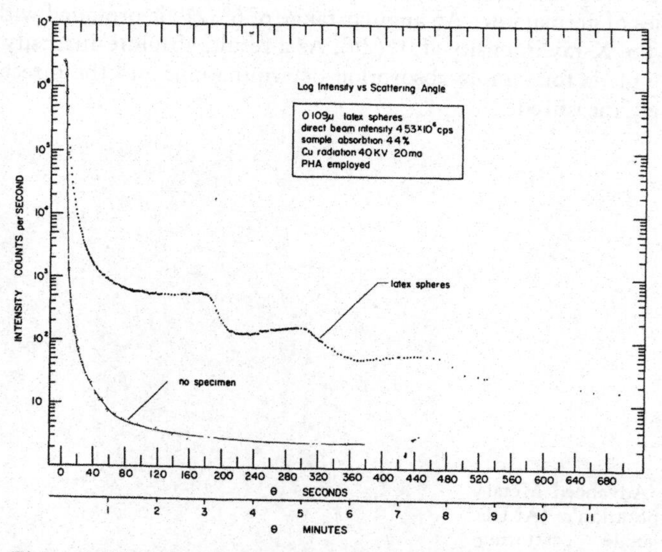

Figure 4. X-ray small-angle scattering from 0.109-μ latex spheres.

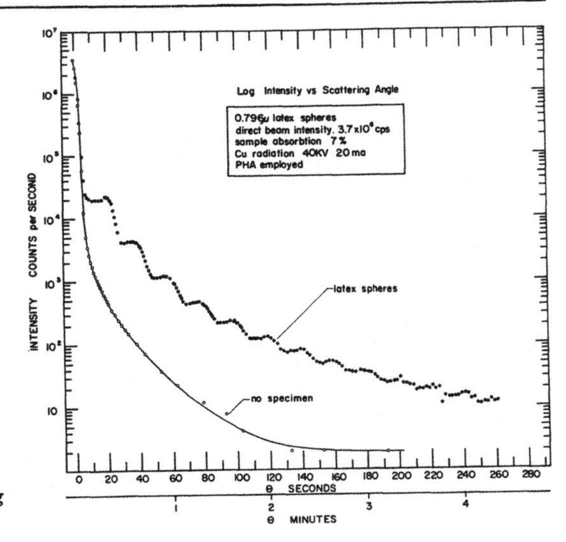

Figure 5. X-ray small-angle scattering from 0.796-μ latex spheres.

These calculations were employed to obtain a measured particle size for comparison with the known particle size. Sizes determined from the high-order maxima (greater than 4) agreed within 2% of the known value. Values obtained from the lower orders deviated by as much as 20%, which indicates that interparticle effects, which are not accounted for by the theory, influence the intensities and positions of these maxima.

The resolution and intensities of the multiple-reflection instrument and freedom from parasitic scattering are also illustrated in Figures 4 and 5. The direct beam exhibits an intensity approaching 4×10^6 counts/sec, with angular resolution of less than 10 sec of arc. In addition, the background tails of the primary beam are two orders of magnitude less than the scattered intensity even at scattering angles of 20 sec of arc (θ).

Figure 5 reveals the ability of the device to resolve and measure precisely structure in the scattering curve which is separated by only 20 sec of arc (θ). This ability allows the investigation of the size and shape of particles as large as 10,000 Å.

SUMMARY

The X-ray small-angle scattering instrument has demonstrated angular resolution which is superior to that of slit systems while at the same time providing high-intensity crystal-monochromated X-rays. The crystals incorporate an even number of reflections, which has the additional advantage of greatly simplifying instrument alignment.

In addition to the usual scattering and diffraction applications, the high resolution and low background intensity near the direct beam should prove extremely valuable in the study of phenomena which greatly influence the scattered intensity in this region, *e.g.*, interparticle-interference and multiple-scattering effects. Moreover, investigations of the size and shape of large particles can be accomplished routinely since the high resolution required for this type of study is an inherent feature of the device.

ACKNOWLEDGMENT

The author wishes to express his gratitude to Drs. R. E. Ogilvie, John T. Norton, and S. H. Moll and Mr. C. Brayton for their efforts in the design of the apparatus and the preparation of this paper.

REFERENCES

1. A. Guinier and G. Fournet, *Small Angle Scattering of X-Rays*, John Wiley & Sons, Inc., New York, 1955, p. 83.
2. W. W. Beeman, P. Kaesberg, J. W. Anderegg, and M. B. Webb, "Size of Particles and Lattice Defects," *Handbuch der Physik* **32**: 321, 1957.
3. P. Kaesberg, W. W. Beeman, and H. N. Ritland, "Double Crystal and Slit Methods in Small-Angle X-Ray Scattering," *Phys. Rev.* **78**: 336, 1950.
4. C. M. Slack, "The Refraction of X-Rays in Prisms of Various Materials," *Phys. Rev.* **27**: 691, 1926.
5. J. W. M. DuMond, "Method of Correcting Low Angle X-Ray Diffraction Curves for the Study of Small Particle Sizes," *Phys. Rev.* **72**: 83, 1947.
6. U. Bonse and M. Hart, "Tailless X-Ray Single-Crystal Reflection Curves Obtained by Multiple Reflection," *Appl. Phys. Letters* **7**: 238, 1965.
7. U. Bonse and M. Hart, "Small Angle X-Ray Scattering by Spherical Particles of Polyvinyltoluene," *Z. Physik* **189**: 151, 1966.
8. M. Renninger, "Messungen zur Röntgenstrahl-Optik des Ideal Kristalls. I. Bestatigung der Darwin–Ewald–Prins–Kohler Kurve," *Acta Cryst.* **8**: 597, 1955.
9. K. Kohra, "An Application for Obtaining X-Ray Beams of Extremely Narrow Angular Spread," *J. Phys. Soc. Japan* **17**: 589, 1962.
10. B. W. Batterman, "Effect of Dynamical Diffraction in X-Ray Fluorescence Scattering," *Phys. Rev.* **133**: A759, 1964.
11. Lord Rayleigh, "The Incidence of Light Upon a Transparent Sphere of Dimensions Comparable with the Wavelength," *Proc Roy. Soc.* (*London*) **A84**: 25, 1911.
12. R. Gans, "Strahlungsdiagramme Ultramikroskopischer Teilchen," *Ann Physik* **76**: 29, 1925.

DISCUSSION

E. L. Gunn (Esso Research and Engineering Company): I notice you show about 12 diffraction orders for your latex spheres. I was wondering about how long it took you to obtain the data shown. I believe these were single point measurements, but you didn't say anything about the time required for measurement.

D. M. Koffman: The data were accumulated by point counting, and the points are shown here. An attempt was made to keep the intensity—the total accumulated counts—above about 2000 to 3000 counts so that you can add them all up to see how much time it took to get 2000 or 3000 counts or more. The lowest counting time we used was 10 sec of time for the higher intensities, and then, for the lower intensities, we just counted until we had accumulated about 2000 counts.

E. L. Gunn: The incident intensity, was it over $1/e$?

D. M. Koffman: It was not. We didn't make any real attempt to maintain the thickness of the sample to that point. Theoretically, the sample should be given attenuation of about $1/e$, which is about 60%, but in practice you will find that, if you do use samples that are that thick, the effects of multiple scattering within particles come into play and this has an effect of shifting the peak positions, filling the wells between the peaks—giving poorer results. Although it is nice to work on a theoretical basis, sometimes it doesn't work out.

STUDY OF EXTENDED X-RAY ABSORPTION FINE STRUCTURE WITH THE USE OF THICK TARGETS

J. T. Hach and E. W. White

Materials Research Laboratory
The Pennsylvania State University
University Park, Pennsylvania

ABSTRACT

A novel method has been investigated for the study of extended X-ray absorption-edge fine structures. The method takes advantage of the absorption of continuum X-rays that emerge at low takeoff angles from an electron-bombarded target. The extended fine structure is observed as undulations in the continuum X-ray intensity in the region of the *self-absorption edge.* Spectra from conventional X-ray diffraction tubes have been recorded for various values of kilovolts and takeoff angles. Spectra obtained by this method are compared with published data. Nearly all the features of the extended fine structures can be clearly resolved by this technique. In the case of iron and copper targets, it is found experimentally that the best contrast is obtained when using a 50-kV accelerating voltage. A takeoff angle of 3° yielded best results for copper, while an angle of 10° was best suited for iron. The Philibert absorption correction successfully used in electron microprobe analysis has been extended to account for the observed self-absorption effect where $k_{\Delta\lambda} = f(\chi)\lambda_1/f(\chi)\lambda_2$. The primary advantage of the thick-target technique is in the ability to obtain absorption spectra without having to prepare thin films. The technique is essentially nondestructive in that the sample need not be pulverized or thinned.

INTRODUCTION

X-ray-absorption fine structure on characteristic absorption edges as well as extended fine structure can be used to help establish how a particular element is chemically and structurally combined in a given specimen. Studies of X-ray absorption structure have seldom been used as a routine tool for the characterization of unknown materials because the necessity of preparing very thin absorbing films has generally made routine study of large numbers of samples very tedious.

The purpose of this paper is to explore the feasibility of observing extended X-ray absorption fine structure directly in the emission spectrum from infinitely thick specimens, which circumvents the need to prepare thin absorbing films. It will be demonstrated that, in the case of the first-transition metal elements, one can observe the extended fine structures as undulations in the intensity of the continuous spectrum of an electron-bombarded target. The experimental conditions that affect the observations will be discussed. This study yields results that further support the recent work of Baun and Fischer[1] who find that the L and M characteristic emission spectra of many elements are strongly affected by overlapping absorption edges.

Figure 1 presents a schematic summary of the features observable in the primary-

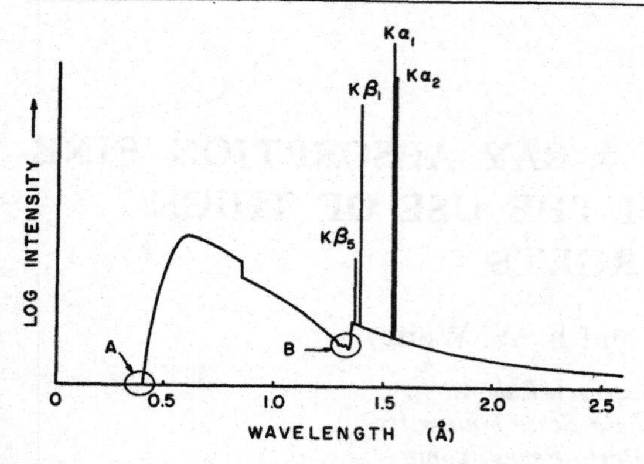

Figure 1. Schematic summary of details observable in primary-emission spectra.

emission spectrum from a copper target. For clarity, the satellite lines and L and M diagram lines have not been included. One can ascertain the following from such a spectrum:

1. The peak acceleration potential of the electron beam.
2. An identification of the target material.
3. The purity of the target.
4. Identification of the detector (in this case, the edge at 0.86 Å indicates the use of a krypton-filled counter).
5. Detector-window-material identification, usually requiring data at much longer wavelengths than shown here.
6. How elements are combined in the target, often established from the relative intensities, shapes, and wavelength positions of X-ray lines.

In addition to the above, two areas in the continuum contain fine structures that yield information on the chemistry and structure of the target. For example, under conditions of moderately high resolution ($\lambda/\Delta\lambda \cong 5000$), one can observe a fine structure in the continuum intensity at the short-wavelength cutoff (circle A in Figure 1). The *isochromat* technique is generally used to make measurements of this structure. The present study will be concerned entirely with moderately high resolution measurements of the extended X-ray absorption fine structure (EXAFS) in the vicinity of circle B in Figure 1.

INSTRUMENTAL

Details of the instrumentation used in this study have been given by White and McKinstry.[2] Modifications of the usual thin-film technique for direct studies of EXAFS in emission spectra are shown in Figure 2. The main difference between conventional EXAFS studies and this one is in the relation of the sample to the X-ray beam. In the conventional method, the sample is a thin film or foil which is alternately inserted into and withdrawn from the X-ray beam for measurement of I_0 and I at increments over the wavelength range of interest. Then, using the familiar relation,

$$I = I_0 \, e^{-\mu x} \tag{1}$$

where μ is the mass absorption coefficient and x is the thickness in centimeters, one plots the μx values as a function of wavelength or energy.

In the present method, the EXAFS is resolved as undulations in the continuum intensity of the electron-bombarded target. Although one could use demountable X-ray

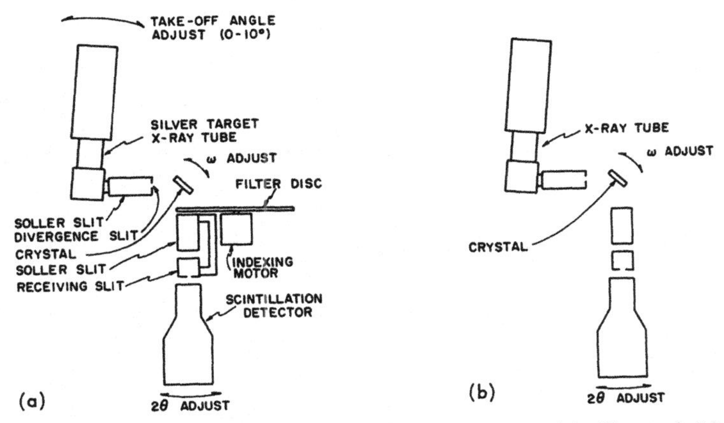

Figure 2. Comparison of instrumentation for the (a) thin-film and (b) thick-target experiments.

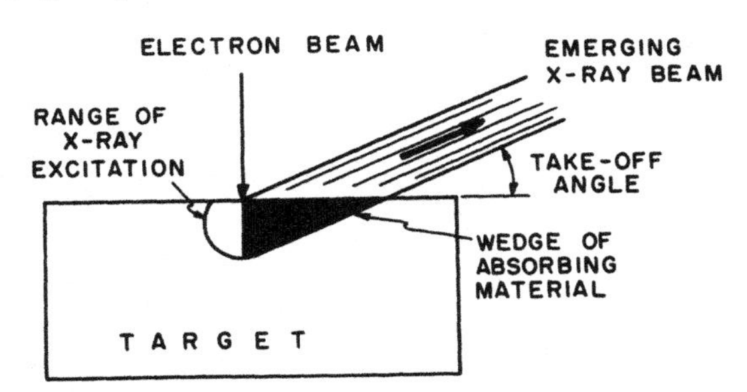

Figure 3. Absorption process for thick targets.

tubes, or electron microprobe samples for that matter, we have only studied the spectra of the Coolidge type of X-ray diffraction tubes.

Figure 3 shows schematically the source of absorption contrast for the thick-target experiment. Primary continuum X-rays generated in the specimen are absorbed to a certain extent as they emerge from the target. The amount of absorption will depend upon (1) the mass absorption coefficient of the target at a given λ, (2) the depth distribution of X-ray generation in the target, and (3) the takeoff angle which determines the absorption-path length of X-rays emerging from a given point below the surface. The depth distribution of X-ray generation will depend upon the acceleration potential of the electron beam and the electron-stopping characteristics of the target.

Spectrometer parameters were a 0.003-in. divergence slit, LiF crystal (4.0272 Å $2d$), 0.002-in. receiving slit, scintillation detector, and pulse-height analyzer. Spectra were step scanned at 0.01° 2θ increments by using a fixed time of 400 sec. Data are calculated as μx *versus* 2θ increment, where μx is an arbitrarily calculated log intensity ratio. The thick-target experiment obviously precludes measurement of a true I_0 value. Instead, an intensity value twice that at a point 250 eV beyond the edge (high-energy side of the edge) is used for "I_0."

RESULTS AND DISCUSSION

Figures 4 and 5 summarize the data obtained for iron and copper. The upper curve in each figure is for thin foils taken according to the setup shown in Figure 2(a). The copper EXAFS compares favorably with the published results of Azároff.[3] The remaining

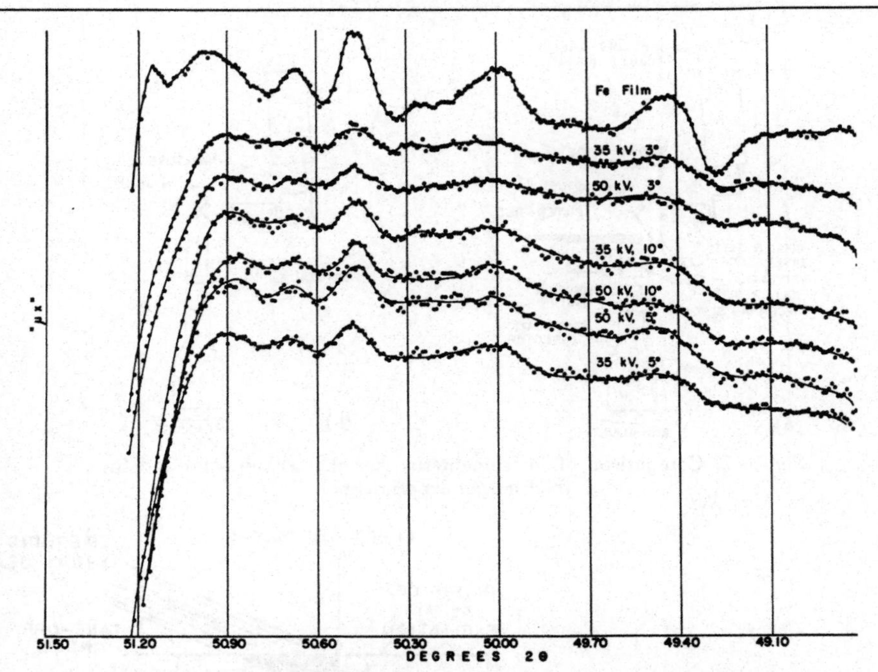

Figure 4. The Fe K-absorption spectrum for thin film and thick target.

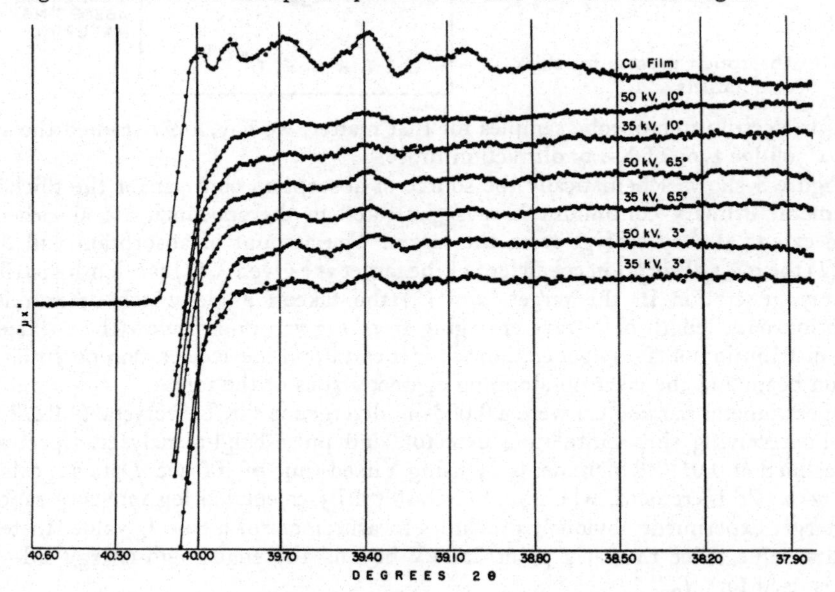

Figure 5. The Cu K-absorption spectrum for thin film and thick target.

curves in the figures are for thick-target data taken at 35 or 50 kV and takeoff angles of 3, 6.5, or 10°. Most of the features observed in the thin foils are also observed in the thick-target spectra. However, the main absorption edge is somewhat masked by the tail of the $K\beta$ emission band.

The 50-kV spectra appear to be slightly better than those for 35 kV. Contrast in the iron spectrum appears to be favored by a large takeoff angle (10°), while a low takeoff angle (3°) favors higher contrast in the copper spectrum. Masking of the main edge by the $K\beta_5$ tail has prevented an evaluation of temperature effects on the Kossel structure. The so-called thickness effect described by Parratt *et al.*[4] is probably responsible for the masking of the Kossel structure since the $K\beta_5$ band is quite intense for the elements studied.

The familiar Philibert absorption equation[5] which is used in correcting microprobe data can be extended for quantitative calculation of the magnitude of the change in continuum intensity across the self-absorption edges as well as for calculation of contrast in the EXAFS.

$$f(\chi) = \frac{1 + h}{[1 + (\chi/\sigma)]\{1 + h[1 + (\chi/\sigma)]\}} \tag{2}$$

where $\chi = \mu/\rho \csc \psi$; ψ is the takeoff angle measured from the plane of the specimen surface; $h = 1.2\bar{A}/\bar{z}^2$; and σ is the Leonard coefficient.

The $f(\chi)$ values for microprobe analysis are used in the following formula to calculate unknown compositions with reference to a pure-metal standard:

$$k_{\mathrm{A}} = C_{\mathrm{A}} \frac{f(\chi)_{\mathrm{unknown}}}{f(\chi)_{\mathrm{pure\ A}}} \tag{3}$$

where C_{A} is the weight fraction of element A in unknown and k_{A} is the measured intensity ratio $I_{\mathrm{unknown}}/I_{\mathrm{pure\ A}}$.

In the case of the self-absorption experiment, the composition is fixed; then, for a given value of ψ and kilovolts, only the μ/ρ varies so that

$$k_{\Delta\lambda} = \frac{f(\chi)\lambda_1}{f(\chi)\lambda_2} \tag{4}$$

where the intensity ratios $k_{\Delta\lambda}$ can be calculated from the ratio $f(\chi)\lambda_1/f(\chi)\lambda_2$. Inherent in this relation is the assumption that the depth distribution of X-ray continuum is the same as for the characteristic radiation of the same wavelength. Although this assumption cannot be rigorously justified, it will be used here for the sake of convenience. That equation (4) appears to hold is illustrated by two examples:

Example 1: For a copper-target X-ray tube operated at 35 kV using a 10° takeoff angle, the calculated intensity ratio on the low- (λ_1) *versus* high-energy side (λ_2) of the self-absorption edge is 2.23. This compares with a value of 2.15 found for the data used to calculate the curve in Figure 5.

Example 2: A pure titanium specimen was studied in an electron microprobe having a 52.5° takeoff angle. At 40 kV, the calculated $k_{\Delta\lambda}$ across the *K*-absorption edge was 1.79. The observed value was 1.80.

SUMMARY

It has been shown, apparently for the first time, that it is feasible to observe extended X-ray absorption fine structure directly in the emission spectra of electron-bombarded thick targets. Curves obtained from iron and copper show that, except for the Kossel structures, one can resolve all the features of the EXAFS. Theoretical justification of the observations are found in an extension of the familiar Philibert equation.[5] The new equation takes the form

$$k_{\Delta\lambda} = \frac{f(\chi)\lambda_1}{f(\chi)\lambda_2}$$

The chief implication of this study is that one should be able to use the thick-target technique to obtain EXAFS for samples which are particularly difficult to prepare as thin films.

ACKNOWLEDGMENT

This work is being sponsored by Air Force Contract F33615-67-C-1047.

REFERENCES

1. W. L. Baun and D. W. Fischer, "Sample Self-Absorption of Electron-Excited Soft X-ray Spectra," paper presented at the Second National Conference on Electron Microprobe Analysis, Boston, Mass., June, 1967.
2. E. W. White and H. A. McKinstry, "Chemical Effect on X-ray Absorption-Edge Fine Structure," in: W. M. Mueller, G. R. Mallett, and M. J. Fay (eds.), *Advances in X-Ray Analysis, Vol. 9*, Plenum Press, New York, 1966, pp. 376–392.
3. L. V. Azároff, "Theory of Extended Fine Structure of X-Ray Absorption Edges," *Rev. Mod. Phys.* **35** (4): 1012, 1963.
4. L. G. Parratt, C. F. Hempstead, and E. S. Jassem, "Thickness Effect in Absorption Spectra near Absorption Edges," *Phys. Rev.* **105**: 1228, 1957.
5. J. Philibert, "A Method for Calculating the Absorption Correction in Electron-Probe Microanalysis," in: H. H. Pattee, V. E. Cosslett, and A. Engstrom (eds.), *X-Ray Optics and X-Ray Microanalysis*, Academic Press, New York, 1963, pp. 379–392.

DISCUSSION

H. W. Pickett (General Electric Co.): How smooth would you say your target was?

E. W. White: This is one of the dangers of working with the Coolidge type of tube. You can't evaluate the target. My experience with X-ray diffraction-tube targets is that they are very rough.

H. W. Pickett: This was a standard commercial X-ray tube?

E. W. White: The copper and iron results were from commercial X-ray tubes. Because we didn't really have quite enough resolution in our electron-probe spectrometer to really see this absorption fine structure, we merely evaluated our sort of theory in terms of just measuring the intensity jump across the absorption edge. But we couldn't, of course, pick up the fine structure quite well enough.

Chairman H. S. Peiser: I wonder if you would like to say a word on the possibility of using this type of technique for the study of the fine structure itself rather than for the characterization of the target material.

E. W. White: I think, in terms of the difficulties, especially in the soft, or very long, wavelength regions of the spectrum, the preparation of uniformly thin films is quite a challenge. Since we seem to be able to observe the contrast which one would expect, it may be that this would be a technique for studying extended fine structure, but I think the best answer is always to have a good thin film and measure the fine structure in the conventional manner.

DETERMINATION OF LATTICE PARAMETERS BY THE KOSSEL AND DIVERGENT X-RAY BEAM TECHNIQUES

A. Lutts

Centre National de Recherches Métallurgiques
Liège, Belgium

ABSTRACT

The principal aim of this article is to develop in a clear and orderly manner a general relationship and show how it can be used to determine with a high degree of precision lattice parameters of tetragonal and hexagonal as well as cubic crystals. The introduction and extensive use of electron-probe microanalyzers provides a ready-made means of obtaining both Kossel and divergent X-ray beam patterns which could previously be produced only with specially constructed X-ray tubes. The present ease of their production as well as the continuing need for precise lattice parameters for the study of many problems associated with crystallized solids has stimulated a renewed interest in these two techniques. As has been recently shown by several experimental results limited to cubic crystals, these techniques are capable of giving lattice parameters with the same degree of precision as those obtained by the more classical means. The development of the general relationship is preceded by a brief historical review, a discussion of the relative merits of the methods, a short description of the nature of the diffraction patterns, and the geometrical conditions necessary for realizing precision parameter measurements. In conclusion, the advantages and disadvantages of the Kossel and divergent-beam methods compared with those of the classical powder techniques are enumerated and discussed.

INTRODUCTION

One of the most characteristic features of a given crystalline solid is the size of its unit cell. A precise value of its dimensions is of great interest in that it can yield much useful information on the physical and chemical nature of the solid state. From a theoretical point of view, such measurements are of primary importance in the study of bonding energies. Interstitial and substitutional solid solutions represent, of course, an extensive field of theoretical as well as practical interest since measuring changes in their lattice parameters is one of the most effective means of studying many of the modifications which they can undergo.

It is thus not surprising that a considerable amount of effort has been expended over the years in the development of more and more precise and accurate methods for the determination of lattice constants. Stimulated by the requirements of present-day research, this work is being actively continued. For example, the last few years has witnessed a renewed interest in the use of both the Kossel and divergent X-ray beam diffraction methods. This has been due in part to the development and extensive use of the electron-probe microanalyzer, which offers a ready-made source of the highly divergent point X-ray source necessary for the most effective use of these techniques. The most recent contributions have dealt with the possibilities of, and have reported on the results obtained by, these methods for the determination of lattice parameters as well as crystal orientation and perfection of cubic materials.

It is the purpose of the present paper to consider the possibilities of the Kossel and divergent-beam methods for the determination of lattice parameters of crystals of lower symmetry. This will be done by the detailed development of a relationship from which parameters of tetragonal and hexagonal in addition to cubic crystals can be determined.

THE KOSSEL AND DIVERGENT-BEAM TECHNIQUES

The divergent-beam X-ray method was first employed by Seemann[1] for the study of rock-salt crystals in the period of 1916 to 1919. Previously, Rutherford and Andrade[2] had studied the spectra of small γ-ray sources, using a rock-salt crystal as the analyzer.

In 1935 and the following years, Kossel and his co-workers[3,4] and Voges[5] developed the technique which bears Kossel's name. It was used by him to determine the parameter of copper[6] and later by van Bergen to measure the lattice constants of aluminum and alpha iron[7] and also copper.[8,9]

Although the divergent-beam X-ray method was employed at intervals over the years,[10,11] it was not until 1947 that its possibilities for the precise determination of lattice parameters and X-ray wavelengths were clearly revealed by Lonsdale[12] in her beautifully written paper. In addition to describing under what conditions divergent-beam patterns could best be recorded and developing a mathematical means for constructing their projections, she also reported the results of her parameter measurements of various diamond crystals and a determination of the Zn $K\alpha_1$ wavelength. This paper stimulated interest which has resulted in considerable development along directions corresponding to each of these two techniques. Imura[13] has considered theoretically in detail the geometry of the divergent-beam patterns. In addition, he redesigned the capillary X-ray tube developed in 1941 by Fujiwara and Takesita[14] and has used it to study the deformation of aluminum single crystals by means of back-reflection patterns. The precision of this method has since been considerably increased and applied for the measurement of strains in single crystals of copper and aluminum,[15] in the age-hardening alloy Al-Cu,[16] in the alloys Cu-Al and Cu-Ge,[17] and in the metal wolfram as well as for the precise measurement of the lattice parameter of the latter.[18]

The Kossel technique has attained its extensive use as a result of the development of the electron-probe microanalyzer. Although this instrument was primarily designed to study the X-ray emission spectra of small volumes, Castaing[19] clearly recognized and showed experimentally that its focused electron beam could produce an X-ray source small enough to reveal Kossel patterns of crystals as perfect as quartz. Several workers have profited from this ready-made means of producing high-quality patterns.

Both Hanneman *et al.*[20] and Heise[21] have developed their respective methods of obtaining precise lattice parameters and have demonstrated their possibilities with nickel single crystals. Potts and his co-workers[22] have measured the parameters of semiconducting materials, using Kossel's original method,[6] while Peters and Ogilvie[23] have shown the advantages of the electron-probe microanalyzer for the determination of crystallographic orientation and parameters in a two-phase Cu-Si alloy.

More recently, Gielen *et al.*[24] have developed a theory of oblique intersection and have discussed the precision to be expected in lattice parameter determinations. They have also reported the advantages which can be realized in the use of a computer to decrease the labor and avoid the uncertainty involved in the search for suitable oblique interactions as well as give the results of parameter measurements of an Fe-Si alloy.

Both these techniques employ a stationary single crystal which is irradiated by a highly divergent X-ray beam emitted from as small a source as possible. Except for several details, the diffraction patterns produced by these two methods are the same. They can be differentiated, however, by the physical origin of the X-ray source. The X-ray beam employed in the divergent-beam method is produced in a thin foil anode of a specially constructed tube which is used in the end-on position. That is, the X-ray beam passes through the thin anode, which is frequently the window of the tube. Such a geometrical arrangement has the obvious advantage that the specimen, located outside the tube, can be placed very near the X-ray source. In the Kossel technique, the specimen itself is made the anode of the X-ray tube. Thus, the divergent beam has its origin at the surface of the crystal being studied. This method is characterized by the fact that quite small, single crystals can be employed. In fact, it is frequently possible to study a grain in a relatively small-grained polycrystalline foil.[23] Despite this advantage, the Kossel technique suffers from several standpoints when compared with the divergent-beam method. First of all, generation of the X-ray beam in the specimen itself subjects the latter to a certain degree of localized heating and the establishment of temperature gradients. The resulting temperature rise can be difficult to measure or evaluate and thus becomes troublesome when highly precise lattice parameters are desired. Thermal expansion can also cause a displacement of the group—consisting of specimen and X-ray source—in relation to the film, which can result in line broadening of the pattern. In addition, the radiation generated will be characteristic of the chemical composition of the specimen. As will become evident below, the choice of the incident radiation to be employed for a given crystal is of primary importance. It rarely happens that the radiation characteristic of the atomic species comprising the specimen will be the most useful. This inconvenience can, of course, be eliminated by evaporating, electroplating, or otherwise attaching to the surface of the specimen the required target material.* Such a procedure does not, however, eliminate localized specimen heating.

Even though the divergent-beam method requires larger specimens, it offers certain important advantages in that both the specimen and the film are outside the X-ray tube. Since it is irradiated only by the divergent X-ray beam, the specimen is not subjected to localized heating. In addition, the specimen being outside the X-ray tube and in air, its temperature can be easily measured and maintained at a known and constant value by circulating a thermostatically controlled fluid. Changing the temperature of a given specimen between successive patterns also makes it possible to establish with great precision the coefficient of thermal expansion in the temperature range considered. Since both the specimen and the film are outside the X-ray tube, they can be easily oriented to each other. This is quite useful as it greatly simplifies the recording of the restricted part of the divergent-beam pattern which is of interest for parameter determination. Such an operation is more difficult when using the Kossel technique since the specimen and frequently the film itself are inside the tube. The most important advantage of the divergent-beam technique is, without doubt, that the operator has complete control of the characteristic radiation to be employed.

THE NATURE OF PATTERNS AND GEOMETRICAL CONDITIONS NECESSARY FOR PRECISE PARAMETER DETERMINATIONS

Although it is not the purpose of the present article to describe in detail all the aspects involved in the formation of these patterns,† a brief description using Figure 1 will be useful in what will follow.

* Such a solution is required in the case of nonconducting crystals.

† The interested reader is referred to the text by James[25] as well as to the papers of Lonsdale[12] and Imura.[13]

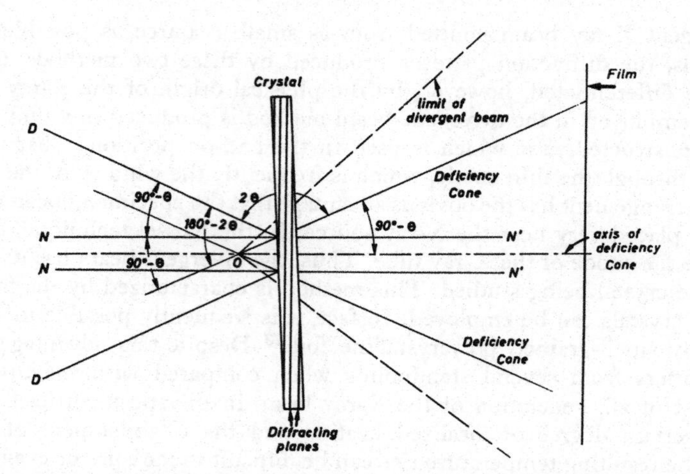

Figure 1. Schematic representation of the formation of divergent X-ray beam patterns. The thickness of the single-crystal specimen is exaggerated.

Suppose, as shown schematically, a certain family of planes is so oriented that it receives a part of the highly divergent X-ray beam emitted by the source 0, at an angle satisfying the Bragg diffraction relationship. Here, we need not apply the dynamical theory developed by von Laue[26] since it is the angular position of the conic sections and not the fine structure or intensity distribution which is of interest. In fact, in order to render the geometrical considerations clearer, we are considering the divergent-beam technique in which the X-ray beam is produced outside the specimen. Except for a few modifications, the following is true for the Kossel method in which the X-rays are generated at the specimen surface.

In the small angular region about the true Bragg angle, whose value is determined essentially by the perfection of the crystal and the size of the X-ray source, the planes in question diffract the incident characteristic component into the back-reflection region along the direction D. This diffracted beam will make an angle of $180° - 2\theta$ with the incident part of the divergent beam and an angle of $90° - \theta$ with the plane normal $N - N'$. Such diffraction removes nearly all of the X-ray energy from that transmitted through the crystal in the corresponding angular region. All other portions of the incident beam not satisfying the Bragg relationship will not be diffracted by this family of planes. They will pass through the specimen and emerge from the far (right) side with a decrease in intensity which will be a function of the normal absorption factors: the chemical nature of the specimen, its effective thickness, and the wavelength distribution of the incident beam. If the lateral dimensions of the crystal and the angular divergence of the X-ray beam are sufficiently large and the planes properly oriented, the diffraction effects can take place in the back-reflection region through an angle of 360° about the plane normal. A flat film placed on the transmission side perpendicularly to the plane normal will exhibit a slightly exposed or deficiency circle on a more intensely exposed background. The circle is, in fact, the intersection with the film of the deficiency cone of semiapex angle $90° - \theta$ whose axis coincides with the normal of the diffracting planes. A flat film placed at the left of the crystal (back-reflection region) will exhibit the inverse effect, a dark line on a less intensely exposed background.

The crystal is, of course, composed of many families of planes so that, depending upon its orientation, the size of its unit cell, the divergence of the incident beam, and the characteristic radiation employed, the actual pattern will be composed of many intersecting lines or conic sections.

Imura[15] has studied in detail the form of these conic sections when the patterns are recorded on flat films placed in both the transmission and back-reflection regions.

In the former, where nearly all* are of the light or deficiency type, these curves are circles, ellipses, parabolas, or hyperbolas depending upon the orientation of the planes with respect to the film. The dark curves on back-reflection patterns are related to the ellipse but are of higher orders.

As Lonsdale has shown, if the approximate lattice parameter and the nature of the crystal are known, it is possible to construct the stereographic projection of the transmission pattern to be expected for a given crystal orientation and a given characteristic radiation. In the case of cubic crystals, the plane of projection is usually the cube plane, while, for hexagonal crystals, it is frequently the basal plane.

Such a projection for the cube plane of alpha-iron (body-centered-cubic) for both Fe $K\alpha$ and Mn $K\alpha$ radiations is shown in Figure 2(a) and (b) respectively. Notice the fourfold symmetry as well as the change in angular position of a given plane as the characteristic radiation wavelength is changed. Stereographic projections of diamond and ice for copper and zinc radiations have been given by Lonsdale and of nickel for nickel radiation by Hanneman *et al.*

It can be shown that certain deficiency conic sections of cubic crystals intersect in the same manner independently of both the wavelength and the unit-cell dimension. Others, on the contrary, as can be seen by comparing Figure 2(a) and (b), are displaced relative to each other for a given lattice constant when the wavelength is changed or for a given wavelength when the lattice parameter is changed.

As shown by Lonsdale, a very sensitive measure of the lattice parameter can be obtained from the oblique intersection of two conic sections of this second type. The

* Dark lines diffracted from planes nearly parallel to the cone axis of the divergent beam are sometimes present on transmission patterns.

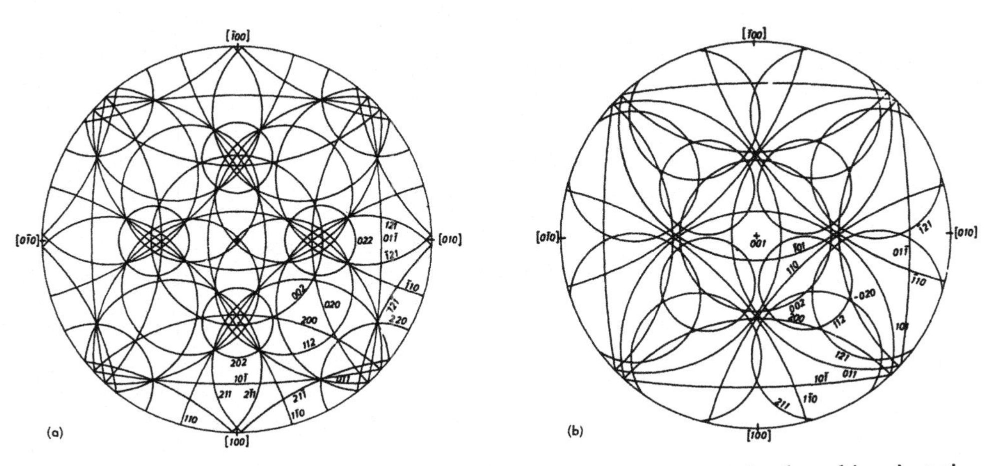

Figure 2. Stereographic projections showing the divergent-beam patterns for the cubic orientation of alpha-iron ($a = 2.866$ Å) produced by the diffraction of (a) Fe $K\alpha$ unresolved doublet ($\lambda = 1.936$ Å) and (b) Mn $K\alpha$ unresolved doublet ($\lambda = 2.102$ Å).

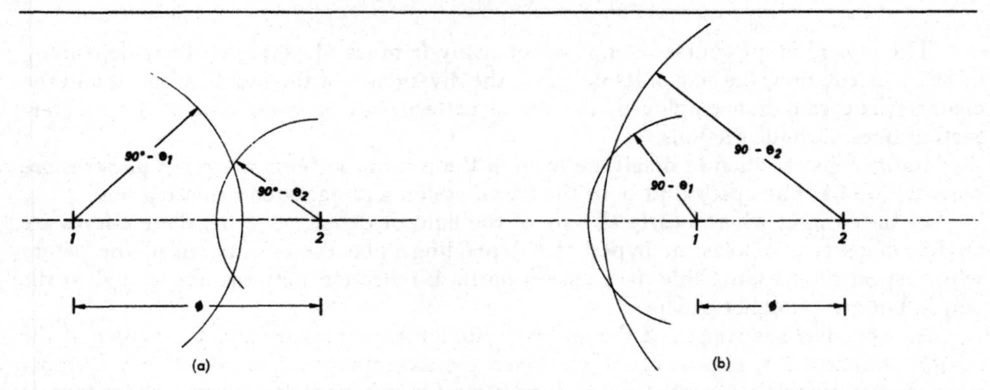

Figure 3. Schematic representation of the production of two types of oblique intersections; unresolved $K\alpha$ doublet.

displacement of such an intersection with a change in the lattice parameter at constant wavelength is much more pronounced than the changes in diameter of the corresponding conic sections. In her treatment, she measured the displacement of such an intersection with respect to another much less sensitive to the parameter change. Present-day workers, as will be shown below, measure the chord length of the overlap which takes into account the displacement two times.

One type of oblique intersection is shown in Figure 3(a). In this case, it is necessary that the sum of the semiapex angles $(90° - \theta_1) + (90° - \theta_2)$ is slightly larger than ϕ, the angle between the two planes. A second possibility frequently encountered is produced when the poles of the planes are on the same side of the intersection of their conic sections. As shown in Figure 3(b), intersection will take place when $(90° - \theta_1) + \phi > (90° - \theta_2)$. Of these two possibilities, that shown in Figure 3(a) will usually be the most sensitive since the change in *radii* of the two conic sections are in opposite directions. It must be stressed that these diagrams show the angular relationships involved.

For simplicity, the intersections shown in Figure 3 have taken into account the unresolved $K\alpha$ doublet. Precise lattice parameter determinations require that this doublet be resolved since the individual wavelengths are known with greater precision than for the unresolved doublet. The angular separation between the doublet components becomes increasingly greater as the interplanar distance decreases.

Since $\lambda K\alpha_2 > \lambda K\alpha_1$, the latter will have the largest semiapex angle and will lie outside the $K\alpha_2$ component on transmission patterns. Another advantage gained in resolving the doublet is that it is possible, as pointed out by Heise,[21] to determine lattice parameters without knowing nor measuring the distance between the X-ray source and the film. In other words, all the information required—except the specimen temperature and the X-ray refraction correction—can be obtained directly from measurements on the film.

In the case of alpha-iron with iron and manganese radiations* [Figure 2(a) and (b)], no such low-angle or oblique intersections appear to be present. One must remember, however, that the $K\beta$ lines are also recorded except when the specimen has an absorption edge at slightly larger wavelengths, as is the case for alpha-iron with cobalt radiation. The presence of the β component thus increases the chance for such an oblique intersection. This condition can also be favored by using a shorter characteristic wavelength

* The stereographic projections were constructed by assuming a reasonable value of the lattice parameter and the unresolved doublet.

to increase the number of diffracting planes. The most effective means is, without doubt, the use of an alloy anode of the proper chemical composition.

As might be expected, the construction of a different stereographic projection for each possible anode material is a laborious and time-consuming task. Gielen *et al.*[24] have shown, however, that the time and effort involved in the search for these intersections can be greatly shortened by using a computer program. Once the proper experimental conditions have been established, it is still useful to construct at least that portion of the stereographic projection which is of interest. This is especially true when dealing with patterns rich in lines, such as those encountered for crystals with large unit cells, or when short wavelengths or alloy anodes are used.

DEVELOPMENT OF THE RELATIONSHIP FOR PARAMETER DETERMINATIONS

We shall now develop in detail a relationship and show how it can be applied for the determination of the lattice parameters of cubic, tetragonal, and hexagonal crystals.

Let us consider the intersection of the type shown in Figure 3(a) produced by two planes $(h_1k_1l_1)$ and $(h_2k_2l_2)$ whose interplanar spacings are $d(h_1k_1l_1)$ and $d(h_2k_2l_2)$, respectively. These will be referred to hereafter as planes 1 and 2. In general, $d_1 \neq d_2$. Let us select the planes such that $d_2 < d_1$. As indicated in Figure 4, which shows the angular relationships involved only, these planes and their respective normals are separated by the angle ϕ. The lines 0–1 and 0–2 are the axes of the conic sections (deficiency lines) and the normals to the planes 1 and 2, respectively. Thus, the reference planes I and II, intersecting at the angle ϕ, are parallel to the crystal planes 1 and 2. The line 0–C, lying in the plane formed by the normals of the two crystallographic planes, will intersect at right angles

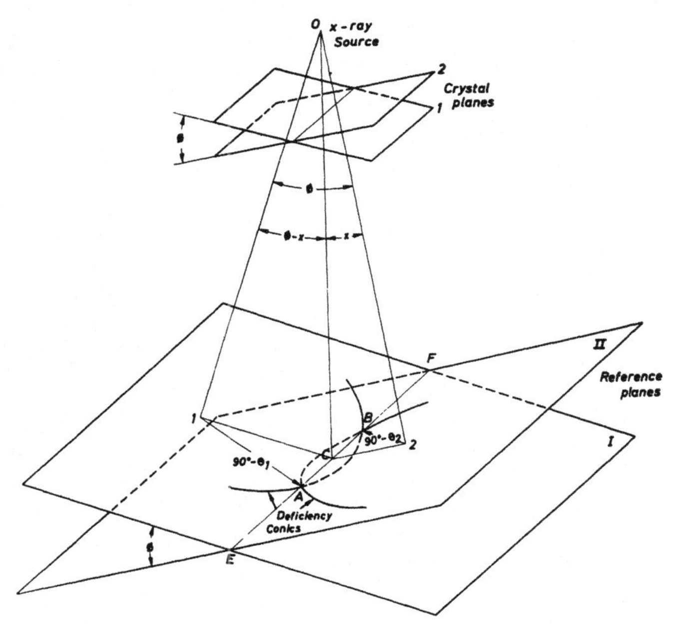

Figure 4. Schematic representation in perspective showing the angular relationship involved in the production and intersection of two deficiency conic sections; unresolved $K\alpha$ doublet.

the line EF, formed by the intersection of planes I and II, midway between the intersections A and B of the deficiency conic sections. The line 0–C thus divides the angle ϕ into two parts $\phi - x$ and x.

In the two right spherical triangles $1AC$ and $2AC$, letting $AC = b$, it can be shown that

$$\cos b = \frac{\cos(90° - \theta_1)}{\cos(\phi - x)} = \frac{\cos(90° - \theta_2)}{\cos x} \tag{1}$$

Using the Bragg relationship and introducing the interplanar distances d_1 and d_2, we obtain

$$\frac{\cos(\phi - x)}{\cos x} = \frac{d_2}{d_1} \tag{2}$$

The angle x is thus independent of the wavelength when both the deficiency conics are produced by the same characteristic radiation. For this reason, both the $K\alpha_1$ and $K\alpha_2$ intersections can be shown to lie on the line EF of Figure 4. In the special case when $d_1 = d_2$,

$$\frac{\cos(\phi - x)}{\cos x} = 1 \tag{2a}$$

and

$$x = \phi/2 \tag{2b}$$

Equation (2) can be rewritten as

$$\tan x = \frac{1}{\sin\phi}\left(\frac{d_2}{d_1} - \cos\phi\right) \tag{3}$$

from which it can be shown that

$$\cos x = \frac{d_1 \sin\phi}{\sqrt{(d_2 - d_1\cos\phi)^2 + d_1{}^2\sin^2\phi}} \tag{4}$$

Introducing the above result into the right side of equation (1), we obtain, for the $K\alpha_1$ component,

$$\cos b\alpha_1 = \frac{\lambda K\alpha_1\sqrt{(d_2 - d_1\cos\phi)^2 + d_1{}^2\sin^2\phi}}{2d_1d_2\sin\phi} \tag{5}$$

This equation, upon further rearrangement, becomes

$$\tan b\alpha_1 = \frac{\sqrt{4d_1{}^2d_2{}^2\sin^2\phi - \lambda\alpha_1{}^2[(d_2 - d_1\cos\phi)^2 + d_1{}^2\sin^2\phi]}}{\lambda K\alpha_1\sqrt{(d_2 - d_1\cos\phi)^2 + d_1{}^2\sin^2\phi}} \tag{6}$$

A similar relationship, equation (6a), can be obtained for the $K\alpha_2$ component.

As shown schematically in Figure 5, let a line from the X-ray source 0 intersect the flat film at right angles at the point C; the distance 0–C being D. *The distances measured on the flat film* between the $K\alpha_1$ and between the $K\alpha_2$ intersections of the deficiency lines produced by planes 1 and 2 are, as indicated, $l\alpha_1$ and $l\alpha_2$, respectively.

The corresponding semiapex angles are $b\alpha_1$ and $b\alpha_2$.

Thus,

$$\tan b\alpha_1 = \frac{l\alpha_1}{2D} \tag{7}$$

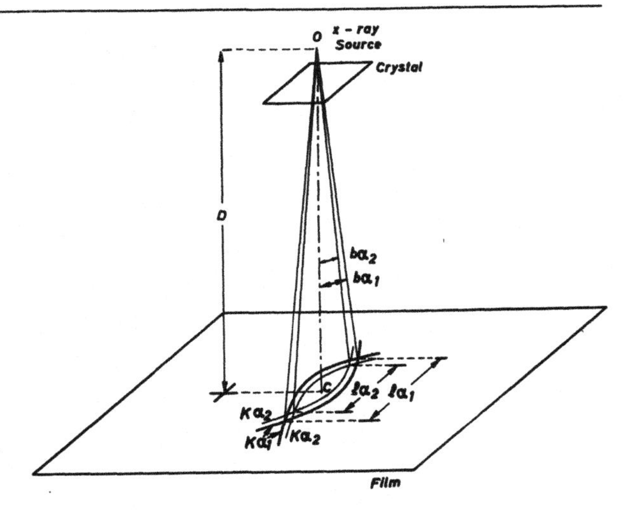

Figure 5. Schematic representation in perspective showing linear and angular relationships needed for lattice parameter determinations in the Heise or symmetrical method; resolved $K\alpha$ doublet.

and

$$\tan b\alpha_2 = \frac{l\alpha_2}{2D} \tag{8}$$

from which we can obtain

$$\frac{l\alpha_1}{l\alpha_2} = \frac{\tan b\alpha_1}{\tan b\alpha_2} \tag{9}$$

Replacing the right side of equation (9) by the values obtained in equations (6) and (6a) and squaring both sides of the resulting relationship, it can be shown that

$$\frac{l^2\alpha_1}{l^2\alpha_2} = \frac{\lambda^2\alpha_2\{4d_1{}^2d_2{}^2\sin^2\phi - \lambda^2\alpha_1[(d_2 - d_1\cos\phi)^2 + d_1{}^2\sin^2\phi]\}}{\lambda^2\alpha_1\{4d_1{}^2d_2{}^2\sin^2\phi - \lambda^2\alpha_2[(d_2 - d_1\cos\phi)^2 + d_1{}^2\sin^2\phi]\}} \tag{10}$$

Equation (10) is the relationship from which the lattice parameters can be obtained by using the method proposed by Heise.[21] In order that this equation can be used in its above form, it is necessary that the line 0–C intersect the film at right angles midway along the line passing through the oblique intersections. As was mentioned previously, this operation is relatively easy to perform with the divergent-beam method, in which both the specimen and film are outside the X-ray tube.

The above equation is, in its present form, also limited to the case where the deficiency lines from both planes are produced by diffraction of $K\alpha_1$ and $K\alpha_2$ radiations from the same target element. If an alloy anode is used and the intersections from two different sets of doublets are measured, this equation might have to be modified.

From equation (2), we can see that the angle x will now depend upon the ratio of the wavelengths employed and, thus, the value of D in equations (7) and (8) will not be the same. There will, in fact, be two distances of source to film, D and D' where $D' > D$. An eventual correction will depend, of course, upon the difference between D and D' and upon the desired precision of the parameter determination.

The parameter obtained from equation (10) is also influenced by two other factors, the temperature of the specimen and the refraction of X-rays in the crystal. Since the lattice parameter depends upon the temperature of the crystal, precise and comparative measurements require that the temperature be maintained constant and that, in addition, it is accurately known during the exposure of the film. As pointed out by Straumanis,[27] the measurement of the parameter of aluminum with a precision of ± 0.00001 Å requires that the specimen temperature be kept constant to at least within ± 0.1°C or less. As previously mentioned, temperature measurement and control is much more easily performed in the divergent-beam method since the specimen is outside the X-ray tube and irradiated only by the primary divergent X-ray beam.

The X-ray wavelengths used in equation (10) are those measured in vacuum or in air. The effective wavelength in the crystal is increased by a very small amount, which, however, becomes measurable when precise lattice parameters are desired. In the Heise or symmetrical case considered here, the deficiency conic sections shift in a direction corresponding to a decrease in their radii. Depending upon the nature of the crystal and the wavelength employed, the measured parameter can be as much as a few parts in 10^4 or 10^5 too low. Lonsdale[12] has discussed the refraction correction to be applied to diamond when using copper radiation.

Cubic Crystals

In order to simplify the development which follows, let us rewrite equation (10) as follows:

$$\frac{l^2\alpha_1}{l^2\alpha_2} = \frac{\lambda^2\alpha_2\,(A - \lambda^2\alpha_1 B)}{\lambda^2\alpha_1\,(A - \lambda^2\alpha_2 B)} \tag{11}$$

in which

$$A = 4d_1{}^2d_2{}^2(1 - \cos^2\phi) \tag{12}$$

and

$$B = d_2{}^2 - 2d_1d_2\cos\phi + d_1{}^2 \tag{13}$$

For cubic crystals,

$$d(hkl) = \frac{a}{\sqrt{h^2 + k^2 + l^2}} \tag{14}$$

and

$$\cos\phi = \frac{h_1h_2 + k_1k_2 + l_1l_2}{\sqrt{h_1{}^2 + k_1{}^2 + l_1{}^2}\,\sqrt{h_2{}^2 + k_2{}^2 + l_2{}^2}} \tag{15}$$

Letting

$$\beta_1 = \sqrt{h_1{}^2 + k_1{}^2 + l_1{}^2} \tag{16}$$

$$\beta_2 = \sqrt{h_2{}^2 + k_2{}^2 + l_2{}^2} \tag{17}$$

and

$$\gamma = h_1h_2 + k_1k_2 + l_1l_2 \tag{18}$$

equations (12) and (13) become, respectively,

$$A = \frac{4a^4(\beta_1{}^2\beta_2{}^2 - \gamma^2)}{\beta_1{}^4\beta_2{}^4} \tag{19}$$

and

$$B = \frac{a^2(\beta_1{}^2 - 2\gamma + \beta_2{}^2)}{\beta_1{}^2\beta_2{}^2} \tag{20}$$

Introducing the results of (19) and (20) into the numerator of the right side of equation (11), we obtain

$$A - \lambda^2\alpha_1 B = \frac{4a^4(\beta_1{}^2\beta_2{}^2 - \gamma^2)}{\beta_1{}^4\beta_2{}^4} - \frac{\lambda^2\alpha_1 a^2(\beta_1{}^2 - 2\gamma + \beta_2{}^2)}{\beta_1{}^2\beta_2{}^2} \tag{21}$$

A similar relationship, equation (21a), exists for the corresponding denominator in (11). Equation (10) can thus be rewritten as follows for cubic crystals:

$$\frac{l^2\alpha_1}{l^2\alpha_2} = \frac{\lambda^2\alpha_2\{4a^2(\beta_1{}^2\beta_2{}^2 - \gamma^2) - \lambda^2\alpha_1[\beta_1{}^2\beta_2{}^2(\beta_1{}^2 - 2\gamma + \beta_2{}^2)]\}}{\lambda^2\alpha_1\{4a^2(\beta_1{}^2\beta_2{}^2 - \gamma^2) - \lambda^2\alpha_2[\beta_1{}^2\beta_2{}^2(\beta_1{}^2 - 2\gamma + \beta_2{}^2)]\}} \tag{22}$$

By referring to equations (16), (17), and (18), it can be seen that all terms on the right side of equation (22) are known except, of course, the lattice parameter a. This relationship is the generalized version of that shown in equation (7) by Heise.[21*] As proposed by this author, it is best to solve graphically for a. Owing to difficulties involved in drawing such curves, a certain error will be involved in such a solution. For this reason, it would be best to use the graphically determined lattice parameter as the starting value from which the more precise parameter can be obtained with the aid of equation (22) by means of successive approximations.

Tetragonal Crystals

For tetragonal crystals,

$$d(hkl) = \frac{ac}{\sqrt{c^2(h^2 + k^2) + a^2l^2}} \tag{23}$$

and

$$\cos\phi = \frac{d_1d_2[c^2(h_1h_2 + k_1k_2) + a^2l_1l_2]}{a^2c^2} \tag{24}$$

As before, let us simplify by writing

$$\epsilon_1 = h_1{}^2 + k_1{}^2 \tag{25}$$

$$\epsilon_2 = h_2{}^2 + k_2{}^2 \tag{26}$$

and

$$\eta = h_1h_2 + k_1k_2 \tag{27}$$

With the aid of these relationships, it can be shown that equations (12) and (13) become, respectively,

$$A = \frac{4a^4c^4[(c^2\epsilon_1 + a^2l_1{}^2)(c^2\epsilon_2 + a^2l_2{}^2) - (c^2\eta + a^2l_1l_2)^2]}{(c^2\epsilon_1 + a^2l_1{}^2)^2(c^2\epsilon_2 + a^2l_2{}^2)^2} \tag{28}$$

and

$$B = \frac{a^2c^2[c^2(\epsilon_1 + \epsilon_2 - 2\eta) + a^2(l_1{}^2 + l_2{}^2 - 2l_1l_2)]}{(c^2\epsilon_1 + a^2l_1{}^2)(c^2\epsilon_2 + a^2l_2{}^2)} \tag{29}$$

As can be seen, these two equations contain both the parameters a and c. It is possible to determine first the parameter a by selecting the proper (h_1k_10) and (h_2k_20) intersections. All terms containing l_1 and l_2 are then zero, while the parameter c cancels out on the right side of equation (10).

* This author gave as an example the determination of the lattice parameter of a nickel crystal with Ni $K\alpha$ radiation and the (004) to (222) intersections.

Once the parameter a is known, c can be determined by selecting a new pair of intersections of the type $(h'_1k'_1l'_1)$ and $(h'_2k'_2l'_2)$ in which l'_1 and l'_2 are not both zero. It may, of course, be necessary to change the radiation employed.

Hexagonal Crystals

For hexagonal crystals,

$$d(hkl) = \frac{\sqrt{3}ac}{\sqrt{4c^2(h^2 + hk + k^2) + 3a^2l^2}} \tag{30}$$

and

$$\cos\phi = \frac{d_1d_2\{4c^2[h_1h_2 + k_1k_2 + \frac{1}{2}(h_1k_2 + k_1h_2)] + 3a^2l_1l_2\}}{3a^2c^2} \tag{31}$$

Again simplifying by letting

$$\rho_1 = h_1^2 + h_1k_1 + k_1^2 \tag{32}$$

$$\rho_2 = h_2^2 + h_2k_2 + k_2^2 \tag{33}$$

and

$$\sigma = h_1h_2 + k_1k_2 + \tfrac{1}{2}(h_1k_2 + k_1h_2) \tag{34}$$

Equations (12) and (13) now become

$$A = \frac{36a^4c^4[(4c^2\rho_1 + 3a^2l_1^2)(4c^2\rho_2 + 3a^2l_2^2) - (4c^2\sigma + 3a^2l_1l_2)^2]}{(4c^2\rho_1 + 3a^2l_1^2)^2(4c^2\rho_2 + 3a^2l_2^2)^2} \tag{35}$$

and

$$B = \frac{3a^2c^2[4c^2(\rho_1 + \rho_2 - 2\sigma) + 3a^2(l_1^2 + l_2^2 - 2l_1l_2)]}{(4c^2\rho_1 + 3a^2l_1^2)(4c^2\rho_2 + 3a^2l_2^2)} \tag{36}$$

The lattice parameters a and c of the hexagonal crystals can be determined with the aid of equations (35), (36), and (10) in a manner similar to that for tetragonal crystals described above.

CONCLUDING REMARKS

It thus appears that both the Kossel and divergent-beam X-ray methods can be employed for the precise determination of lattice parameters of tetragonal and hexagonal as well as cubic crystals.

What then are their major advantages and disadvantages compared with the powder method? The latter, whether the classical film techniques or the more modern and sophisticated diffractometers are used, has two important advantages: a nearly universal application due to many years of development, and the use of a readily available or easily prepared specimen. Simultaneously, however, the powder techniques suffer from two principal inconveniences: the need for more or less complicated and, therefore, expensive cameras or diffractometers and associated electronic circuits and the necessity to correct for several sources of experimental errors. The latter are now well known and can be effectively reduced or eliminated to the point where it is possible, in certain cases, to obtain a precision of better than 1 part in 10^5. Such corrections are usually performed by a graphical or analytical means which require a series of measurements on a film or with a diffractometer over an extensive angular range.

As several recent results have shown, both the Kossel and divergent X-ray beam methods are also capable of attaining the same degree of precision. These methods require, as we have already mentioned, single crystals or large grains of a certain size. Unfortunately, such specimens are not always as common as are polycrystals or powder. Once the specimens and a suitable X-ray source are available, however, these techniques offer the decided advantage of requiring only relatively simple and inexpensive auxiliary equipment: a thermostatically controlled bath, thermally insulated and orientable specimen support, and a flat film holder. Because both the specimen and film are immobile in relation to the X-ray source, such complications as arise from strictly controlled specimen movement in powder cameras and 2:1 detector–specimen rotation in diffractometry are totally eliminated.

Provided that a sufficiently sensitive oblique intersection is used, the precise determination of a lattice parameter can be obtained from only two length measurements in a quite limited region of the entire diffraction pattern. Except for specimen temperature and refraction effects, no other corrections are required.

The precise determination of lattice parameters is at best a detailed and more or less time-consuming operation. In a certain number of cases, the parameter changes between successive experiments are small; thus, the need for precision measurements. Good examples are the influence on the unit-cell dimensions of interstitial solid solutions with solute–atom concentration and the study of radiation damage and its recovery. For this reason, the initial choice of operating conditions—characteristic radiation to be employed, *etc.*—will probably be valid for a given series of experiments and will represent the major part of the preliminary work. Thus, the effort involved in determining the best combination of radiation and oblique intersections required by both the Kossel and divergent-beam methods might well, in light of their advantages, be justified. Such a possibility appears to be even more promising now that much of the time and effort involved can be greatly reduced by computers.

REFERENCES

1. H. Seemann, *Z. Physik* **20**: 169, 1919.
2. E. Rutherford and E. N. da C. Andrade, *Phil. Mag.* **28**: 263, 1914.
3. W. Kossel, V. Loeck, and H. Voges, *Z. Physik* **94**: 139, 1935.
4. W. Kossel and H. Voges, *Ann. Physik* **23**: 677, 1935.
5. H. Voges, *Ann. Physik* **27**: 694, 1936.
6. W. Kossel, *Ann. Physik* **26**: 533, 1936.
7. H. van Bergen, *Ann. Physik* **39**: 533, 1941.
8. H. van Bergen, *Naturwiss.* **25**: 415, 1937.
9. H. van Bergen, *Ann. Physik* **32**: 737, 1938.
10. W. Gerlach, *Z. Physik* **12**: 557, 1927; *Verk. Phys. Med. Ges. Wurzburg* **56**: 55, 1927.
11. B. Hess, *Z. Krist.* **97**: 197, 1937, and **104**: 294, 1942.
12. K. Lonsdale, "Divergent-Beam X-Ray Photography of Crystals," *Phil. Trans. Roy. Soc. London* **A240**: 219, 1947.
13. T. Imura, "The Study of Deformation of Single Crystals by the Divergent X-Ray Beams," *Bull. Naniwa Univ.* **A2**: 51, 1954.
14. T. Fujiwara and I. Takesita, *J. Hiroshima Univ.* **11**: 93, 1942.
15. T. Imura, "A Study of the Deformation of Single Crystals by Divergent X-Ray Beams (Part III)," *Bull. Univ. Osaka Prefect.* **A5**: 99, 1957.
16. T. Imura, S. Weissmann, and J. Slade, Jr., "A Study of Age-Hardening of Al–3.85% Cu by the Divergent X-Ray Beam Method," *Acta Cryst.* **15**: 786, 1962.
17. J. Slade, Jr., S. Weissmann, K. Nakajima, and M. Hirabayashi, "The Effect of Low-Temperature Anneal in Copper-Base Alloys," *Rept.* DA-ARO(D)-31, 164, G, 300 (Dept. Army), 1963 (unpublished).

18. T. Ellis, L. Nanni, A. Shirier, S. Weissmann, and N. Hosokawa, "Strain and Precision Lattice Parameter Measurements by the X-Ray Divergent Beam Method," *J. Appl. Phys.* **35**: 3364, 1964.
19. R. Castaing, "Application des sondes électroniques à une méthode d'analyse ponctuelle chimique et cristallographique," Thesis, University of Paris, 1957.
20. R. Hanneman, R. Ogilvie, and A. Modrzejewski, "Kossel Line Studies of Irradiated Nickel Crystals," *J. Appl. Phys.* **32**: 1429, 1962.
21. B. Heise, "Precision Determination of the Lattice Constant by the Kossel Line Technique," *J. Appl. Phys.* **32**: 938, 1962.
22. H. Potts, G. Pearson, and V. Macres, "Precise Measurements of Semiconductor Single-Crystal Lattice Parameters by the Kossel Line Technique," *Bull. Am. Phys. Soc.* **28**: 593, 1963.
23. E. Peters and R. Ogilvie, "X-Ray Orientation and Diffraction Studies by Kossel Lines," *Trans. AIME* **232**: 89, 1965.
24. P. Gielen, H. Yakowitz, D. Ganow, and R. Ogilvie, "Evaluation of Kossel Microdiffraction Procedures: The Cubic Case," *J. Appl. Phys.* **36**: 773, 1965.
25. R. James, "The Optical Principles of the Diffraction of X-Rays," in: L. Bragg (ed.), *The Crystalline State*, G. Bell & Sons, Ltd., London, 1950, p. 448.
26. M. von Laue, "Die Fluoreszenzröntgenstrahlung von Einkristallen," *Ann. Physik* **23**: 705, 1935.
27. M. Straumanis, "Parameters of Crystal Lattices: II. High Precision Measurements by the Asymmetric Diffraction Method," in: G. L. Clark (ed.), *The Encyclopedia of X-Rays and Gamma Rays*, Reinhold Publishing Corp., New York, 1963, p. 705.

DISCUSSION

H. W. King (Imperial College of London): How long does it take you to set up your experiment if you have all your apparatus ready?

A. Lutts: Unfortunately, I cannot answer that because this is a preliminary study of the method. We have not done any experimental work.

H. W. King: You say that there is a lot of time involved.

A. Lutts: In establishing the correct operating conditions, yes. Determining which planes give the required oblique intersections. Once these operating conditions are known, however, and the pattern obtained, it is necessary to measure only two lengths on the film. The exposure time required could vary from a few minutes on. In the case of the Kossel technique, exposure times are usually shorter than for the divergent-beam method.

H. W. King: With a powder determination, you can do, say, a complete diffraction pattern in about 1 hr and feed the results into a computer, so I think you are slightly overstating the time involved in conventional diffraction techniques. I think you have a case, but you are overstating it by trying to run down the other side.

A. Lutts: Yes. Eventually. Right now they are feeding the results from a powder method into computers. That is true.

H. W. King: I think both techniques are obviously improved by having a computer anyway.

A. Lutts: That is true.

FULLY AUTOMATED HIGH-PRECISION X-RAY DIFFRACTION

T. W. Baker, J. D. George, B. A. Bellamy, and R. Causer

Atomic Energy Research Establishment
Harwell, Didcot, Berkshire, England

ABSTRACT

X-ray diffraction angles are measured precisely, conveniently, and automatically by a specially designed instrument, the automatic precision X-ray goniometer connected on line to an Elliott 903B data processor. A monitor program controls two such instruments and two diffractometers simultaneously and allows a comprehensive set of experiments to be performed. The sensitivity is such that, when translated into terms of changes in the crystal-lattice parameter, a precision of 1 part in 10,000,000 is being attained, and indications are that absolute measurements are almost as good. The procedures, instrument, and automatic control are described, and the results of performance tests and some applications are given.

INTRODUCTION

The precision measurement of X-ray diffraction angles has applications in the measurement of crystal lattice parameters and changes in these with irradiation, temperature, pressure, *etc.*; the study of crystal imperfections; and the measurement of X-ray wavelengths and their distribution in the emitted beam.

This paper describes how X-ray diffraction angles are being measured precisely, quickly, and conveniently by using certain procedures which are embodied in a specially designed instrument, the automatic precision X-ray goniometer, and how this instrument's usefulness has been increased by its on-line connection to a data processor.

This experimental facility has been developed because of the continually increasing demand at Harwell for very high precision X-ray measurements, particularly in support of various programs of work concerning investigations into the defect structure of neutron-irradiated materials. Its use, as reported to this conference, has been largely influenced by such considerations since that is where most of the effort has been directed to date, although it has much wider applications than this.

Some aspects of this work have already been reported in a highly summarized form.[1]

THE PROBLEM

Figure 1 shows the top part of a diffraction peak. The problem is to describe and locate this profile with sufficient precision, and this entails the measurement of both intensity and angle.* The measurements of these two quantities are treated separately. Special treatment has also to be given to systematic errors that occur in the measurement of diffraction angles.

*In this work, angles are measured in millionths of a revolution for reasons of technical convenience, and this unit is dubbed the *micrev*, an abbreviation for microrevolution. Thus, 1 micrev (μR) = 10^{-6} revolution = 1.296″.

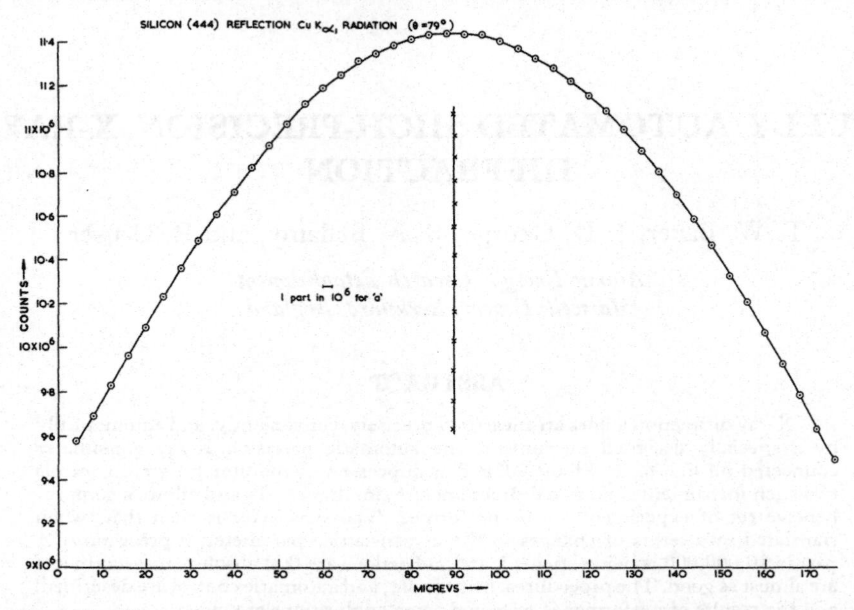

Figure 1. The top part of a diffraction profile showing midchord peak construction.

THE MEASUREMENT OF INTENSITY

The measurement of intensity requires that sufficient counts should be accumulated to ensure satisfactory statistics and that the X-ray generating and counting systems should be sufficiently stable. The generator used with this instrument is a Philips PW1010 connected to the mains by an Advance Volstat constant-voltage transformer. The counting system consists of a 20th Century Electronics PX28e/Xe xenon-filled proportional counter connected to the Harwell 2000 Series electronics. Tests were carried out to assess the stability of the equipment by setting the instrument on a diffraction peak and counting for a long period. These showed that, provided variations in atmospheric conditions are

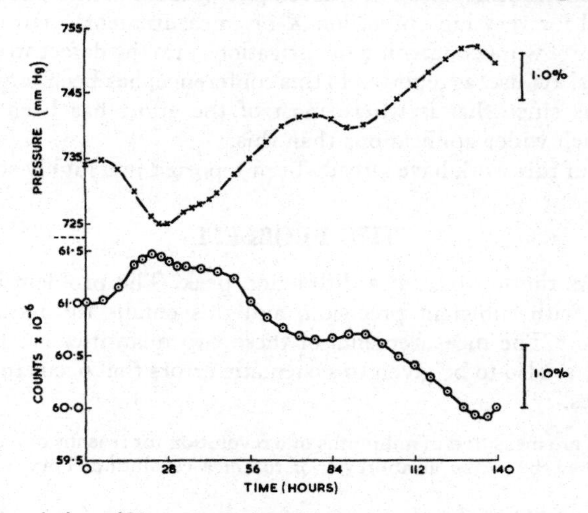

Figure 2. The variation of X-ray intensity with barometric pressure (typical conditions).

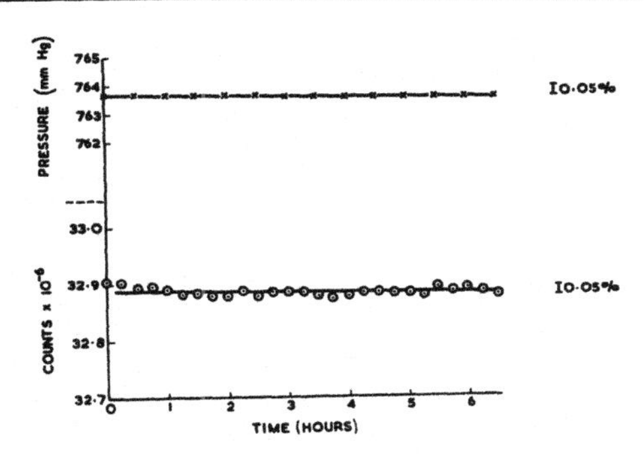

Figure 3. The variation of X-ray intensity with barometric pressure (steady conditions).

guarded against, satisfactory intensity stability is being attained. The dependence of the intensity on barometric pressure is demonstrated in Figure 2, which shows the variation in the peak intensity over 5 days during December, 1965, plotted with the variation in the barometric pressure. Figure 3 shows similar plots for steady barometric conditions when the variation in intensity is statistical. This intensity variation is due to the changing X-ray absorption of the atmosphere as its density varies with the pressure. Variations in humidity also affect the intensity.

In many applications, this effect causes no trouble, and, in those where it does, it can be avoided by enclosing either the whole instrument or the beam path in an air-tight enclosure. Alternatively, the experiment can often be planned so that it is not affected by factors that vary slowly with time.

THE MEASUREMENT OF ANGLES

Angles are measured by counting pulses to a stepping motor connected to a gear and worm, as shown in Figure 4. The particular property of this system that is made use of is the remarkable fidelity with which the gear follows the worm. Originally, around the gearwheel, there is an angular error of about $\pm 15''$, and this can be corrected by means of

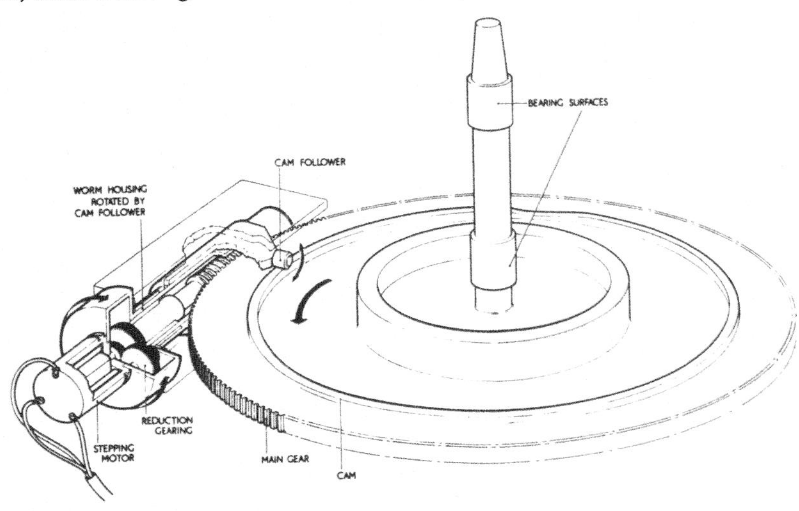

Figure 4. Method of angle setting and measurement.

a cam so that the residual error is reduced to about $\pm 5''$. (This cam correction is a refinement that is a convenience in some applications but is not fundamental to the method.)

TREATMENT OF SYSTEMATIC ERRORS

The Procedure Described by Bond

In the measurement of angles, the eccentricity error is one of the most frequent. Bond,[2] in a notable paper, pointed out that, if measurements are made with the use of a single crystal and if the crystal is rotated through equivalent reflecting positions on each side of the collimator as shown in Figure 5 taken from Bond's paper, then the angle through which the crystal is rotated yields θ free from specimen eccentricity and absorption errors as these are associated with the diffracted beam. This method also avoids zero errors. This is the procedure that is adopted in this work. The angle that is measured is that through which the crystal is rotated between equivalent reflecting positions on each side of the collimator.

For a lattice-parameter measurement to be correct to 1 part in 10 million it is necessary to measure this angle to approximately $0.25''$ (0.2μR) at $\theta = 80°$.

Toleration of Nonparallel Conditions

At the degree of precision dealt with here, there is a further error that can occur in the measurement of diffraction angles, which requires treatment. If the plane of diffraction is not parallel to the plane of angular measurement, the angle measured is only a projection of the true angle and, hence, an incorrect value is obtained. The plane of diffraction is defined by the direction of the incident beam (*i.e.*, the collimator) and the normal to the crystal plane. For a measuring circle in the horizontal plane, errors in the measured angle will occur if either the collimator or the crystal has a vertical tilt. For the angle to be measured correctly, the plane of diffraction must be perpendicular to the axis of rotation of the gear. A simple procedure has been devised, however, that allows the nonparallel conditions due to a collimator tilt of small but unknown magnitude to be tolerated and that at the same time adjusts the tilt of the crystal to the optimum value. The procedure is to tilt the crystal progressively in the vertical plane when it is found that the angle between the two reflecting positions of the crystal passes through a maximum value and that this maximum value is very close to the correct one, for, although nonparallel conditions exist, the projected value is very close to the true value. This fact was originally discovered empirically, but the explanation is that, in these circumstances, the projection function

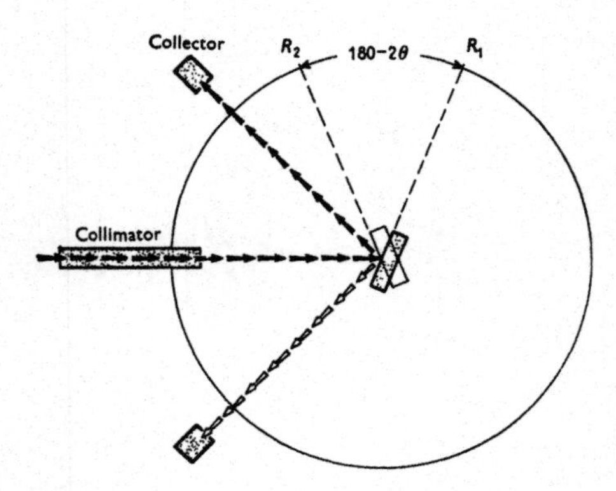

Figure 5. The method of diffraction-angle measurement described by Bond.

is changing only very slowly with the angle of tilt. This can be described analytically by the following expression:*

$$\phi' = \cos^{-1} \frac{\cos(\phi/2) - \sin\beta \sin\gamma}{\cos\lambda \cos\beta} + \cos^{-1} \frac{\cos(\phi/2)\cos 2\beta + \sin\beta\sin\gamma}{\cos\beta\cos^2\gamma - 4\cos(\phi/2)\sin\beta\,[\cos(\phi/2)\sin\beta - \sin\gamma]^{\frac{1}{2}}}$$

where ϕ' is the measured angle between the two reflecting positions of the crystal; ϕ is its true value; and β and γ are the crystal and collimator tilts, respectively, *i.e.*, the angles that the normal to the crystal plane and the collimator make with the plane of angular measurement.

Its solution by computer shows that fairly large collimator tilts can be tolerated. For example, at a Bragg angle of 80°, a collimator tilt of 10′ introduces an error of only 0.25″ in the measured angle, which is equivalent to an error of 1 in 10 million in the crystal-lattice parameter. A collimator tilt of 30′ is necessary to introduce an error of 1 in 1 million in the lattice parameter.

In practice, this method is very convenient; the crystal is merely set on the side of the peak farthest from the collimator, and its tilt is varied until the maximum intensity occurs, when the angle is also at its maximum and the crystal is adjusted for optimum conditions.

The Closed Error Loop Method

Errors in angular measurement due to the gearwheel, notably eccentricity, can be dealt with by employing the closed error loop method. This method is based on the fact that 0° and 360° occupy the same position on the smooth continuous surface of the gear and that the gear follows the worm with very high fidelity. This enables the gear to be used to the limit of its basic sensitivity, which is the fidelity with which it follows the worm when a pulse is sent to the stepping motor.

The method consists of using each part of the gear in turn to measure the angle and averaging the results. For example, if the angle measured is 21°, the value obtained for the silicon (444) reflection, with Cu $K\alpha_1$ radiation, this is measured on one part of the gear, the specimen is then rotated through 21° with reference to the gear, the measurement is repeated, and so on, until each part of the gear has, as far as is possible, been used in the measurement. In this case, the angle would be measured 360/21 to the nearest integer, or 17 times. The rotation of the specimen with reference to the gear is referred to in this work as offsetting.

If necessary, the method can be extended by further repeats to take account of the gaps or overlaps that will occur owing to the angles not dividing exactly into 360.

It should be emphasized that this method is only necessary when absolute measurements of angle are required. In most applications where small changes in angle are all that is needed, there is no need to carry out this procedure since measurements are always being made on about the same part of the gear.

THE AUTOMATIC PRECISION X-RAY GONIOMETER

The automatic precision X-ray goniometer is an instrument that has been designed to carry out the experimental procedures necessary to measure X-ray diffraction angles accurately and free from systematic errors. It is shown in Figure 6 with its sides removed so that the main gear can be seen. Only one counter is used, and the diffracted beam is

* We are indebted to Dr. A. J. E. Foreman of the Metallurgy Division, AERE, for this expression.

Figure 6. The automatic precision X-ray goniometer with sides removed to show the main gear.

tracked by this so that the same part of the counter is always presented to the beam. Beyond this, it plays no part in the angle measurement and has no slits. Details of the specimen tilting and offsetting mechanisms, which are also controlled by stepping motors, are shown in Figure 7.

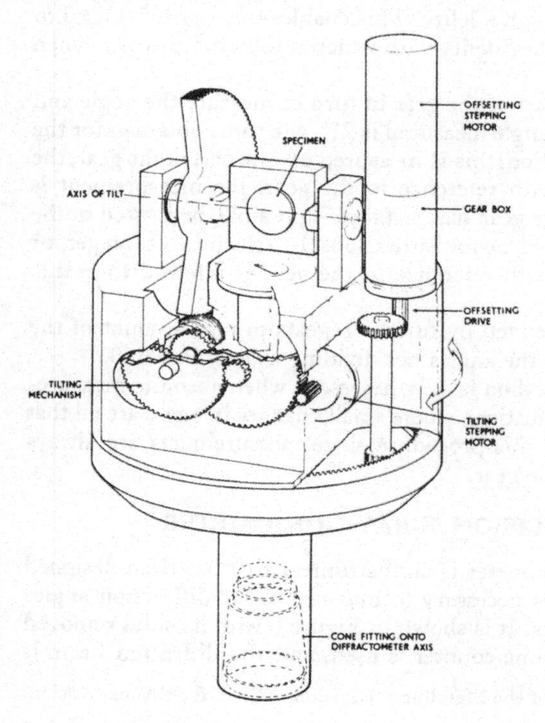

Figure 7. The automatic precision X-ray goniometer showing specimen tilting and offsetting mechanism.

Figure 8. The automatic precision X-ray goniometer with double-crystal spectrometer mode of operation.

As well as its main use, the instrument can be used as a double-crystal spectrometer (see Figure 8) and adapted as a normal powder diffractometer. Clamping pads are provided on its top and sides, to which various additional equipment can be rigidly attached. It has several accessories such as automatic specimen changers and scanners and pressure and temperature attachments. In use, it is contained in a constant-temperature enclosure and operated by remote control.

In this paper, it is described as it is used on line to a data processor, but it can of course be used with a less sophisticated form of control. Arrangements have been made for its manufacture* under license from the United Kingdom Atomic Energy Authority.

THE AUTOMATIC CONTROL

Two precision X-ray goniometers together with two Philips diffractometers are simultaneously controlled on line by an Elliott ARCH 9000B system (based on the 903 central processor). This is a small data processor with an 18-bit word length and a fast store of 8192 words. It has both ALGOL and FORTRAN compilers and a very useful data-processing capability.

A comprehensive monitor program has been written with a large number of input options, so that practically any diffractometer experiment can be undertaken. It is possible to check on the progress of the experiments by means of a display unit, and this can be used to bring up intermediate results without confusing the printed-out log with unnecessary detail.

A typical automatically controlled precision measurement would be as follows: The goniometer rotates quickly to a region where a reflection peak is expected; it then moves on, assessing whether it is on a peak or the background. When a peak is found, the goniometer counts over it and computes roughly the peak position. It then reverses back off the peak and moves forward again until it is five-eighths of the way up the side of the peak; it then tilts the crystal and adjusts it for nonparallel conditions, using the method of the maximum angle and finding the position of maximum intensity by fitting a quartic to the

* McLean Research Engineering Company, Limited, 59A High Street, Hungerford, Berkshire, England.

points. When this is completed, it counts slowly over the peak to obtain adequate precision and fits a quartic to the points so obtained. This is to find the position of the peak. A quartic is used because it allows for some degree of asymmetry in the profile. The goniometer then moves quickly to the position where the second peak is expected, and the procedure (with the exception of the specimen adjustment) is repeated. Then the goniometer is reset to its original position and the following can take place:

1. The procedure is repeated.
2. The specimen is offset with reference to the gear and the procedure repeated until the closed error loop strategy has been completed.
3. Pulses are sent to another motor to change the specimen, scan the specimen, *etc.*, and the procedure is repeated.

Typical outputs are:

1. The angle between the peaks.
2. The average of the angles between the peaks, when offsetting.
3. The position, intensity, breadth at a specified intensity, and a skewness factor of each peak.
4. The standard deviation of the quartic fit to the points.
5. Various check sums.

The information in items 3 and 4 is required for assessing whether the experiments have functioned satisfactorily.

It may be remarked that the data processor has another very marked advantage. It provides immediate data processing of experimental results, so that an intelligent assessment of the progress of the experiments can be made at each occasion and before conditions are altered. This is not usually the case when recourse has to be made to large computers, which, although they compute faster, frequently have a long turnaround time.

PERFORMANCE TESTS ON THE SYSTEM

Basic Sensitivity

The basic sensitivity of the angle-measuring system is the fidelity with which the gear follows the worm when the pulses are sent to the stepping motor. A clear indication of this can be obtained from an examination of measurements on diffraction profiles, an example of which is Figure 1. The original size of this figure was 3ft 6in. by 2ft 6in., where it could be seen that the gear followed the worm with very high precision. Examination of portions of peaks on a more enlarged scale further indicated that the gear was being stepped onward correctly to about 0.1″. Owing to their large scale, these results cannot be conveniently reproduced graphically in this paper.

Further measurements were carried out by staff of the National Physical Laboratory (NPL) on two parts of the gear about 180° apart and over regions of about 80″. The instrument was stepped onward by 2μR and the NPL measured the angle to 0.1″. These results also cannot be reproduced conveniently in graphical form. However, Tables I and II show a least-squares fit of a straight line to the points, and the third column shows the deviations from the calculated straight line in micrevs. Calibration 1 shows only 2 of the 33 points deviating by as much as 0.1μR. Calibration 2 has 5 points deviating by 0.2μR. These measurements confirm those made on the diffraction profiles regarding the high degree of precision with which the gear follows the worm, and since the direct measurement of angles to 0.1″ is no easy matter, the measurements on the diffraction profile itself probably give the best indication of the precision. To the limit which it has been examined at the present, this is of the order of 0.1″.

Table I. NPL Calibration 1

Fit of the points $X(I)$, $Y(I)$ to the straight line $Y = MX + C$ by the method of least squares*

$X(I)$, sec	$Y(I)$, μR	Y(calc.) − Y(obs.), μR
92.380	2.000	0.089
89.850	4.000	0.086
87.430	6.000	−0.004
84.900	8.000	−0.007
82.400	10.000	−0.034
79.920	12.000	−0.076
77.310	14.000	−0.016
74.880	16.000	−0.098
72.320	18.000	−0.077
69.820	20.000	−0.104
67.180	22.000	−0.020
64.760	24.000	−0.110
62.030	26.000	0.045
59.510	28.000	0.034
56.900	30.000	0.094
54.380	32.000	0.083
51.840	34.000	0.088
49.390	36.000	0.022
46.770	38.000	0.090
44.360	40.000	−0.008
41.700	42.000	0.092
39.290	44.000	−0.006
36.800	46.000	−0.041
34.300	48.000	−0.067
31.570	50.000	0.087
29.170	52.000	−0.018
26.670	54.000	−0.045
24.020	56.000	0.047
21.640	58.000	−0.074
19.020	60.000	−0.006
16.500	62.000	−0.017
13.980	64.000	−0.028
11.410	66.000	0.000

* Slope, $M = -0.7893$; standard deviations of coefficients are $M = 0.00046$ and $C = 0.026$ μR.

Tables I and II show that the standard deviation on the intercept C is better than 0.03 μR in each case, and this gives some indication of the error in fixing the peak position when a curve such as a quartic is fitted to this number of points.

Peak Description and Location

As well as the comment in the previous paragraph, the precision of the peak description and location is indicated by measurements of the profile as shown in Figure 1. The midchord peak construction[2] is shown, and, at the graph's original size, it was estimated that the peak was being fixed to 0.2 μR, which would give a precision of 1 part in 10 million in the lattice parameter.

Table II. NPL Calibration 2

Fit of the points $X(I)$, $Y(I)$ to the straight line $Y = MX + C$ by the method of least squares*

$X(I)$, sec	$Y(I)$, μR	Y(calc.) − Y(obs.), μR
90.100	6.000	0.164
87.740	8.000	0.025
85.080	10.000	0.121
82.520	12.000	0.139
80.120	14.000	0.031
77.620	16.000	0.002
75.040	18.000	0.036
72.590	20.000	−0.033
69.980	22.000	0.024
67.460	24.000	0.016
65.020	26.000	−0.066
62.460	28.000	−0.048
59.960	30.000	−0.077
57.550	32.000	−0.178
54.770	34.000	0.014
52.440	36.000	−0.150
49.930	38.000	−0.171
47.270	40.000	−0.074
44.710	42.000	−0.056
42.190	44.000	−0.070
39.570	46.000	−0.005
37.060	48.000	−0.026
34.560	50.000	−0.056
31.930	52.000	0.017
29.340	54.000	0.059
26.940	56.000	−0.049
24.290	58.000	0.040
21.810	60.000	−0.005
19.130	62.000	0.107
16.760	64.000	−0.025
14.210	66.000	−0.015
11.660	68.000	−0.005
8.810	70.000	0.242
6.480	72.000	0.079

* Slope, $M = -0.7883$; standard deviations of coefficients are $M = 0.00062$ and $C = 0.034$ μR.

Reproducibility of Angle Measurements

These tests were all carried out by using the silicon (444) reflection with Cu $K\alpha_1$ radiation, for which $\theta = 79°$, and 2μR steps.

The top part of the peak was used, and, as a result of experience, between 50,000 and 100,000 counts were collected per point for a precision of 1 part in 1 million, and between 5 million and 10 million for 1 part in 10^7 million.

Test on the Same Part of the Gear. Table III shows the results of 32 measured angles averaged by groups of eight so that the counts on the peak were 6×10^6. The deviation from the mean of these four results is about 0.1μR. Before the results were so

Table III. Reproducibility Test Using the Same Part of the Gear

Angle, μR	Deviation from mean,* μR
59357.43	0
59357.56	+0.13
59357.42	−0.01
59357.31	−0.12

* Mean = 59357.43 μR and the standard deviation = 0.1 μR.

averaged, the standard deviation of the 32 measurements was 0.7μR and the maximum deviation from the mean was 1.5μR.

Tests of Offsetting Round the Gear. These indicate the reproducibility of absolute measurement of angle when the closed error loop method is used.

The angle of 21° is offset 17 times round the gear. The average of these 17 values should be free from gear errors. The crystal is reset at each position. The standard deviation is for each of these 17 values and represents the residual error round the gear. The test is the agreement between the final average values.

Table IV gives the results for a test of three offset experiments, each starting at a different position on the gear, made semiautomatically, and computed graphically. It shows that, whereas the standard deviation for each of the 17 readings is about 3.8μR, the maximum spread on the final results is 0.4μR.

Table IV. Test of Closed Error Loop Method (Semiautomatic and Graphical Methods)

Measurement	Final value, mean of 17 measurements, μR	Standard deviation, μR
1	59371.7	3.8
2	59371.3	3.8
3	59371.7	3.6

Table V. Test of Closed Error Loop Method (Fully Automatic Operation)

Measurement	Final value, mean of 17 measurements, μR	Deviation from mean,* μR
1	59368.68	+0.27
2	59368.68	+0.27
3	59367.94	−0.47
4	59368.18	−0.23
5	59368.32	−0.09
6	59368.68	+0.27

* Standard deviation of the six values is 0.32 μR.

Table V shows the results of a similar test carried out entirely automatically, the final values being those printed out by the data processor. During this test, the crystal, which is reset for each measurement, was, for programming reasons, not being set with its full precision, and this could have had an adverse effect on the results. Nevertheless, the results given in Table V do show good agreement, with a maximum deviation from the mean of 0.47 μR and a standard deviation of 0.32 μR (0.4″), whereas the standard deviation for the average of the 17 values was about 4.0 μR in each case.

The crystal setting has since been improved by a programming alteration.

Pressure on the Specimens

An early reproducibility test of the goniometer consisted of making measurements on eight different single crystals of silicon. It was found that, whereas the measurements on the different specimens always agreed with each other about to 1 part in 10 million, the values obtained varied from day to day by amounts greater than 1 part in 100,000. It was not possible to correlate this variation with any environmental conditions that varied from day to day. However, when the adjustment of the specimens by the method of the maximum angle was devised, it became possible to scan across the specimen without concern for tilting that might occur in the process and that might be greater than could be tolerated. The results of scans in the horizontal and vertical directions are shown in Figures 9 and 10. There is a marked decrease in the measured angle at the center of the specimen, which was a disc 20 × 1 mm. The variations turned out to be due to a small, weak leaf spring on the back of the specimen, holding it in position. Different operators on different days were fitting the specimen and spring in different positions in an unsymmetrical specimen holder, according to their ideas of correctness. Such specimens are now held in position by silicone grease, and scans on specimens held in this manner do not show a change of angle across them.

The effect of the pressure of the spring was confirmed by compressing a coil spring against the back of the specimen with a screw. Figure 11 shows how the angle decreased with revolutions of the compressing screw.

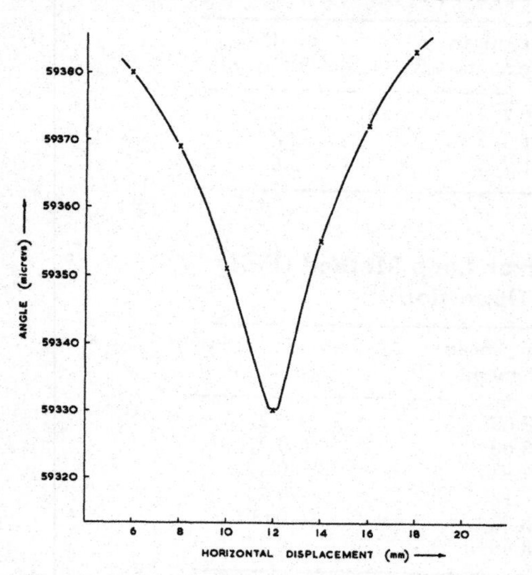

Figure 9. Horizontal scan of the specimen.

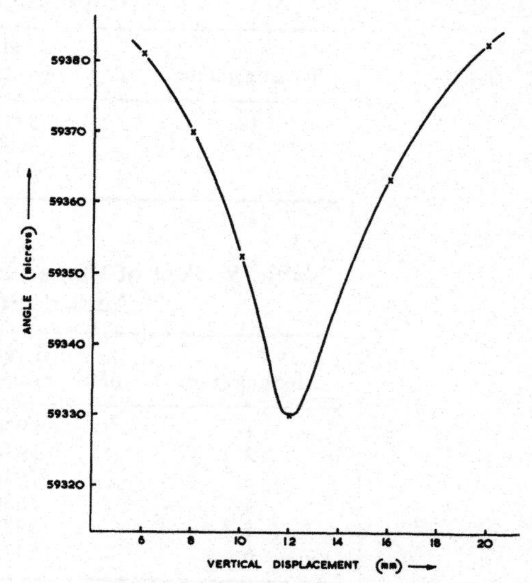

Figure 10. Vertical scan of the specimen.

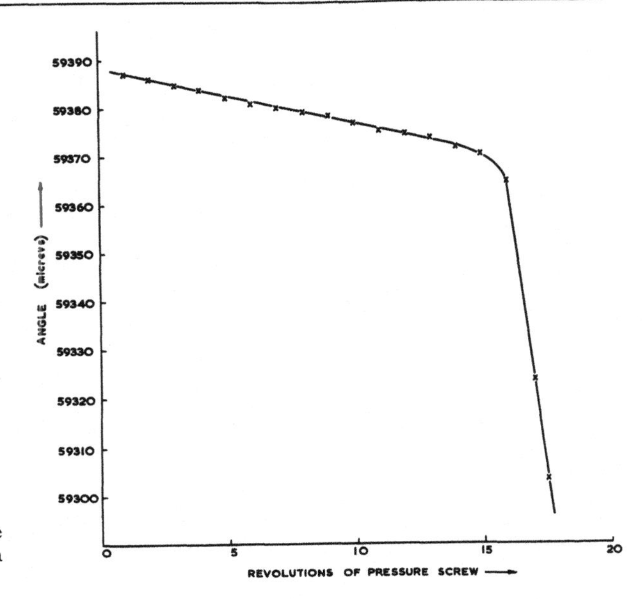

Figure 11. Effect of tightening the spring on the back of the specimen by turning a pressure screw.

ANNEALING CURVES FOR IRRADIATED MAGNESIUM OXIDE

The results obtained in these experiments were the first in this type of application for which the instrument had been principally designed. Interest was therefore centered around the confidence which could be placed in the results, as well as in the results themselves. They follow the change in the lattice parameter of irradiated magnesium oxide as the radiation damage is annealed out.

Figure 12 shows a typical experiment with the use of standard accurate powder methods of lattice-parameter measurement. Fine detail is obscured by the scatter of points. Figure 13 shows the results of two such experiments in which the automatic precision X-ray goniometer was used. The results of the two experiments on separate specimens are in good agreement.

Points A and B are of interest. At temperatures above 1100°C, it was found that the anneal was affecting the surface of the specimen, so that, to overcome this, the specimens were cleft after anneal and the inner surfaces were used for measurement. Point A was a measurement made on such a surface, and it lay off the curve. Accordingly, the corresponding cleft surface on the other piece of the specimen was measured, and point B was

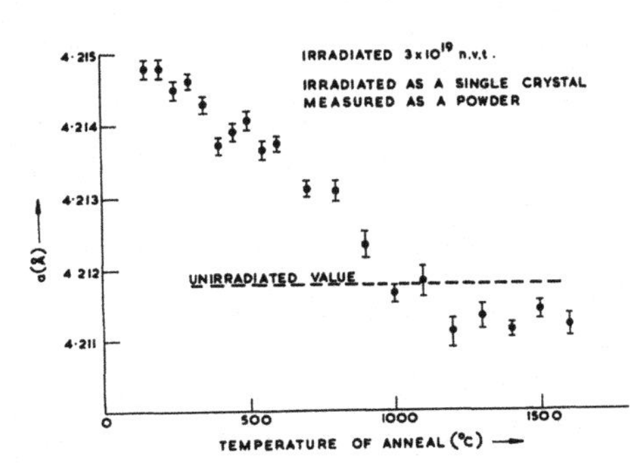

Figure 12. The variation of the lattice parameter of irradiated magnesium oxide with annealing temperature, measured by accurate powder methods.

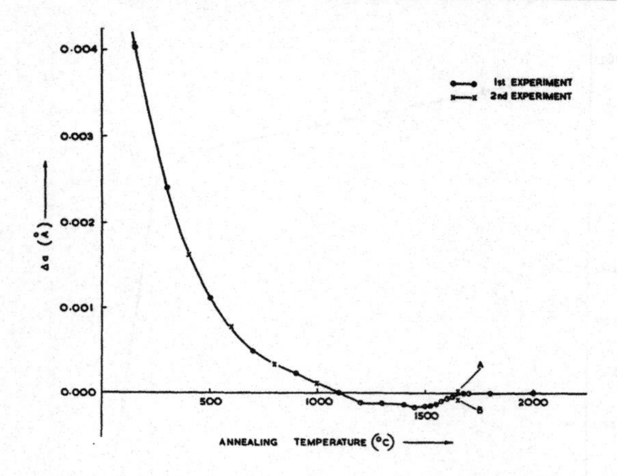

Figure 13. The variation of the lattice parameter of irradiated magnesium oxide with annealing temperature by using the automatic precision X-ray goniometer (irradiated 8.8×10^{20} nvt > 1 MeV).

obtained, again off the curve but on the other side of it. If confidence in these measurements was to be maintained, it was necessary to postulate strain in the specimen to account for these results. Examination of the specimen under a polarizing microscope confirmed the presence of strain, which turned out to be due to hydrogen leaking into the furnace during this anneal. It is known that annealing in hydrogen can cause strain in magnesia. The instrument was thus vindicated.

THE STATIC METHOD

When all that is required is the measurement of changes in angle of an *in situ* specimen with, say, temperature for the purpose of measuring thermal expansion, a simpler and quicker method can be employed. Angle and intensity measurements are made up the side of a diffraction peak, and the intensity is thus calibrated with angle. The instrument is then set in the middle of this region and the temperature altered. The change in angle can then be obtained by intensity measurements alone. A high degree of intensity stability is required for this method, and care must be taken to check that the intensity and shape of the peak have not altered with the change of conditions. If necessary, such changes can be allowed for. This procedure is referred to as the *static method.*

This method was tested during a determination of the coefficient of thermal expansion of magnesium oxide at room temperature. By using the normal method, the changes in angle were measured over the ranges 20–30, 25–30, and 24.5–25.5°C. The static method was then employed, and intensity measurements were made with the temperature increasing and again with it decreasing at a constant rate. The increases and decreases were part of checks to establish that the temperature of the diffracting surface was in fact being measured. The temperature was measured by a thermistor in good thermal association with the specimen. The result of one set of measurements is shown in Figure 14. This is regarded as a reliable determination, and it was in good agreement with measurements made by using the normal method. It is given here because it shows clearly the sensitivity of the method.

The sensitivity of this method, which is high, depends on the accuracy of the intensity measurement and the rate of change of intensity with angle, *i.e.*, the steepness of the peak. Thus, it can be increased by using pre-crystal-reflected radiation with the instrument in its double-crystal mode of operation. It is hoped that this method may be used to follow the lattice-parameter changes in a piezoelectric crystal with applied voltage.

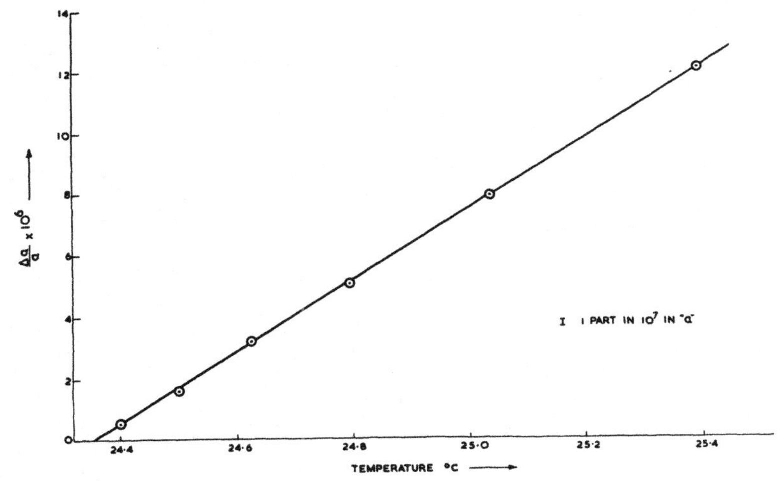

Figure 14. The thermal expansion of magnesium oxide over 1.0°C, showing the sensitivity obtained by the static method.

CONCLUSIONS

An instrument has been built which measures X-ray diffraction angles precisely, conveniently, and automatically. The high precision is attained by making use of the following points:

1. The high fidelity with which the gear follows the worm.
2. The closed error loop strategy.
3. The remote control which permits the measurements to be carried out in an enclosure free from environmental variations.
4. The method described by Bond.
5. The toleration of nonparallel conditions by the method of the maximum angle.

It should be emphasized that, with this instrument, the measurement of diffraction angles involves no dimension of the instrument, the crystal does not have to be exactly centered, the collimator does not have to be exactly perpendicular to the axis of rotation, and errors in the gear can be tolerated. Thus, for its ultimate precision, dependence is not placed on any engineering construction of dimension, angle, or scale. The engineering requirements are for an instrument that does not distort under the movements that take place during the measurement, a well-surfaced gear, and a bearing that does not stick or wobble.

The automatic control allows the measurements to be made conveniently. Since the time taken to carry out precision measurements depends on the collection of sufficient counts, the on-line connection of the instrument to a data processor is of particular value, as it allows the nights and weekends to be used for data collection and, thus, the time available is increased by a factor of about 4. It also frees the experimenter from a lot of tedious computation, and the organizational control it affords permits the experiments to be carried out in improved ways.

The tests that have been described, although not completely assessing the precision of the system since time has not yet permitted this, do give confidence that the diffraction angles are being measured with high precision. The estimate of this precision, when translated into terms of changes in crystal-lattice parameter, is 1 part in 10 million, with absolute measurements approaching this.

Before the installation of this system, lattice-parameter measurements were made by accurate powder methods, to about 1 part in 20,000, with considerable drudgery. The output was one per day. With the present system, measurements can be made with a precision of 1 in 1 million at the rate of one every half hour, with relatively little effort on the part of the experimenter.

ACKNOWLEDGMENTS

It is a pleasure to record the valuable assistance we have had from many colleagues at Harwell. In particular, thanks are due to Mr. F. D. Seymour for advice concerning the Harwell 2000 series electronics and to Mr. C. N. Davey for additional electronics instrumentation. Mr. N. S. Tobin and Mr. F. F. Freeman gave particularly valuable experimental assistance. Special thanks are due to Dr. J. Williams, Head of the Ceramics Division, for his support, encouragement, and useful discussion.

The monitor program was written by Mr. B. Pearce of Computer Analysts and Programmers Limited, London.

REFERENCES

1. T. W. Baker, J. D. George, B. A. Bellamy, and R. Causer, "Very High Precision X-Ray Diffraction," *Nature* **210**: 720, 1966.
2. W. L. Bond, "Precision Lattice-Constant Determination," *Acta Cryst.* **13**: 814, 1960.

DISCUSSION

Z. W. Wilchinsky (Enjay Polymer Labs.): Do these precisions that you quoted—high precisions —hold for powders as well as single crystals?

T. W. Baker: No, it is fundamentally a single-crystal method because it is based on Bond's method, which is based on the Bragg spectrometer. It is this that allows you to deal with some of the errors. If it were adapted as a powder diffractometer, it would do better than some commercial ones but only marginally so. However, these days, more and more serious work is being done on single crystals, and I think that this is one of the reasons why it is proving so useful.

B. S. Sanderson (National Lead Co.): When you operate all of these instruments simultaneously, is this working in a time-sharing method or do they each have to wait for each other?

T. W. Baker: It is time-shared. The control of all the instruments is going on simultaneously. It's really a fairly sophisticated form of control and shows what a small computer like this is capable of. Incidentally, the computer itself cost about £12,000, and, with the additional hardware necessary for the control, it cost about, £20,000. The automatic precision goniometer costs about £5000.

G. Martin (Stanford University): You mentioned the Nelson–Riley plot. You don't use that as a correction factor for single-crystal work, do you?

T. W. Baker: No, it's not necessary in this method. It doesn't come into it at all. We built up this automatic facility because of the precision side of things. Because we've got the computer for the precision side, we are controlling our other diffractometers as well, and then, of course, they just behave in the normal way. For the powder work, a Nelson–Riley plot might be used. It is used in our photographic powder work.

H. W. King (Imperial College, London): You have taken a lot of care over your goniometer. What care have you taken to stabilize your X-ray source, because you don't measure both peaks at the same time, *i.e.*, you have a time lag?

T. W. Baker: We find that, if we take care of the barometric variations, then we can keep intensity correct to 0.025% for a very long time and better than that as we go over the two peaks. But having the computer gives you the organizational control of the experiment that you do not normally have. For instance, supposing you want to measure a diffraction peak which is a very weak one and, for some purpose, you want to know its shape very accurately. It could take you a weekend to collect the necessary counts, and, during this time, the barometric pressure could have changed and affected the intensity by several per cent; also, over this length of time, you might not feel too sure of the stability of your X-ray generator—although I think you can if you use a Philips PW 1010 generator or an equivalent. Thus, you are probably faced with a long-term intensity drift while you collect sufficient counts. However, using the computer, you can go over the peak in a time that is short compared with this drift and then repeat a sufficient number of times to give the necessary counts. On each repeat, the counts at each specific angle are put in the same store location as on the previous occasion. Thus, sufficient counts can be collected to give the required accuracy, with the effect of the long-term intensity drift being taken care of. Let us say, for instance, it takes several days to collect the necessary counts; then you might arrange to go over the peak every hour, always putting the intensity data in the appropriate store locations. This is the "enhancetron" technique.

H. W. King: Is the barometric pressure affecting your crystals or your X-ray unit?

T. W. Baker: It's just due to the fact that there are more molecules of air to absorb the X-rays at higher barometric pressures as they go through the path length of a diffractometer. It is purely absorption in air. For Cu $K\alpha_1$, barometric-pressure variations can cause several per cent of intensity variations, and this fact is not generally brought out in the literature.

H. W. King: This means that you make no attempt to improve on the commercial instrument for controlling the voltage or the current stability?

T. W. Baker: I think the Philips PW 1010 generator is probably good enough. We have a constant-voltage transformer in front of this, but this is because, at Harwell, we have large accelerating machines that cause a lot of trouble in the electricity mains. I imagine most of the error results from barometric-pressure variations. It's the barometric pressure that causes most of the trouble.

COMPUTERIZED MULTIPHASE X-RAY POWDER-DIFFRACTION IDENTIFICATION SYSTEM

G. G. Johnson, Jr., and V. Vand

Materials Research Laboratory
The Pennsylvania State University
University Park, Pennsylvania

ABSTRACT

A computer search system utilizing the Powder Diffraction File compiled by the Joint Committee on Powder Diffraction Standards, originally developed for an IBM 7074 in FORTRAN II and reported at the Pittsburgh (1966) conference,[1] has been revised and extended to run on IBM 360/50. The system is now written in FORTRAN IV. This search system, which uses all the lines of the reference patterns, has successfully identified up to six standard reference patterns from a multiphase unknown X-ray diffraction pattern in less than 2 min running time. No chemical information is necessary for the system to run. In the revised program, the chemical composition of the patterns is now available from the magnetic tapes in immediate conjunction with the printout of the "most likely" components of the mixture. However, this chemical information is not used by the program itself in the search procedure since, if the unknown pattern is absent from the file, it is helpful to know those compounds which are isostructural with the unknown pattern. With the immediate use of chemical information, these patterns would be eliminated. An estimation of the relative concentration of each of the components, based on absolute intensities, is also calculated by the program. This identification system has been run on experimental data both of the Guinier type and of a less reliable type, with the present Powder Diffraction File on the search tape. Although the number of false matches was increased with the poorer quality of input data, the programs yielded excellent results for both single- and multiple-phase patterns even with poor data and the absence of any chemical information. A series of results from the Materials Research Laboratory of The Pennsylvania State University, illustrating the system in operation with increasingly difficult mixtures, will be given. With such a system in operation at such a small cost, the diffractionist can concentrate on the results and meaning of the identification rather than on the method of identification itself.

INTRODUCTION

The field of analytical chemistry has many methods of identification of unknown materials; however, the use of X-ray diffraction as a tool has often been neglected. Since many of the specimens considered by chemists can often be obtained in polycrystalline samples, it seems that this analytical tool should be used for quick and nondestructive testing on extremely small samples. The sample preparation for X-ray diffraction is quite simple and has not proved a problem in the use of this method in the chemical laboratory. The main difficulty until now has been the identification of the standard from the resulting X-ray diffraction powder patterns of a mixture, but now a computer-based information retrieval system has been developed which identifies all the components present in a multicomponent mixture.

We have developed a retrieval written in FORTRAN II for an IBM 7074, which we described at the Pittsburgh (1966) conference.[1] This system is available and has been described in the literature[2] and is also available through the Joint Committee on Powder Diffraction Standards.[3] Since the introduction of this original system, several significant improvements have been introduced and the system will be tested in practice. This new system is also available through the Joint Committee on Powder Diffraction Standards.[4] One improvement was the introduction of the chemical information in the output. Another was a significant shortening of computer running time by optimalization of the program.

A table of times for identification (using a tape-oriented system with both the direct and inverted files on magnetic tape) of a three-component problem given by Frevel,[5] which contained selenium, arsenic, and As_2O_3 is as follows:

*IBM 7074	FORTRAN II	PENN STATE Operating System		300 sec
†IBM 360/50	FORTRAN IV(H) OPT = 02	OS/360	Version 10	139 sec
‡IBM 360/50	FORTRAN IV(H) OPT = 02	OS/360	Version 10	112 sec
‡IBM 360/65	FORTRAN IV(H) OPT = 02	OS/360	Version 10 HASP	84 sec
‡IBM 360/67	FORTRAN IV(H) OPT = 02	OS/360	Version 10 HASP	57 sec

It may be thought that, for X-ray diffraction to be used by analytical chemists, special experimental techniques are necessary. It has been found in actual practice with the system in operation that ordinary laboratory diffraction techniques, *i.e.*, ordinary collection of data with no corrections for absorption, sample preparation, particle size, or other experimental parameters, can be used. Data from Guinier cameras (Hagg, AEG, DeWolff–Nonius), Debye–Scherrer cameras (114 and 57 mm) and diffractometers have been tested with the computer system. Some data have been both internally calculated and read with the latest microdensitometers (Mann, Jarrell–Ash, and Joyce–Loebl). Although the number of false drops (the patterns incorrectly retrieved) increases with the poorer quality of the data, the correct results are properly retrieved. For the *poorest* data with a 50-line unknown test pattern, the ratio of false drops to correct results is about 20 : 1 (*i.e.*, a 2-component pattern retrieves about 40 possible standard patterns). The list of possible patterns is then greatly reduced by the user with the knowledge of possible elements present and the resulting four to five standard patterns are then examined in greater detail by the diffractometer.

The measured pattern is used to determine the Bragg reflection peaks and a simple estimation of the intensities (the area under the peak or the peak-height value if all peaks show the same profile). From these peaks, it is then possible to record the ordered sets $d_i I_i$.

It is a well-established fact that every polycrystalline substance is characterized by its own powder pattern $d_i I_i$. Although the spacings are determined solely by the size and shape of the unit cell, the intensities are determined by the arrangement of the atoms within the unit cell. Since, basically, all atoms scatter the incident X-ray differently, these three conditions make every pattern "unique" (except for very rare exceptions).

The identification of laboratory samples would be impossible if a file of reference patterns were not available since there is no *a priori* method of identifying the substance

* Requires 8000 storage plus intermediate tape storage, version 1 of search system.
† Requires 40,000 storage, version 2 of search system.
‡ Requires 40,000 storage, version 3 of search system.

from its powder pattern. Such a file was established in 1938 by the Joint Committee, has been continually updated and improved (at the rate of about 2000 standard patterns per year), and currently (in 1967, the Powder Diffraction File) consists of approximately 13,000 standard patterns (about 10^7 characters). The File, prepared for computer use by the authors, is now available for the first time on magnetic tape, and it is this magnetic tape that is used in the system described in this paper.

Now the question arises of the ease of retrieval of these standard reference patterns. It is a simple problem to retrieve an exact pattern of one substance no matter what problem you are doing. But let a few practical complications be introduced as follows:

1. Errors in both the measurement of the standard and the unknown.
2. Overlap of identifying lines.
3. More than one component present.
4. Instrumental resolution.

Since both the standard reference patterns and the laboratory experimental pattern contain certain errors, the information retrieval system considers the possibility of such errors. This is described later. It is because of these difficulties that the identification of specimens by X-ray diffraction methods has only been attempted by the professional diffractionists (metallurgists, crystallographers, mineralogists, and physicists).

The traditional methods of identification as published by the Joint Committee on Powder Diffraction Standards, using the various book indices (Hanawalt–Davey, Fink, and KWIC), and peek-a-boo cards (Matthews) are well known. However, the use of these methods for complex or multiphase unknowns becomes a very difficult problem. It was this impetus which brought about the application of high-speed electronic computers as the solution to the problem.

Now let us just look at what this pattern recognition system will do. In essence, it allows people who do not wish to become professional diffractionists to use the knowledge of both professional programmers and crystallographers to solve their own problems. Just as the Busing–Martin–Levy program has allowed those interested in single-crystal analysis to solve the crystal structure of complex substances, this system has proved to be successful in the solution of multiphase (multicomponent) samples of polycrystalline specimens.

The two assumptions made in the computer search system are the same as those used in any other method of identification of powder diffraction patterns. They are:

1. A multiphase (multicomponent) pattern is simply the weighted sum of the individual standard patterns.
2. The intensities of the peaks from that component are proportionate to the amount of each individual component present.

A brief summary of the other assumptions which are described in detail in a previous paper[2] are as follows:

3. The effect of solid solution can be neglected.
4. The patterns of the component substances of a mixture are present in the ASTM file.
5. The maximum error allowable between the positions of the lines and intensities of the file patterns and the unknown pattern is given.
6. An inverted file instead of a direct file is used for the first stage of the search.
7. Packed representation of data is used within the computer system.

8. Integer arithmetic, which is considerably faster than floating point arithmetic, is used.
9. The data of the file should be transformed into a system having a constant error window.
10. For the powder diffraction file, this means the use of reciprocal spacing and logarithms of intensities.
11. The search system should work successfully without introducing at the outset the chemical knowledge of elements present.
12. The search should give maximum possible concentrations of file substance compatible with criteria 1 and 2 of additivity of patterns present.

The input data to the information retrieval system consists of the measured d_iI_i properly normalized from the unknown pattern. The search system makes the comparison of all 13,000 standard patterns, each of which can consist of $99d_iI_i$, in less than 2 min. The manner in which this computer-based system operates will now be described.

The ASTM Powder Diffraction File consists of 13,000 standard patterns each with the respective decimal interatomic spacings d and integer relative intensities I. In order to best utilize the speed and storage capability of today's modern high-speed electronic computers, certain modifications are made. The first modification is obtained by converting by the computer the ordered pair d_iI_i into a characteristic integer number which represents the pair. This transformation has been described in earlier papers.[6,7]

Then the use of integer packing gains in speed and increases the effective storage of the computer. While this characteristic number is being constructed, it is being transformed into a number set with a constant bandpass, or error window. Thus, it is possible to examine the entire range of experimental values with a constant bandpass, a practice which should be used in all data processing systems. The second modification is the construction of both direct and inverted tape files. The inverted file consists of only the characteristic integer numbers which represent the strongest lines of the standard patterns. If a component is present in a mixture, then the strongest lines of that component are present. If a line is not present, then the standard component with that line is also missing.

The input data d_iI_i are also transformed into characteristic integer numbers and compared with the inverted file. All reference patterns which have characteristic integer numbers within the bandpass are considered as possibly correct. Certain standard patterns are thus retrieved and are then used with the direct file. The direct file compares the entire reference pattern, starting with the most intense lines, with the unknown pattern. Thus the difficulty of line coincidence does not present a problem.

Two other computer search systems—those of Frevel[5] and Nichols[8,9]—seem to have the problem of producing negative intensities as the search proceeds. Both these systems, Frevel's in ALGOL 60 for the Burroughs 5000 and Nichols' in FORTRAN IV for the IBM 7094, seem to run faster than the original system which we developed,[1] but this is because their systems use only a subset of the ASTM reference file. The increased speed of the improved system described in the file makes this system at least an order of magnitude faster than their systems. However, during the summer of 1967, both Frevel and Nichols were collaborating at Dow Chemical Company on a much improved system.

An estimation of the relative amount of each component present, based on absolute intensities, is obtained in this section of the search system. As each component of the mixture is identified, the number of successful line matches, relative concentrations, and chemical formulas of the standard is printed out for the user.

It should be noted that no chemical information has been initially used in the solution of the pattern. Since the chemical composition of each of these standards is printed as it is retrieved, the information is not used before the results are printed. Those retrieval patterns (about ten) can then be examined with the chemical knowledge of the unknown mixture. Even if the unknown sample is not present in the ASTM Powder Diffraction File, the printed information (chemically incorrect) gives the user an idea of the substances in the file which have a similar arrangement of atoms and size of unit cell. Thus the system gives more information than the standard reference file.

SOME RESULTS OF THE SYSTEM

The search procedure has been run by the authors on over 150 problems at the Materials Research Laboratory of The Pennsylvania State University and on about a dozen problems at the University of Michigan by J. D. Hanawalt. Two selected results will be shown here to illustrate the system in operation.

The first example is known to contain arsenic and selenium in major proportions and antimony as a minor component. Figures 1 and 2 show the results of the new computer system. Eleven standard patterns are retrieved from the Powder Diffraction File in 136 sec. Of these 11 patterns, the chemical knowledge of the sample eliminates many of the substances. It can be seen that $(Sb,As)_2O_3$, As_2O_3, As, and Se are identified as possibly being in the sample.

The second example is known to contain silver, yttrium, lead, selenium, arsenic, and potassium as those elements which can be found from fluorescence analyses. The input is

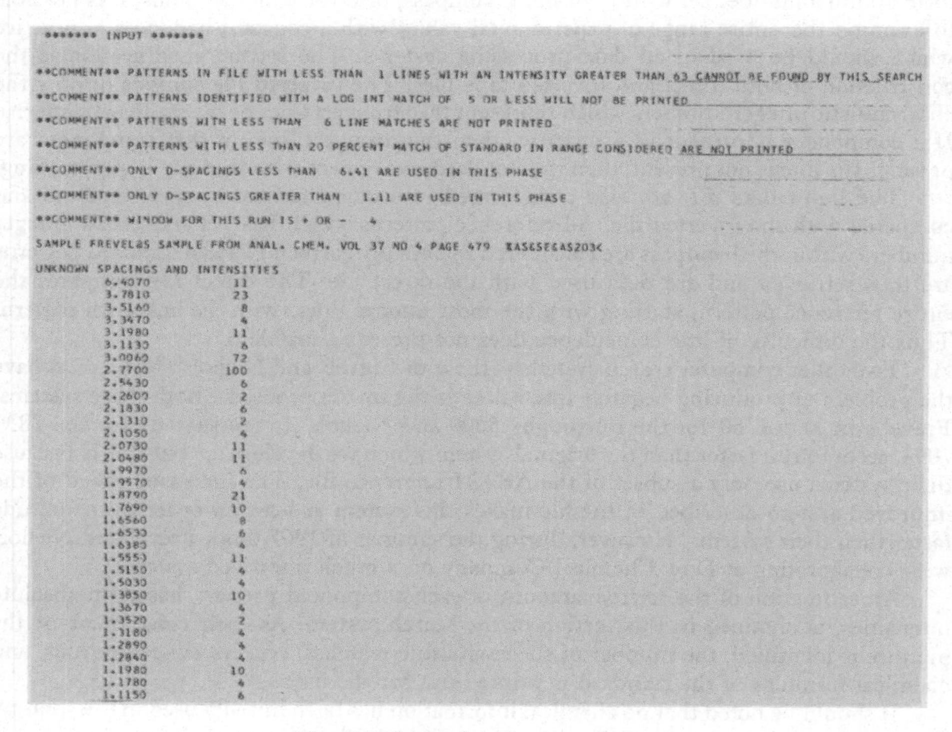

```
******* INPUT *******

**COMMENT** PATTERNS IN FILE WITH LESS THAN  1 LINES WITH AN INTENSITY GREATER THAN 63 CANNOT BE FOUND BY THIS SEARCH
**COMMENT** PATTERNS IDENTIFIED WITH A LOG INT MATCH OF  5 OR LESS WILL NOT BE PRINTED
**COMMENT** PATTERNS WITH LESS THAN   6 LINE MATCHES ARE NOT PRINTED
**COMMENT** PATTERNS WITH LESS THAN 20 PERCENT MATCH OF STANDARD IN RANGE CONSIDERED ARE NOT PRINTED
**COMMENT** ONLY D-SPACINGS LESS THAN   6.41 ARE USED IN THIS PHASE
**COMMENT** ONLY D-SPACINGS GREATER THAN   1.11 ARE USED IN THIS PHASE
**COMMENT** WINDOW FOR THIS RUN IS + OR -   4
SAMPLE FREVEL@S SAMPLE FROM ANAL. CHEM. VOL 37 NO 4 PAGE 479  ꓘAS&SE&AS2O3<
UNKNOWN SPACINGS AND INTENSITIES
        6.4070        11
        3.7810        23
        3.5160         8
        3.3470         4
        3.1980        11
        3.1130         6
        3.0060        72
        2.7700       100
        2.5430         6
        2.2600         2
        2.1830         6
        2.1310         2
        2.1050         4
        2.0730        11
        2.0480        11
        1.9970         6
        1.9570         4
        1.8790        21
        1.7690        10
        1.6560         8
        1.6530         6
        1.6380         4
        1.5550        11
        1.5150         4
        1.5030         4
        1.3850        10
        1.3670         4
        1.3520         2
        1.3180         4
        1.2890         4
        1.2840         4
        1.1980        10
        1.1780         4
        1.1150         6
```

Figure 1. Input data for example 1.

```
******* OUTPUT *******            F3

NUMBER    PERCENT    MINIMUM    LOG    ASTM                      FORMULA                          RELIABILITY
  OF        OF      INTENSITY   INT    NUMBER                                                       FACTOR
 LINE    STANDARD      OF      MATCH
MATCHES    LINE       LINES
          MATCHES   COMPARED

   6        31          6       15     10758    ✓   ( SB AS )2 O3                                      21
   6        27         15        6     40566    ✓   AS2 O3                                              9
  15       100          0       25     50632    ✓   AS                                                135
   8        42          6       15     60362    ✓   SE                                                 28
   6        39         15        6    100319        MN FE2 O4                                           9
   6        29          6       15    100455        PB BI S                                            21
   7        31          6        6    120403        BA BI2 NB2 O9                                      16
   7        27         10       10    130408        U O3 .2H2 O0                                       17
   6        46         10        6    160241        MN2 TI O4                                          12
   6        29         15        6    160551        /- BA2 ZN ( P O4 )2                                 9
   6        27          6       10    160598        H2 O (A9ODEGK )                                    18

IF THE ABOVE RESULTS DO NOT RESULT IN 'AN IDENTIFICATION', IT IS SUGGESTED THAT THE PARAMETERS BE CHANGED AND THIS PHASE REPEATED

THE EXECUTE TIME FOR THIS RUN WAS 136 SECONDS

RUN COMPLETE
```

Figure 2. Computer output of results for example 1.

shown in Figure 3, while the output is shown in Figure 4. Of the 22 possible standards which could make up these unknown patterns, from the chemistry, it can be seen that 6 standards are correct. The time given in the figure of 192 sec reflects the slow speed of an IBM 360/50. On an IBM 360/67, this pattern took 58 sec.

```
******* INPUT *******

**COMMENT** PATTERNS IN FILE WITH LESS THAN  1 LINES WITH AN INTENSITY GREATER THAN 63 CANNOT BE FOUND BY THIS SEARCH

**COMMENT** PATTERNS IDENTIFIED WITH A LOG INT MATCH OF 11 OR LESS WILL NOT BE PRINTED

**COMMENT** PATTERNS WITH LESS THAN   8 LINE MATCHES ARE NOT PRINTED

**COMMENT** PATTERNS WITH LESS THAN 40 PERCENT MATCH OF STANDARD IN RANGE CONSIDERED ARE NOT PRINTED

**COMMENT** ONLY D-SPACINGS LESS THAN  50.00 ARE USED IN THIS PHASE

**COMMENT** ONLY D-SPACINGS GREATER THAN   1.44 ARE USED IN THIS PHASE

**COMMENT** WINDOW FOR THIS RUN 15 + OR -   3

SAMPLE FHEVEL, IBID., SAMPLE U-3.6 KY2O3, AS2O3, AG3ASO4, KH2ASO4, PBSEO4<

UNKNOWN SPACINGS AND INTENSITIES
        7.0660       10
        6.7880        9                    2.1310        14
        6.3970       66                    2.1000        10
        5.2580       24                    2.0780         9
        5.2240       24                    2.0000        21
        5.0700        5                    1.9880         9
        5.0010        9                    1.9660         7
        4.8160        7                    1.9580        17
        4.4770       12                    1.8740        31
        4.4300       12                    1.8630         5
        4.3100       12                    1.8520         7
        4.1460       10                    1.8450         5
        3.8140       30                    1.8170         5
        3.7050        5                    1.7890         5
        3.4820       17                    1.7690         7
        3.3820       17                    1.7190         5
        3.2690       26                    1.7150         5
        3.2430        5                    1.7040        19
        3.2120       19                    1.7000        19
        3.1980       78                    1.6920         5
        3.1520       17                    1.6680        10
        3.0790       21                    1.6380        28
        3.0610      100                    1.6200         7
        2.9840       59                    1.5980        22
        2.9380        5                    1.5630         7
        2.9190        5                    1.5490         9
        2.8520       14                    1.4910         7
        2.7670       19                    1.4420         9
        2.7410       78                    1.5320        10
        2.6960        7                    1.9400         9
        2.6500       17
        2.6220        7
        2.6120        5
        2.5870        7
        2.5760        5
        2.5400       23
        2.5020       48
        2.3160        5
        2.3080        5
        2.2590       14
        2.2110        5
```

Figure 3. Input data for example 2.

******* OUTPUT ******* FS

NUMBER OF LINE MATCHES	PERCENT OF STANDARD LINE MATCHES	MINIMUM INTENSITY OF LINES COMPARED	LOG INT MATCH	ASTM NUMBER	FORMULA	RELIABILITY FACTOR
11	100	0	15	10604	BA S O3	66
9	40	6	15	10624	NA H S O4	32
15	71	6	15	10810	K2 SE O3	54
8	100	0	25	11113	SR O2	70
10	76	6	15	40230	U O2 F2	36
13	92	1	39	40566	AS2 O3	93
8	66	2	15	50534	SB2 O3	38
16	100	0	39	50574	Y2 O3	128
8	100	0	25	60493	AG3 AS O4	70
8	88	3	15	70234	SR O2	33
11	68	6	15	80044	K RE O4	39
10	76	6	15	100084	CA3 FE2 (GE O4)3	36
10	71	6	15	100090	CA3 CR2 (GE O4)3	36
13	76	2	39	100194	HO2 O3	83
10	52	6	15	110023	CA3 GA2 (GE O4)3	36
10	90	6	15	110237	B P O4	36
13	100	0	15	110309	(H3 O)2 (U O2)2 V2 O8Q	78
10	50	6	15	120767	K H2 AS O4	36
9	64	3	25	130007	CD3 SC2 GE3 O12	44
9	59	6	15	130008	CD3 CR2 GE3 O12	32
8	42	6	15	130010	CD3 FE2 GE3 O12	28
11	64	3	25	130426	DY3 GA2 (GA O4)3	53
12	57	3	15	140026	SN3 (O H)4 (N O3)2	50
8	50	6	15	140836	C3 H3 SM O6	28
26	68	6	15	150375	PB SE O4	93
8	72	3	15	160728	CA P	33

IF THE ABOVE RESULTS DO NOT RESULT IN 'AN IDENTIFICATION', IT IS SUGGESTED THAT THE PARAMETERS BE CHANGED AND THIS PHASE REPEATED

THE EXECUTE TIME FOR THIS RUN WAS 192 SECONDS

RUN COMPLETE

Figure 4. Computer output of results for example 2.

CONCLUSIONS

The search system is written in both FORTRAN II for the IBM 7074 and in FORTRAN IV for the IBM 360/65. It gives the chemist an additional analytical tool (using X-ray diffraction) which can be used to identify small polycrystalline samples. This method of identification, previously used only by diffractionists, can be used with minimal experimental techniques and yields correct results of six component mixtures in less than 2 min of computer time.

Note: A limited number of card decks can be obtained by writing to the authors. Anyone using the program does so at his own risk. The authors would appreciate any comments to improve or, if necessary, to correct this program.

REFERENCES

1. G. G. Johnson, Jr., and V. Vand, Twenty-fourth Pittsburgh Diffraction Conference, Paper No. B-1, 1966.
2. G. G. Johnson, Jr., and V. Vand, "A Computerized Powder Diffraction Identification System," *Ind. Eng. Chem.* **59**: 18, 1967.
3. G. G. Johnson, Jr., and V. Vand, "FORTRAN IV Programs for the Identification of Multiphase Unknown Powder Diffraction Patterns Using the Joint Committee Powder Diffraction File," *ASTM* (35 pp.), 1967.
4. The American Society for Testing and Materials, The American Crystallographic Association, The (British) Institute of Physics, and the National Association of Corrosion Engineers.
5. L. K. Frevel, "Computation Aids for Identifying Crystalline Phases by Powder Diffraction," *Anal. Chem.* **37**: 471, 1965.
6. V. Vand, Fourteenth Pittsburgh Diffraction Conference, Paper No. 8, 1965.
7. G. G. Johnson, Jr., and V. Vand, "A FORTRAN II System for the Identification of Unknown Multiphase Powder Diffraction Patterns," The Materials Research Laboratory, Pennsylvania State University, University Park, Pa. (52 pp.), 1967.
8. M. Nichols, Twenty-fourth Pittsburgh Diffraction Conference, Paper No. B-3, 1966.
9. M. Nichols, "A FORTRAN II Program for the Identification of X-ray Powder Diffraction Patterns," University of California Research Laboratory, UCRL–70078, 1966.

DISCUSSION

B. S. Sanderson (National Lead Co.): Can you give us some ideas for those of us who have rather small computers? What size core memory would you need for each of these programs?

G. G. Johnson: The FORTRAN II program requires about 6000 storage. It requires two storage devices such as discs or tapes. The programs are written so that they are device independent. In other words, if you have a disc, it is going to work exactly the same as with tape. The FORTRAN IV program takes about 45,000 words of core. I am presently in the process of writing it for a remote station so that I can sit in my office and use the search system and I won't have to walk to the computation center.

B. S. Sanderson: Can you also comment on what would happen, for instance, with that pattern that had about 96 lines. What would happen as you successively cut out the lower intensity lines? Would that increase the number significantly?

G. G. Johnson: As you cut out the lower intensity lines?

B. S. Sanderson: Yes, in other words, if you only take the higher intensity lines.

G. G. Johnson: The latest program has what is known as the background correction. Supposing you thought your background radiation was an intensity of about 5 and you only went down measuring those lines to about 5, you substantially get the same results. You do have to apply this value. Suppose you can't measure intensities of 2. You put a parameter in, known as the background of 2, and what the system does is, when it goes through the standard and if there is a whole series of lines in the standard with an intensity of 1, the program just ignores them. It doesn't try to fit those lines which are less than the background. The system doesn't base the identification on those lines which are less than the background which you state.

B. S. Sanderson: You did make a comment which you did in your paper in *Analytical Chemistry* about the assumption that the most intense lines all had equal intensity. As I understand it, you are making the assumption that the 100% lines on each of the cards all have equal intensity.

G. G. Johnson: This is one reason why we put this in—I didn't mention this, but the intensities are packed on a logometric scale and the reason we've done this is because we do not have absolute intensities such that, in one pattern, the line of 100 would really be a line of 50 in another pattern. But, if you take the logs of these and base them all on 100, these are really quite the same. There isn't much difference. Just Wednesday, I was talking to Don Hannawalt in Michigan. He really feels that you should use these numbers exactly. He doesn't feel that you should use logs. We feel that because these patterns are not on an absolute basis, we have to use logs. You can use what you want.

B. S. Sanderson: This point is well taken, and I think the comment that you made in your paper about the necessity perhaps of one additional number on each card which would represent an absolute intensity measurement would certainly be a very important thing to have.

G. G. Johnson: Dr. Scott of the Joint Committee is working on this, and he is referencing all lines to aluminum oxide. There are only about 400 or 500 patterns which have this on it right now, and they aren't in the file. It may be a future development.

B. S. Sanderson: Thank you. That was a very interesting paper.

A. Giamei: (Pratt and Whitney Aircraft): I have a suggestion, a cautioning remark, and a question. The suggestion is that you might be able to reduce your execution time if you first eliminate all the chemical compounds which are absolutely possible. I should think that this would cut down the possibilities by quite a bit. For example, if you are looking at a steel, you might eliminate all compounds of uranium, *etc.* This is the suggestion. The cautioning remark is that, in the use of the ASTM card file, one has to be very careful in using the quoted intensities for several reasons. I am sure you are familiar with these. For example, changes in wavelength or changes in type of measuring instrument could lead to discrepancies. Consider monochromator *versus* filter radiation, texture problems which the experimenter may have had, or grain-size problems. Also, you have to keep in mind that you may have severe texture or grain-size problems yourself. The question that I have is: What are the criteria for rejecting patterns? You have discussed "reliability factors." I should like to hear some of the details as to how you arrived at those factors. For example, if you just had a pure material, but, let's say, due to texture, one line happened to be absolutely missing which should have been present, what would the conclusion be?

G. G. Johnson: Let me answer your questions in reverse. If it were a line which would probably be one of the three strongest lines which was completely missing, and this is one of the problems that I was discussing with Hannawalt—he was missing lines of 80—, it rejects them. The search system works in the following manner. If you have lines of 100 there and you have the lines of 80 missing, that substance may not be there. The chemical questions we've already considered, and we found that it doesn't shorten the execute time at all. You have to read the serialized device which contains the whole ASTM file. So, even if it shortened it from, say, 57 sec down to 50 sec, you will be losing some information. Supposing you have a pattern in the file of rare earth and you also took another rare earth that you took the X-ray pattern of. If you eliminated these on chemical grounds, you would never obtain the isostructural compound. So, suppose you start pulling out all rare earths, all X_2O_3. You would have eliminated all these X_2O_3's because yours isn't in the file. So we feel the shortening of time for putting the chemical information in is not really necessary—you lose information.

In the ASTM file, in regard to the intensities being bad, yes, I'll agree some of the intensities are quite bad and this is the reason that we've gone to this logometric scale. Suppose you measure one intensity as 80 and someone else measures it as 100. If you take the log of both these numbers, 5 times the log to the base 10 of intensity, you will find that both of these have the same scaled intensity; they both have the intensity of 9. This is the reason that, in this paper, I describe that we feel we should be using a logometric intensity scale because it tends to wash out the differences in experimental procedures. This is described in the writeup that has just been published, and we feel that, for the present ASTM file where the intensities are differing by $\pm$ 10 to 20%, the logometric scale is the best scale. If the intensities get down to the point where they are differing by less than 10%, then I feel that a straight 10 representing 100,9 representing 90 to 100, 8 representing 80 to 89 would be the best scale. So we feel that the ASTM intensities are taken care of this way.

Chairman H. S. Peiser: I should like to say that, while this would obviously give the right answers in most cases, there would be some glaring examples where you would undoubtedly go wrong by this program. You gave the rare earth example. I think it is the second strongest line of garnet which is practically missing in some of the rare earth garnets, for instance, in some of the hydrogarnets. So, problems do exist.

G. G. Johnson: It would give us some problems.

H. S. Peiser: I think this is a very exciting method, but one which, in the case of difficult mixtures, would fall down relative to detailed interpretation by an expert. However, there is a lot to be said for a method which can be put in the hands of every analyst. I think that the work Dr. Johnson and Dr. Vand are doing is really a great step forward.

W. Ostertag (Corning Glass Works): Dr. Johnson, how can you handle solid solutions? Solid solutions occur quite frequently in inorganic chemistry.

G. G. Johnson: The problem of solid solution is taken care of by one of two methods. One is to open the window up wider; the second is—supposing it is a binary compound—to have a series of compounds across the tie line. We have found that, for that one example that Frevel had, there was a solid solution, and we were able to retrieve it because the change in the lattice parameter was quite small. If the change is larger, we will have to have a series of compounds in the ASTM file representing the solutions from A to B to give a series of compounds represented in the file of A–B compounds.

W. Ostertag: Can your program identify the components of a mixture if one component is present in trace amounts?

G. G. Johnson: Yes, it can. A pattern was sent to me by Tem-Pres Corporation at State College, I think it was something like 93% quartz and a small trace of vanadium. We were able to retrieve out of the vanadium compound. That's 93 to 7. I don't know how low you want to go.

W. Ostertag: Thank you very much.

MEASUREMENT OF ELASTIC STRAINS IN CRYSTAL SURFACES BY X-RAY DIFFRACTION TOPOGRAPHY

Brian R. Lawn*

Division of Engineering
Brown University
Providence, Rhode Island

ABSTRACT

The use of X-ray topographic techniques for studying elastic strains in crystals deformed at their surfaces is becoming widespread, especially in the field of silicon semiconductor devices. Although the broad features of the phenomenological processes involved in producing the strain patterns on the X-ray micrographs are understood, little attention has been devoted to evaluating the detailed nature or range of the strain fields in the crystal. In this paper, an elastic model is proposed for cases in which a region of crystal surface is uniformly deformed over a thin layer. With this model, the associated strain field in the surrounding crystal, which is readily computed from elasticity theory, may be characterized by a single parameter. The model is in accord with observed strain patterns on topographs of abraded diamond surfaces and silicon surfaces onto which a strip of metal film has been evaporated. From the spatial range of the diffraction contrast, an estimate of the parameter characterizing the strain field may be made.

INTRODUCTION

In recent years, the methods of X-ray topography have found increasing application in the study of elastic strains in crystal surfaces. Such strains may be introduced by any process which deforms the crystal surface and thus leaves it in a state of residual stress. Partial relief of the residual stress is then manifested as an elastic distortion of the crystal matrix surrounding the deformed region of surface, provided, of course, no plastic flow occurs. The edges of metal strips evaporated onto silicon substrates[1] and the boundaries of parts of crystal surface damaged by mechanical abrasion,[2] particle or photon irradiation,[3] or chemical effects such as surface diffusion and oxidation[1,4] are examples of regions where elastic strain fields have been detected by X-ray diffraction contrast. The existence of this type of strain field is of interest from a practical as well as an academic standpoint. For instance, the presence of the elastic strains at the edges of evaporated thin films is believed to be an instrumental factor in the breakdown of semiconductor devices.

Most attention in the studies performed to date has been directed toward a basic understanding of the physical processes causing the elastic strains and to a qualitative interpretation of the associated diffraction contrast on the topographs. This paper is concerned more with the nature and magnitude of the elastic strain field itself. An elastic model is first proposed in which the strain field can be calculated and characterized by a single parameter. The diffraction contrast on X-ray topographs is then examined and used to verify the essential predictions of the model. Finally, from a measurement of the

* Present address: School of Physics, University of N.S.W., Kensington, N.S.W., Australia.

range of observable diffraction contrast, an estimate of the characterizing strain parameter is made.

ELASTIC MODEL

In the interest of simplicity, we specify the following assumptions: (1) The strains in the matrix crystal are taken to be representable by the equations of linear isotropic elasticity; (2) the depth δ of the deformed surface layer of crystal is assumed small compared to the surface dimensions of the layer, which are in turn assumed small compared to the dimensions of the crystal itself; (3) the deformation is considered uniformly distributed throughout the deformed region of crystal at the surface. Thus, we have as our model an otherwise perfect semi-infinite crystal matrix with a thin isotropic layer of "bad" crystal embedded in its surface.

The strain field in crystal surrounding a deformed surface layer of arbitrary shape may now be computed in the following manner. Let us consider the residual stress in the bad layer compressive (tension will simply reverse the sign of the strains) so that the layer must expand to relieve the stresses. Since δ is small, this tendency to expand will give rise to an outwardly directed normal pressure p, very nearly parallel to the crystal surface, on the matrix at the peripheral boundary of the layer. By Saint-Venant's principle, as long as we do not concern ourselves with the strain situation within distances δ of the periphery, we may consider the pressure p equivalent to a line force F per unit of length of the periphery. Writing $\mathbf{n}$ as a unit vector lying in the crystal surface and directed perpendicularly outward from the periphery of the bad layer (Figure 1), we then have $\mathbf{F} = p\delta\mathbf{n}$. Defining a system of cartesian coordinates by the unit vectors $\mathbf{x}_1$, $\mathbf{x}_2$, $\mathbf{x}_3$, we can now compute the displacements at any point (x_1, x_2, x_3) in the crystal owing to an effective point force $F dl$ at $(x_1', x_2', 0)$ and, by integrating around the entire periphery of the bad layer, the total displacement field may be determined. Following Landau and Lifshitz,[5] we arrive at an expression of the following type for the ith component of the displacement vector:

$$u_i = \frac{F}{E} \oint G_{ik}(x_1 - x_1', x_2 - x_2', x_3, \nu) n_k \, dl \tag{1}$$

where E is Young's modulus; ν is Poisson's ratio; n_k is the component of $\mathbf{n}$ in the direction $\mathbf{x}_k$; the G_{ik} are given by Landau and Lifshitz;[5] and k is a repeated suffix. In equation (1), F is assumed constant everywhere on the periphery (this assumption demands further consideration in special instances). From equation (1), the strain field may be established for any shape of periphery of bad crystal. A high-speed computer is generally required to perform the calculations. Such calculations have been done for elliptical microabrasion patches on surfaces of diamond.[2]

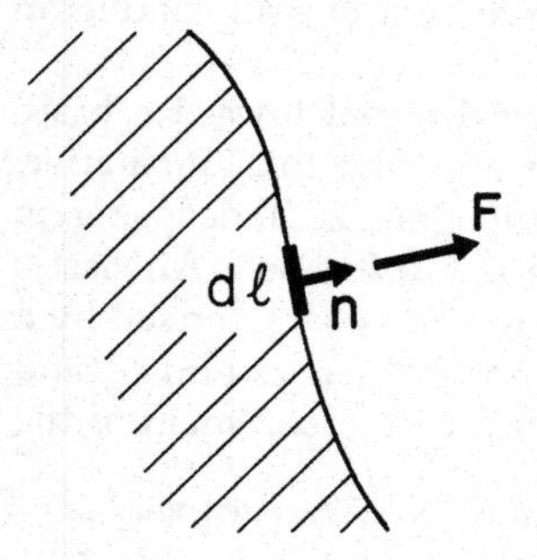

Figure 1. The force exerted by a deformed surface layer on a crystal matrix.

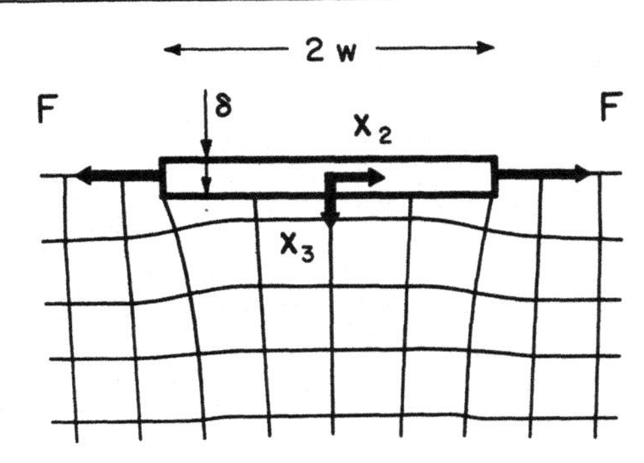

Figure 2. Distortion of the crystal lattice due to residual compressive stress in the deformed layer.

In the present paper, we can choose a particularly simple peripheral geometry without loss in generality. We take the bad layer to be a long strip parallel to $\mathbf{x}_1$ with edges at $x_2 = \pm w$ and thickness δ measured in direction x_3. Such a description probably applies reasonably well to the strips of metal evaporated onto silicon surfaces.[1] Solutions for the strain field may be computed from equation (1) or, alternatively (for this particular case) from the two-dimensional solution for straight-line forces in a surface given by Timoshenko and Goodier[6] and others. The expressions for the displacements and strains are too cumbersome to be given here. Instead, the displacements are displayed in schematic form in Figure 2 as a distorted lattice for an arbitrary value of F.

SOME VERIFICATIONS OF THE MODEL

The object of this section is to provide evidence supporting the model outlined in the previous section.

Optical Examination of the Crystal Surface

One of the predictions of the elastic model is readily investigated by optical means. As seen in Figure 2, the crystal surface surrounding the embedded bad layer is flat, a prediction which turns out to be independent of the peripheral shape of the layer. The crystal surfaces surrounding elliptical abrasion patches on diamond and evaporated metal strips on silicon are indeed found to be flat (allowing for original surface roughness) within the limits of detection of interference microscopy. The model similarly predicts that the central, deformed region of the surface will be raised or lowered slightly according to whether the direction of $\mathbf{F}$ is outward or inward, but the very nature of the deforming mechanism precludes any observation of such an effect in the cases studied here. In the case of the abraded diamond, some material is removed from the surface, and, in the case of the metal strip on the silicon substrate, the thickness of the evaporated strip obscures any measurement. However, in the latter case, the top surface of the strip is found to be flat, as the model would predict for a uniformly thick layer.

X-Ray Topographs of the Crystals

The optical examinations mentioned above provide no indication that the crystal might be in a state of residual strain. However, the extremely strain-sensitive techniques of X-ray topography[7] reveal quite plainly the presence of the strain field, as seen in Figures 3 and 4 (taken under the usual experimental arrangement in which the angular divergence of the incident X-ray beam greatly exceeds the range of reflection of the perfect crystal). In Figure 3 are seen topographic images of the elastic strains around the edges

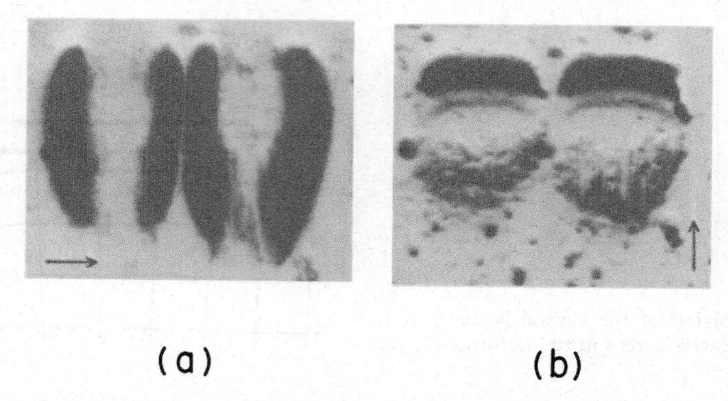

Figure 3. X-ray transmission topographs of elliptical abrasion patches on the (111) face of a natural diamond; Mo $K\alpha$ radiation; (a) $2\bar{2}0$ and (b) 220 reflections; the arrows denote **g** vectors; the mean diameter of patches is 220 μm; and $\delta \sim 1$ to 10 μm.

Figure 4. X-ray transmission topographs of a silicon wedge with an aluminum strip evaporated onto the X-ray exit surface; the arrows indicate values of μt at each end of wedge; the surface plane is closely parallel to (111); Mo $K\alpha$ radiation, $2\bar{2}0$ (top) and $\bar{2}20$ (bottom) stereo-pair; width $2w = 720$ μm; and $\delta \simeq 0.25$ μm.

of two abrasion patches on a natural octahedron surface of diamond.[2] Natural percussion damage gives rise to the extra specks. The second set of topographs (Figure 4) shows a stereopair[7] of a silicon wedge onto which a strip of aluminum has been evaporated. The thickness of the wedge varies roughly linearly from 0.15 mm at the left-hand side of the figure to about 1.60 mm at the right-hand side. The thickness of the wedge may be inferred from the Pendellösung fringe pattern[8] which depicts contours of equal thickness. The "fading" of the pattern every fourteenth or fifteenth fringe is a diffraction effect which has been described elsewhere.[9] Further diffraction contrast, *e.g.*, the vertical white lines[10] and the specks due to accidental surface damage, are merely pointed out here and will receive no further attention.

The nature of the diffraction contrast observed on transmission topographs depends on the thickness of the specimen. For $\mu t < 1$ (μ is the linear absorption coefficient and t is the thickness of the specimen), the contrast is always positive and insensitive to the sign of the strains. The contrast seen in Figure 3 and on the very left of Figure 4 falls into this category. Now it is well established that this type of contrast becomes zero when

g · **u** (**g** is the reciprocal lattice vector) itself becomes zero, so that, with proper choice of reflecting planes, indications of the direction (but not the *sense*) of **u** may be obtained. A survey of this kind reveals that, in all diamond and silicon specimens studied, the contrast reduces to zero at all positions on the peripheral boundaries of the bad layers where **g** · **n** is zero and reaches its maximum where **g** · **n** is a maximum (see especially Figure 3). This is in accord with the model, which predicts a state of very nearly plane strain in the plane defined by **n** and $\mathbf{x}_3$, so that **u** is always very nearly perpendicular to the periphery.

When $\mu t \gg 1$, the contrast may be either positive or negative, or both. This condition is realized on the extreme right of Figure 4. Meieran and Blech[1] have empirically established a rule for determining the sign of the strains, and thus of force **F**, from the asymmetry of the diffraction contrast. Their rule may be stated as follows: The peripheral contrast has net positive value where **F** · **n** and **g** · **n** have the same sign. From Figure 4, we thereby deduce that **F** is directed outward from the strip edges, so that the bad layer is in this case in a state of residual compression. (A similar deduction is made for the diamond specimens.[2])

INTERPRETATION OF CONTRAST MECHANISMS

Before we can make an estimate of the magnitude of the elastic strain field giving rise to the observed diffraction contrast on the topographs, we must first understand the contrast mechanisms themselves. In the previous section, we distinguished between contrast at low absorption (commonly termed *extinction* contrast) and contrast at high absorption (*Borrmann* contrast). A tentative inspection of Figure 4 would indicate that the distinction between these two contrast regimes is not always too clear. The edges of the strip for which **g** · **n** is positive give rise to a band of positive contrast invariant of crystal thickness, while the opposite edges give rise to a band of similar width within which the contrast changes progressively from positive to negative as the crystal becomes thicker. The striking implication to be made from this observation is that the *spatial range* of contrast is determined uniquely by the strain level and not at all by the absorption level, absorption accounting only for the *distribution* of intensity *within* the bands of contrast. Thus, if we can establish the strain situation at the extremities of the bands where the contrast becomes zero, we have a means by which the entire strain field may be specified, and this without having to place any restrictions on specimen thickness.

Unfortunately, the diffraction of X-rays in nearly perfect crystals is a very complicated process when the beam incident on the crystal is widely angularly divergent. The reason for this is that, in general, the diffracted intensity reaching each point on the crystal exit surface is composed of the sum of contributions from rays contained within the *energy-flow-triangle* delineated by the directions of the direct and diffracted beams. The relatively large angle subtended by these two directions, together with the relatively thick specimens used, generally precludes such practical simplifications as the "column" approximation made in electron microscopy. But, in certain instances, such as at very low and very high absorption, and with the strain mainly localized near the X-ray exit surface of the crystal, at least a semiquantitative analysis of the contrast can be developed. The treatment is further aided by considering only cases of symmetrical transmission so that **g** is parallel (or antiparallel) to **n**.

Penning and Polder[11] were the first to present a formal treatment of the behavior of X-ray beams in *slightly* distorted crystals. A slightly distorted crystal is interpreted as one in which the allowed wave field at each point can be completely specified by wave points on the various branches of the dispersion surface construction in reciprocal space.

Each wave point corresponds to a ray (energy-flow vector for a pair of incident and diffracted waves) whose direction of propagation is normal to the dispersion surface at that point. The Penning and Polder theory indicates that, in an undistorted crystal, the wave points retain their location on the dispersion surface as the rays pass through the crystal, while, in a slightly distorted crystal, they "migrate" along their respective branches. The physical interpretation of the wave-point migration is that the rays propagate along curved paths rather than in straight lines as they do in perfect crystals and that there is an associated redistribution of the energy flow in the incident and diffracted directions. The amount of migration of a given wave point as the ray propagates an incremental distance is proportional to a deformation parameter η defined in terms of the component of $\mathbf{u}$ parallel to $\mathbf{g}$: for $\mathbf{g}$ parallel to $\mathbf{x}_2$, we have,[11,12,13] with θ the Bragg angle and with $\mathbf{x}_2$ and $\mathbf{x}_3$ defining the plane of incidence,

$$\eta = \cos^2\theta \frac{\partial^2 u_2}{\partial x_3{}^2} - \sin^2\theta \frac{\partial^2 u_2}{\partial x_2{}^2} \tag{2}$$

Reversing the sign of u_2 or $\mathbf{g}$ reverses the sign of η. The total migration of a ray when it reaches the exit surface is then proportional to the integrated value of η over the ray path between the entrance and exit surfaces. The terms $\partial^2 u_2/\partial x_3{}^2$ and $\partial^2 u_2/\partial x_2{}^2$ [computed from equation (1)] are shown as functions of relative depth x_3/w below the crystal surface in Figure 5. It is noted that, in all cases, these terms reverse their sign just beneath the crystal surface and that they attain their maximum value *at* the crystal surface.

We now investigate the application of the above concepts to the strain field at the edges of the evaporated aluminum strip on the silicon substrate (Figure 4). At both high and low absorption, it will be found that the observed contrast can only be explained in terms of a breakdown of the Penning and Polder theory.

High-Absorption Case ($\mu t \gg 1$)

When the absorption is high, Borrmann transmission occurs; *i.e.*, rays which belong to that branch of the dispersion surface with the longest wave vectors and which travel closely parallel to the Bragg planes show anomalously high transmission, while all remaining rays are rapidly attenuated. For the weakly absorbed rays, Penning and Polder show that, when the deformation parameter η is positive, the wave points migrate along the dispersion surface in such a way that the flow of energy into the diffracted beam becomes increased (at the expense of the direct beam); similarly, for a negative parameter, energy is transferred into the direct-beam direction. The integration of η is now performed by making the assumption that the ray-propagation vector remains antiparallel to $\mathbf{x}_3$; since the strongest distortion is at the X-ray exit surface, this assumption should not lead to much error owing to spreading of the rays. When this is done, it is found that the integrated value of η at the exit surface is zero for all values of x_2. This is not altogether

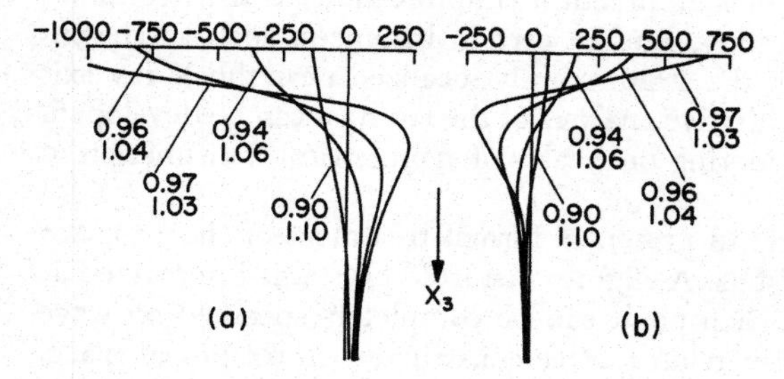

Figure 5. Plots of (a) $\partial^2 u_2/\partial x_3{}^2$ and (b) $\partial^2 u_2/\partial x_2{}^2$ (plotted on abscissas in units of F/Ew^2) as a function of relative depth x_3/w (plotted down to the depth $x_3 = 0.12\,w$); labels on the curves denote particular values of x_2/w.

surprising since we saw in Figure 5 that the terms in η reversed their signs below the crystal surface. The physical interpretation of the zero integral is best seen with reference to the first of the two terms in equation (2). The integration of this term over the thickness of the specimen along x_3 gives, for small θ, the difference in lattice tilt at the exit and entrance surfaces. Both these tilts must be zero according to the model of Figure 2 in which the surfaces are flat and stress free except at the peripheries of the bad layer. Thus the Penning and Polder theory would predict zero contrast, which is, of course, in direct contradiction with the observation in Figure 4.

The explanation of this apparent discrepancy lies in the fact that we have assumed the lattice distortion to be small enough that the behavior of X-rays in the crystal may be represented by a migration of wave points along the branches of the dispersion surface. However, Penning[12] shows that, when $|\eta|$ locally exceeds some critical value $|\eta_c|$, the propagating ray can no longer adjust its curvature sufficiently rapidly to satisfy the above dynamical description (see later). Bearing this in mind, we now investigate the possible behavior of a ray approaching the exit surface in Figure 2. We find from Figure 5 that, at the edge for which **g** is parallel to **n**, the integrated value of η will first become increasingly positive, reaching a maximum value where the curves cross the ordinate, and will subsequently decrease again toward zero. If $|\eta_c|$ is exceeded before reaching the exit surface, the ray can no longer be dynamically scattered and will pass out of the crystal without further reflection. This will prevent the integrated η from attaining its zero value, so that the *net* η will be effectively positive at the exit surface and will thus give rise to extra intensity in the diffracted-beam direction. Similarly, with **g** antiparallel to **n**, the sign of η becomes reversed, so that, for the opposite edge of the strip, extra intensity will appear in the direct- rather than diffracted-beam direction. This explains the asymmetric contrast on the right-hand side of Figure 4.

Low-Absorption Case ($\mu t < 1$)

The low-absorption case is complicated by the presence of rays from *all* points on *all* branches (including those branches due to the polarization of the beam) of the dispersion surface. However, the situation is simplified by the fact that the upper and lower branches have an equal and opposite effect on the integrated intensity at the exit surface. Therefore, we should again expect to observe zero contrast at the periphery of the bad layer. But, when absorption is low, we must take cognizance of the divergent incident beam. Only that portion of the beam falling within the angular range of reflection of the perfect crystal is dynamically diffracted according to the dispersion-surface representation, the remainder of the beam passing through the crystal almost unattenuated. It is this latter part of the incident beam which can give rise to diffraction contrast. Those regions of crystal for which $|\eta| > |\eta_c|$, although not able to diffract the X-rays dynamically, will be suitably oriented to diffract some part of the intense direct beam, which gives rise to positive contrast at *both* edges of the strip on the left-hand side of Figure 4.

ESTIMATE OF PARAMETER F

In the previous section, we saw that the diffraction contrast at high and low absorption could be explained if we postulated that the lattice distortion near the periphery satisfied the condition $|\eta| > |\eta_c|$. (The argument can be extended to the case of intermediate absorption.[2]) Penning[12] interprets the breakdown of the dynamical scattering concept in the following way. If the distortion of the crystal is too severe, there are insufficient

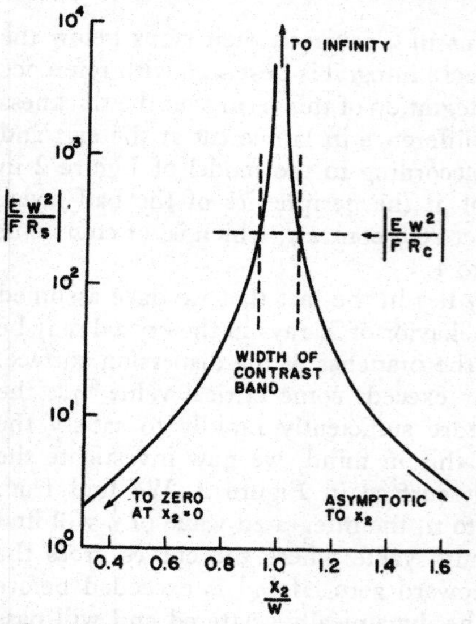

Figure 6. Plot of $|Ew^2/FR_s|$ as a function of the relative length x_2/w. The vertical dotted lines represent the limits of observed diffraction contrast and the corresponding horizontal dotted line then represents the cutoff value of $|R_c^{-1}|$.

reflecting planes to generate the necessary curvature of the ray path. Using an argument of this kind, one obtains[2]

$$|\eta_c| \simeq 1/gt_e^2 \tag{3}$$

with t_e an extinction distance.

For the diffraction conditions used in Figure 4, $\theta = 11°$, so that the first term in equation (2) is the dominant one. We may therefore approximate η_c to R_c^{-1}, where R is the radius of curvature of the Bragg planes. Now, we saw in Figure 5 that this curvature attained its maximum value R_s^{-1} at the X-ray exit surface. Thus, for diffraction contrast to appear, the curvature $|R_s^{-1}|$ should exceed the critical value $|R_c^{-1}|$ computed from equation (3). In Figure 6, $|R_s^{-1}|$ is plotted as a function of x_2 (the quantities E, F, and w appear when the curvature is plotted in dimensionless form). From the position of the contrast band at the aluminum strip edges marked on this diagram, we find

$$\left|\frac{1}{R_c}\right| = 2.5 \times 10^2 \left|\frac{F}{Ew^2}\right| \tag{4}$$

as the critical curvature relevant to Figure 4. Equating the right-hand sides of equations (3) and (4) and inserting $E = 13 \times 10^{11}$ dyne/cm^2 for silicon, $2w = 0.072$ cm, $g = 5.2 \times 10^7$ cm^{-1}, and $t_e = 37 \times 10^{-4}$ cm for Mo $K\alpha$, 220 reflection in silicon, we arrive at a value of 1×10^4 dyne/cm for F. This may be compared with the value $F = 5 \times 10^5$ dyne/cm obtained for the abrasions on the diamond surface in Figure 3.[2]

CONCLUSION

We have presented an elastic model from which the strains around damaged layers of crystal surface may be calculated, and it has been shown that the single parameter F characterizing the strain field can be determined from X-ray topographs by simply observing the spatial range of diffraction contrast. It is, however, pointed out that several approximations attend the treatment given above. For instance δ is considered effectively infinitesimally small so that we can introduce the concept of the line force $\mathbf{F}$ in the surface. In the cases discussed above, the width of the contrast bands greatly exceeds δ, so that little error is introduced. We also considered θ small so that η_c might be replaceable by R_c^{-1}; for θ larger, Figure 6 would require R_s^{-1} to be replaced by η_s, and this plot would then only be valid for the one value of θ. It was further assumed that the curvature of the rays through the distorted crystal was not severe; the small θ and the localization of the strain field at the exit surface permitted this assumption to be made. Finally, the Penning criterion predicting the breakdown of dynamical scattering [equation (3)] is not exact and is subject to some degree of uncertainty. Thus, in applying the above treatment to the evaluation of crystal surface strains, one must ensure that the various assumptions are adequately satisfied, and the deduced value of F must be regarded as having order of magnitude accuracy only.

ACKNOWLEDGMENTS

The author wishes to point out that parts of this work were carried out in collaboration with Professor F. C. Frank, Dr. M. Hart, Dr. A. R. Lang, all of Bristol University, and Dr. E. M. Wilks, of Oxford University. He also wishes to thank Professor Barton Roessler, of Brown University, for useful criticisms of the manuscript. He is indebted to Industrial Distributors and the Advanced Research Projects Agency for financial support.

REFERENCES

1. E. S. Meieran and I. A. Blech, "X-Ray Extinction Contrast Topography of Silicon Strained by Thin Surface Films," *J. Appl. Phys.* **36**: 3162, 1965.
2. F. C. Frank, B. R. Lawn, A. R. Lang, and E. M. Wilks, "A Study of Strains in Abraded Diamond Surfaces," *Proc. Roy. Soc.* (*London*) **A301**: 239, 1967.
3. K. Martin, "Irradiation Damage in Lithium Fluoride," Thesis, Bristol University, 1966.
4. J. M. Fairfield and G. H. Schwuttke, "Precipitation Effects in Diffused Transistor Structures," *J. Appl. Phys.* **37**: 1536, 1966.
5. L. D. Landau and E. M. Lifshitz, *Theory of Elasticity*, Pergamon Press, London, 1959, pp. 26–29.
6. S. Timoshenko and J. N. Goodier, *Theory of Elasticity*, McGraw-Hill Book Company, New York, 1951, pp. 85–91.
7. A. R. Lang, "Studies of Individual Dislocations in Crystals by X-ray Diffraction Microradiography," *J. Appl. Phys.* **30**: 1748, 1959.
8. N. Kato and A. R. Lang, "A Study of Pendellösung Fringes in X-ray Diffraction," *Acta Cryst.* **12**: 787, 1959.
9. M. Hart and A. R. Lang, "The Influence of X-ray Polarization on the Visibility of Pendellösung Fringes in X-ray Diffraction Topographs," *Acta Cryst.* **19**: 73, 1965.
10. M. Hart and A. R. Lang, "Direct Determination of X-ray Reflection Phase Relationships through Simultaneous Reflection," *Phys. Rev. Letters* **7**: 120, 1961.
11. P. Penning and D. Polder, "Anomalous Transmission of X-rays in Elastically Deformed Crystals," *Phillips Res. Rept.* **16**: 419, 1961.
12. P. Penning, "Theory of X-ray Diffraction in Unstrained and Lightly Strained Perfect Crystals," Thesis, Technological University of Delft, 1966.
13. M. Hart, "Dynamic X-ray Diffraction in the Strain Fields of Individual Dislocations," Thesis, Bristol University, 1963.

X-RAY DOUBLE-CRYSTAL METHOD OF ANALYZING MICROSTRAINS WITH BeO SINGLE CRYSTALS

Jun-ichi Chikawa*

University of Denver
Denver, Colorado

and Stanley B. Austerman

Autonetics
Anaheim, California

ABSTRACT

A double-crystal arrangement was employed in the symmetrical Laue arrangement [$(+n, -n)$ setting]. A perfect BeO crystal was used for the first crystal of the double-crystal spectrometer. To obtain a high X-ray intensity, the thickness of the crystal was made to correspond to a maximum of Pendellösung interference. A slit was placed between the first and second (specimen) crystals to select the X-rays which precisely satisfy the Bragg condition. The slit was adjusted to avoid significant Fraunhofer diffraction. In this method, the incident beam for the specimen crystal was parallel enough to obtain intrinsic rocking curves of the specimen crystal. As an application, the method was used for determination of the senses of slight strains in BeO crystals.

INTRODUCTION

Lattice imperfections and lattice distortions have been observed by X-ray diffraction topography.[1] Images of imperfections by X-ray topography are more complicated than those by electron microscopy[2] and, thereby, inherently contain more information. For the X-ray case, Bragg diffraction-peak widths are very narrow, and the incident beams have a large divergency, so that the wave fields corresponding to the entire Bragg-peak width are produced. The X-rays in the incident beam deviated from the Bragg condition produce so-called "kinematical images," and the X-rays which nearly or exactly satisfy the Bragg condition make "dynamical images."[3] If the distortion in crystals is intermediary between both kinematical and dynamical images, new wave fields are produced.[3] These different images are superimposed so that the net images are very complicated. Therefore, we attempted to simplify this problem with a highly parallel incident beam with which we can separate these images.

Such highly parallel beams may be obtained by a double-crystal method in Laue arrangement which has been used by Authier.[4] In the present experiment, BeO crystals have been used for the first and second crystals in the double-crystal method. Crystals with a very low absorption coefficient such as BeO provide a parallel beam with a fairly high intensity.

THEORY

According to the dynamical theory of X-ray diffraction for perfect crystals, the energy flow of X-rays which satisfy the Bragg condition passes exactly parallel to the

* Present address: Broadcasting Science Research Laboratories of NHK, Kinuta-machi, Setagaya-ku, Tokyo, Japan.

reflecting plane, and the energy flow of X-rays deviated slightly from the Bragg condition are radiated into the X-ray fan from the entrance position of the incident beam on the crystal surface,[5] as shown schematically in Figure 1. When the glancing angle of the incident beam changes slightly, there is a large change in the direction of the energy flow, *i.e.*, the crystal acts as an angular amplifier. This effect is central to the present method of strain detection. The amplification factor A is expressed by

$$A = \frac{d\alpha}{d\theta}$$

where α is the angle between the energy path and the reflecting plane; and θ, the glancing angle of the wave vector on the reflecting plane. Since the paths are perpendicular to the dispersion surfaces of the wave fields,[5] the amplification factor is obtained from the curvatures of the dispersion surfaces. The amplification factor takes a maximum value at $\theta = \theta_B$, where θ_B is the Bragg angle. For a symmetric Laue case,

$$A = \left(\frac{d\alpha}{d\theta}\right)_{\theta=\theta_B} = \frac{2\sin^2\theta_B}{C|\phi_h|}$$

where C is the polarization factor and ϕ_h is related to the crystal structure factor F_h by the well-known formula[6]

$$\phi_h = \frac{\lambda^2}{\pi V}\left(\frac{e^2}{mc^2}\right)|F_h|$$

where V is the volume of the unit cell and e, m, and c have the usual meanings.

For example, for the $2\bar{1}\bar{1}0$ reflection of BeO crystals with Mo $K\alpha_1$ radiation, $A = 0.9 \times 10^5$ at $\theta = \theta_B$. Therefore, if a slit is placed behind the crystal as shown in Figure 1 (second slit), we can, in principle, obtain a very parallel beam diffracted from the perfect crystal, which serves as an incident beam for the specimen crystal.

Here we should consider Fraunhofer diffraction from the slit. With a slit width w, the first maximum of the diffraction appears at the angle

$$\delta \sim \frac{0.7\lambda}{w}$$

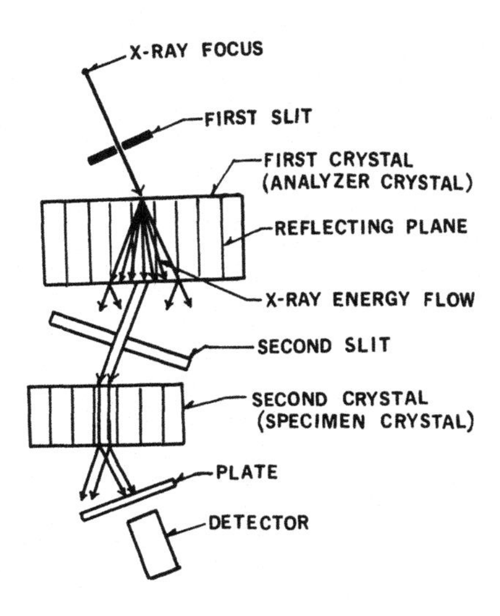

Figure 1. Double-crystal method in the Laue arrangement.

relative to the primary beam.[7] In order to attain a divergency of the beam one order less than rocking-curve widths (a few seconds of arc), the slit width should be as wide as 100 μ. To select the energy flows with such a wide slit, a very thick crystal is necessary. For most kinds of crystals, the necessary thickness is in a range of anomalous transmission. For BeO crystals, however, the very low absorption coefficient ($\mu_0 = 2.86$ cm^{-1} for Mo $K\alpha_1$ and $\mu_0 = 23.9$ cm^{-1} for Cu $K\alpha_1$) allows strong Pendellösung interference to be observed for a thick crystal. Therefore, a high intensity of the parallel beam can be obtained for such a thickness as gives a maximum of Pendellösung[8] interference.

EXPERIMENT

The section topographs in Figure 2 were taken for a selected BeO crystal, grown by the flux method,[9] by placing a plate between the crystal and the second slit in Figure 1. At the center of this section topograph for the diffracted beam in Figure 2(a), a dark line with a width of about 70 μ can be seen, which is due to the interference maximum. In the section topograph for the direct beam in Figure 2(b), the reverse contrast is seen, *i.e.*, the center line is very light. This observation shows that almost all X-ray energy satisfying the Bragg condition is directed toward the diffracted direction. Rough estimation indicates that the intensities at the interference maxima are four or more times higher than the intensities in the range of anomalous transmission.

Moreover, there is another advantage in using Pendellösung interference effects in crystals with low absorption coefficient. In Figure 2(a), a very light line can be seen on each side of the interference maximum at the center of the topograph. These two lines correspond to interference minima for X-rays which are slightly deviated from the Bragg condition.[8] Therefore, even with a width of the second slit that is equal to the separation of the two interference minima, the effective slit width is much smaller.

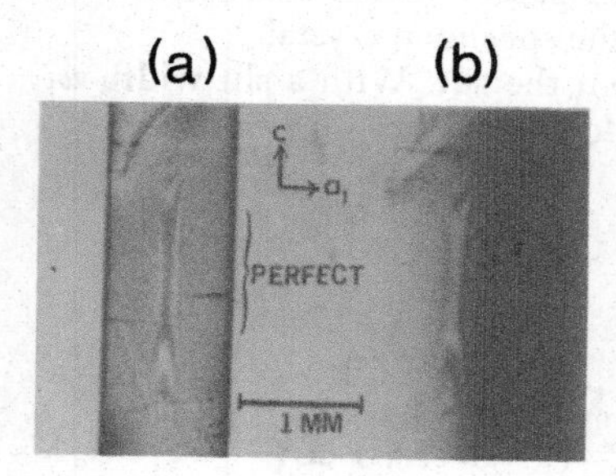

Figure 2. Section topograph of the first crystal (BeO) of the double-crystal method. The diffracted beam from the perfect region was used as the incident beam for the specimen crystal; (a) and (b) were taken for the diffracted and transmitted beams, respectively; BeO $2\bar{1}\bar{1}0$ reflection, Mo $K\alpha_1$; crystal thickness, 2 mm.

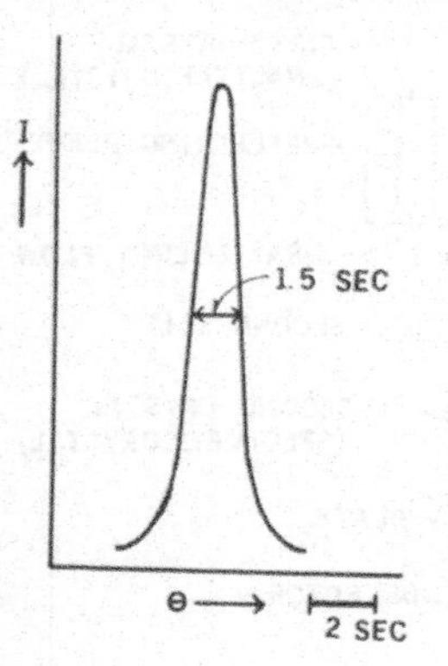

Figure 3. A rocking curve of a specimen BeO crystal obtained by the method in Figure 1, with the incident beam shown in Figure 2 and a $2\bar{1}\bar{1}0$ reflection ($+n$, $-n$) setting. The rocking curve was measured for a small area of the specimen crystal to minimize vertical divergency of the X-ray beam.

That is, the divergency of the very parallel beam is defined by the peak width of the central interference maximum, while the second slit, which blocks out all other maxima, has a greater width that produces a negligibly small angle δ in the Fraunhofer diffraction.

To examine parallelism of the beam obtained, rocking curves were measured for another selected BeO crystal used as the specimen crystal. The crystals were in the $(+n, -n)$ setting of the Laue arrangement. A rocking curve thus obtained is shown in Figure 3. The rocking-curve width is 1.5″ of arc, which is very close to the theoretical width of 1.4″ of arc calculated with the assumption of an ideally perfect crystal and a completely parallel incident beam.[6] The narrow rocking curve shows that the beam which enters the second crystal is parallel enough that it does not contribute significantly to the rocking curve of the specimen crystal.

AN APPLICATION

According to Penning and Polder's theory,[10] the energy flows of X-rays are bent in a strain field. This bending can be seen by this method. Figure 4(a) is a conventional Lang topograph of a BeO crystal which was sliced perpendicular to the c axis to form a thin plate. The two black parts indicated by the arrows are inversion twin cores.[11] The absolute sign of the c-axis polarity of the two parts is opposite to the rest of the crystal. Figure 4(b) is a section topograph of the same crystal which was taken by the method shown schematically under the topograph. The position of the section topograph is indicated by the dotted line in (a). In the central part of (b), hook-shaped fringes can be seen which indicate a slight lattice strain. The fringes can be explained by the spacing

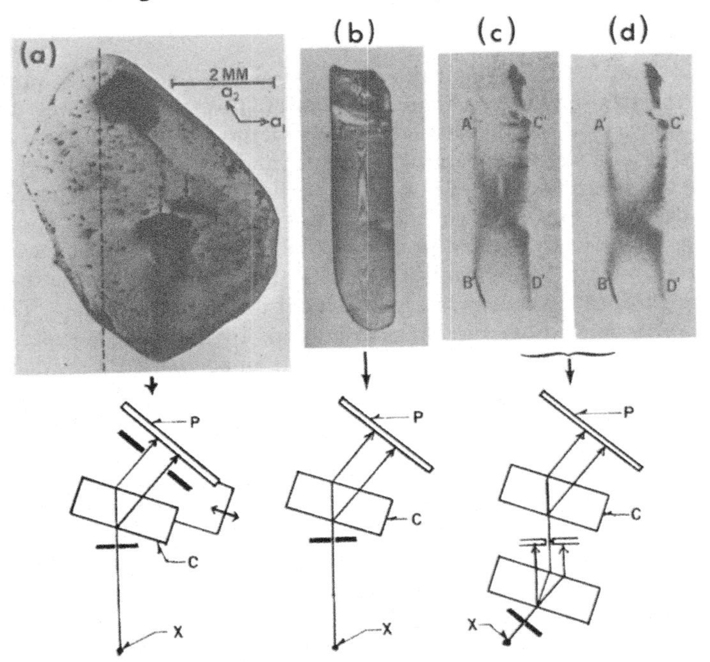

Figure 4. Topographs of a BeO crystal with local strains. The crystal thickness is 2.3 mm. The methods of taking topographs are illustrated under each topograph: (a) Lang topograph; (b) section topograph at the position indicated by the dotted line in (a); (c) and (d) topographs taken at the same position as (b) by the double-crystal method.

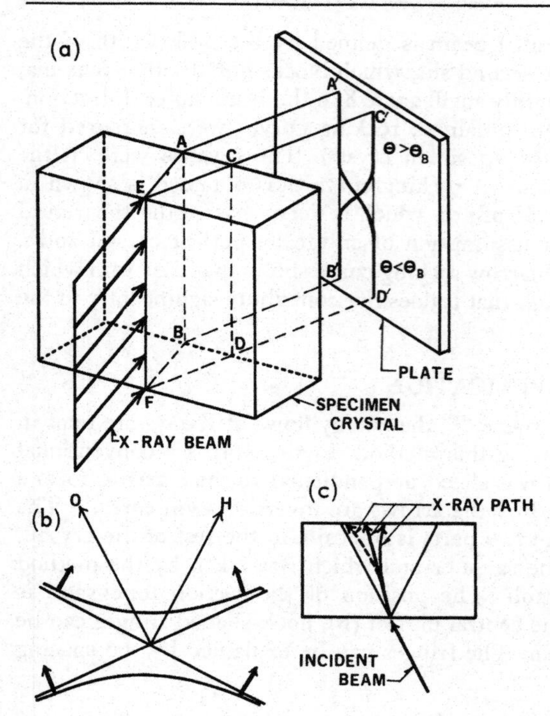

Figure 5. Line images and X-ray energy flow in the double-crystal method: (a) specimen crystal and line images on the plate; (b) dispersion surfaces in the reciprocal lattice and directions of X-ray energy flows (arrows); (c) bending of the X-ray energy flow due to strains.

contraction of Pendellösung fringes due to the strain.[12] The strain is presumed to be produced by the twin boundary. Figures 4(c) and (d) were taken by the double-crystal method. The first crystal was that crystal used for Figure 2. A plate was placed behind the specimen crystal, as shown schematically under (c) and (d). Of course, the incident beam is parallel in the horizontal plane but has a vertical divergency perpendicular to the sheet of the drawing. Furthermore, the reflecting planes of both the specimen crystal and the first crystal usually are not precisely parallel owing to almost inevitable instrumentation misalignment. Consequently, the incident beam can satisfy the Bragg condition only for a small area of the specimen crystal even if the crystal is perfect. The upper and lower parts are deviated from the Bragg condition. In the present experiment, it was found by rocking-curve peak position that, for the upper part, the glancing angle θ is larger than the Bragg angle θ_B [Figure 5(a)] and the wave fields are excited in the right-hand side of the dispersion surfaces. For the lower parts, θ is smaller than θ_B. First, note the lower branch of the dispersion surfaces in Figure 5(b). For the upper part of the crystal, the X-ray path is in the direction of EC in Figure 5(a) because the path is perpendicular to the dispersion surface,[5] and the diffracted beam reaches the line $C'D'$ on the plate. The value of $\theta - \theta_B$ decreases from the upper part to the lower part. This decrease corresponds to the moving of the wave points in the dispersion surfaces from right to left in Figure 5(b). That is, the direction of the X-ray path changes from EC to EA, and, therefore, the image by the diffracted beam shifts from $C'D'$ to $A'B'$, as seen in the schematic drawing. Similarly, for the upper branch of the dispersion surfaces, the image should be a line curved in the opposite way. The line images from both the branches should intersect each other in the central part where the Bragg condition is precisely satisfied.

Now consider the effect of local strain in the specimen crystal. The X-ray paths in a perfect crystal would be straight. In a slightly distorted crystal, however, the X-ray paths are bent, as shown schematically in Figure 5(c),[10] and the sense of this bending depends on the sense of the strain. This bending effect is superimposed on the image shift. When the path is bent in the same direction as the shift of the line image due to nonparallelism of the reflecting planes in the two crystals, this bending makes the shift abrupt. If the path is bent in the opposite direction, the shift occurs more slowly. As mentioned above, the specimen crystal has a local strain near the twin cores, which causes the hook-shaped fringes in Figure 4(b). In a comparison of (b) and (d), it can be seen that, in (d), the crystal was placed in such an orientation that both the line images intersect in that strain region. In (c), the part where the line images intersect is more perfect than that in (d). The abrupt shift can be seen on the line image $A'D'$ in (d). Therefore, one can conclude that the strain has the sense which bends the X-ray path in the same direction as the gradual change in $\theta - \theta_B$. In this way, we can find the sense of slight strains relative to the direction of nonparallelism of the reflecting planes in the two crystals, which in turn can be determined by the rocking-curve measurement.

SUMMARY

A double-crystal method in the Laue arrangement was described. This method is appropriate for crystals with very low absorption coefficient such as BeO. The beam from the first crystal has a fairly high intensity and is parallel enough to obtain essentially intrinsic rocking curves from a perfect second crystal. As an application, the sense of a local strain in a BeO crystal was determined. This method is expected to be useful for further minute study of crystal imperfections.

ACKNOWLEDGMENT

The authors express their sincere gratitude to Professor J. B. Newkirk for his valuable discussion.

REFERENCES

1. J. B. Newkirk and J. H. Wernick (eds.), *Direct Observation of Imperfections in Crystals*, Interscience Publishers, Inc., New York, 1962.
2. N. Kato, "Wave Optical Theory of Diffraction in Single Crystals," in: G. N. Ramachandran (ed.), *Crystallography and Crystal Perfection*, Academic Press, London, 1963, p. 153.
3. A. Authier, "Contrast of Dislocation Images in X-Ray Transmission Topography," in: J. B. Newkirk and G. R. Mallett (eds.), *Advances in X-Ray Analysis, Vol. 10*, Plenum Press, New York, 1967, p. 9.
4. A. Authier, "Mise en evidence experimentale de la double refraction des rayons X," *Compt. Rend.* **251**: 2003, 1960.
5. N. Kato, "The Flow of X-Ray and Material Waves in Ideally Perfect Single Crystals," *Acta Cryst.* **11**: 885, 1958; also, P. P. Ewald, "Group Velocity in X-Ray Crystal Optics," *Acta Cryst.* **11**: 888, 1958.
6. R. W. James, "The Dynamical Theory of X-Ray Diffraction," in: F. Seitz and D. Turnbull (eds.), *Solid State Physics, Vol. 15*, Academic Press, New York, 1963, pp. 55–220.
7. M. Born and E. Wolf, *Principles of Optics*, Pergamon Press, London, 1964, p. 393.
8. N. Kato and A. R. Lang, "A Study of Pendellösung Fringes in X-Ray Diffraction," *Acta Cryst.* **12**: 787, 1959; also, N. Kato, "A Theoretical Study of Pendellösung Fringes, Part I, General Considerations," *Acta Cryst.* **14**: 526, 1961; and N. Kato, "A Theoretical Study of Pendellösung Fringes, Part II, Detailed Discussion Based on a Spherical Wave Theory," *Acta Cryst.* **14**: 627, 1961.

9. S. B. Austerman, "Growth and Properties of Beryllium Oxide Single Crystals," *J. Nucl. Mater.* **14**: 225–236, 1964.
10. P. Penning and D. Polder, "Anomalous Transmission of X-Rays in Elastically Deformed Crystals," *Philips Res. Rept.* **16**: 419, 1961.
11. S. B. Austerman, J. B. Newkirk, and D. K. Smith, "Study of Defect Structures in BeO Single Crystals by X-Ray Diffraction Topography," *J. Appl. Phys.* **36**: 3815, 1965.
12. N. Kato, "Pendellösung Fringes in Distorted Crystals," *J. Phys. Soc. Japan* **19**: 971, 1964; also, M. Hart, "Pendellösung Fringes in Elastically Deformed Silicon," *Z. Physik* **189**: 269, 1966.

DISCUSSION

H. S. Peiser (National Bureau of Standards): This is a very important paper, and I do congratulate the authors. There are many interesting points raised. It brings the BeO crystals into the very small elite of crystals which come very close to the theoretical Darwin width. There are a number of other points I should like to raise in private discussion, but could the author say something about the magnitude of the strain, and possibly also why it is you think that Fraunhofer diffraction applies when the slit does not accurately geometrically define the beam.

J. Chikawa: I can't say anything about the magnitude of the strain.

Chairman W. C. Hagel: Are these of the order of 1000 psi?

H. S. Peiser: No, 1 ppm. Is it of the order of 1 ppm or 1 part per hundred million? Because I think other methods could give the indication.

J. Chikawa: The sense of the strain can be determined, but the magnitude was not determined. The magnitude may be estimated from the spacing–contraction of Pendellösung fringes. The second question?

S. B. Austerman (Atomics International): The question relates to the effectiveness of the mechanical slits in producing Fraunhofer diffraction when the slits do not actually define the width of the central fringe. The calculations relating to 100-μ-width slit assumed that there was definition of the beam by that slit, but, when the fringe is narrower than the slit, there is no diffraction from the slit, and the value of the slit blades is to block off the Pendellösung fringes other than the central peak.

X-RAY STRESS MEASUREMENT BY THE SINGLE-EXPOSURE TECHNIQUE

John T. Norton

Massachusetts Institute of Technology
Cambridge, Massachusetts

and

Advanced Metals Research Corporation
Burlington, Massachusetts

ABSTRACT

X-ray techniques employing film recording continue to play an important role in the routine determination of residual stresses in metals, particularly where heavy and bulky specimens are involved. The single-exposure technique in which the two diffraction patterns required are obtained simultaneously in a single X-ray exposure has several practical advantages over the more conventional two-exposure technique for many routine applications. Details of the method are discussed and an analysis is made of the factors influencing the reliability of the stress determination. An apparatus for stress measurement having some unique features is described and a practical application presented, illustrating the use of the single-exposure technique for the determination of the stress constant of a specimen of hardened steel. The specimen was subjected to a series of known stresses in bending, both in tension and compression. The measured diffraction-line displacements were plotted against the applied stress, and a line was fitted to the points by a least-squares calculation. From its slope, the stress constant was determined. The standard deviation of individual stress determinations from the "true" values given by the least-square line was calculated and was in excellent agreement with the value calculated from the errors of measurement. It is concluded that the single-exposure technique with its reduced time required for measurement and less demanding requirement of precise instrumental adjustment provides a reliable and practical method for many routine stress measurement applications.

INTRODUCTION

X-ray measurements of surface-stress components on metal specimens have been employed for many years and have been applied to many practical problems. A number of modifications of the basic technique have been developed, each with its advantages and limitations. In making the decision as to which technique can best be employed for a particular problem, it is necessary to consider several factors. These include the purpose for which the stress information is needed, for this determines the precision with which the results must be obtained; the inherent precision of the several techniques; the errors introduced by the measuring process in each; and the time and effort needed to make an individual measurement. Another important consideration has to do with the type of specimen involved since this determines whether the alignment can be carried out easily and accurately in a laboratory environment or whether it is necessary to work under what might be called field conditions. The latter often involves bulky specimens of complex shape and restricted working conditions which may well pose serious problems in achieving the precise alignment which the X-ray method requires.

X-RAY FILM TECHNIQUES

The X-ray method measures lattice strains in the surface layer of the specimen. If two strain components are measured whose directions with respect to the surface normal are known and which are coplanar with the surface normal and the desired surface stress component, their difference multiplied by a suitable elastic constant gives the magnitude of the stress component, independent of the other components of the stress system.

Two techniques involving recording the X-rays on film are in common use. The first, shown schematically in Figure 1, is called the *single-exposure technique* (SET). The incident X-ray beam is directed toward the specimen surface at a fixed angle β from the surface normal, and the diffracted beams corresponding to the two measuring directions which bisect the angle between the incident and diffracted beams are recorded simultaneously on two separate films or on the two sides of the same film. In this technique, the directions of the measured lattice strains are fixed by the chosen β angle and by the Bragg diffraction angle θ_0, of the specimen material.

The second, known as the *double-exposure technique* (DET) and shown in Figure 2, requires two separate exposures on separate films, one with the incident beam inclined so that the direction of measurement is parallel to the surface normal and the second with the incident beam tilted in relation to the specimen so that the measuring direction makes an angle ψ with the surface normal. In this technique, the investigation can select, within limits, the desired value of the angle ψ independent of the diffraction angle θ_0.

By making some reasonably acceptable assumptions, one may write

$$\text{Stress component } \sigma = K(S_2 - S_1) \tag{1}$$

where the quantity $(S_2 - S_1)$ is the relative diffraction-peak displacement as measured from the two films. It is convenient to divide the stress constant K into two parts

$$K = K_E \cdot K_G \tag{2}$$

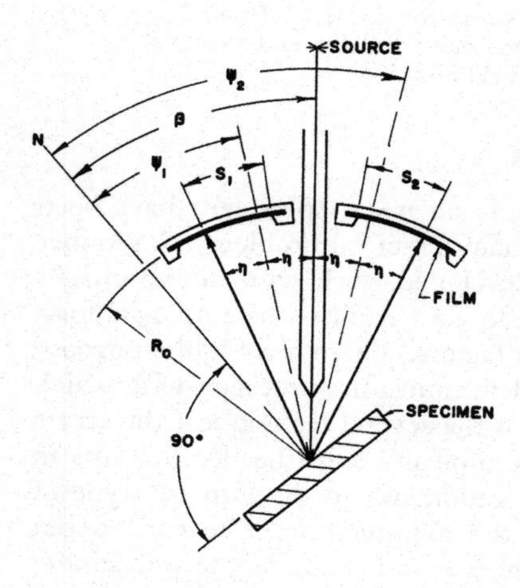

Figure 1. Schematic diagram of single-exposure technique.

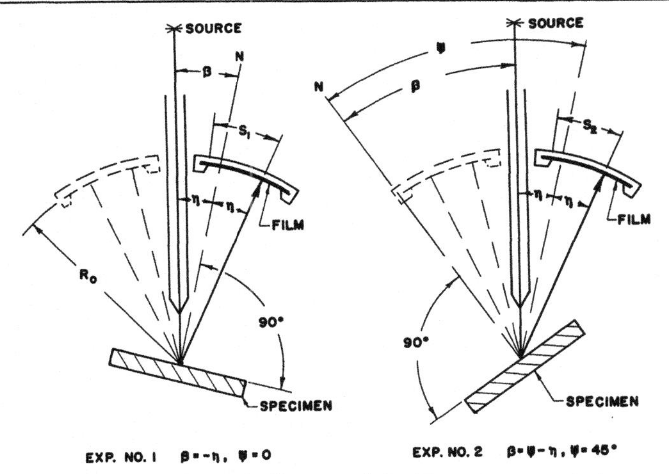

Figure 2. Schematic diagram of double-exposure technique.

where K_E is equal to $E/(1 + \nu)$ which is the appropriate elastic constant of the material and K_G is the geometric constant depending only on the geometrical aspects of the particular technique employed. With reference to Figure 1 for the SET,

$$K_G = \frac{1}{4R_0 \sin^2 \theta_0 \sin 2\beta} \tag{3}$$

If β is 45°, the sin 2β term drops out.

Again, with reference to Figure 2 for the DET,

$$K_G = \frac{\cot \theta_0}{2R_0 \sin^2 \psi} \tag{4}$$

If the same specimen and radiation are used, the value of K_E will be the same for the two techniques and they may be compared by considering the numerical values of K_G. In Figure 3, a plot of K_G is shown as a function of θ_0 for a camera of 70.0-mm radius. In the case of the SET, curves are plotted for β angles of 30, 35, and 45°, while for the DET the curves are for ψ angles of 30, 35, and 45°.

An inspection of these curves shows at once the important difference between the two techniques. In the useful range of values of θ_0 extending from about 75 to 85°, the value of K_G for the SET is nearly independent of θ_0 and changes only slightly for variations in the β angle in the vicinity of 45°. On the other hand, the value of K_G for the DET varies markedly with both θ_0 and ψ. Since the value of K_G can be taken as a measure of the inherent sensitivity of the method, assuming that there are no errors of measurement involved, it is seen that, while the sensitivity of the SET is essentially constant in the practical range of diffraction and inclination angles, that of the DET can be improved by increasing θ_0 or ψ or both. On the other hand, in the SET, it is possible to make significant errors in the values of the diffraction angle or inclination angle without introducing corresponding errors in the value of K_G and consequently in the calculated stress.

In Table I, the values of K_G and K are calculated for an iron specimen with the use of both chromium and cobalt radiation, which shows a comparison of the two techniques. From the point of view of the inherent sensitivity, the advantage of the DET, as well as

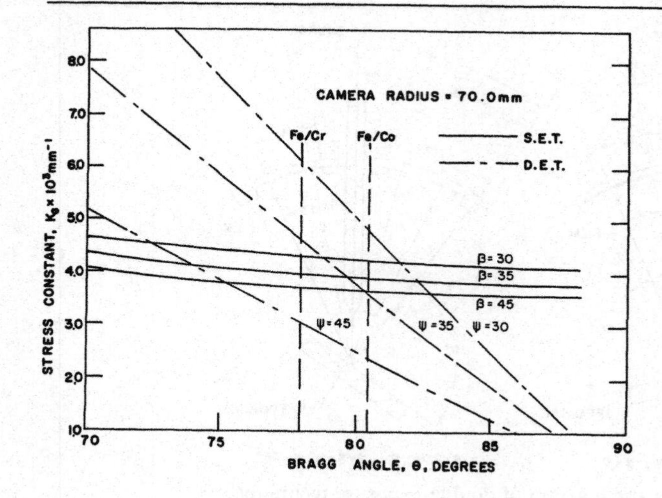

Figure 3. Stress constant K_G as a function of θ_0.

Table I. Evaluation of Stress Constants

Camera radius, 70.0 mm

	Iron with chromium radiation		Iron with cobalt radiation	
	211 Planes, SET, $\beta = 45°$	$\theta_0 = 78.0°$, DET, $\psi = 45°$	310 Planes, SET, $\beta = 45°$	$\theta_0 = 80.5°$, DET, $\psi = 45°$
K_E*	27,900 ksi	27,900 ksi	20,000 ksi	20,000 ksi
K_G	0.00373 mm^{-1}	0.00308 mm^{-1}	0.00367 mm^{-1}	0.00239 mm^{-1}
K	104.1 ksi/mm	85.9 ksi/mm	74.1 ksi/mm	48.3 ksi/mm

* Values taken from Donachie and Norton.[1]

the advantage of the larger diffraction angle, is clear. It is important, then, to consider how the errors associated with individual determinations resulting from misalignment and inherent in the film measurement influence the two techniques.

EXAMPLE OF THE SINGLE-EXPOSURE TECHNIQUE

As an example of the application of the SET, results are presented on the experimental determination of the stress constant of a steel specimen by using chromium radiation. The specimen was a bar of SAE 01 steel heat-treated to a hardness of Rockwell C55. The bar was mounted in a bending fixture such that either tensile or compressive stress could be applied to the upper surface, and SR-4 strain gages were attached to upper and lower surfaces close to the center. The loading cycle extended to 45 ksi in tension and compression, and, in this range, the strains measured by the gages on the two surfaces agreed at all stress levels within 10 μin./in. The bar proved to be elastically very stable with no evidence of creep at the highest loads and no shift of the strain gage readings on the unloaded bar after the loading cycles. Two separate runs were made over the complete stress range. The camera was realigned after each stress increment.

The X-ray unit employed was the Advanced Metals Research Corporations portable residual stress analyzer shown in Figure 4. This unit has several unique features. The

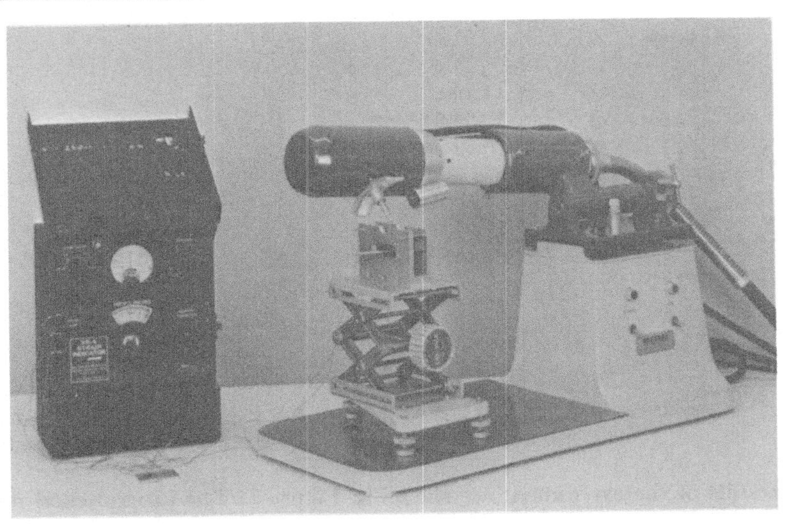

Figure 4. Stress unit set up for stress-constant determination.

film camera is cylindrical with a radius of 70.0 mm. The films used are standard dental-film packets which slide into curved slots in the film holder, which permits the films to be inserted and removed for processing without disturbing the camera alignment. Knife edges machined in the film holder cast sharp X-ray shadows at each end of the film, which serve as fiducial marks for measurement of diffraction-peak positions and for film-shrinkage correction.

The X-ray beam passes through the hollow shaft of the camera and is collimated by an aperture located 15 mm from the specimen surface. Pinhole or slit apertures may be used. A retractable pointer which replaces the collimator aperture locates the center of the cylindrical camera and the axis of the X-ray beam. It is used to adjust the specimen to film distance, and this adjustment is facilitated by a micrometer screw. In order to set the incident beam at the desired inclination angle, the unit tilts about a horizontal axis which can be adjusted to coincide with the axis of the cylindrical camera. The angle of tilt can be adjusted to $\pm 60°$, and the value of the angle is read from a graduated drum.

The film reading was carried out on a conventional film reading box by a scale and vernier reading directly to 0.05 mm. The diffraction lines from the specimen were reasonably sharp with partial resolution of the alpha doublet. On each film, four readings were made and averaged on the fiducial mark and on the diffraction peak. The difference of these two averages gave the value of S. This value was corrected in all cases for film shrinkage. The experimental conditions are summarized in Table II.

Table II. Experimental Conditions for Stress-Constant Determination

Specimen	SAE 01 steel; hardness C55
Radiation	Cr $K\alpha$, 211 planes
Camera radius	70.0 mm
β angle	40.0°
Diffraction angle θ_0	78.0°
Stress constant K_G (calc)	0.00380 mm^{-1}
Young's modulus for strain-gage readings	30,000 ksi

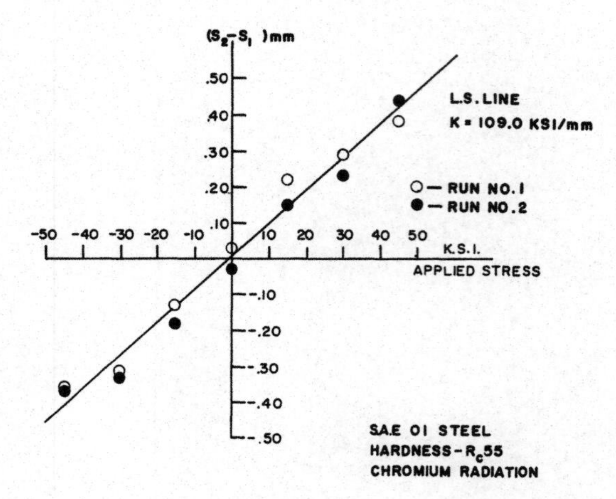

Figure 5. Graph of experimental results of stress-constant determination.

The results of the experiment are shown in Figure 5. The experimental points for both runs were plotted, and a least-square line calculated for all of the points. From the slope of this line, the stress constant K was calculated and found to be 109.0 ksi/mm. Using the value of K_G in Table II, the value of K_E was found to be 28,700 ksi. This compares favorably with the value of 27,900 ksi found on ingot iron by Donachie and Norton[1] using the diffractometer technique. In addition, the rms deviation of individual points from the least-square line was calculated and proved to be 0.041 mm, or 4.49 ksi. A straight line appears to be a very satisfactory representation of the data points, and its passage through the origin shows that the unloaded specimen is essentially stress free.

ERRORS IN THE STRESS-MEASUREMENT TECHNIQUE

The errors which are inherent in the stress-measurement process and influence the absolute values of the measured stress as well as the reproducibility of individual measured values may be divided into three kinds. In the first group are the random errors involved in determining the value of $(S_2 - S_1)$. Their effect on the calculated stress is independent of the stress level in the specimen. The second group of errors arise from the fact that the camera and specimen must be aligned for each measurement and the quantities R_0, β, and ψ, as actually used, will differ in a random fashion from the values used in the calculation of the constant K_G which is applied to a complete series of measurement. The effect of these errors on the calculated stress is proportional to the magnitude of the stress level in the specimen. The third group comprises the errors in θ_0 and K_E. These are constants for a series of experiments involving a specific material and radiation, and it is important to understand how precisely these constants must be known for the two techniques being discussed. In the treatment of the errors to follow, numerical comparisons of the two techniques are based on the specific and comparable situation of an iron sample with chromium radiation under the conditions shown in Table I.

Perhaps the largest error in the measuring process is associated with the location of the peak of the diffraction line on the film. This error depends on the method of measurement and on the sharpness of the diffraction line. In order to obtain a measure of this error, one of the films from the stress-constant experiment described above was selected and measured repeatedly in the same manner as previously described. In this way, averages of many readings were obtained for the positions of the fiducial mark and the diffraction peak. The rms deviation of individual measurements was found, and the rms deviation of the quantity $(S_1 - S_2)$ was calculated by the expression

$$\Delta(S_2 - S_1) = \sqrt{2(\Delta F^2 + \Delta L^2)} \tag{5}$$

where ΔF is the rms deviation of the fiducial-mark position and ΔL is the rms deviation of the diffraction-peak positions.

Table III gives the results, which show that the value of $\Delta(S_1 - S_2) = 0.040$ mm, which is in excellent agreement with the value 0.041 mm observed in the stress-constant experiment described above. It is an important indication that the film-measurement error is the principal random error in the SET.

In the DET, there is an added error in the quantity $(S_2 - S_1)$ due to the fact that the sample-to-film distance is not identical in the two exposures. If the rms deviation of the specimen-to-film distance is ΔR, the error is equal to $2\eta\Delta R$, where $\eta = 90° - \theta_0$. The rms deviation is then equal to

$$\Delta(S_2 - S_1) = \sqrt{2[\Delta F^2 + \Delta L^2 + (2\eta\Delta R)^2]} \tag{6}$$

Experiments were made to determine the value of ΔR by using the mechanical method of the pointer, a dial indicator being used to measure the true variation in the distance. The value of ΔR was found to be 0.045 mm. On this basis, $\Delta(S_2 - S_1) = 0.047$ mm, as shown in Table III.

Another method of correcting for the variations in specimen-to-film distance is the use of a reference powder on the specimen surface. This means an additional diffraction peak to measure on each film, with its associated random errors. If the rms deviation of the reference powder line is ΔP, then,

$$\Delta(S_2 - S_1) = \sqrt{2(\Delta F^2 + \Delta L^2 + \Delta P^2)} \tag{7}$$

Since the reference powder is chosen to give sharp lines, P will have a value only slightly larger than that found for the fiducial mark. If the value is taken as 0.020 mm, then

Table III. Random Error in Film Measurement

Single-exposure technique

$\beta = 40°$ $K = 109.0$ ksi/mm
$\Delta F = 0.018$ mm $\Delta L = 0.021$ mm
$\Delta(S_2 - S_1) = \sqrt{2(\Delta F^2 + \Delta L^2)} = 0.040$ mm
$\Delta\sigma = 109.0 \times 0.040 = 4.36$ ksi

Double-exposure technique

$\psi = 45°$ $K = 88.5$ ksi/mm
Pointer: $\Delta F = 0.018$ mm $\Delta L = 0.021$ mm $\Delta R = 0.045$ mm
$\Delta(S_2 - S_1) = \sqrt{2[\Delta F^2 + \Delta L^2 + (2\eta\Delta R)^2]} = 0.047$ mm
$\Delta\sigma = 88.5 \times 0.047 = 4.16$ ksi
Reference powder: $\Delta F = 0.018$ mm $\Delta L = 0.021$ mm $\Delta P = 0.020$ mm
$\Delta(S_2 - S_1) = \sqrt{2(\Delta F^2 + \Delta L^2 + \Delta P^2)} = 0.048$ mm
$\Delta\sigma = 88.5 \times 0.048 = 4.25$ ksi

Δ = rms deviation of measurement
F = fiducial mark
L = diffraction line from specimen
R = specimen to film distance
P = reference powder

$\Delta(S_2 - S_1)$ becomes 0.048 mm, again summarized in Table III. Thus, the mechanical method, if carefully carried out, and the reference powder method give essentially the same results.

A word should be said about the employment of a reference powder. Its use adds significantly to the time required to make a measurement and, unless applied with considerable care, can introduce errors greater than those it corrects. For the SET, where the distance setting is less critical, the reference powder appears to offer no advantages to balance against the longer time required for a measurement. For the DET under conditions where alignment adjustments are difficult, the reference powder offers advantages which may well compensate for the longer time required.

The random error in K_G is due to the fact that the actual values of R_0, β, or ψ in a particular measurement may be different from the values used in the calculated value of this constant. It can best be evaluated for a particular technique if expressed in terms of the error in each of these, which will produce an error of 1% in the value of K_G and hence in the calculated value of the stress. In both the SET and the DET, as seen in Table IV, a change of 1% in R_0 results in a change of K_G of 1%. In the SET, a change of 4° in β results in the 1% change in K_G, while, in the DET, a change in ψ of 0.3° produces the same error. This illustrates the fact that the alignment is less critical in the SET.

The problem in adjusting the ψ angle in the DET, even where the unit is capable of precise tilt-angle adjustment, is that of establishing the direction of the true surface normal of the specimen. It is especially difficult where the surface at the measuring point is curved or of irregular shape. Alignment for the normal exposure is not critical, but it is critical for the oblique exposure, as the above figures show.

The value of θ_0, the Bragg diffraction angle for the stress-free specimen, can seldom be measured precisely because of the absence of a specimen known to be free from stress. Thus a value must be assumed. In order that K_G may be correct to within 1% in the SET, θ_0 must be correct to within 1.3°, while, for the DET, it must be within 0.11°. Again, the more critical character of the DET is illustrated.

The appropriate value of K_E to be used in the residual-stress calculation presents perhaps the most serious problem of all in so far as absolute stress values are concerned. In spite of extensive discussion, there is at present no satisfactory theoretical basis for its calculation. It has been shown[2,3] that, if the mechanically determined values of Young's modulus and Poisson's ratio are employed, errors as great as 30% may result. Its true value appears to depend not only on the composition of the specimen but upon its structure and heat treatment as well as the X-radiation and diffraction planes involved. The only practical solution at the present time seems to be an experimental determination of the stress constant by using a specimen and experimental conditions identical with those of the actual measurements.

The most common problem in stress measurement is concerned with a comparison of similar specimens with different treatments or the variations in the stress level at

Table IV. Changes to Produce 1% of Change in Stress Constant

	Single-exposure tech.	Double-exposure tech.
R_0	0.70 mm	0.70 mm
β, at 45°	4°	
ψ, at 45°		0.30°
θ_0, at 78°	1.3°	0.11°

various points on the same specimen. Under these circumstances the investigator is primarily concerned with the reliability of an individual measurement as compared to other individual measurements. In other words, he would like to keep the random errors at a minimum and at the same time know their probable magnitude. Except in a few special cases, the absolute magnitude of the stress is of lesser importance, and an error of the order of 10 to 20% in the absolute magnitude would seldom modify the interpretation of the overall results. For this reason, many investigators employ a value of K_E calculated from the mechanical elastic constants with a clear understanding of the uncertainty this involves.

CONCLUSIONS

On the basis of the discussion given above, it is clear that the choice of the proper technique to use is a judgment question involving the particular reasons for measuring the stress in the first place. Because of the lower value of the constant K_G, the best precision obtainable with the DET is superior to the best obtainable by the SET by a factor of 10 to 30% depending upon the material and experimental conditions. However, this advantage may not be realized if the DET is not carried out under conditions which permit very precise alignment for each exposure. Also, the advantage is realized only at the expense of a double or triple time requirement for each individual measurement.

The SET is quicker and simpler and makes less demands upon the skill of the operator. Its speed permits duplicate measurements to be made in a time which does not exceed that required for a single measurement by the DET, and duplicate measurements are a practical necessity in this area. This technique is well adapted to many practical problems of stress measurement, and experience over the years has amply demonstrated its usefulness.

REFERENCES

1. M. J. Donachie, Jr., and J. T. Norton, "X-ray Studies of Lattice Strains under Elastic Loading," *Trans. ASM* **55**: 51, 1962.
2. D. A. Bolstad and W. E. Quist, "The Use of a Portable X-Ray Unit for Measuring Residual Stresses in Aluminum, Titanium, and Steel Alloys," in: W. M. Mueller, G. R. Mallett, and M. J. Fay (eds.), *Advances in X-Ray Analysis, Vol. 8*, Plenum Press, New York, 1965.
3. H. R. Woehrle, F. P. Reilly, III, W. J. Barkley, III, L. A. Jackman, and W. R. Clough, "Experimental X-Ray Stress Analysis Procedures for Ultrahigh-Strength Materials," in: W. M. Mueller, G. R. Mallett, and M. J. Fay (eds.), *Advances in X-Ray Analysis, Vol. 8*, Plenum Press, New York, 1965.

DISCUSSION

P. R. Morris (Armco Steel Corp.): I think it is only fair to point out that these X-ray techniques for stress measurement depend on a homogeneous stress state in the surface of the sample and also that the beam size which is used in these measurements is typically very small and, in order to get any kind of reasonable results, the grain size in the material must be very small. These are conditions which are often not satisfied in commercial materials of interest.

J. T. Norton: This is very true, of course. I think this is well recognized. In this particular arrangement, the beam size is a millimeter in diameter for the majority of the work, but it can be made larger or smaller and, if you make it smaller, you pay the price of longer exposures. Then, of course, you get into the trouble of grain size, but it is possible in this film technique to oscillate the ψ angle or the β angle of the camera through $\pm 2°$ while the exposure is being made. This does wonders in a coarse-grained sample to give you a measurable line. Incidentally, the camera itself oscillates about the axis of the X-ray beam during exposure to help this coarse-grained size effect.

D. Bolstad (Boeing Airplane Co.): What kind of exposure time are you talking about?

J. T. Norton: With this sample of steel with the hardness of C55, the X-ray diffraction lines were quite sharp and there was a suggestion, at least, of resolution of the doublet. These exposures were of the order of 3 min, with 35 kV and 15 mA. But, as the lines get broader, the exposures get longer.

RESIDUAL STRESS AND SHAPE DISTORTION IN HIGH-STRENGTH TOOL STEELS

L. B. Gulbransen and A. K. Dhingra

Materials Science
Washington University, St. Louis, Missouri

ABSTRACT

One of the major problems that has plagued the tool and die maker for many years and more recently has come to the attention of the manufacturer of missiles and high-performance aircraft is the problem of shape distortion which occurs during heat treatment in the high-strength tool and die steels. Not only is shape distortion a problem in the heat treatment and use of these materials, but the origin of shape distortion has been a controversial issue among metallurgists for many years. The quantitative measurement of shape distortion on heat-treated steels is simply carried out by machining standard shape samples, in this case, an L-shaped sample, and making a measurement of the variation after heat treatment from the 90° of the original 90° angle of the L. It is usually assumed that relief of residual stresses in heat-treated parts will occur by the shape changes which have been described above; however, it has been demonstrated that elastic residual stresses may still be present in heat-treated parts that have been tempered and theoretically should be stress free. By a very straightforward and simple application of the back-reflection X-ray diffraction method for residual-stress determination, a very striking relationship has been demonstrated between the shape (angular) distortion of both A_2 tool steel (air hardening) and O_1 tool steel (oil hardening) and the residual-stress pattern of these steels. Conversely, one could presumably utilize residual-stress data at changes in cross section to estimate semiquantitatively the amount of shape distortion which occurs in rather complex parts.

INTRODUCTION

The theory of shape distortion has been discussed in detail by B. S. Lament.[1] Shape distortion is mainly caused by (locked-in) residual stresses. Residual stresses are generally created by nonuniform heating or cooling or because of the development of temperature gradients between the surface and interior of the part. In addition, nonuniform phase transformations may result in a misfit between the various phases at their common interface and generate residual stresses.

METALLURGICAL FACTORS AFFECTING SHAPE DISTORTION

Austenitizing

Shape distortion may be caused by both elastic and plastic deformation. Usually, tool-steel parts contain residual stresses before being heated to the austenitizing temperature. These residual stresses may be created during the machining and mechanical working of the material. Because the yield strength of steel decreases with an increase in temperature, most of the (angular) shape distortion that occurs during heating to the austenitizing temperature is due to the resultant residual stresses. As the material is heated, the residual stress becomes greater than the elastic strength, which results in plastic flow in order to relieve a part of the residual stress.

During heating, thermal stresses are created because of nonuniform temperature distribution in the various sections of the part. In addition, the temperature gradient between the exterior and core of the part will produce thermal stresses, and, wherever the thermal stresses are greater than the elastic limit of the material, plastic flow will take place and result in shape distortion.

If one portion of the part reaches the critical temperature earlier than another during heating transformation, stresses will also result. For example, the surface of a thick part will reach the critical temperature before the center; therefore, the surface of the part will tend to contract (because of the formation of austenite) while the center is still expanding. As a result of this volume expansion, high stresses will be set up in the material and cause considerable shape distortion. Shape distortion, in general, increases with the rate of heating, section size, and nonuniformity of the part.

Quenching

When steel is quenched from above the critical temperature, face-centered cubic austenite begins to transform to body-centered tetragonal martensite wherever the temperature of the part reaches the M_s temperature. Now, as the austenite begins to cool, it undergoes normal thermal contraction. However, the martensitic transformation is accompanied by an expansion, an increase in specific volume, which is initiated at the M_s temperature. The extent of transformation to martensite and the amount of retained austenite may also result in residual stresses.

Nonuniform rate of cooling causes considerable shape distortion. The magnitude, sign, and the resulting stress pattern in a quenched tool-steel part is a function of nonuniformity of cooling in addition to the chemical composition of the steel, its size and shape, and the severity of quench.

Tempering

One of the main objectives of tempering a quenched steel is to relieve the high microstresses in the martensitic structure and to decrease the level of macroresidual stresses. Although tempering generally lowers the overall macrostress level, microresidual stresses may be intensified by thermal and transformation stresses that are introduced during the tempering operation. In general, the level of residual stresses decreases with an increase in tempering temperature and longer tempering times.

SPECIMEN PREPARATION

In order to compare the effect of quenching media on residual-stress distribution and distortion, air-hardening tool steel, type A_2, and oil-hardening tool steel, type O_1, were selected. A typical analysis of type O_1 tool steel is

Carbon	0.9%
Chromium	0.50%
Manganese	1.30%
Tungsten	0.50%

and the analysis of type A_2 tool steel is

Carbon	1.00%
Manganese	0.70%
Silicon	0.20%
Chromium	5.00%
Molybdenum	1.00%
Vanadium	0.20%

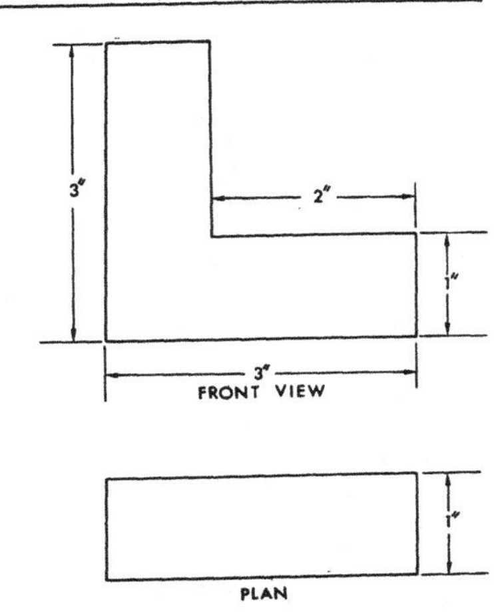

Figure 1. Machined specimen.

The specimens were machined in the form of angle sections with approximate dimensions of 3 by 3 by 1 in., as illustrated in Figure 1. All specimens were carefully finish ground on all sides and were checked for flatness and parallelism of the sides. Special precautions were taken while grinding the sides to ensure that the angle section was 90° and that all corners were square.

HEAT TREATMENT

The O_1 tool-steel specimens were heat treated in a neutral salt-bath furnace. Tempering was carried out in circular blower furnaces for 2 hours according to the schedules given in Table I. The A_2 specimens were heat treated in an atmospheric furnace with a dewpoint of 20 to 30° and were air cooled to room temperature from the austenitizing temperature of 1775°F. Tempering was carried out for 2 hours over a range of temperatures according to the schedules given in Table II.

MEASUREMENT OF SHAPE DISTORTION

A universal shadowgraph was used to measure the L angle of each specimen before and after heat treatment. The angles were measured to an accuracy of 1′ of arc. The angle through which the two sides of the specimen distorts depends, among other factors, on its position during austenitizing in the heat-treating furnace, the angle at which

Table I. Heat-Treatment Data for O_1 Tool-Steel Specimen Treated in Neutral Salt-Bath Furnace

	Preheat		Austenitize			
Specimen No.	Temp., °F	Time, min	Temp., °F	Time, min	Quench media	Tempering temp., °F
1	1200–1300	30	1475	30	Oil	300
2	1200–1300	30	1475	30	Oil	400
3	1200–1300	30	1475	30	Oil	500
4	1200–1300	30	1475	30	Oil	600
5	1200–1300	30	1475	30	Oil	700
6	1200–1300	30	1475	30	Oil	800

Table II. Heat-Treatment Data for A_2 Tool-Steel Specimens

Specimen No.	Preheat Temp., °F	Preheat Time, min	Austenitize Temp., °F	Austenitize Time, min	Quenching media	Tempering temp., °F
1	700–900	120	1775	60	Air*	No temper
2	700–900	120	1775	60	Air*	No temper
3	700–900	120	1775	60	Air	300
4	700–900	120	1775	60	Air	600
5	700–900	120	1775	60	Air	900

* These specimens were deep frozen at −120°F for 2 hours in order to transform retained austenite to martensite.

the sample is quenched, and the uniformity of quench. The amount of angular distortion of the specimen will be minimized if all of the previous factors listed are held uniform for each specimen. In general, the angular distortion of the two legs of the L specimen will not be equal. One leg may distort in a clockwise direction, while the other may distort in a counterclockwise direction, or both legs may distort in the same direction. The angular distortion of the two legs and also the resultant change in right angle are given in Table III for O_1 tool steel and in Table IV for A_2 tool steel.

Table III. Angular-Distortion Results of O_1 Tool-Steel Specimen Heat-Treated in Neutral Salt-Bath Furnace

Specimen	Condition, °F	Angular distortion* One side, min	Angular distortion* Other side, min	Resultant change* in right angle, min
1	Tempered 300	−12	+22	+10
2	Tempered 400	−16	−10	−26
3	Tempered 500	+45	−2	+43
4	Tempered 600	−12	−3	−15
5	Tempered 700	+73	−3.5	+69.5
6	Tempered 800	−10	−10	−20

* The positive sign indicates increase, and the negative sign, decrease, in right angle.

Table IV. Angular Distortion of A_2 Tool-Steel Specimen on Tempering

Specimen No.	Tempering temp., °F	Angular distortion* One side, min	Angular distortion* Other side, min	Resultant change in right angle, min
1	No draw	−10.5	−4	−14.5
2	No draw†	−4	−4	−8
3	300	+20.5	−11	+9.5
4	600	−28	−8	−36
5	900	+10	+3	+13

* The positive sign indicates increase and the negative sign decrease in right angle.
† This specimen was deep frozen at −120°F.

MEASUREMENT OF RESIDUAL STRESSES

Residual stresses after heat treatment were determined by standard back-reflection X-ray diffraction techniques with film as the recording medium. In the measurement of residual stress with X-rays, the interatomic spacing of the (211) reflecting lattice planes of iron were used as the gage length for measuring strain.[2] The elastic strain normal to the surface of the specimen was calculated by the equation

$$e_3 = \left(\frac{d_1 - d_0}{d_0}\right) \tag{1}$$

where e_3 is the strain normal to the surface of the specimen; d_1, the lattice spacing for the stressed surface; and d_0, the lattice spacing for the stress-free surface.

From the theory of elasticity, the elastic strain normal to the free surface was expressed by the equation

$$e_3 = -(\sigma_1 + \sigma_2)\frac{v}{E} \tag{2}$$

where $(\sigma_1 + \sigma_2)$ is the sum of principal stresses in the surface of the specimen; E is Young's modulus for steel; and v is Poisson's ratio for steel.

Finally, the sum of the principal stresses lying in the specimen surface were calculated by the equation

$$(\sigma_1 + \sigma_2) = -\frac{E}{v}\left(\frac{d_1 - d_0}{d_0}\right) \tag{3}$$

A standard back-reflection camera was used on a Norelco X-ray generator. The setup for the specimen and the camera is schematically represented in Figure 2. In order to determine the film-to-specimen distance very accurately, a thin film of stress-free silver powder was applied at the junction of the L specimen for each stress determination. The X-ray generator was operated with a chromium target and vanadium filter at 30 kV and 10 mA. The silver powder on the specimen surface was exposed to X-rays for about 2 hours, after which the powder was removed from the specimen surface with care taken not to disturb the film-to-specimen distance. The specimen was then further exposed to X-rays for about 3 hours. After development of the film, the mean diameter of the silver and iron rings was determined by using an illuminator and recording the diameter readings at several locations on the rings.

By using the interplanar spacing of the reference material (silver), the film-to-specimen distance was calculated and, finally, also the lattice spacing for the steel specimens. From this data, the residual stresses were computed for each steel specimen. Tables V and VI and Figures 3 and 4 illustrate the residual stress data obtained for O_1 and A_2 tool steels at various tempering temperatures.

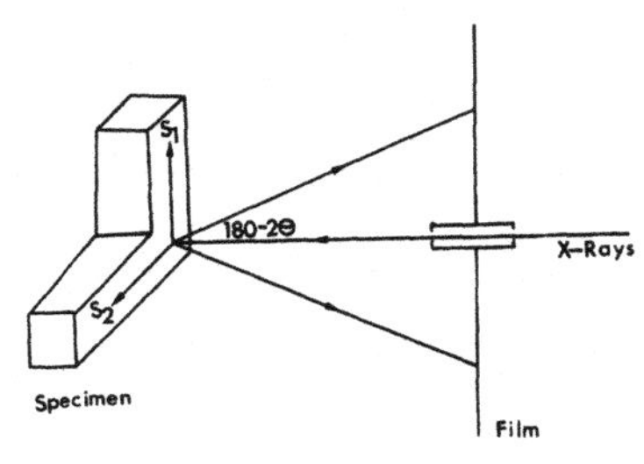

Figure 2. X-ray back-reflection method for the determination of surface residual stresses.

Table V. Surface Residual Stresses (Sum of Two Principal Stresses) of O_1 Tool-Steel Specimen Heat-Treated in Neutral Salt-Bath Furnace

Specimen No.	Condition, °F	Residual stress, 10^3 psi
1	Tempered 300	+36.5 ± 3.0
2	Tempered 400	+29.5 ± 3.0
3	Tempered 500	+45.8 ± 3.0
4	Tempered 600	−24.1 ± 3.0
5	Tempered 700	+13.2 ± 3.0
6	Tempered 800	+23.5 ± 3.0

Table VI. Results of Surface Residual Stresses of A_2 Tool Steel (Sum of Two Principal Stresses) as Determined by X-Ray Diffraction Technique*

Specimen No.	Condition	Residual stress 10^3 psi
1	As hardened	−20.8 ± 3.0
2	After deep freezing	−30.0 ± 3.0
3	Temper 300°F	−21.6 ± 3.0
4	Temper 600°F	−60.8 ± 3.0
5	Temper 900°F	−7.7 ± 3.0

* Data assumed: Cr $K\alpha_1$ = 2.2850 Å; d(Ag) = 1.231 Å.

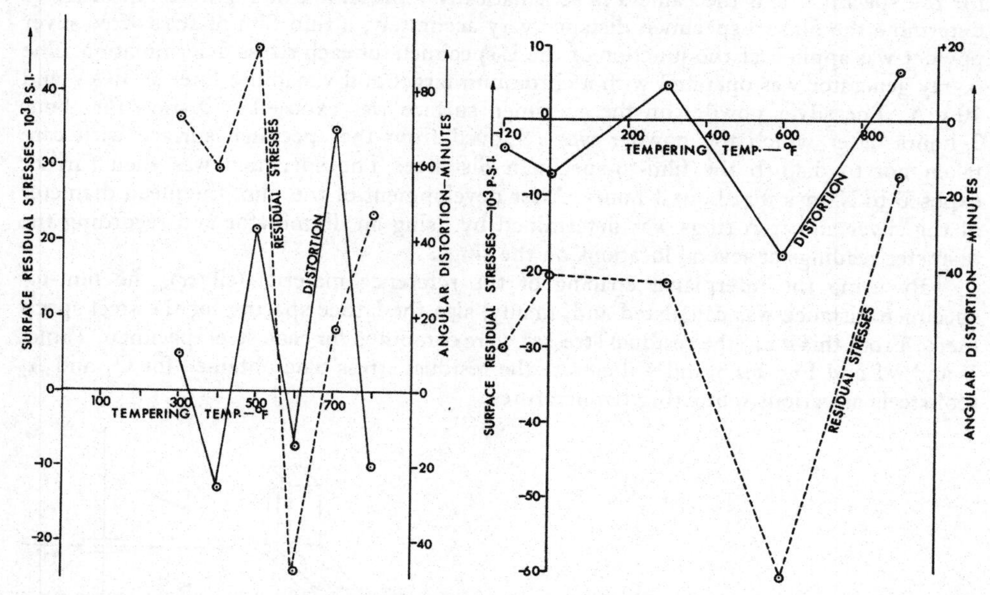

Figure 3. Surface residual stresses and angular distortion *versus* tempering temperature in O_1 tool steel.

Figure 4. Surface residual stresses and angular distortion *versus* tempering temperature in A_2 tool steel.

DISCUSSION

The relationship between angular distortion and residual stress is clearly illustrated in Figures 3 and 4. Angular distortion and residual stress for both oil-hardening and air-hardening steels exhibit a striking correspondence in relative magnitude over a wide range of tempering temperatures. As would be expected, the level of residual stress in O_1 tool steel tempered at 300°F, where little stress relief might occur, is considerably greater than in the A_2 tool steel tempered at 300°F, and it is also tensile rather than compressive. Angular distortion follows much the same pattern in both A_2 and O_1 tool steel. As the temperature is increased, an inversion in residual stress from tensile to compressive occurs in O_1 tool steel, and the residual-stress level in both A_2 and O_1 is highly compressive at 600°F. The rather high residual stress present in both A_2 and O_1 tool steels tempered at 500 to 600°F may result from the transformation of retained austenite, which is known to occur over this temperature range. The expansion resulting from this transformation would be expected to result in rather high compressive residual stresses. Above 600°F, the residual-stress level and angular distortion in both types of tool steel decreases appreciably.

CONCLUSIONS

The determination of residual-stress levels in an air hardening, high-strength tool steel of type A_2 and in an oil-hardening, high-strength tool steel of type O_1, in conjunction with measurements of angular distortion in these steels, indicates a rather good degree of correlation of residual stress and shape distortion in these steels. This correlation exists over a rather wide range of tempering temperatures for both types of steel. Length distortion in steel has been shown to be the result of many factors involving phase transformations, thermal expansion and contraction, and others,[3] but it seems clear that angular distortion is very closely related to residual-stress patterns in the class of tool steels utilized in these experiments.

It may be possible in certain instances to use the residual-stress measurement as a semiquantitative predictive estimate of the angular or shape distortion in machine parts and related equipment. In any event, a qualitative analysis of shape distortion can be the result of rather simple residual-stress measurements in machine parts.

REFERENCES

1. B. S. Lament, *Distortion in Tool Steel*, American Society for Metals, Metals Park, Ohio, 1959.
2. C. S. Barrett, *Structure of Metals*, McGraw-Hill Book Company, New York, 1943.
3. A. K. Dhingra, "Distortion and Residual Stresses in Tool Steel at Various Severities of Quench and Tempering Temperatures," MS thesis, Washington University, St. Louis, Mo., June, 1966.

IRRADIATION EFFECTS IN SOME CRYSTALLINE CERAMICS*

W. V. Cummings

General Electric Company
Vallecitos Nuclear Center
Pleasanton, California

ABSTRACT

Irradiation effects that have been observed in the structures of a number of ceramic materials are reviewed. Results of X-ray diffraction studies indicate that, to a great extent, the magnitudes of the crystallographic changes depend upon the type of crystal structure. However, the nature of the atomic bonding and the type of radiation can be the predominant factor in radiation stability in some materials. Damage mechanisms that have been investigated include: (1) fast-neutron and high-energy gamma-ray effects, (2) transmutation effects in high-neutron cross-section materials, and (3) the effects of the (n, α) reaction in various boride-containing structures. Some crystallographic changes observed include lattice parameter changes and structure damage of various magnitudes, the appearance of a transmutation product structure, and changes from the crystalline to the amorphous state.

INTRODUCTION

Lattice defects produced by radiation can be of several general types. Frenkel defects or interstitial-vacancy pairs can be formed by fast-neutron–lattice-atom collisions and by atom–atom knock-on collisions. These defects are of an isolated nature in contrast to "spike" effects that are characterized by the disordering of a large number of atoms in a localized volume. In addition to these fast-neutron effects, the slow-neutron nuclear reactions that result in the release of a high-energy alpha particle or fission fragment also produce gross lattice defects. Another type of nuclear reaction that can be of importance is neutron capture and transmutation. The foreign atoms that are produced in some high-capture cross-section materials can be incompatible with the parent structure either chemically or physically. A number of excellent reviews on the theories of radiation damage are available[1–5] and will not be repeated here.

In order to evaluate the fundamental nature of radiation damage by X-ray diffraction techniques and to classify the effects of various types of damage, materials should be studied in which a single reaction will be maximized relative to the other possible reactions. This paper is a summary of such studies that have been made over a period of time at the Vallecitos Nuclear Center. The radiation effects that have been observed in ceramic materials include (1) point defects, (2) transmutation effects, and (3) the (n, α) reaction that occurs in boron and beryllium.

POINT DEFECTS IN TiC AND TiB_2

An experimental study of Frenkel defects (interstitial-vacancy pairs) and other multidefect agglomerates must be made in a material that has no (n, α) or fission reactions,

* A portion of this work was supported by the U.S. AEC, Contract No. AT(04-3)-189, Project Agreement No. 4.

a low cross section for neutron capture, and a minimum of displacement spike effects. Since these defects anneal quite readily, a high-melting-point is desired. A cubic crystal structure is desirable for experimental simplification. Of all the materials considered on the basis of these criteria, titanium carbide seems to be a logical choice for this study. Titanium diboride also would be a good candidate material except for the (n, α) reaction in the B^{10} isotope. Since this reaction has been studied in a number of borides (including TiB_2), as will be described later in this report, it was decided to irradiate a second set of TiB_2 crystals in such a way that the (n, α) reaction would be minimized. This was accomplished by shielding the crystals from thermal neutrons with blankets of cadmium and B^{10}. By this technique, the (n, α) reactions were essentially eliminated and the irradiation was limited to neutrons of relatively high energies. Two advantages were gained by this approach. First, another material in addition to TiC would be studied for the formation of Frenkel defects, and, second, the same material (TiB_2) would be investigated with and without the (n, α) reaction.

After preirradiation diffraction data were obtained, single crystals of TiC and TiB_2 were irradiated to three exposure levels of 1.4×10^{19}, 6.4×10^{19}, and 7.9×10^{20} nvt at temperatures of about 100°C. Lattice parameters and line breadths were measured and are plotted as a function of exposure in Figures 1 and 2. It can be seen that the lattice parameter increased to a maximum of 0.3% at the highest exposure for the cubic TiC. At the same time, the increase in line breadth was almost immeasurable. The diffraction peaks remained very sharp, and the lattice parameter could be measured to a high degree of precision for all samples. The lattice parameters also increased for TiB_2 but not in an isotropic manner. This could be expected since the crystal structure is hexagonal. The a_0 value increased by 0.33% and the c_0 value by 0.21% at the highest exposures. The

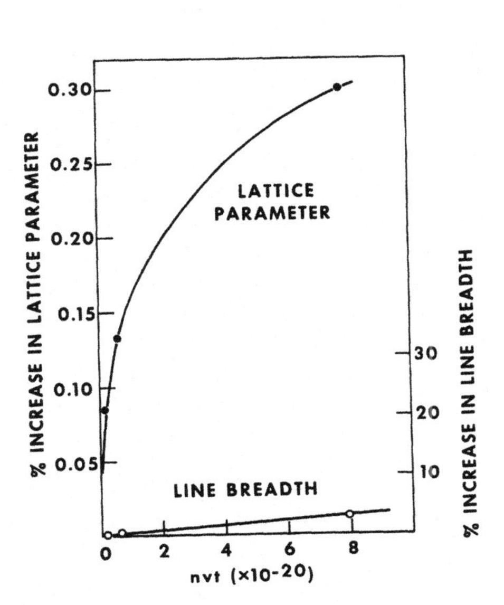

Figure 1. Lattice expansion and line-breadth changes in neutron-irradiated titanium carbide.

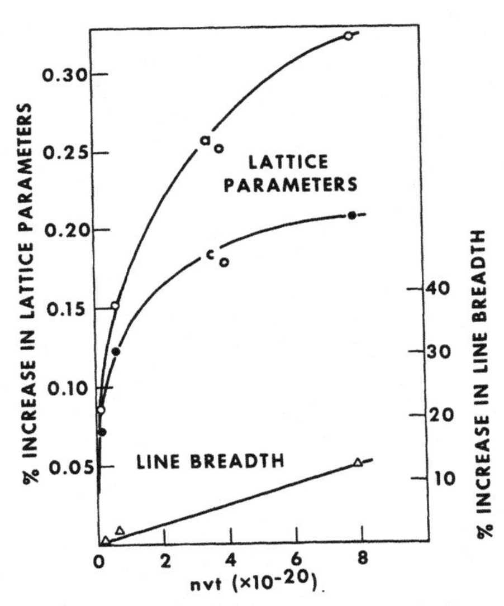

Figure 2. Lattice expansion and line-breadth changes in neutron-irradiated titanium diboride. The specimens were shielded from thermal neutrons with blankets of cadmium and B^{10}. Line-breadth increases were essentially equal for all reflections.

diffraction lines broadened slightly but almost insignificantly when compared with the broadening found in the unshielded TiB_2, as will be shown later. Since it is impossible to eliminate the thermal neutrons completely, the small amount of broadening that is found is probably due to the few (n, α) reactions that are bound to occur.

The observed diffraction effects in TiC and shielded TiB_2 are undoubtedly caused by point defects in the structure. Interstitial-vacancy pairs will cause changes in the lattice parameter but will not cause line broadening.[6,7] Frenkel defects also can be produced by a high-energy gamma-atom recoil mechanism. Crystals of the two materials were irradiated in a Co^{60} facility to an exposure in excess of 10^{10} R to measure the magnitude of this effect. No changes in the diffraction patterns could be measured. It is concluded that the high-energy gamma effect is negligible compared to the neutron irradiation effect.

TRANSMUTATION EFFECTS IN RARE-EARTH OXIDES

The number of foreign atoms that are produced by transmutation can be calculated with fair accuracy and is quite small for most materials. However, in high-neutron cross-section elements, the effect can be significant. If the effects of the transmuted atoms are to be measurable by X-ray diffraction, a new phase must be formed in a significant concentration. For this to occur, not only must the cross section for neutron capture be large, but also the transmutation product must be predominantly of one stable species. Dysprosium meets these requirements with a cross section of 940 barns and a rather simple transmutation scheme predominantly to holmium. Europium has a cross section considerably higher (4300 barns) but its transmutation and decay scheme is quite complicated. The oxides of these two rare earth elements were irradiated to integrated thermal neutron exposures of 2.4×10^{21} and 1.9×10^{21}, respectively.

The diffraction data obtained from the Dy_2O_3 indicated that the material retained good crystallinity during irradiation. Structural changes were in evidence, however, from an examination of peak positions and shapes. The structure of this oxide is large body-centered cubic with 80 atoms per unit cell. Inherently, diffraction patterns from structures of this type are characterized by numerous peaks, most of which have low intensities, and by small angular separation of these peaks even at low diffraction angles. Following irradiation, this poor dispersion and low intensity together with the increased background from radioactivity resulted in difficult, but not serious, problems in analyzing the data. Discriminating procedures reduced background levels to less than 10 counts/sec.

A large number of low-intensity lines in addition to those from an expanded Dy_2O_3 parent lattice and a considerable amount of line broadening characterized the post-irradiation diffraction patterns. The majority of these extra lines could be indexed as a body-centered cubic cell with a slightly smaller lattice size than the original Dy_2O_3. The interplanar spacings for this cell are shown in Table I, along with those that are indexed as the original Dy_2O_3. It can be seen from the data that these values are very close to those calculated for Ho_2O_3. The observed values are slightly larger than the calculated values, but this percentage increase is much less than that measured for the irradiated Dy_2O_3. Three unidentified lines also appear in the diffraction pattern from the irradiated sample.

Thermal neutron capture by dysprosium results in short-lived isotopes that decay to holmium by beta emission. The high cross-section isotope (2700 barns) in the parent material is the 28.2% abundant Dy^{64}. Neutron capture forms Dy^{165}, which decays to Ho^{165}, a stable isotope with a relatively low cross section. Also, Dy^{165} has a high cross section and despite its short half-life, some Dy^{166} is produced, which in turn decays to

Table I. X-Ray Data for Irradiated Dy_2O_3

		Calculated		I/I_1 (obs.)		(*hkl*)	
d (obs.)	*I*	Dy_2O_3	Ho_2O_3	Dy_2O_3	Ho_2O_3	Dy_2O_3	Ho_2O_3
3.77	3	3.77	3.75	–	–	(220)	(220)
3.56	3	–	–	–	–		
3.34B	3	3.37	3.35	–	–	(310)	(310)
3.09	50	3.08		100		(222)	
3.06*	15		3.06		100		(222)
2.98*	4	–	–	–	–	–	–
2.88	4	2.85		8		(321)	
2.83*	3		2.83		20		(321)
2.74*	3	–	–	–	–	–	–
2.68	20	2.67		40		(400)	
2.65*	5		2.64		33		(400)
2.53	4	2.51		8		(411)	
2.50*	3		2.49		20		(411)
2.29	4	2.27		8		(332)	
2.25*	3		2.25		20		(332)
2.19	5	2.18		10		(422)	
2.10	8	2.09		16		(510)	
2.08*	2		2.08		15		(510)
2.02*	2	–	–	–	–	–	–
1.96	2	1.95		4		(521)	
1.95*	2		1.94		15		(521)
1.89	25	1.89		50		(440)	
1.88*	2		1.88		15		(440)
1.83	2	1.83		4		(530)	
1.81*	2		1.82		15		(530)
1.80	4	1.78		8		(600)	
1.78*	2		1.77		15		(600)
1.76	5	1.73		10		(611)	
1.73*	2		1.72		15		(611)
1.65	5	1.65	1.64	?	?	(541)	(541)
1.61	15	1.61		30		(622)	
1.60	4		1.60		27		(622)
1.58	5	1.57		10		(631)	
1.54	5	1.54		10		(444)	

* Additional lines—not found in unirradiated material.

Ho^{166} and finally to the stable Er^{166}. When half-lives and cross sections are considered together, however, the rate of formation of Ho^{165} should be much greater than that for Er^{166}.

The exposure of dysprosium to an integrated thermal neutron flux of 2.4×10^{21} nvt will result in the formation of a substantial amount of holmium. If the dysprosium is in the chemical form of Dy_2O_3, then the irradiated product should be Ho_2O_3 since the two rare earth oxide structures are isomorphous. It would seem that the transmutation product should be a solid solution of the three compounds $(Dy, Ho, Er)_2O_3$; however, the concentration of erbium should be small enough to be almost negligible.

The increase in the interplanar spacings in Dy_2O_3 and also the diffraction line broadening in this structure is caused by the fast neutrons in the flux spectrum. In the case of the Ho_2O_3, it is noticed from Table I that the observed data agree quite well with the calculated values. If pure Ho_2O_3 had been formed during irradiation, then it is reasonable to expect the same type of damage to this structure as occurred in Dy_2O_3, *i.e.*, an expansion of the lattice. The magnitude of this effect is smaller than would be expected. However, if only a small amount of erbium is present in solid solution, then a noticeable decrease in lattice parameter would result if Vegard's law approaches linearity in this case. The lattice parameter in an undamaged structure of this type should be somewhat smaller than that for Ho_2O_3. Fast-neutron damage to this solid-solution structure would expand the lattice, and the final measured value would be somewhere between that expected for damaged Ho_2O_3 and that for an undamaged solid-solution structure of these rare earth oxides. The observed values in Table I fall within this range.

Since neutron capture is random in time and space, it would seem that a homogeneous solution of $(Dy, Ho)_2O_3$ should be formed. The diffraction pattern should then show only a single phase with a lattice parameter value somewhere between those for the two phases Dy_2O_3 and Ho_2O_3. It seems that the only way two distinct phases could exist in the final product would be the result of self-shielding, which would cause a gradient with distance in the sample in the number of transmutations. The center portion would remain essentially free of Ho atoms or at least would contain a much lower concentration than the outside. The two concentrations would then be mixed during preparation for X-ray studies, and two distinct patterns would be recorded.

No peaks were observed in the diffraction pattern obtained from Eu_2O_3, which indicates that gross damage occurred to the structure during irradiation. The reason for this extensive damage compared to the moderate damage to Dy_2O_3 cannot be postulated from the experimental data available. Even though europium has a thermal neutron cross section of 4300 barns compared with 940 barns for dysprosium and, in addition, has somewhat unique nuclear properties in that it has a series of high cross section daughter products, it is difficult to theorize a mechanism that would result in such gross structure damage. A series of exposures beginning with very low nvt values and extending to approximately 10^{21} nvt should prove fruitful in defining the nature of this phenomenon.

EFFECT OF (n, α) IN BORON-CONTAINING MATERIALS

Crystal Structures

The irradiation of a boron-containing material with neutrons can cause structural changes in several ways. First, the fast neutrons produce damage by bumping collisions with atoms in the crystalline lattice. Second, thermal neutrons are captured by the high cross section B^{10} isotope and produce an (n, α) nuclear reaction. This reaction can be considered similar to a nuclear fission since the products are He^4 and Li^7 nuclei, which dissipate approximately 2.4 MeV of energy by collisions and ionization as they travel through the crystal lattice. The final form of damage to the structure is caused when these relatively large atoms of helium and lithium come to rest in nonequilibrium positions and produce lattice strains.

The various borides crystallize as hexagonal diborides, tetragonal tetraborides, simple cubic hexaborides, and face-centered cubic dodecaborides. The transition metals titanium, zirconium, hafnium, and vanadium are some of the elements that combine with boron to form the diborides. The C^{32} type of structure is essentially hexagonal-close-packed, with the metal atoms situated at the points of the hexagonal lattice or in the 0, 0, 0 positions and with the boron atoms occupying the $\frac{1}{3}$, $\frac{2}{3}$, $\frac{1}{2}$, and $\frac{2}{3}$, $\frac{1}{3}$, $\frac{1}{2}$ positions. This arrangement gives one formula weight per unit cell. The resulting structure is characterized by alternating layers of metal and boron atoms and a small $c : a$ ratio. It can be seen that the basic structure shown in Figure 3 is strictly a layered structure.

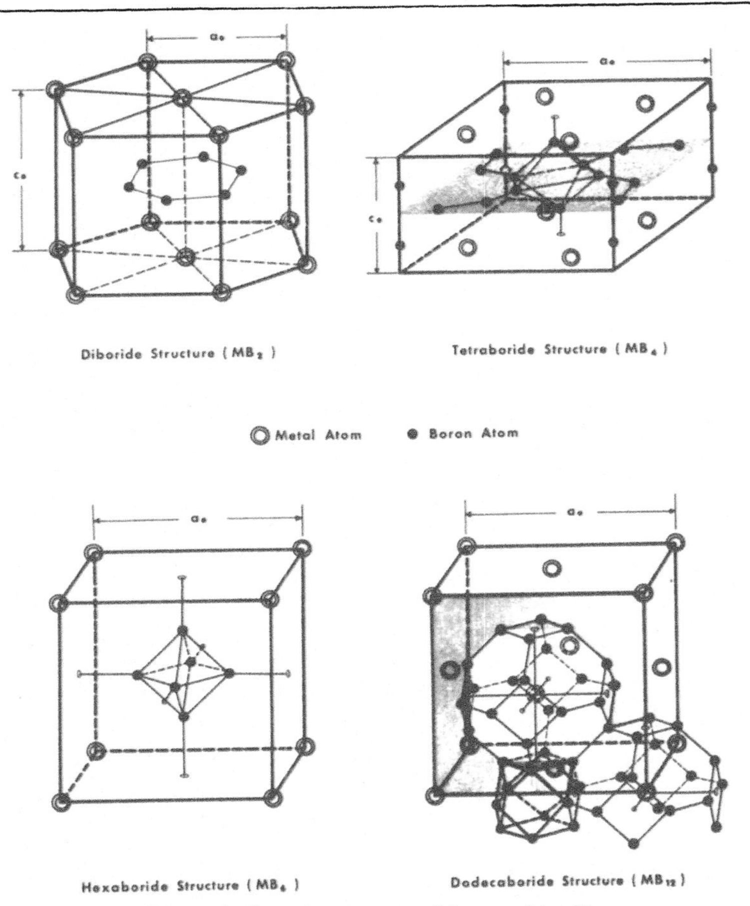

Figure 3. Crystal structures of the metal borides.

The hexaborides form a cubic structure of the CaB_6 $D2_1$ type but also can be visualized as a CsCl type. This structure, as shown in Figure 3, is basically a framework of three-dimensional boron octahedra with metal atoms occupying the relatively large interstitial positions that exist between the octahedra. Atom positions are one metal atom at 0, 0, 0 and six boron atoms at $\frac{1}{2}$, $\frac{1}{2}$, x and like positions, where $x \cong 0.207$. This unit cell also contains one formula weight. The rare earths and a few other atoms such as calcium, yttrium, barium, and thorium combine with boron to form this hexaboride structure.

The tetraborides form a tetragonal $D1_e$ type of structure that exhibits some of the geometry found in both the diborides and hexaborides. The boron atoms in the hexagonal diborides form a layered structure and in the cubic hexaborides, a three-dimensional structure in the form of octahedra. A close study of the structures in Figure 3 reveals these basic features of each in the transitional tetraboride structure. The octahedra formed by the boron atoms can be seen in the center of the distorted cube outlined by the metal atoms. Other remaining boron atoms in the structure are in the form of layers between the metal atoms.

Table II. Crystallographic Data for Some Borides

	Lattice parameters	r(M–M), Å	r(B–B), Å	r(M–B), Å
*TiB_2	Hex. $a = 3.028$ $c = 3.228$	3.028	1.77	2.38
*HfB_2	Hex. $a = 3.14$ $c = 3.47$	3.14	1.81	2.51
*ZrB_2	Hex. $a = 3.17$ $c = 3.53$	3.17	1.83	2.54
*YB_4	Tetr. $a = 7.09$ $c = 4.01$	3.31	1.60	2.59
DyB_4	Tetr. $a = 7.23$ $c = 4.09$	3.50	1.64	2.64
*YB_6	Cubic, P $a = 4.08$	4.08	1.69	3.01
CaB_6	Cubic, P $a = 4.153$	4.15	1.72	3.06
DyB_6	Cubic, P $a = 4.13$	4.13	1.71	3.04
SmB_6	Cubic, P $a = 4.129$	4.129	1.71	3.04
*EuB_6	Cubic, P $a = 4.175$	4.175	1.73	3.08
*YB_{12}	Cubic, F $a = 7.50$	5.30	1.77	2.80

* Postirradiation study of crystallographic changes.

The MB_{12} dodecaboride structure is a $D2_f$ type and is also characterized by a three-dimensional arrangement of boron atoms. The metal atoms have a face-centered cubic coordination and are surrounded by clusters of boron atoms that can be visualized as forming a 14-sided polyhedron (truncated octahedron). Perhaps the most fundamental nature of this structure is represented if the boron atoms are visualized in their closest arrangement. This is shown in Figure 3 by the heavily outlined group of atoms that is midway between the metal atoms. This grouping represents the structure in a manner more nearly in line with structural tendencies of the other higher borides and also of elemental boron, *i.e.*, a network composed basically of a three-dimensional array of boron atoms and, in the case of the compounds, with the metal atoms occupying the large interstitial positions. This method of representation gives the dodecaborides a sodium chloride type of structure in the same sense that the hexaborides are a cesium chloride type of structure.

Structure data for the borides of basic interest are given in Table II. It can be seen that the metal-to-metal and metal-to-boron distances are least in the hexagonal diborides, intermediate in the tetragonal tetraborides, and greatest in the cubic hexaborides. This increasing interatomic spacing with boron content also continues for the metal-to-metal distances when the lone dodecaboride YB_{12} is included, but the metal-to-boron spacing for this compound is less than for the hexaborides. The boron-to-boron distances do not vary greatly with structure and are not significantly different from that found in elemental boron.

Material Selection

Very few investigations of the effects of irradiation on the structure of boron-containing materials have been conducted. However, Tucker and Senio[8] have made a rather extensive study of the effects of neutron irradiation on the structure of boron carbide. An increase in a_0 and a decrease in c_0 was measured, but in general the stability of the material was quite good. It was theorized that this stability was due to the ability of the relatively open boron carbide structure to accommodate the irradiation-generated helium atoms in interstitial positions. On the basis of these results, it has been rather generally concluded that the most radiation-resistant borides would be those with both an open crystal structure and isotropic properties.

The borides that were studied for structure damage include the diborides TiB_2, ZrB_2, and HfB_2; the tetraboride YB_4; and the hexaborides YB_6 and EuB_6. The dodecaboride YB_{12} was also included, but the examination was primarily of a cursory nature. Three crystal systems, hexagonal, tetragonal, and cubic, are included in this study. Most of the samples were polycrystalline, but some single crystals of TiB_2 were included.

Experimental Procedures

All of these borides were irradiated in the General Electric test reactor in the range of 200 to 500°C. The radioactivity levels of some of the samples were in the range of 100–200 mR/hr at 10 in., and shielded X-ray diffraction facilities were necessary for their examination. In order to decrease the background in the diffraction patterns and to provide personnel shielding, a pulse-height selector was employed in the detector system and a lead cell with a maximum thickness of 2 in. was placed around a modified sample holder. These modifications are shown in Figure 4. This arrangement reduced the background in the diffraction patterns by a factor of 10 and provided adequate shielding for laboratory personnel. Because of the minute size of the single crystals (0.2 by 0.4 mm), the radioactive intensity from these specimens was quite low. Little trouble was encountered in mounting the crystals on quartz fibers by conventional techniques for examination. A single-crystal orienter with direct intensity recording and both Laue and rotation cameras was used for postirradiation examination of these crystals.

The X-ray diffraction patterns obtained from the irradiated samples show that the hexagonal structure of the diborides of zirconium and hafnium were distorted rather severely. The unit cells in general expanded in the a_0 direction and at first expanded in the c_0 direction but then contracted as the exposure increased. The diffraction patterns from the polycrystalline specimens were characterized by very broad lines that showed displacement to both higher and lower angles when compared with patterns from unirradiated controls. These patterns are shown in Figure 5(a). Even though the background intensities were not excessive and did not cause appreciable interference in the measurements, the peaks were too broad and diffuse to be discernible at 2θ angles greater than 90°.

The structure of the tetragonal YB_4 was disrupted to a greater extent than that of the diborides. One specimen of this type received an exposure of 62% burnup of the B^{10} atoms, and a lattice shrinkage of 1.5% was measured in the c_0 direction. A change of +0.40% occurred in the a_0 direction. The diffraction peaks were very broad, which indicates that rather extensive microstrains are present or crystallite fragmentation has occurred. In another specimen exposed to a burnup of 94% of the B^{10} atoms, the lattice distortion was even greater. Shrinkage in the c_0 direction increased in magnitude to a value of 1.7%, and expansion in the a_0 direction increased to 0.76%. An example of the

Figure 4. Instrument modifications: lead cell on diffractometer and pulse-height selector in detector instrument rack.

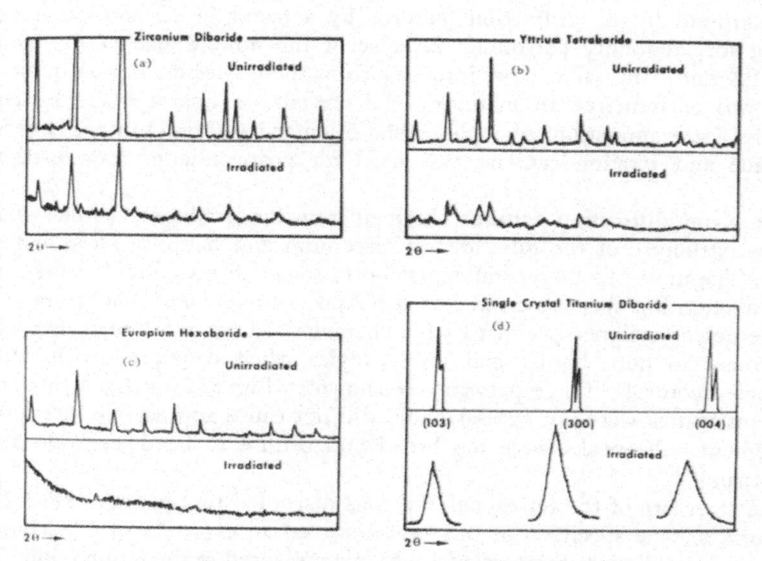

Figure 5. X-ray diffraction patterns of metal borides before and after irradiation.

diffraction patterns obtained from the tetraborides before and after irradiation is shown in Figure 5(b).

All samples of the cubic hexaborides suffered such extensive structure damage during irradiation that no peaks were discernible in the diffraction pattern. This effect indicates that the structures are approaching an amorphous state. The structure damages quite rapidly since one of the samples reached an exposure of only 37% burnup of the B^{10} atoms. These effects are illustrated by the diffraction patterns of EuB_6 shown in Figure 5(c).

Two borides were irradiated that were enriched to 92% in the fissionable B^{10} isotope. In B^{10} enriched YB_6 and YB_{12} samples, gross self-shielding from thermal neutrons occurred. The outer portions of the solid specimens were damaged to such an extent that no crystalline structure was detectable by X-ray diffraction; however, only minor damage to the crystalline structure was measured in the inner portions of the specimens. Since these inner portions were shielded from thermal neutrons, no (n, α) reactions occurred and consequently no foreign atoms were introduced. The damage in this region, then, was caused by fast-neutron bombardment alone. Radiation damage from fast neutrons undoubtedly produces lattice defects. However, the damage from this mechanism is almost negligible when compared with the damage caused by the (n, α) reaction and the presence of foreign atoms in the lattice.

The single crystals of TiB_2 fractured during irradiation even though the exposure level was relatively low (14% burnup of B^{10}). The separate pieces were still sufficiently large for examination, and no experimental difficulties were introduced. The crystals were first oriented by use of a single-crystal orienter, and diffraction data were obtained by direct recording with the diffractometer. Very good diffraction data were obtained compared with those from the polycrystalline specimens. The peak intensities diffracted from these crystals were increased by at least an order of magnitude over those from polycrystalline samples, and background interference from radioactivity was only a few counts per second. Peaks were recorded over the full 2θ range, and reflections with either identical or nearly identical 2θ positions were readily separated. A portion of the diffractometer tracings obtained from these crystals compared with the tracings obtained before irradiation can be seen in Figure 5(d).

Data obtained from the single crystals revealed that the TiB_2 lattice had expanded in both the a_0 and c_0 directions. The expansion of a_0 was 1.49%, and the expansion of c_0 was 0.74%. The $c : a$ axial ratio has changed from 1.063 to 1.058. Laue and rotation photographs of the TiB_2 single crystals show general damage to this hexagonal structure. Both a_0 and c_0 axis photographs are shown in Figures 6 to 9. During the alignment process, it was observed that the plane of the fracture mentioned previously was within a few degrees of the c_0 axis. It can be seen that not only have the reflections broadened and shifted in position but also the relative intensities have changed. Diffractometer tracings of the reflections give quantitative measurements of these changes. Table III shows the large changes in relative intensities and some anisotropic broadening of the reflections. These measurements were made on both a set of reflections of the type $(00l)$ and the type $(h00)$. In addition, reflections from planes of the [100] zone were included so that a number of crystallographic directions would be represented. It is seen that the broadening increases more rapidly with the order of the reflection in the a_0 direction or $(h00)$ reflections than in the c_0 direction or $(00l)$ type of reflections. Broadening of the X-ray reflections can indicate damage of two general types, fragmentation of the crystallites or microstrains (distortion) within the lattice, or both may be found. Line broadening caused by microstrains or distortion of the lattice is dependent upon the order of the reflection, but that caused by fragmentation is independent of order.[9] This relationship

(a) Unirradiated

(b) Irradiated

Figure 6. Photographs of TiB_2 crystal. Rotation about *a* axis.

(a) Unirradiated

(b) Irradiated

Figure 7. Photographs of TiB_2 crystal. Rotation about *c* axis.

and the observed broadening strongly suggest that distortion of the lattice is much greater in the a_0 direction than in the c_0 direction. Since the largest interstices are in the (100) faces, nucleation and clustering of the atoms along these planes would cause the observed effect. In addition, this could be a factor in the causes for the weakening and fracture in this direction during irradiation.

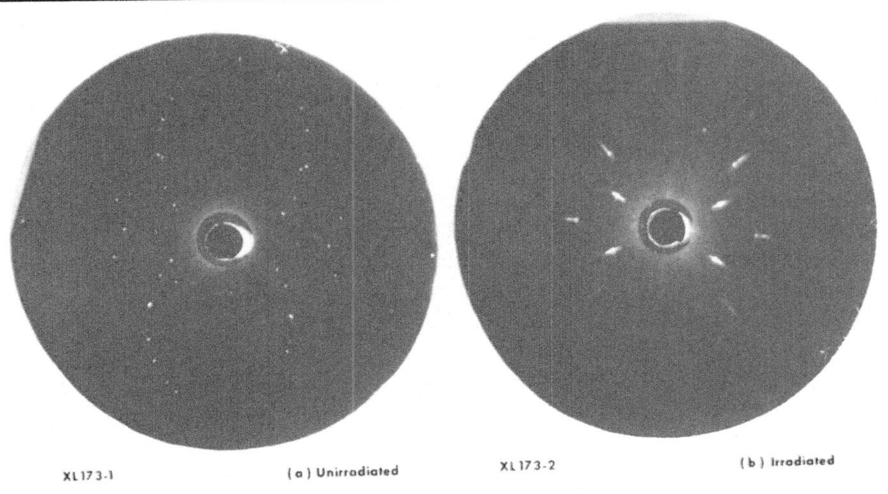

Figure 8. Transmission Laue photographs of TiB_2 crystal. Incident beam parallel to a axis.

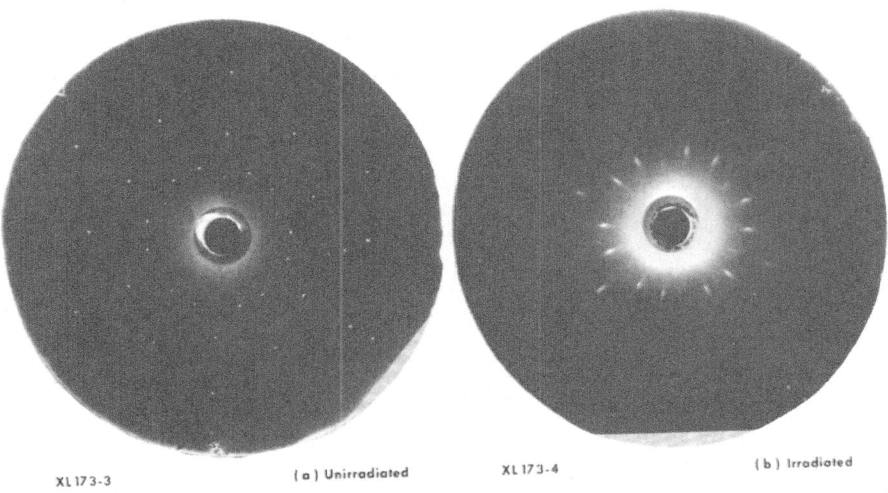

Figure 9. Transmission Laue photographs of TiB_2 crystal. Incident beam parallel to c axis.

Rather than express burnup as a per cent of the B^{10} isotope that undergoes an (n, α) reaction, it is sometimes more descriptive to use the per cent of burnup of total atoms. These values are shown in the third column of Table IV (with a review of irradiation conditions and results in the other columns) and represent the per cent of the original atoms that has been converted to helium and lithium by the nuclear reaction. It follows that the per cent of foreign atoms that have been introduced into the material is twice this total burnup value. These atoms must come to rest in interstitial positions, grain boundaries, or at some defect site in the lattice. Of course, only one vacancy is generated at the original lattice site of the B^{10} atom, and the ratio of interstitials to vacancies is 2 : 1.

Table III. Line-Broadening and Intensity Data from Irradiated TiB_2 Single Crystals

(hkl)	I_0/I_i	B_i/B_0
100	12	6.4
200	17	9.7
300	40	13
001	2.2	3.4
002	1.7	3.4
003	2.4	3.0
004	7.1	4.9
201	2.9	8.3
101	3.8	3.2
102	1.3	5.3
103	14	5.2
110	10	4.3
111	28	2.3

Table IV. Irradiation Data for Borides

Material	% Burnup		Structure	Dimensional changes, %			Helium release, %	
	B^{10}	Total		Δa_0	Δc_0	ΔV	During irrad.	During anneals to 1000°C
ZrB_2	83	10.4	Hex.	+1.6	−0.30	+2.9	3.6	–
HfB_2	87	10.9	Hex.	+1.6	−0.20	+2.9	0.3	31.6
YB_4	62	9.3	Tetr.	+0.40	−1.5	−0.7	4.9	55.9
YB_6	42	6.8	Cubic	–	–	–	14	100
EuB_6	37	6.0	Cubic	–	–	–	19	70
ZrB_2	99	12.4	Hex.	+1.7	−1.2	+2.2	9	
HfB_2	93	11.6	Hex.	+1.9	−0.60	+3.2	13	
YB_4	94	14.1	Tetr.	+0.76	−1.7	+0.50	28	
YB_6	91	14.6	Cubic	–	–	–	47	
EuB_6	68	10.9	Cubic	–	–	–	61	
YB_4	9	1.35	Tetr.	+0.35	+0.80	+1.5	–	
TiB_2	14	1.75	Hex.	+1.49	+0.74	+4.01	–	

Both of the atoms that are produced by the B^{10} reaction are larger than the original boron atom. If these atoms are retained within the lattice as interstitials, the misfit will cause lattice strains, the size of which depends upon crystal structure.

If only the filling of the interstitial positions are considered (size and number), it would seem that the hexaboride structure would be able to accommodate the helium and lithium atoms with the least amount of distortion. The nature of the cubic structure would indicate that any distortion that would occur would be isotropic in nature.

The tetraboride structure has a number of interstitial positions in the lattice, but these are considerably smaller than those for the hexaborides and cannot readily accommodate the helium and lithium atoms. The inherently anisotropic nature of the tetragonal structure also would be expected to be reflected in any distortion of the lattice. Again, based only upon the disposition of the helium and lithium atoms in available interstitial positions, the radiation stability of the tetraborides should be less than that for the hexaborides.

The hexagonal structure of the diborides is a tightly packed structure with very small interstitial positions. The interstices vary in relative size from those in the close-packed structures of pure elements because of the large differences in size between the metal and boron atoms. The largest interstitial site and the only one of significance is at $\frac{1}{2}$, 0, $\frac{1}{2}$ and like positions. Since this is the center of the (100) face, two such positions exist per unit cell. These interstices are smaller than those in the hexaborides and tetraborides, and any trapped atoms in these sites would cause large lattice strains. The degree of distortion in this structure caused by the presence of the helium and lithium atoms in interstitial sites should be greater than for either the tetraborides or the hexaborides.

The concentration of foreign atoms that is retained within the lattice can be deduced by measuring the amount of helium that is released during irradiation. The pertinent values have been measured independently[10] and are shown in Table IV along with the total gas that could be released by heating to 1000°C. In spite of the fact that the hexaborides have the largest and most numerous interstices in which the foreign atoms may be trapped, this structure has retained the smallest amount of helium gas both during irradiation and during annealing treatments to 1000°C. On the other hand, the diborides have retained nearly all of the helium atoms during irradiation, and, even after the series of annealing treatments, well over half of the gas remains within the material. Again, the tetraborides occupy an intermediate position in this respect.

Obviously, damage to the various structures cannot be explained from the effects caused by the final disposition within the lattice of the interstitial atoms. The mechanism by which the structure is damaged must be pursued further. When the B^{10} isotope "fissions," the helium and lithium nuclei are propelled through the lattice in opposite directions with a total energy of approximately 2.4 MeV. The magnitude of the structure damage would be dependent upon both the number of such fissions and the nature of the structure. As pointed out by Samsonov,[11] the bonding between the atoms in the various boride structures differs to a large degree. The layered structures of the hexagonal diborides form strong metal-to-metal and metal-to-boron bonds. These bonds are much weaker in the tetragonal tetraborides and cubic hexaborides primarily because of the three-dimensional nature of the boron clusters. Insufficient numbers of valence electrons are available to form both the covalent boron–boron links and the metal–boron bonds. As a result, the strengths of the structure bonds are greatest for the diborides, as might be surmised from the bond lengths, as tabulated in Table II. The strongest bonds are undoubtedly most resistant to breakage by the passage of high-energy particles through the lattice. Based upon this criterion, resistance to radiation damage should be greatest for the diborides and least for the hexaborides, with the tetraborides again in an intermediate position.

From the above considerations of structural characteristics and the observations of both the relative damage to the various structures and the ability of these structures to retain radiation-generated foreign atoms within the lattice, it becomes apparent that the release of fission-generated atoms does not depend primarily upon the size or number of trapping sites but rather upon the damage to the structure caused by the (n, α) reaction. In turn, the magnitude of structure damage is shown to be inversely proportional to the strength of the interatomic bonding. The weakly bonded cubic structure of the hexaborides is damaged extensively by the (n, α) reaction at relatively low exposures. The structure is damaged to such an extent that, very early in the exposure, interstitial sites inherent in the structure are destroyed and diffusion barriers are drastically lowered. Relatively speaking, the strongly bonded diboride structure, though highly distorted, has remained intact and interstitial sites have not been destroyed.

As mentioned previously, no quantitative measurements could be obtained from the cubic hexaboride diffraction data, and the data obtained from the tetragonal tetraboride analysis are of such poor precision that a mechanism of radiation damage in these struc-

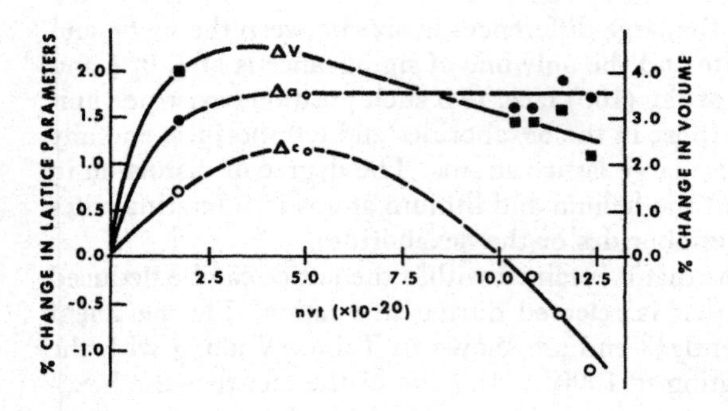

Figure 10. Lattice-parameter and unit-cell-volume changes in neutron-irradiated metal diborides.

tures cannot be proposed with any degree of confidence. However, sufficient data are available from the diborides for a radiation-damage mechanism to be surmised with some degree of credibility.

A study of the measured lattice dimensional changes in the metal diborides shows that the expansion in the a_0 direction occurs at all exposures. On the other hand, the c_0 dimension initially expands, then reverses and contracts as the exposure increases. The net effect at all exposures is an increase in volume. The data points that describe the changes in lattice parameter as a function of exposure are shown in Figure 10. This evidence indicates that, initially, an expansion of the hexagonal lattice in both directions occurs as a result of the trapping of helium and lithium atoms in interstitial positions. During the early stages of the irradiation, damage to the lattice is insufficient to disrupt an appreciable number of trapping sites, and very little diffusion of the impurity atoms occur at the temperature of irradiation. The lattice strains that are produced by the misfit of the relatively large interstitials provide nucleation sites for clustering of the lithium atoms along the (100) planes. As irradiation continues, the structure is disrupted to the extent that trapping sites are gradually destroyed and the diffusion of the interstitials is enhanced. The removal of interstitials from the structure will permit the structure to collapse in the c_0 direction. Figure 10 shows that this diffusion and collapse begins at about 5.0×10^{20} nvt. Beyond this point, the absolute number of interstitials diminishes because of the increasing damage to the structure.

The limited amount of data available from the tetraborides indicates that a radiation-damaging mechanism of this general type also would explain the observed behavior in this structure. The rate of damage with boron burnup has been shown to be greater for this structure than for the diborides. For this reason, diffusion barriers will be diminished at a lower exposure, and the concentration of interstitials and therefore the initial expansion of the lattice will be less than for the diborides. It also is felt that the radiation effects in the cubic hexaborides would follow this model, but, since the rate of damage to this structure is much greater than for the other structures, it would be possible to obtain a damage *versus* exposure function only by irradiating specimens to much lower burnups.

SUMMARY AND CONCLUSIONS

It has been shown that, by a judicious choice of materials, the various types of radiation effects can be separated to a great extent. (1) Point defects are maximized in materials containing light atoms, and, in general, retention of these defects is greatest in materials with high melting points. X-ray diffraction studies have shown that by far the most significant radiation effect in TiC is Frenkel defects. This is also true of TiB_2 when the (n, α) reaction is suppressed by shielding the sample from thermal neutrons. (2) Transmutation products can be observed and are the major irradiation effect in materials containing atoms with high cross sections for neutron capture and having relatively simple transmutation schemes to a predominant stable isotope. The transmutation product Ho_2O_3 has been observed in irradiated Dy_2O_3. (3) The (n, α) reaction is the primary cause of structure damage in boron-containing materials. It has been established that irradiation damage to the metal borides is related to crystal structure. This relationship, however, does not depend primarily upon the crystal system, *i.e.*, cubic, tetragonal, or hexagonal, as such, but upon the strength of the structural bonding. The diborides with their tightly bonded structure are the most resistant to radiation damage. The hexaborides with their open structure and weak bonding are extensively damaged by radiation. Retention of the interstitial atoms within the structure depends upon the integrity of the interstitial sites and varies inversely with irradiation damage.

REFERENCES

1. F. Seitz, "On the Disordering of Solids by Action of Fast Massive Particles," *Discussions Faraday Soc.* **5**: 271, 1949.
2. F. Seitz, "Radiation Effects in Solids," *Phys. Today* **5**: 6, 1952.
3. J. A. Brinkman, "On the Nature of Radiation Damage in Metals," *J. Appl. Phys.* **25**: 961, 1954.
4. F. Seitz and J. S. Koehler, "Displacement of Atoms During Irradiation," in: F. Seitz and D. Turnbull (eds.), *Solid State Physics, Vol. 2*, Academic Press, New York, 1956, pp. 307–442.
5. G. J. Dienes and G. H. Vineyard, *Radiation Effects in Solids*, Interscience Publishers, Inc., New York, 1957.
6. C. W. Tucker and J. B. Sampoon, "Interstitial Content of Radiation Damaged Metals from Precision X-Ray Lattice Parameter Measurements," Knolls Atomic Power Laboratory 1037, Jan. 26, 1954.
7. D. L. Gray and W. V. Cummings, "An X-Ray Diffraction Study of Irradiated Molybdenum," *Acta Met.* **8**: 446, 1960.
8. C. W. Tucker and P. Senio, "X-Ray Scattering by Neutron Irradiated Single Crystals of Boron Carbide," *Acta Cryst.* **7**: 456, 1954.
9. B. E. Warren and B. L. Averbach, "The Separation of Cold-Work Distortion and Particle Size Broadening in X-Ray Patterns," *J. Appl. Phys.* **23**: 496, 1952.
10. E. W. Hoyt and D. L. Zimmerman, "Radiation Effects in Borides, Part I. Helium Release and Swelling in Irradiated Borides," General Electric Atomic Power 3743, Feb. 13, 1962.
11. G. V. Samsonov, "Rare Earth Metal Borides," *Usp. Khim.* **28**: 189, 1959.

DISCUSSION

R. Baro (Université de Strasbourg, France): May I ask you, did you observe any preferential orientations between the phases which were formed?

W. V. Cummings: No, I did not. The intensities in the diffraction pattern fit the theoretical intensities very nicely, relatively speaking. Of course, there is some variation, but not enough to definitely attribute to orientation effects.

W. Ostertag (Corning Glass Works): Did you ever observe the structure of samarium oxide when irradiating the oxides of europium and gadolinium?

W. V. Cummings: No, we did not.

W. Ostertag: I should like to bring to your attention some very recent work on the phase stability diagram of rare-earth oxides by Georg Brauer. Professor Brauer reported at the Sixth Rare Earth Conference held in Gatlinburg that the C type of structure is not stable from La_2O_3 through Gd_2O_3 at any temperature.

W. V. Cummings: Thank you.

Chairman W. C. Hagel: In the abstract, I noticed that you talked about crystalline to amorphous changes. You didn't mention any in your talk that I know of, but presented a slide showing no peaks. Could you comment on that?

W. V. Cummings: The hexaborides we studied gave no diffraction peaks, just diffuse scattering at low angles. On the basis of these data, it was concluded that the structure had become amorphous. We did not observe this effect in either the tetraborides or the diborides, but there were indications of reduced crystallinity in both.

AN X-RAY DIFFRACTION STUDY OF THE AGING REACTION IN TWO AUSTENITIC ALLOYS

J. K. Abraham and T. L. Wilson

Republic Steel Corporation
Research Center
Cleveland, Ohio

ABSTRACT

Aging processes exhibiting cluster to precipitate transitions were studied in polycrystalline austenitic iron-base alloys with a Siemens' Guinier camera. This camera combines the Seemann–Bohlin focusing geometry with a curved-crystal monochromator and thus maximizes the resolution of observed sidebands and the weak precipitate lines. Growth studies encompassing a cluster-size range of 15 to 70 unit cells were followed. For the systems of interest, this coincided with a variation from detectable hardness increase to a stage of maximum hardness immediately preceding precipitation. Cluster sizes were calculated on the basis of the Guinier model; variation with time and temperature permitted calculations of an apparent activation energy in the one system where decomposition was spontaneous. An iron–nickel–titanium alloy was used to study aging in a ternary system. Behavior was classic in that the cluster size present on quenching grew with aging coincident with a simultaneous hardness increase. Calculation of activation energies indicated strongly that transportation of nickel to, or iron from, the cluster was rate determining. Upon overaging, the nickel–titanium enriched clusters gave way to the hexagonal Ni_3Ti phase. An iron–nickel–chromium–niobium quaternary, in addition to presenting a clustering system similar to the above ternary, showed two rather interesting phenomena. First, chromium was necessary for precipitation; the ternary iron–nickel–niobium did not age. Secondly, a stable Fe_2Nb Laves phase present upon quenching from 2200°F disappeared on aging in favor of nickel–niobium clusters; an incubation time for the formation of these clusters existed, and its duration was about 4 hr. An asymmetry was noted in the diffraction intensities about the $(311)_\gamma$ line in both systems. In the iron–nickel–titanium case, the asymmetry was only in intensity, whereas, with the iron–nickel–chromium–niobium alloys, the asymmetry existed in both intensity and position. Interpretation of these observations is made on the basis of anticipated variations in scattering factors, lattice spacings, and cluster sizes.

INTRODUCTION

Since the earliest recognition of aging phenomena, a classical metallurgical problem has been the determination of precipitation kinetics. Depending on the system in question, metallography, resistance measurements, internal friction, and X-ray techniques have all been used in studies. The literature on this subject is voluminous and would be impossible to review here. What will be discussed are the results of using one technique particularly applicable to a problem in which we are concerned—the kinetics of aging in iron-base austenitic alloys.

Aging studies in austenitic alloys are interesting for two particular reasons. First, the problem of strengthening austenitic alloys exists *per se*. Aside from cold working, precipitation is the most effective means of increasing tensile properties. Secondly, there is some indirect evidence that precipitation processes responsible for the strengths of "high-strength steels" are initiated in the austenitic condition during cooling; *i.e.*, the rapid hardening which occurs during the first few minutes of aging in maraging steels and the periodic arrangement of the precipitate particles are indicative of the occurrence of a solute-clustering, preprecipitation process.[1] It was this latter thought that proved intriguing and served as a stimulus for this work.

Since the earliest stages of precipitation were to be followed, it is evident that a technique sensitive enough to detect small fluctuations in composition and structure is required for the studies in mind. Studies utilizing a focusing Guinier camera satisfied this requirement. Fluctuations in measured a_0 values of the order of 0.01 Å can be detected routinely and Manenc *et al.*,[2,3] for instance, have used this type of camera to detect clustering in many systems, including Ni–Ti, Ni–Si, and Fe–Ti–Si; interpretation was based on composition fluctuations in the matrix.

The purpose of this study was to follow the aging processes in austenitic iron-base alloys by use of techniques applicable to Guinier analysis, discussed in detail by Guinier,[4] Hillert,[5] and de Fontaine.[6] Particular emphasis was placed on the preprecipitation processes indicated by sideband development. Two examples are presented, the results of which are amenable to compositions used in high-strength steels. An FeNiTi alloy was studied which eventually produced an Ni_3Ti precipitate in the overaged condition. Similarly, an FeNiCrNb alloy producing an Ni_3Nb overaged precipitate was studied. This latter system was chosen when it became evident that the FeNiTi ternary was rather unique in presenting age-hardening precipitation; chromium seemingly is an element essential to causing the general precipitation of many nickel compounds in iron-rich matrices.

PROCEDURE

Material

The compositions of the alloys selected for this study were:

Fe–31.1Ni–3.5Ti (wt. %)

and

Fe–33Ni–15Cr–4Nb

The high nickel concentration was chosen so as to have an austenitic (fcc) structure at room temperature, whereas the amount of titanium and niobium was selected to form a sufficient amount of Ni_3X. A chromium level of 15% was found necessary to ensure the precipitation of Ni_3Nb.

The alloys were melted in a 5-lb vacuum induction furnace and then cast into 2-in.-diameter ingots. Forging, followed by hot rolling, was performed under a protective atmosphere of argon before homogenizing for 4 hr at 1950°F for the titanium alloy and 2100°F for the niobium material. Cold rolling reduced the thickness to 0.03 in.

The titanium material was solution treated at 1850°F for 1 hr, then water quenched to room temperature to produce a complete supersaturated austenite prior to the aging treatment. Solution treating the niobium specimens at 2050°F for 1½ hr resulted in a supersaturated austenite plus some Fe_2Nb at the grain boundaries. Aging was performed in salt pots at temperatures and times in accordance with results of hardness values taken on material treated in a gradient furnace. Two to four specimens per treatment were used to determine reproducibility.

Examination

A coherent scale formed during the aging process. This was removed by polishing on 320-grit emery paper and then electropolishing. Three hardness readings were taken per sample on the cleaned surface, and their averages are reported in the results. The morphology of the structure was observed by optical microscopy taken across the cross section of the samples. This also gave a quantitative determination of the depth of the scale. A nital etchant was used on the titanium specimens, whereas an alcohol–5% HCl electrolyte was used for niobium samples.

X-Ray Diffraction

The X-ray study employed a Siemens' Guinier type of camera with a cobalt tube operated at 45 kV and 10 mA. The Guinier type of camera consisted of a 1-radian-diameter Seemann–Bohlin camera and a Johannson type of curved quartz crystal monochromator. The Seemann–Bohlin focusing geometry increased the resolution of the 1-radian camera to that of a 2-radian Debye–Scherrer camera. As a result of the high-intensity focusing of the diffracted beam and in spite of the reflection losses in the monochromator crystal, the exposure times for Guinier patterns are of the same order as for Debye–Scherrer patterns. The monochromator, which could be adjusted to give $K\alpha_1$, $K\alpha_2$, or both, nearly eliminated the background due to white radiation and provided both the high resolution and the very low background needed to measure sidebands arising from the preprecipitation state. In this study, the monochromator was set to give only $K\alpha_1$ radiation. To keep the resolution as fine as possible, the back side of the film was masked with tape before developing; thus, the emulsion of that side was removed during fixing. Exposure times ranged from 5 to 62 hr, depending on the speed of the film. For the very fine grain Ilford CX film, an exposure time of at least 48 hr was required; for the fast Ilford G film, 5 hr was sufficient.

With this camera, the specimen could be mounted in a wide range of positions, symmetrical or asymmetrical, back-reflection or transmission. For this study, the asymmetrical ($\alpha = 45$ and $70°$) back-reflection methods were used. The specimens were positioned tangent to the inner circumference of the film in the camera. The X-ray beam was collimated with a fixed vertical-divergence slit and a 4° horizontal divergence slit ahead of the monochromator. Two more sets of adjustable vertical and horizontal slits were positioned between the monochromator and the sample. These slits were set to minimize the vertical divergence and to reduce the scattered radiation. The sample area irradiated by the monochromatic incident beam was approximately $\frac{3}{16}$ by $\frac{1}{8}$ in. (for asymmetrical back reflection). By oscillating the sample in the beam, the total area exposed to the X-ray was increased to $\frac{7}{16}$ by $\frac{1}{8}$ in.

The sidebands flanking the $(311)_\gamma$ line in the back-reflection patterns were the sharpest and most intense, and the aging process was followed by measuring the variation of position of the sidebands about this reflection.

The films were read on a standard Norelco optical film reader. Each film was read a minimum of four times. Preprecipitation zones sizes were calculated from the equation derived by Daniel and Lipson,

$$Q = \frac{h \tan \theta}{N \,\delta\theta} \tag{1}$$

In this relation, Q is the zone size or the wavelength of the periodic structure in dimensions of unit cells; N is the sum of the squares of the Miller indices of the reflection plane; h is the index for the direction in which the periodic variation takes place; θ is the diffraction angle of the matrix diffraction line; and $\delta\theta$ is the angle between the sideband and the matrix line in radians.

The reproducibility of the results was checked throughout the study. This was accomplished by repeating the diffraction measurements on a second and third specimen given the same treatment as the first. In most instances, this difference was less than the deviation, which was up to 10%, obtained on the four readings of each film. The largest uncertainty in the measurement occurred at the two extremes, when the zone sizes were very small and the sidebands very weak and diffuse or when the zone sizes were large and the sidebands exceedingly close to the main diffraction line.

RESULTS

Fe–Ni–Ti

The aging treatment used for the Fe–31Ni–3.5Ti alloy along with the hardness response and X-ray results are listed in Table I; hardness results are plotted for the sideband in Figure 1. Of particular interest here are the kinetics of the strengthening process. Observations indicate that both the precipitation sequence and initial hardness response are related. Figure 2 illustrates that the resultant hardness is associated with the zone size responsible for the sidebands, independent of the temperature at which the samples were aged.

Table I. X-Ray and Hardness Data of Fe–31Ni–3.5Ti

Aging treatment		Hardness, R_N30	X-ray observation	
Temp., °F	Time		Zone size	Other phases
1150	20 min	20	18.0	
	80 min	33	19.6	
	2 hr, 46 min	35	21.7	α'
	8 hr, 20 min	43	24.3	α'
1200	5 min	14		
	17 min	33	17.6	
	1 hr	41	21.2	
	2 hr, 46 min	46	24.2	α'
1250	3 min	17	19.3	
	8 min	28	21.1	
	17 min	37	22.1	
	33 min	43	23.6	α'
	1 hr, 7 min	47	26.7	α'
	1 hr, 57 min	51	31.4	α'
	2 hr, 46 min	53	32.4	α'
	5 hr, 32 min	56	39.3	α'
	13 hr, 53 min	58	43.6	α'
	27 hr, 46 min	68		$Ni_3Ti + \alpha'$
1300	2 min	13.5		
	5 min	35	26.4	
	17 min	45	28.4	
	30 min	47	29.5	α'
	1 hr	50	32.2	α'
1380	2 min	28	25.6	
	5 min	49	36.2	
	17 min	53	38.3	
	1 hr	56	52.0	α'
	2 hr, 46 min	58	69.5	α'

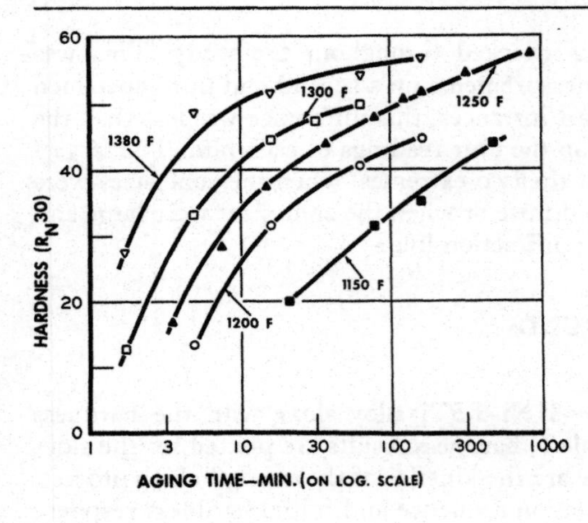

Figure 1. Hardness *versus* aging time for the Fe–31Ni–3.5Ti alloy.

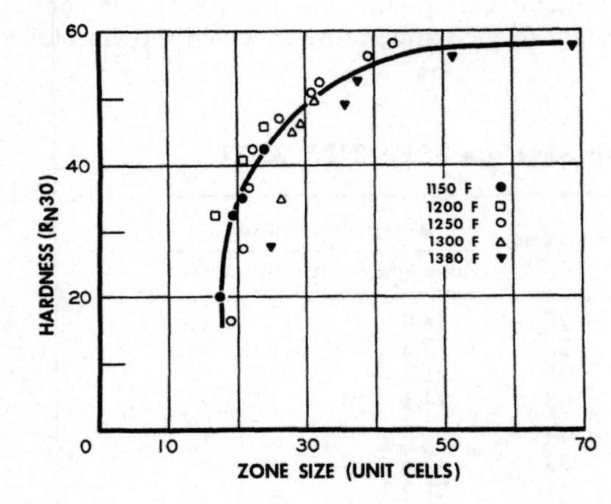

Figure 2. Hardness *versus* zone size for the Fe–31Ni–3.5Ti alloy.

Two complicating factors which enter the aging kinetics are the formation of a cellular Ni_3Ti (hcp) phase at the grain boundaries and the appearance of the α' martensitic phase. The volume per cent of the cellular reaction, however, was small for the aging region. Broadening of grain boundaries observed optically and substantiated by transmission electron microscopy, indicates an order of 1 to 3 vol. % at peak hardness. Presumably, the martensite forms upon cooling in the nickel-denuded regions formed by the cellular reaction, but it must also occur in some of the denuded regions of the matrix resulting from the clustering phenomenon.

This is based on the martensite formation prior to the general precipitation of Ni_3Ti and its occurrence in larger volume percentages than the Ni_3Ti phase ($\sim$10% at the cluster-to-precipitate transition). The morphology of the martensite was extremely fine in scale since metallographic examination did not indicate the presence of a bulk martensitic phase until much later in the aging treatment. It should be emphasized that a martensitic transformation is occurring only in some of the denuded regions since sidebands remain visible concurrent with transformation. The lack of a decrease in hardness as the overaged condition is approached is undoubtedly due to a matrix strengthening resultant from the shear transformation. Some surface martensite unquestionably formed on the sample during polishing and, hence, contributed to the large volume per cent of martensite obvious on the X-ray patterns, but the hardness results indicate that a matrix martensite was truly present. Secondary evidence leading to the same conclusion was that Knoop hardness determinations closely paralleled superficial R_N measurements.

This system provided an opportunity to conduct a detailed analysis of zone formation as indicated by sideband developments inasmuch as such a large time and temperature range were covered. A sequence indicating the sideband development about the $(311)_\gamma$ line as determined from film exposure and densitometer analysis is shown in Figure 3. It should be indicated in this case that the asymmetry in intensity is a manifestation of differences in the size of the denuded and enriched regions. The intensity of the sidebands as explained by Hargreaves,[7] is governed by modulation of either scattering factor or spacing. As in most systems, the difference in scattering power is very small, and, hence, it is the difference in spacing that causes the observed sidebands. This is undoubtedly the case for Fe–Ni–Ti.

The treatment of kinetics from sideband information in this system has been treated in detail elsewhere,[8] but an outline of the pertinent procedure may be given here. By assuming that the growth of the clusters are diffusion controlled, it can be shown that the following growth equation is applicable:

$$G = G_0 + \alpha(Dt)^m \tag{2}$$

where G is the size of the growing cluster; G_0 is the size at zero aging time; D is the

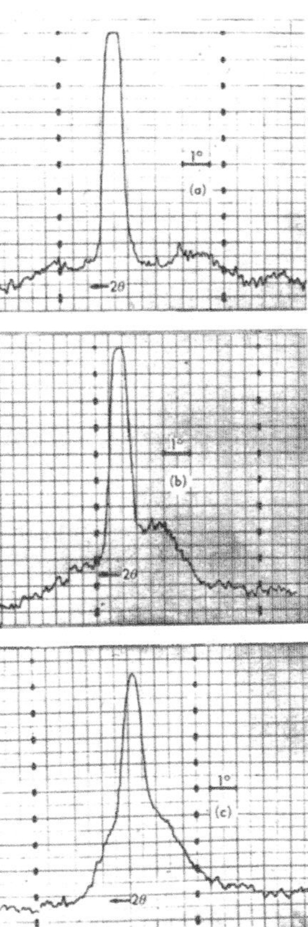

Figure 3. Microphotometer traces of the $(311)_\gamma$ sidebands of the titanium alloy; aged at 1250°F for (a) 17 min; (b) 2 hr, 46 min; and (c) 13 hr, 53 min.

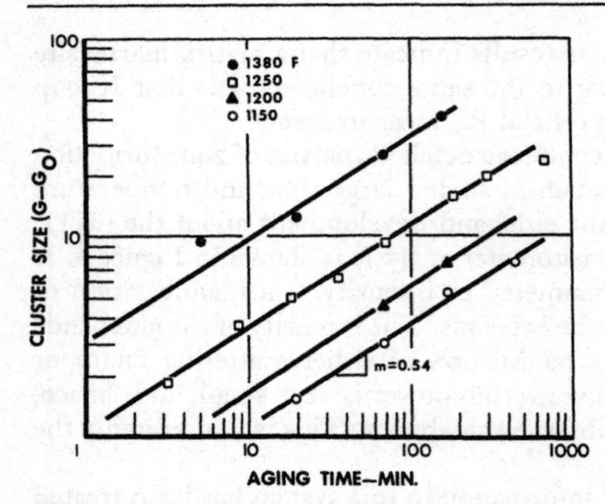

Figure 4. Zone size increment ($G - G_0$) *versus* aging time for the Fe–31Ni–3.5Ti alloy.

diffusion coefficient; t is the aging time; α is the growth coefficient; and m is an exponent which can be determined from the slope of a plot of log ($G - G_0$) *versus* log t. Such a plot is shown in Figure 4, where m is found to equal 0.54. The values of G_0 were obtained by extrapolating to zero time on the plots of G *versus* $t^{\frac{1}{2}}$.

By coupling equation (2) with the empirical equation that relates the diffusion coefficient to the activation energy Q and the temperature T,

$$D = D_0 \exp\left(-\frac{Q}{RT}\right) \tag{3}$$

Table II. Comparison of Observed and Reported *d* Values* of γ- and α-Iron and Ni_3Ti (hcp)

	d Values from ASTM card file for					
	γ		α		Ni_3Ti	
d Observed†	*d*	*hkl*	*d*	*hkl*	*d*	*hkl*
1.013 M			1.013	220		
1.034 VW	1.037	222				
1.046 VW					1.046	207
1.081 W	1.083	311				
1.089 W					1.087	224
1.170 VS			1.170	211	1.173	206
1.267 MS	1.270	220				
1.278 M					1.276	220
1.330 MW					1.330	205
1.433 VS			1.433	200		
1.512 VW					1.51	204
1.728 M					1.72	203
1.789 MS	1.80	200				
1.952 MS					1.95	202
2.027 VS			2.027	110		
2.078 W	2.08	111			2.07	004
2.138 M					2.13	201
2.213 M					2.21	200

* The zero position was extrapolated from the martensite lines.
† Notation: VS, very strong; MS, medium strong; M, medium; MW, medium weak; W, weak; VW, very weak.

the activation energy was determined to be 59 kcal/mole. This value is probably low and may be off by a factor of 10 to 15% since no account can be taken of the temperature dependence of the free-energy change.[5] However, it is near the magnitude of the reported values for the activation energy for diffusion of nickel and self-diffusion of iron in austenite, which are 67,500 and 67,800 cal/mole, respectively. It is believed that the activation energy for diffusion of titanium in austenite will be substantially higher. Hence, it is concluded that the process is controlled by the transportation of nickel to, or iron from, the clusters.

Upon the collapse of the sidebands into the matrix's reflections, the Ni_3Ti (hcp) lines appear. In addition to the cellular precipitation, it would thus appear that the precipitates form generally from the existing clusters. The *d* values of the lines present at this stage are given in Table II.

Fe–Ni–Cr–Nb

For the austenitic Fe–Ni–Nb material to become an age-hardening alloy, Weiner and Irani[9] have found that the addition of chromium is necessary. We also discovered this to be true and that the level of chromium needed for the aging phenomenon seemed to be related to the electron-to-atom ratio $e:a$ of the alloy. As the $e:a$ ratio of the matrix approaches 8.25,* given as the necessary formation ratio for γ', the fcc Ni_3Ti, or $Ni_3(Al,Ti)$ transition phase,[10] a hardness response is noted upon aging along with the appearance of sidebands on the X-ray patterns. The hardness values, X-ray results, and aging treatments for an Fe–33Ni–15Cr–4Nb alloy are given in Table III.

Hardness as a function of aging time is shown in Figure 5. Unlike the titanium alloy, which had a finite cluster size in the quenched state, there is an apparent incubation period of the order of 3 to 4 hr before the hardness starts to increase. The reason for the long incubation time is probably not simple. It suffices to say that two complicating factors are present. First, niobium is obviously not clustering at the solution-treatment temperature, which appears to be quite likely for the titanium alloy, and, secondly, a Laves phase is present after solution treatment. If nothing else, this phase lowers the supersaturation driving the clustering reaction in the matrix. Either of the above factors would lead to the observed incubation behavior.

* For purposes here, this value was calculated for the simplest alloy chemistry assumptions. These may be considered only a first approximation for such calculations, but they are consistent with methods used in the referenced work.

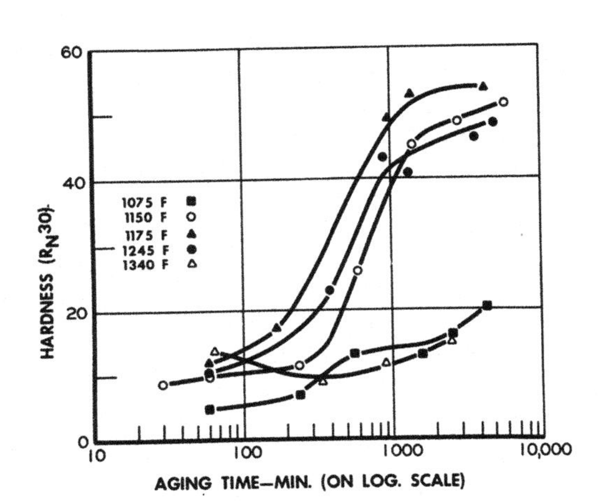

Figure 5. Hardness *versus* aging time for the Fe–33Ni–15Cr–4Nb alloy.

Table III. X-Ray and Hardness Data of Fe–33Ni–15Cr–4Nb

Aging treatment		Hardness,	X-ray observations	
Temp., °F	Time, hr	R_N30	Zone size	Other phases
1075	1	5		Fe_2Nb
	4	7		Fe_2Nb
	9.5	13		Fe_2Nb
	26	13		Fe_2Nb
	45	17		Fe_2Nb
	76	20		Fe_2Nb
1150	0.5	9		Fe_2Nb
	1	10		Fe_2Nb
	4	11		Fe_2Nb
	10	26	37	Fe_2Nb
	24	45	28	Fe_2Nb
	50	48	40	Fe_2Nb
	100	51	35	Fe_2Nb
	145	44	43	Fe_2Nb
	366	54	55	Fe_2Nb
1175	1	11		Fe_2Nb
	3	17		Fe_2Nb
	16.3	49	50	Fe_2Nb
	24	53	49	Fe_2Nb
	72	54	72	Fe_2Nb
1245	1	11		Fe_2Nb
	6.8	23		Fe_2Nb
	15	43		Fe_2Nb
	22.5	40		Fe_2Nb
	63.5	46	55	Fe_2Nb
	86	49	66	Fe_2Nb
1325	15.5	41	50	Fe_2Nb
	146	31		$Fe_2Nb + Ni_3Nb$
1340	6	9		Fe_2Nb
	15	11		Fe_2Nb
	44	16		Fe_2Nb
	65	14		Fe_2Nb

Aside from the long incubation time, the aging kinetics are rather routine. Hardening increases to a plateau maximum at 1150, 1175, and 1245°F. Overaging occurs at 1325°F with the formation of the Ni_3Nb orthorhombic phase; at 1340°F, no effect is observed.

Optical microscopy indicates that the structure of the aged material is identical to those that were solution treated. No precipitates are seen, but a bulky second phase is present mainly at the grain boundaries in amounts proportional to the niobium content. If anything, the amount of this phase decreases with increasing aging time. It has been identified as Fe_2Nb (hcp), a Laves phase. Spectroscopic examination of extracted particles reveals that some nickel and chromium are present; hence, the formula should be written as $(FeNiCr)_2Nb$. The *d* values obtained from the extraction are given in Table IV. This phase may contribute to the solution-treated hardness by a dispersion-hardening mechanism, but its role is only secondary in the aging kinetics.

Shortly after the initiation of the hardness increase, very weak sidebands are seen on the X-ray patterns. Both the intensity and position of the sidebands about the matrix's

Table IV. Comparison of Observed and Reported *d* Values of Fe_2Nb

Observed*		Fe_2Nb†		Fe_2Nb‡
2.85	W			2.86
2.40	MS	2.403	MW	2.41
2.21	S	2.214	S	2.22
2.08	W	2.074	W	2.085
2.045	S	2.047	S	2.055
2.010	S	2.009	S	2.015
1.955	W	1.961	MW	1.965
1.835	W	1.840	MW	1.845
1.765	W	1.780	W	1.780
1.462	W			1.465
1.386	W	1.390	W	1.391
1.347	MW	1.351	M	1.353
1.309	MW	1.312	M	1.313
1.250	MW	1.254	M	1.257
1.226	VW			1.229
1.200	M	1.207	M	1.206
1.110	W			1.111
1.056	M			1.059

* Notation: S, strong; MS, medium strong; M, medium; MW, medium weak; W, weak; VW, very weak.
† From ASTM Card No. 15-316.
‡ Calculated value.

reflections are asymmetrical, as illustrated in Figure 6. Similar asymmetrical maxima were seen in the Cu–Ti system by Bückle and Manenc[11] and Nesterenko and Chuistov.[12] Observations of such maxima are not so well documented and are worthy of note. As stated before, sidebands arise basically because of modulations in lattice spacing or scattering factor in intimate zones. Two factors are necessary to provide an asymmetry in position. These are (1) a spectrum in cluster size to provide multiple sidebands and (2) a factor to provide intensity differences on sideband pairs. De Fontaine[6] has analyzed the problem of sideband formation theoretically and used an elecronic analog to simulate the development of intensity maxima under various conditions of modulation in scattering factor, spacing, and the combination of both spacing and scattering factor. It was shown that, when there is a wide spectrum of sidebands (corresponding to different cluster sizes), only the envelopes of the sidebands are seen. The intensity of each individual sideband pair will be asymmetrical when modulation of both scattering factor and spacing are operating. Therefore, the center of gravity of the two envelopes need not be positioned at a symmetrical distance from the matrix reflection but, rather, is expected to be asymmetrical.

For the case considered here, experimentally, the scattering factor for niobium is approximately 1.6 times greater than those for iron, nickel or chromium. A variation in spacing, due to the large size difference between niobium and iron, nickel, or chromium, also exists. Combining these two modulations of scattering power and spacing, which are out of phase, explains the asymmetry in the observed sidebands intensity. Another complicating factor in the intensity arises, as was the case for Fe–Ni–Ti, from the difference in the volume of the enriched and denuded regions. The asymmetry in position indicates, from condition (1), that a variation in cluster size is present. Thus, the sizes determined from equation (1) represent, at best, only an *indication* of the average size.

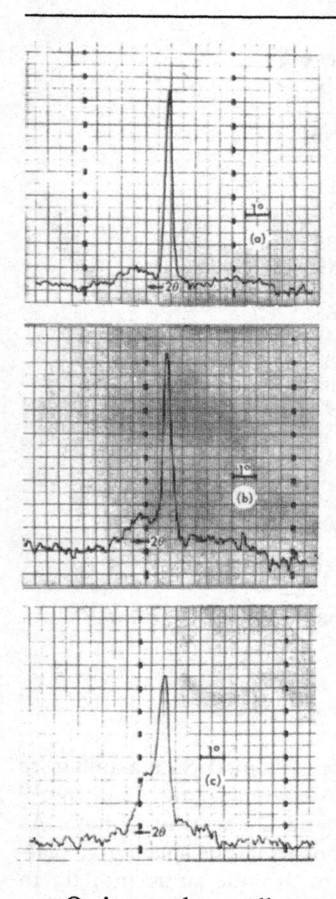

Figure 6. Microphotometer traces of the $(311)_{\gamma}$ sidebands of the niobium alloy; aged at 1150°F for (a) 24 hr, (b) 100 hr, and (c) 366 hr.

Owing to the small temperature range over which the aging process is occurring and the wide range of cluster sizes existing at a given treatment, no attempt was made to calculate the activation energy for growth of the clusters.

DISCUSSION AND CONCLUSIONS

The aging kinetics of the two systems are dissimilar even though their aging sequences are identical in that a coherent transition phase forms prior to the development of the equilibrium precipitates. There is an immediate aging response in the Fe–Ni–Ti system which could be accounted for by either (1) spinodal decomposition during the transformation that is occurring or (2) the formation of nuclei during the solution treatment. There is indirect evidence based on the $c:a$ ratio of the martensite phase[13] that ordering of titanium has taken place in the austenite region prior to the martensitic transformation. Thus, the titanium sites could act as the nuclei for clustering of nickel–titanium. In either case, the distribution of the nuclei and, hence, the clusters will be generally periodic. All of the clusters will have a relatively uniform size for a given time and temperature since their growth rate is diffusion controlled and the distance for which the nickel or iron atom must travel is essentially constant for each zone.

For the niobium system, there appear to be no preexisting nucleation sites for the clusters; hence, they must form by the standard nucleation and growth process. Guinier's model of discrete cluster formation would, therefore, be most logical for treatment. The randomness in relationship to time and position at which the clusters develop will result in a spectrum of zone sizes. Coalescence would be expected to occur commensurate with the aging process until a more or less uniform equilibrium size is obtained. The slowness at which overaging is reached is undoubtedly associated with coalescence.

The above arguments are compatible with the experimental observations. Diffuse maxima positioned symmetrically about the matrix reflections are seen immediately upon aging of the titanium alloy. The movement of these maxima and, hence, the growth of the clusters are directly related to the diffusion of nickel or iron. On the other hand, not until a substantial time has elapsed do sidebands appear in the niobium system, and then they are asymmetrical in position and intensity. The asymmetry in intensity increases with time, but that of position seems to decrease, which indicates coalescence.

Thus the precise measurements available with a focusing Guinier camera provide a sensitive means for following kinetics in austenitic alloys. Where the system is favorable, kinetic data can be generated, and, in all cases, observations are consistent with mechanical property observations. The one factor that cannot be measured by this technique which is necessary for quantitative treatment is interparticle spacing. There are, however, few techniques that could give both size and distribution on the scale discussed here.

ACKNOWLEDGMENTS

A discussion with Dr. de Fontaine was helpful in interpreting results. The text was read critically by Messrs. Klems, Arko, and Pascover of this laboratory.

REFERENCES

1. B. G. Reisdorf and A. J. Baker, "The Kinetics and Mechanisms of the Strengthening of Maraging Steel," Air Force Materials Laboratory *Tech. Rept.* AFML-TR-64-390, Wright–Patterson Air Force Base, Ohio, Jan., 1965.
2. C. Bückle, B. Genty, and J. Manenc, "Some Aspects of the Precipitation in a Nickel–Titanium Alloy," *Rev. Met. (Paris)* **56**: 247, 1959.
3. J. -P. Henon, J. Manenc, and C. Crussard, "A Few Results Concerning the Precipitation in Fe–Ti, Fe–Ti–Si, and Fe–Ti–Ni Alloys," *Compt. Rend.* **257**: 671, 1963.
4. A. Guinier, "Heterogeneities in Solid Solutions," *Solid State Phys.* **9**: 293, 1959.
5. M. Hillert, M. Cohen, and B. L. Averback, "Formation of Modulated Structures in Copper–Nickel–Iron Alloys," *Acta Met.* **9**: 536, 1961.
6. D. de Fontaine, "A Theoretical and Analogue Study of Diffraction from One-Dimensional Modulated Structures," in: J. B. Cohen and J. E. Hilliard (eds.), *Local Atomic Arrangements Studied by X-Ray Diffraction*, Gordon & Breach, New York, 1966, p. 51.
7. M. E. Hargreaves, "Modulated Structures in Some Cu–Ni–Fe Alloys," *Acta Cryst.* **4**: 301, 1951.
8. J. K. Abraham, J. K. Jackson, and L. Leonard, "X-Ray Study of the Aging Process in an Austenitic Fe–31Ni–3.5Ti Alloy," *Trans. ASM* **61**: 233, 1968.
9. R. T. Weiner and J. J. Irani, "Intermetallic Precipitation in Austenitic Steels Containing Niobium and/or Tantalum," *Trans. ASM* **59**: 340, 1966.
10. R. F. Decker and S. Floreen, "Precipitation from Substitutional Iron-Base Austenitic and Martensitic Solid Solutions," in: G. R. Speich and J. B. Clark (eds.), *Precipitation From Iron-Base Alloys*, Gordon & Breach, New York, 1965, p. 69.
11. C. Bückle and J. Manenc, "Investigation of the Precipitation in Cu-Rich Cu–Ti Alloys," *Rev. Met. (Paris)* **57**: 436, 1960.
12. E. G. Nesterenko and K. V. Chuistov, "Initial Disintegration of Supersaturated Solid Solution of Titanium in Copper," *Fiz. Metal. i Metalloved.* **9**: 140, 1960.
13. Y. Honnorat, G. Henry, and J. Manenc, "Study of Hardening of Fe–30% Ni Alloys with Titanium," *Mem. Sci. Rev. Met.* **62**: 429, 1965.

DISCUSSION

Chairman W. C. Hagel: With the Fe–Ni–Ti alloy, you made an activation energy analysis and mentioned observing 59 cal/mole. It should be kilocalories, correct? And then you compared it with substitutional diffusion in iron–nickel base alloys. Usually, the number is around 70 rather than 59. I don't know whether that is too good a comparison. What I suspect happens with this first alloy is that the precipitate might be something different than the nickel–titanium phase that you are focusing on. It could be an impurity sequence, or, if aluminum is present, it could be gamma prime, which often happens with these alloys. I just wonder if you know that much about the initial precipitate. You mention that it could be spinodal, and so on. Would you care to elaborate on that?

J. K. Abraham: Yes, I would. As far as impurities present, our alloys were reasonably pure. The level of aluminum present was less than what can be detected by spectrographic analysis, and the total interstitial content was less than 0.02%. The gamma-prime phase which we have observed is quite similar to the aluminum or aluminum–titanium gamma prime. Mihalisin and Decker ("Phase Transformation in Ni-Rich Ni–Ti–Al Alloys," *Trans. AIME* **218**: 507, 1960) have shown that titanium could replace aluminum during the precipitation of alpha in Ni–Ti–Al system; hence it is not surprising to see the similarities. With respect to activation energies, Moll and Ogilvie ("Solubility and Diffusion of Titanium in Iron," published in *Trans. AIME* **215**: 613, 1959) determined that the activation energy for the diffusion of titanium in austenitic steels was 60,000 cal/mole. This compares favorably with the numbers obtained here.

Chairman W. C. Hagel: Right on the button.

J. K. Abraham: Yes, but I think that's fortuitous, of course.

Chairman W. C. Hagel: It's possible that titanium is the rate-recontrolling species.

J. K. Abraham: It seems that way, but as Hillert[5] points out, and stated in the text, the energies derived from the analysis presented here would be low.

Chairman W. C. Hagel: You mentioned that nickel or iron could also be controlling.

J. K. Abraham: I'm not sure which of these it is, but the process appears to be diffusion-controlled and not nucleation-controlled. Also, for the data here, we do not believe coalescence is very important.

Chairman W. C. Hagel: You're really neglecting nucleation in the first alloy, but, in the second alloy, you're forming a complicated Laves structure which requires a long-term nucleation and growth mechanism.

J. K. Abraham: In the second structure, there is a nucleation and growth process definitely occurring, but the Fe_2Nb Laves phase which was discussed is one that was present initially and apparently decomposes with aging.

O. Kimball (University of Michigan): Am I correct in assuming these were powder samples in all cases?

J. K. Abraham: No, they were not. The samples were standard $\frac{1}{4}$ by 1 by approximately $\frac{1}{8}$ in. thick.

O. Kimball: This was the sample you examined by X-ray diffraction in the Guinier camera?

J. K. Abraham: Yes.

O. Kimball: Did you check for problems with grain size, preferred orientation, *etc.*?

J. K. Abraham: Preferred orientation should not be important in the studies made here. As I mentioned, the total area irradiated by the sample was approximately $\frac{1}{2}$ by $\frac{1}{8}$ in.

O. Kimball: In a system where you've got either composition fluctuations or zones this large, they should be easily detectable by microscopy. Has anyone, either yourself or others, tried this as a check? I have another question which is somewhat similar. Have you tried single-crystal techniques, which are more sensitive than powder methods?

J. K. Abraham: As of yet, we have not done single-crystal work on this system. We hope to. Microscopy work, yes. Seeing that this is an X-ray conference, I did not want to mention electron microscopy work that is being done and has been done in this system. In both cases, work that has been done in this system confirms the aging sequence that has been mentioned.

A SIMPLIFIED METHOD OF QUANTITATING PREFERRED ORIENTATION

Michael M. Klenck

Atomics International
A Division of North American Aviation, Incorporated
Canoga Park, California

ABSTRACT

Qualitative evaluation of preferred orientation in 10-mil Hastelloy N seamless tubing was carried out by obtaining conventional pole figures for several samples with the modified Schulz reflection technique. This established the texture as a (110)[112]. Quantitative determination of this texture was achieved for a large number of samples by a radial traverse of an azimuthally averaged (110) stereographic projection. This was accomplished with a Schulz goniometer by rotating the sample rapidly in its own plane while slowly varying the tilt angle ϕ. The first minimum observed in the resultant pattern approximates the pole density of the randomly oriented component of the sample. A scan obtained from a randomly oriented sample and suitably scaled to match pole density at these minima serves to distinguish the random from the oriented component of the sample. A ratio of the oriented to the random component as a function of the tilt angle ϕ is independent of the instrumental effects of defocusing and absorption. This ratio is proportional to the pole density of the oriented component since the pole density of the random component is a constant by definition. The same constant of proportionality yields a pole density of unity for the random component. Integrating the pole density for the random component over the surface of a reference sphere of unit radius gives the value 4π for the number of randomly oriented poles N_R. The number of poles corresponding to the oriented component is obtained by evaluating the integral

$$N_0 = 2\pi m \int_0^{\Phi} R(\phi) \sin\phi \, d\phi$$

In this equation, $R(\phi)$ is the ratio of the oriented to the random component, Φ is the tilt angle corresponding to the first minimum in the sample scan, and m is a multiplicity factor which compensates for integration over just one 110 face of each grain. Evaluation of this integral may be made by plotting it with the aid of a family of sine curves and measuring the area under the curve.

The per cent of volume of the oriented component is then calculated by the relationship

$$\text{Volume \% oriented} = \frac{N_0}{N_0 + N_R} 100$$

This method entails about 30 min of an analyst's time and 40 to 120 min of machine time, depending on the grain size of the sample. Further information on the angular breadth of the texture may be obtained from the plot of the integrand. This technique is subject to the limitations that the material have a single texture and it must be possible to prepare the specimens so that a suitably chosen pole figure contains a central concentration of poles which are delineated from other regions of pole concentration by a surrounding minimum in pole density. This is generally not too difficult in the case of sheet textures but requires that the specimen normal be parallel to the fiber axis for a fiber texture.

TECHNIQUE

The procedure described was generated by a requirement that a large number of samples of Hastelloy N tubing be quantitated with respect to the degree of preferred orientation. The limited time available for this effort precluded preparation of full pole figures for each sample. In addition, the large grain size (ASTM 4) of the samples made it statistically necessary to evaluate a number of specimens for each step in the processing of the tubing.

An initial qualitative analysis of the character of the preferred orientation was carried out for several samples by using the Schulz reflection technique[1] as modified by Chernock and Beck.[2] A copper-target X-ray tube was used and the nickel beta filter was moved to a position between the X-ray tube and the sample to reduce nickel fluorescence in the sample, due to Cu $K\beta$ radiation. A typical pole figure is shown in Figure 1. It is instrumentally limited to a tilt angle of 70° but may readily be recognized as a duplex (110)[$\bar{1}$12],[1$\bar{1}$2] texture. The ideal for this texture is indicated by the solid dots.

The basis for quantitating such a pole figure is the assumption that the grains of the sample may be divisible into two groups. One group represents the random component and, by definition, has a constant pole density over the reference sphere. The other group corresponding to the oriented component has a variable pole density which falls to zero in some areas. Thus, in an actual pole figure, the minimum pole density (after averaging out statistical fluctuations) corresponds to the random component and must be subtracted from the total pole density to obtain the density function for the oriented component. This function may be expressed as a function of the polar coordinates ϕ and α, but, in the method described in this paper, it is forced to be a function of the polar angle ϕ only.

The means of obtaining this function of ϕ is a variation of the Schulz method in which the rate of change in the azimuthal angle α (Figure 2) is very large relative to the rate of change of the tilt angle ϕ. In this investigation, $d\alpha/dt = 120°/\text{sec}$ and $d\phi/dt = 1°/\text{min}$. A time constant for the detection system of 16 sec was selected because it is large with respect to $d\alpha/dt$ and small with respect to $d\phi/dt$. The customary oscillation of the sample back and forth in its own plane was simultaneously carried out. In Figure 3, a simple solution to carrying out these motions is shown. The oscillation motor drives the azimuthal angle, and the high-speed goniometer gears are used with the clutch disengaged to provide the translatory motion. Particularly in the case of coarse-grained samples, it is important that the ratio of the frequencies of the oscillatory motion and the

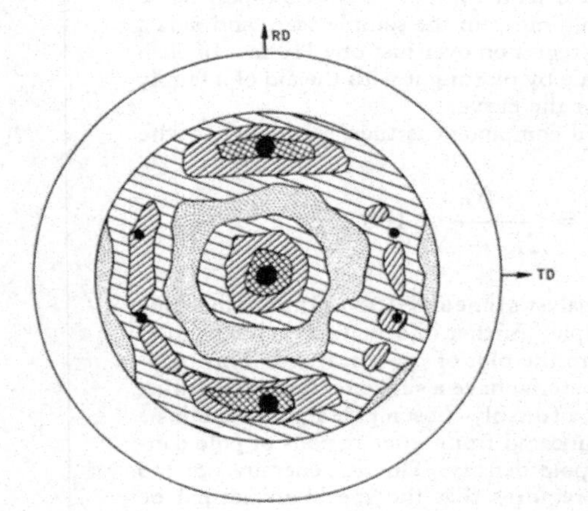

Figure 1. Pole figure (110) of Hastelloy N tubing after grain growth. The dark spots show the ideal (110)[$\bar{1}$12],[1$\bar{1}$2] texture.

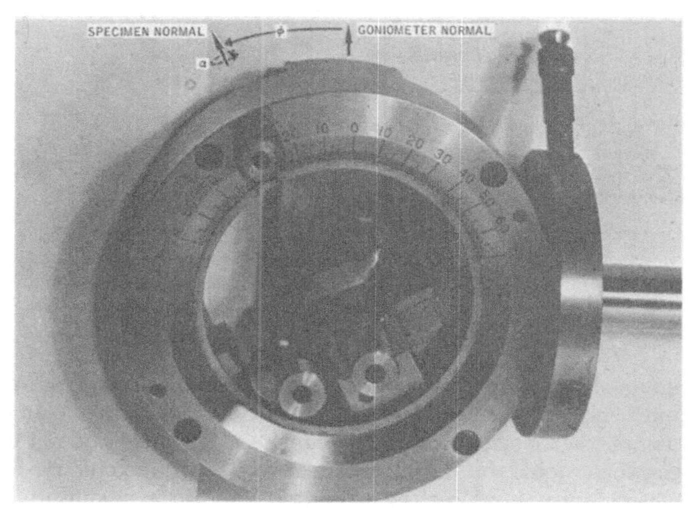

Figure 2. A Schulz goniometer showing the polar angle ϕ and the azimuthal angle α.

azimuthal rotation should not be expressible as a ratio of small integers (*i.e.*, 1 : 3, *etc.*), or the sample will not receive maximum exposure of its area.

The detector output for the motion described constitutes a radial traverse of an azimuthally averaged (110) stereographic projection. This is shown in Figure 4 as a solid line. Owing to the coarseness of the grains in the sample, it was advantageous to scan from 70° ϕ to $-70°$ ϕ and base the results on the average of the two sides. The observed asymmetry about $\phi = 0$ is the result of a combination of imperfection in instrument geometry

Figure 3. Diffractometer with Schulz goniometer in place showing flexible cable connections. The oscillation drive provides $d\alpha/dt$, and the connection to the 2θ drive gears provides the translatory motion.

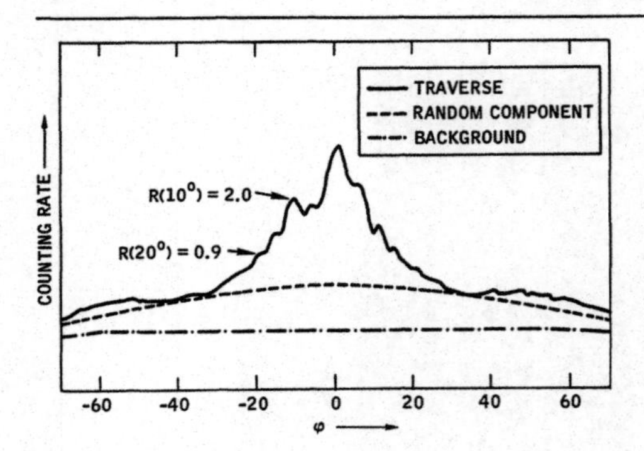

Figure 4. A typical traverse of an azimuthally averaged stereographic projection. Values of the ratio $R(\phi)$ are shown at two points.

and the coarse-grained sample. For a fine-grained material, the scan is symmetric about $\phi = 0$ so that only half would need to be run. The background shown in the figure is obtained by changing the 2θ setting on the goniometer. The magnitude of this change is necessarily larger than would be customary in ordinary diffractometry owing to the deliberate defocusing introduced by the Chernock and Beck modification. Since the background is flat over the range of interest, it is only evaluated at $\phi = 0°$ either before or after the ϕ scan.

A synthetic random sample was prepared by annealing fine filings of Hastelloy N and sprinkling them on a glass slide coated with vacuum grease. A ϕ scan of this random sample serves to characterize the instrumental defocusing effect. The random-component scan shown in Figure 4 is derived from this random-sample scan by scaling it to equal the pole-density minimum which occurs at 30° ϕ in this case. This scaling procedure is easily accomplished by drawing a family of random scans corresponding to various attenuations of the random-sample scan. When the sample scan is laid over the family of random scans and their respective backgrounds are made to coincide, a "random component" curve, such as is shown in Figure 4, may easily be drawn to match the sample curve at its pole-density minimum and conform to the family of random curves.

Since the units of pole density are unspecified, the magnitude of the density of this random component is assigned a value of 1. The purpose of the arbitrary assignment is to make the ratio of the preferred to the random component [denoted by the function $R(\phi)$] equal numerically to the pole density of the preferred component.

$$\text{Pole density of preferred component} \equiv R(\phi) = \frac{\text{sample scan} - \text{random component}}{\text{random component} - \text{background}} \tag{1}$$

The empirical correction usually applied to such functions to correct for defocusing has not been included since it conveniently cancels out in the ratio of the components.

The total number of poles corresponding to the random component will be equal to its unit pole density integrated over the surface of the reference sphere

$$N_R = \int_0^{\pi} 2\pi \sin\phi \, d\phi \tag{2}$$

The total number of poles corresponding to the preferred component will be equal to the integral of the ratio $R(\phi)$ of the preferred to the random components integrated over the same reference sphere. In practice, the function $R(\phi)$ is not evaluated over the entire range, so that integration over the entire sphere cannot be carried out. It is possible, however, to integrate over the one pole which tends to lie normal to the sample surface. This integration covers the range of $\phi = 0$ to $\phi = \Phi$, where Φ is the value of the first minimum in the function $R(\phi)$. In the example shown in Figure 4, this is approximately 30°. Since this only encompasses one pole of each grain in the preferred component, it must be multiplied by the multiplicity m of that plane in order to give the total number of poles that would have been obtained had the integration covered the whole reference sphere. Thus, the number of poles corresponding to the preferred component will be

$$N_0 = m \int_0^{\Phi} R(\phi) 2\pi \sin \phi \, d\phi \tag{3}$$

Evaluation of the integral in equation (3) is accomplished by plotting the integrand and measuring the area under the curve. Plotting this function is facilitated by the use of a family of sine curves such as is shown in Figure 5. An adequate representation of $R(\phi)$ was obtained by measuring it at 5° intervals. The result of a typical plot is shown, and its area can be converted to a number either by use of a planimeter or by cutting the area out of paper of constant areal density and weighing it. In this work, the latter method was more convenient.

On the assumption that the material is single phased and single textured, the sum of the random and the preferred components will add up to 100%. Therefore, the volume per cent which is oriented may be calculated by the equation

$$\text{Volume \% oriented} = \frac{N_0}{N_0 + N_R} 100 \tag{4}$$

The factor of 2π which appears in both equations (2) and (3) has been included for the sake of correctness. In the final calculation of volume per cent, it cancels out; so, in practice, the evaluation of N_0 and N_R does not include it. A number proportional to N_0 is

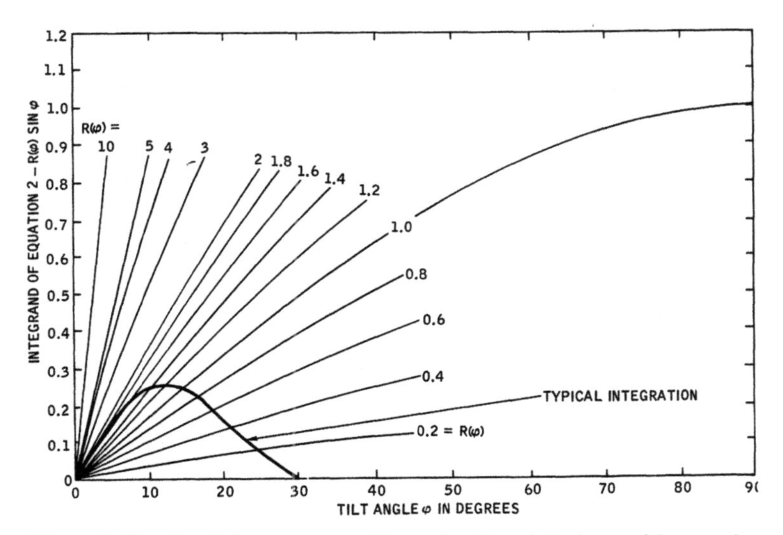

Figure 5. Family of sine curves used in evaluation of the integral in equation (3). The dark line envelops the area associated with a typical integration.

Table I. Sources of Error

	Vol. % oriented	Std. dev.	Variance
Six independent runs of the same sample	31.9	2.7	7.3
Positioning of sample		1.1	1.2
Drawing in random component		1.5	2.2
Evaluating $R(\phi)$		1.7	2.9
Variability of paper density		0.9	0.8
Variability of instrument response		<0.5	<0.2

obtained by multiplying the area under the "typical integration" of Figure 4 by the multiplicity (in this example, 12). The area under the sin ϕ curve in Figure 5 from $\phi = 0$ to $\phi = \pi/2$ is multiplied by 2 to obtain a number proportional to N_R.

For the example used in this paper, the calculated volume is 30.2%. Table I shows the breakdown of the sources of error of the method, based on multiple runs of a different sample.

The two largest sources of error, evaluating $R(\phi)$ and drawing in the random component, are sensitive to the grain size of the sample and would be significantly reduced if fine-grained samples were being analyzed. The values given in the table reflect only the precision of the method as applied to the Hastelloy N samples analyzed. The accuracy of the method will be affected principally by the validity of the assumptions that, at the minimum $\phi = \Phi$, the function $R(\phi)$ is independent of the azimuthal angle α, that only one texture is present in the sample, and that the sample is single phased. A method entailing a far more sophisticated approach was described by J. R. Holland at a previous Denver Conference.[3]

Not including whatever time is needed to prepare the sample surfaces, this approach requires no more than 30 min of an analyst's time and 40 to 120 min of machine time, depending on the grain size of the sample. Additional information on the angular distribution of the texture may be easily obtained by breaking the typical integration curve up into angular ranges before evaluation.

REFERENCES

1. L. G. Schulz, "A Direct Method of Determining Preferred Orientation of a Flat Reflection Sample Using a Geiger Counter X-Ray Spectrometer," *J. Appl. Phys.* **20**: 1030–1033, 1949.
2. W. P. Chernock and P. A. Beck, "Analysis of Certain Errors in the X-Ray Reflection Method for the Quantitative Determination of Preferred Orientations," *J. Appl. Phys.* **23**: 341–345, 1952.
3. J. R. Holland, "Quantitative Determinations and Descriptions of Preferred Orientation," in: W. M. Mueller, G. R. Mallett, and M. J. Fay (eds.), *Advances in X-Ray Analysis, Volume 7*, Plenum Press, New York, 1964, pp. 86–93.

DISCUSSION

H. Pickett (General Electric Co.): As I see it, and I admire your acceptance of the assignment to evaluate different materials in an easily reported and interpretable way, I am prepared to believe that this is adequate for your purpose. But, I want you to verify one thing that I think I understand in this connection. There's a risk that you get the same report for two different materials in terms of per cent oriented where the sharpness of orientation around the central pole varies substantially from one to another as against the broad peak, so that your method implies that you believe that the risk of this is low or that, if it did happen, it wouldn't influence the usability of the material.

M. M. Klenck: The sharpness of the orientation is revealed by the typical integration shown in Figure 5. The tilt angle corresponding to the peak value of the ordinate would provide a simply obtained number inversely proportional to the sharpness of the orientation.

Z. W. Wilchinsky (Enjay Polymer Labs.): If your intensity had dropped down to the background level zero, would that imply that your per cent orientation would be 100%?

M. M. Klenck: Yes, in the definition of this method, it would be so.

CRYSTALLITE ORIENTATION ANALYSIS FOR ROLLED CUBIC MATERIALS

Peter R. Morris and Alan J. Heckler

Research and Technology
Armco Steel Corporation
Middletown, Ohio

ABSTRACT

Roe's method for deriving the crystallite orientation distribution in a series of *generalized spherical harmonics* is applied to the analysis of texture in rolled cubic materials. The augmented Jacobi polynomials, which are the basis of the generalized spherical harmonics, have been derived for cubic crystallographic symmetry and orthotropic physical symmetry through the sixteenth order. Truncation of the series expansions at the sixteenth order should permit treatment of textures having a maximum of 17 times random and a minimum angular width at half maximum of 34°. A numerical technique has been developed which permits approximate evaluation of the integral equations from a finite array of data points. The method is illustrated for commercial steels and is used to elucidate the primary recrystallization texture of a decarburized Fe–3%Si alloy.

INTRODUCTION

Three angles are required to specify the orientation of a crystallite with respect to a physical reference frame. The experimental determination of crystallite orientation distribution by measuring the orientations of a large number of grains by Laue or etch-pit methods is a tedious procedure. In addition, one has to make some estimate of the individual grain volumes and weight the results accordingly or assume that the grains can be treated as equal in size (see, for example, the work of C. G. Dunn[1]).

Because of the experimental difficulties involved in determining the crystallite orientation distribution, most workers have employed experimental techniques which lend themselves either to representing the angular distribution of the poles of a particular type of crystallographic plane in a physical reference frame (pole figure) or to representing the angular distribution of a particular type of physical direction in a crystallographic reference frame (inverse pole figure or axis density figure). Either representation is a function of two angular coordinates.

Intuitive determination of the (three-angle) crystallite orientation distribution from two or more (two-angle) pole figure distributions is extremely difficult except in those cases where the orientation approaches that of a single crystal.

Recently, Roe[2] devised a technique in which the crystallite orientation distribution is determined from several pole figures. In a subsequent paper,[3] he particularized the treatment for cubic-crystal symmetry. In this paper, we shall illustrate the application of Roe's method to commercial flat-rolled steels.

DISCUSSION OF ROE'S METHOD

The set of Euler angles, ψ, θ, and ϕ, used by Roe to specify the orientation of a crystallite in a physical reference frame is illustrated in Figure 1. The axis for each

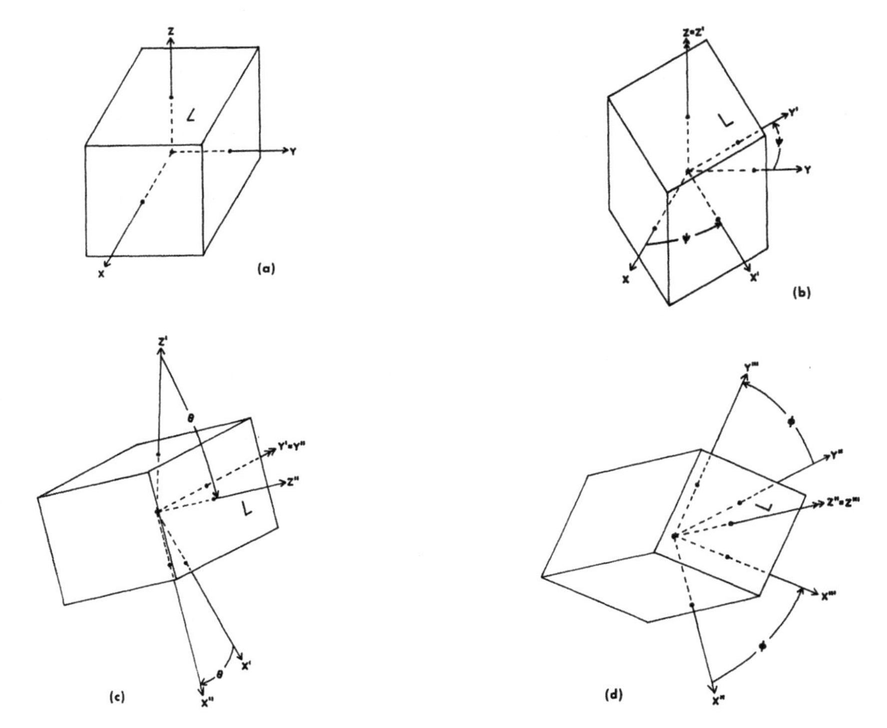

Figure 1. Euler angles. (a) Initial cube orientation; (b) ψ rotation; (c) θ rotation; (d) ϕ rotation.

rotation is denoted by a double arrowhead. The positive sense of each rotation is that of a right-handed screw advancing in the direction of the double arrowhead. The ψ rotation about the **Z** axis is carried out first by rotating **X** into **X**′, **Y** into **Y**′, and leaving **Z** unchanged, *i.e.*, $\mathbf{Z}' = \mathbf{Z}$. The θ rotation about the **Y**′ axis is carried out second by rotating **X**′ into **X**″, **Z**′ into **Z**″, and leaving **Y**′ unchanged, *i.e.*, $\mathbf{Y}'' = \mathbf{Y}'$. The ϕ rotation about the **Z**″ axis is carried out last by rotating **X**″ into **X**‴, **Y**″ into **Y**‴, and leaving **Z**″ unchanged, *i.e.*, $\mathbf{Z}''' = \mathbf{Z}''$. Any specified orientation of the **X**‴, **Y**‴, and **Z**‴ crystallographic axes relative to the **X**, **Y**, and **Z** physical axes (rolling direction, transverse direction, and sheet normal direction, respectively) can be obtained by appropriate ψ, θ, and ϕ rotations.

The orientation of a reciprocal lattice vector $\mathbf{r}_j$ with respect to the crystallographic axes is specified by the colatitude angle Θ_j measured from **Z**‴ and by the longitude angle Φ_j measured from **X**‴ toward **Y**‴. The orientation of this same reciprocal lattice vector $\mathbf{r}_j$ to the physical reference frame is specified by the colatitude angle χ_j measured from **Z** and by the longitude angle η_j measured from **X** toward **Y**.

The distribution $q_j(\chi, \eta)$ of the jth reciprocal lattice vector (pole figure) is first expanded in a series of spherical harmonics

$$q_j(\chi, \eta) = \sum_{l=0}^{\infty} \sum_{m=-l}^{l} Q_{lm}^{j} P_{lm}(\zeta) e^{-im\eta} \tag{1}$$

where $\zeta = \cos \chi$. The coefficients Q_{lm}^{j} of this expansion can be determined by using the orthogonality relations for spherical harmonics [see Roe,[2] equation (9)]

$$Q_{lm}^{j} = \left(\frac{1}{2\pi}\right)\int_{0}^{2\pi}\int_{-1}^{1} q_j(\chi, \eta)P_{lm}(\zeta)e^{+imn}\,d\zeta\,d\eta \tag{2}$$

In the application of equation (2) to experimental data, the integration is replaced by summation over a finite number of data points. If the measured value $q_j(\chi, \eta)$ is assumed to be constant and to represent the interval (χ_k, η_k), we may take it outside the integral, which gives

$$Q_{lm}^{j} = \left(\frac{1}{2\pi}\right)\sum_{k} q_j(\chi_k, \eta_k)\int_{\zeta_{k-\frac{1}{2}}}^{\zeta_{k+\frac{1}{2}}} P_{lm}(\zeta)\,d\zeta \int_{\eta_{k-\frac{1}{2}}}^{\eta_{k+\frac{1}{2}}} e^{imn}\,d\eta \tag{3}$$

where $\zeta_{k+\frac{1}{2}}$, $\zeta_{k-\frac{1}{2}}$, $\eta_{k+\frac{1}{2}}$, and $\eta_{k-\frac{1}{2}}$ designate the upper and lower limits of ζ and η for the kth interval.

The crystallite orientation distribution is next expanded in a series of generalized spherical harmonics

$$w(\psi, \theta, \phi) = \sum_{l=0}^{\infty}\sum_{m=-l}^{l}\sum_{n=-l}^{l} W_{lmn}Z_{lmn}(\xi)e^{-im\chi}e^{-in\phi} \tag{4}$$

where $\xi = \cos\theta$, and the $Z_{lmn}(\xi)$ are the Jacobi polynomials augmented by the square root of the weight function. They form an orthonormal set with respect to the l index on the interval -1, 1. The coefficients W_{lmn} of this expansion are related to the coefficients Q_{lm}^{j} of the pole figure expansion by the system of linear simultaneous equations

$$Q_{lm}^{j} = 2\pi\left(\frac{2}{2l+1}\right)^{\frac{1}{2}}\sum_{n=-l}^{l} W_{lmn}P_{ln}(\Xi_j)e^{in\Phi_j} \tag{5}$$

where $\Xi_j = \cos\Theta_j$ [see Roe,[2] equation (13)].

EFFECTS OF PHYSICAL, CRYSTALLOGRAPHIC SYMMETRY

The restrictions imposed on $q_j(\chi, \eta)$, Q_{lm}^{j}, and W_{lmn} owing to physical and crystallographic symmetry have been discussed by Roe[2] (his Tables 1 and 2). For mirror planes perpendicular to the rolling, transverse, and sheet normal directions, the Q_{lm} are all real, $Q_{lm} = Q_{l\bar{m}}$, and l and m are restricted to even values. For this physical symmetry and cubic crystallographic symmetry, the W_{lmn} are all real, $W_{lmn} = W_{l\bar{m}n} = W_{lm\bar{n}} = W_{l\bar{m}\bar{n}}$, l and m are restricted to even values, and n is restricted to $4k$, where $k = 0, 1, 2, \ldots$. Because of symmetry conditions the nonzero W_{lmn} are not all linearly independent. The relations between linearly independent and linearly dependent W_{lmn} for cubic symmetry are given by Roe's[3] equations (6) and (7) and Table 1. The Z_{lmn} have been derived for orthotropic physical and cubic crystallographic symmetry in multiple-angle form up to $l = 16$ and are tabulated in the Appendix. The Z_{lmn} are given in this form by

$$Z_{lmn} = \frac{(D_{lmn})^{\frac{1}{2}}}{2^k}\sum_{s=0}^{l} E_{lmns}\cos(s\theta) \tag{6}$$

The multiple-angle form results in less error than the power series expansion when the number of digits which may be carried is limited to typical computer usage (seven or eight significant figures). Roe[2] gives relations between Z_{lmn} and $Z_{l\bar{n}\bar{m}}$, between Z_{lmn} and Z_{lnm}, and between $Z_{l\bar{m}\bar{n}}$ and Z_{lmn} [his equations (A10), (A11), and (A12), respectively]. A relation between Z_{lmn} and $Z_{lm\bar{n}}$ may be found by changing the sign of ξ in the differential equation from which the Z_{lmn} are derived [Roe's[2] equation (A1)]. The original differential equation can be recovered by substituting $-\xi$ for ξ since the equation contains products of even powers of ξ by even-order derivatives and of odd powers of ξ by odd derivatives, with the exception of the term $-2mn\xi$. Hence,

$$Z_{lm\bar{n}}(\xi) = Z_{lmn}(-\xi) = Z_{l\bar{m}n}(\xi) \tag{7}$$

or, if the Z_{lmn} are expressed in terms of θ,

$$Z_{lm\bar{n}}(\theta) = Z_{lmn}(\pi - \theta) = Z_{l\bar{m}n}(\theta) \tag{8}$$

Finally, since the W_{lmn} are all real,

$$w(\psi, \theta, \phi) = \sum_{l=0}^{\infty} \sum_{m=-l}^{l} \sum_{n=-l}^{l} W_{lmn} Z_{lmn}(\xi) \cos(m\psi + n\phi) \tag{9}$$

The truncation error σ_w in $w(\psi, \theta, \phi)$ can be obtained by first estimating the average of the square of Q_{lm} over reciprocal space [see Roe's[2] equation (33)]. An estimate of $\sigma_w{}^2$ is then given by plotting

$$\sum_{m=-l}^{l} \langle Q_{lm}^2 \rangle$$

versus l and fitting an equation of the form

$$\sum_{m=-l}^{l} \langle Q_{lm}^2 \rangle = Ae^{-Bl}$$

to the data by a least-squares method. Then,

$$\sigma_w{}^2 \approx \int_{\lambda+1}^{\infty} (l + \tfrac{1}{2}) Ae^{-Bl}\, dl \tag{10}$$

where λ is the order at which the expansion indicated in equation (9) is truncated, *e.g.*, in this work, $\lambda = 16$.

The severity of texture which can be adequately treated for truncation at order l may be estimated in the following manner: The spherical harmonics of order l are homogeneous polynomials of order l. They can fit exactly the homogeneous polynomial $(l + 1) \cos^l \theta$, where the normalization coefficient has been selected so that the average over 4π solid angle is unity. For $l = 16$, this polynomial has a maximum value of 17, and an angular width at half maximum of 34°.

EXPERIMENTAL PROCEDURE

In order to apply Roe's method to materials with cubic crystallographic symmetry and orthotropic physical symmetry, *e.g.*, flat-rolled body-centered-cubic iron, it is sufficient to determine one quadrant of two or more pole figures. The data presented in this paper were obtained by using a composite sampling technique[4,5] which permits determination of a quadrant of the pole figure by the reflection method.[6]

Mo $K\alpha$ radiation was used with a Siemens texture goniometer to obtain the data required for construction of {110}, {200}, and {222} pole figures. A random standard specimen was used to correct for defocusing effects.[7] The X-ray data were collected along a spiral path, and the output was in the form of digital data punched on paper tape.[8]

For the numerical approximation of the integrals used to obtain the coefficients Q_{lm}^j [Roe[2] equation (9)], it is convenient to specify the reciprocal lattice vector density

$q_j(\chi, \eta)$ at grid points formed by the intersections of parallels of colatitude χ and longitude η. A linear interpolation was used to estimate q_j at specific values of χ and η from the three nearest data points along the spiral path scanned by the Siemens texture goniometer. The $q_j(\chi, \eta)$ may also be determined by using a combined transmission[9] and reflection[10] technique.

TEXTURE ANALYSIS OF FLAT-ROLLED STEELS

The primary recrystallization texture of decarburized Fe–3%Si will be used to illustrate the application of Roe's[2] method to flat-rolled steels. This texture has been variously described.[11–13] The texture components are important commercially since a strong {110}⟨001⟩ secondary recrystallization texture develops from this texture during a subsequent anneal. The difficulty of interpreting this texture from pole figures is apparent in Figures 2 to 4. The {200} pole figure is consistent with a {110}⟨001⟩

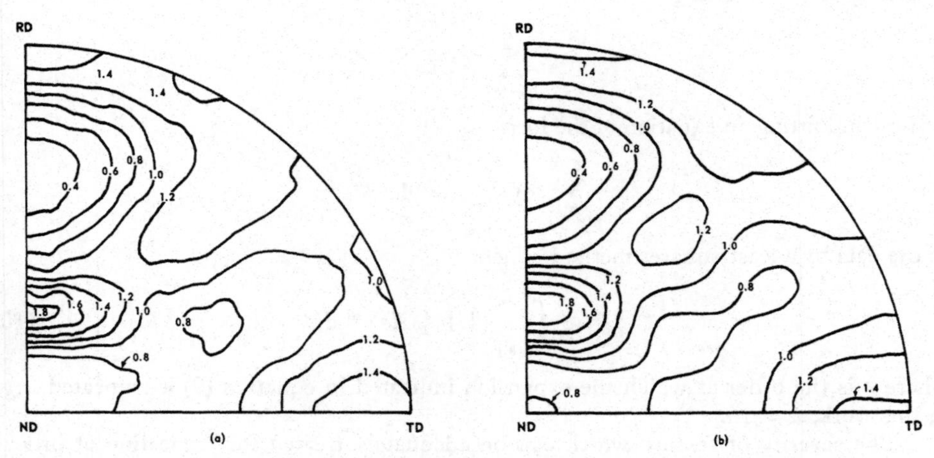

Figure 2. {110} Pole figures of Fe–3% Si. (a) Measured; (b) derived from W_{lmn}.

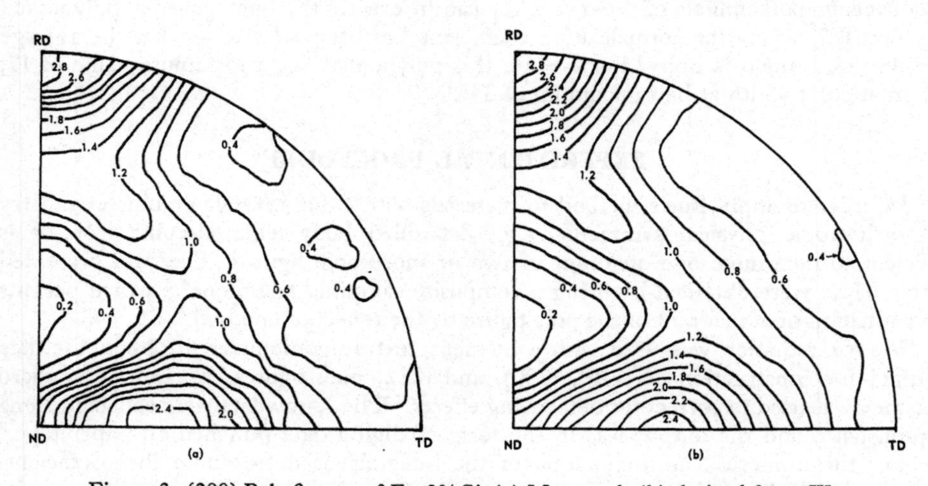

Figure 3. {200} Pole figures of Fe–3% Si. (a) Measured; (b) derived from W_{lmn}.

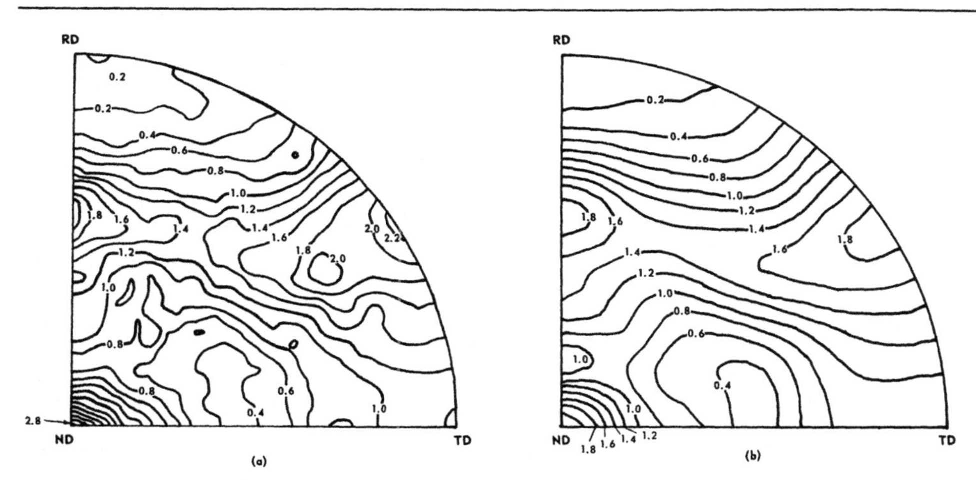

Figure 4. {222} Pole figures of Fe–3% Si. (a) Measured; (b) derived from W_{lmn}.

texture, while the {110} pole figure indicates a less than random intensity normal to the sheet and suggests a {111}⟨211⟩ texture. The {222} pole figure suggests a ⟨111⟩ fiber axis in the normal direction.

Series Expansion of Pole Figures

Coefficients Q^j_{lm} for the series expansion indicated in equation (1) were determined from the {110}, {200}, and {222} pole figure data by using equation (3). The restriction $W_{2m0} \equiv 0$, together with equation (5), requires $Q^j_{2m} \equiv 0$. Values of Q^j_{2m} obtained from experimental data by equation (3) were generally of the same order of magnitude as the fourth- or sixth-order coefficients. No explanation of this apparent paradox will be offered here. The coefficients W_{lmn} for the series expansion of the crystallite orientation distribution indicated in equation (4) were then determined from equation (5) by a least-squares method. To test whether the series expansion indicated in equation (4) was consistent with the {110}, {200}, and {222} pole-figure data, the Q^j_{lm} were then derived from the W_{lmn} by equation (5) and the series expansion indicated in equation (1) was used to construct {110}, {200}, and {222} pole figures. These figures can be compared with figures constructed from measured data in Figures 2 to 4. The crystallite orientation distribution adequately describes the essential features of the measured pole figures. The truncation errors $\sigma_q{}^j$ for the {110}, {200}, and {222} pole-figures [see Roe[2] equation (30)] are 0.2, 0.3, and 0.4, respectively. It should be noted that both $q_j(\chi, \eta)$ and $\sigma_q{}^j$ have been multiplied by 4π according to convention to give an average value of 1 for q_j. Equations (5) and (1) may also be used to construct pole figures for which diffraction data are not available.

Construction of Inverse Pole Figures

The normal direction (ND) inverse pole figure can be calculated from the set of W_{lmn} by averaging equation (4) with respect to the variable ψ to give

$$w(\theta, \phi) = \sum_{l=0}^{\infty} \sum_{n=-l}^{l} W_{l0n} Z_{l0n}(\xi) \cos n\phi \tag{11}$$

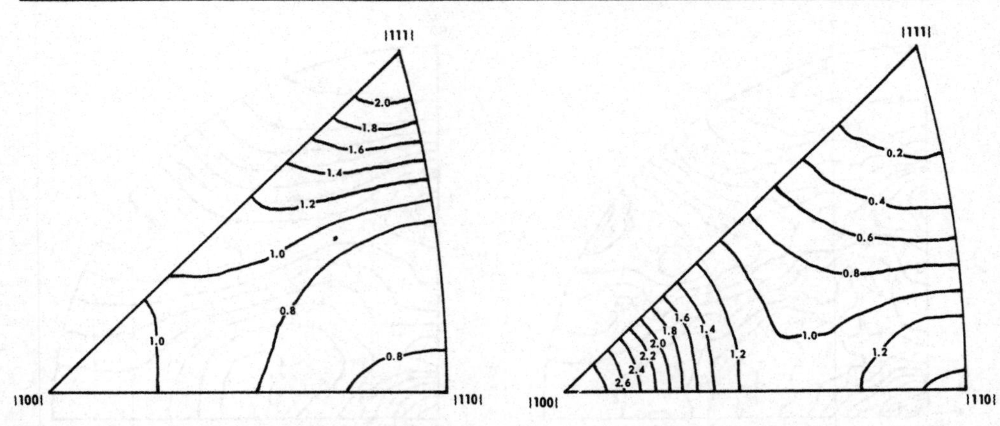

Figure 5. Normal-direction inverse pole figure for Fe–3% Si.

Figure 6. Rolling-direction inverse pole figure for Fe–3% Si.

An inverse pole figure corresponding to any sample direction can be calculated from equation (11) after a new set of W_{lmn} are obtained by a rotation of the sample coordinate system [see Roe[2] equation (17)]. Figures 5 and 6 show the normal-direction and rolling-direction inverse pole figures for the Fe–3%Si.

The normal-direction inverse pole figure shows a {111} component, and the rolling-direction inverse pole figure shows a {200} component and weaker {110} component. Again, interpretation of the crystallite orientation distribution from these figures is difficult.

REPRESENTATION OF CRYSTALLITE ORIENTATION DISTRIBUTION

Once the coefficients W_{lmn} are known, the crystallite orientation is given by equation (9). For the symmetry conditions considered here, *i.e.*, l even, m even, and $n = 4k$,

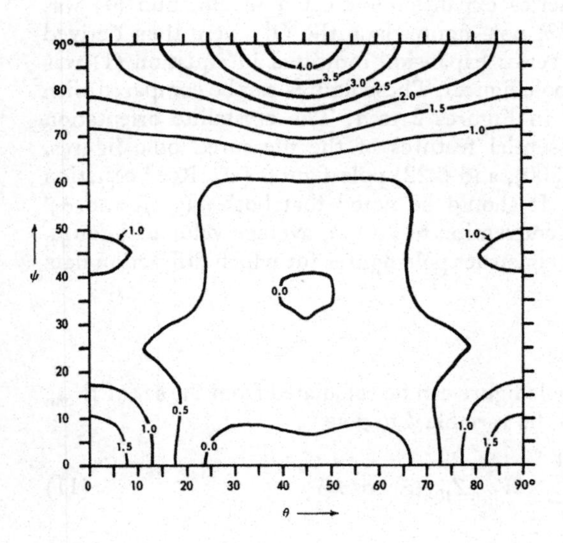

Figure 7. The $w(\psi, \theta, 0°)$ for Fe–3% Si.

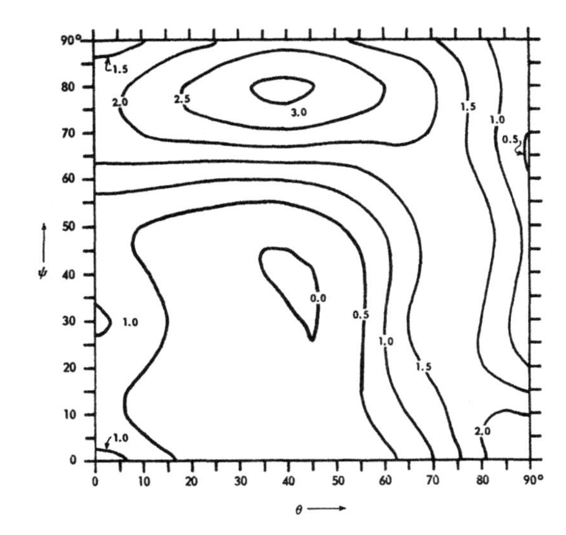

Figure 8. The $w(\psi, \theta, 15°)$ for Fe–3% Si.

where $k = 0, 1, 2 \ldots$, it is sufficient to specify $w(\psi, \theta, \phi)$ for the region

$$0 \leq \theta \leq \frac{\pi}{2} \qquad 0 \leq \psi \leq \frac{\pi}{2} \qquad 0 \leq \phi \leq \frac{\pi}{2}$$

Two-dimensional, constant ϕ, sections of the crystallite orientation distribution corresponding to $\phi = 0$, 15, 30, and 45° are shown in Figures 7, 8, 9, and 10, respectively, for the Fe–3%Si. For this material, the truncation error σ_w is estimated from equation (10) to be 0.2, where w and σ_w have been multiplied by $8\pi^2$.

The ψ, θ, and ϕ coordinates corresponding to some ideal textures of general interest are given in Table I. For $\theta = 0°$, ψ and ϕ rotations are equivalent, and, in this case, it is not possible to specify ψ and ϕ uniquely but only the sum $\psi + \phi$. This accounts for the displacement of the values of w along the $\theta = 0°$ boundary with a change in ϕ such that $w(\psi, \theta, \phi)$ remains unchanged for constant $\psi + \phi$.

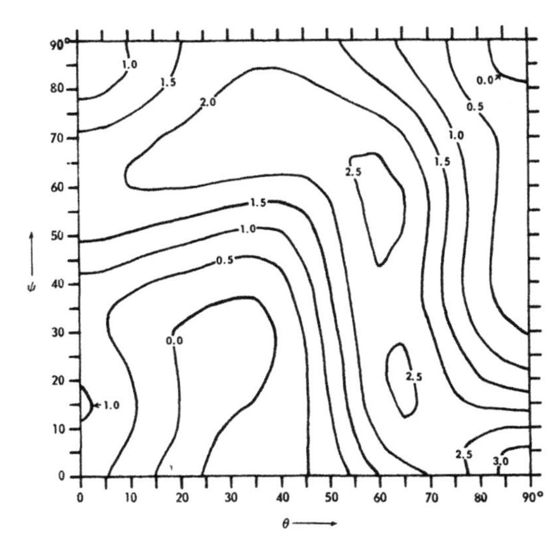

Figure 9. The $w(\psi, \theta, 30°)$ for Fe–3% Si.

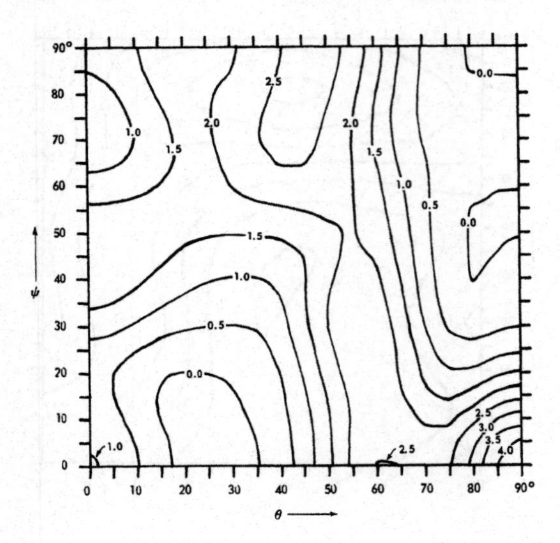

Figure 10. The $w(\psi, \theta, 45°)$ for Fe–3% Si.

Figure 7, corresponding to $\phi = 0°$, displays the expected symmetry, *i.e.*,

$$w(\psi, \theta, 0) = w\left(\psi, \frac{\pi}{2} - \theta, 0\right) \quad w(\psi, 0, 0) = w\left(\frac{\pi}{2} - \psi, 0, 0\right)$$

and

$$w\left(\psi, \frac{\pi}{2}, 0\right) = w\left(\frac{\pi}{2} - \psi, \frac{\pi}{2}, 0\right)$$

The $\phi = 45°$ section, Figure 10, contains all the ideal textures listed in Table I. If other textures are of interest, additional sections may be required. For some purposes, constant θ or constant ψ sections may prove more useful.

Table I. Euler Angles for Ideal Textures

Figure:	7		8		9		10	
ϕ:	0°		15°		30°		45°	
Texture:	ψ	θ	ψ	θ	ψ	θ	ψ	θ
{100}⟨001⟩	0°	0°	75°	0°	60°	0°	45°	0
	0°	90°						
	90°	0°						
	90°	90°						
{100}⟨110⟩	45°	0°	30°	0°	15°	0°	0°	0°
							90°	0°
{110}⟨001⟩	90°	45°					0°	90°
{111}⟨110⟩							30°	54.7°
							90°	54.7°
{111}⟨112⟩							0°	54.7°
							60°	54.7°

Table II. The w for Several "Ideal" Textures Based on a Random Standard Value of Unity for an Fe–3%Si Sample

{100}⟨001⟩	{100}⟨110⟩	{110}⟨001⟩	{111}⟨110⟩	{111}⟨211⟩
1.9	1.1	4.4	2.4	2.1

RESULTS

Fe–3%Si

For the Fe–3%Si sample used to illustrate the method, the strengths of certain ideal textures relative to a randomly oriented sample are given in Table II. The results given in Figures 7 to 10 and Table II can be summarized:

1. All common "ideal" textures listed in Table I are present in amounts greater than random. Other unlisted textures are also present in greater than random amounts. The strongest single texture component revealed by calculation of $w(\psi, \theta, \phi)$ on a 5 by 5 by 5° net (6859-mesh points) is the {110}⟨001⟩.

2. For $\phi = 0°$, $\theta = 45°$, relatively small deviations from $\psi = 90°$ produce appreciable changes in w. A change in ψ for constant θ and ϕ corresponds to textures having a common crystallographic direction parallel to the normal direction. For $\psi = 90°$, deviations of θ and ϕ from $\theta = 45°$, $\phi = 0°$ produce less significant changes in w, which indicates that the preference of ⟨001⟩ directions for the rolling direction is greater than the preference of {110} planes for the plane of the sheet.

3. For given values of θ and ϕ, the average value of w with respect to ψ corresponds to the value of the normal-direction inverse pole figure at θ and ϕ; *e.g.*, for $\theta = 45°$, $\phi = 0°$, the average of w from $\psi = 0$ to 90° is 0.9. This corresponds to the value in the ⟨110⟩ direction of the normal-direction inverse pole figure and to the value in the normal-direction of the {110} pole figure. Thus, while the {110}⟨001⟩ is the strongest single texture component (4.4 times random), the number of {110} planes in the plane of the sheet is less than random.

4. The cube-on-corner group consists of all texture components having a {111} plane in the plane of the sheet. For this group, $\phi = 45°$, $\theta = 54.7°$, and different values of ψ correspond to different crystallographic directions parallel to the rolling direction. Figure 10 reveals no strong preference for a particular member of the group. Values of w for $\theta = 55°$, $\phi = 45°$ vary from a minimum of 2.0 to a maximum of 2.4, the maximum corresponding to {111}⟨110⟩, which indicates an approximate ⟨111⟩ fiber axis in the normal direction.

Low-Carbon Aluminum-Killed Steel

As an additional example, the technique was applied to a commercial low-carbon aluminum-killed steel. The texture of this material is often described as {111}⟨110⟩. A {200} pole figure and a $\phi = 45°$ section of the crystallite orientation distribution for this material are shown in Figures 11 and 12, respectively. For $\phi = 45°$, $\theta = 55°$, $w(\psi, \theta, \phi)$ varies from a minimum of 5.8 corresponding to {111}⟨112⟩ to a maximum of 8.1 corresponding to {111}⟨110⟩. All texture components not associated with the cube-on-corner group are small in this material. The truncation error for this sample is estimated to be $\sigma_w = 0.9$.

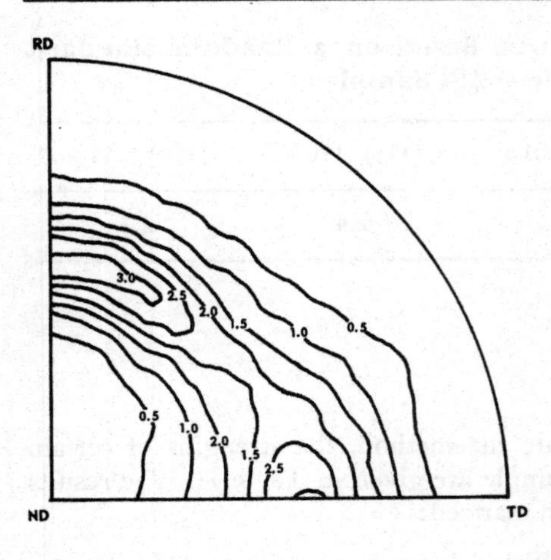

Figure 11. {200} Pole figure for low-carbon aluminum-killed steel.

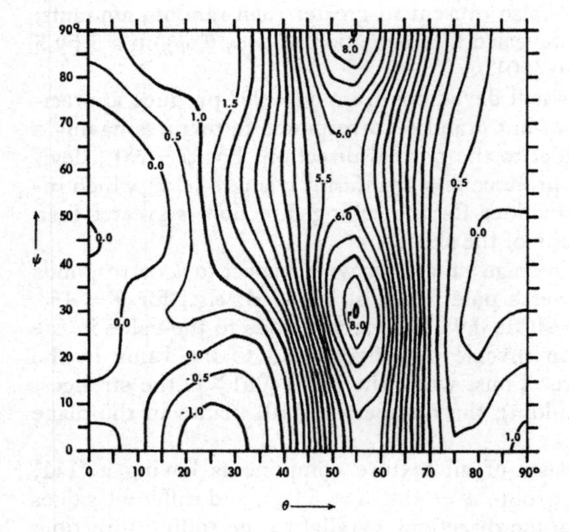

Figure 12. The $w(\psi, \theta, 45°)$ for low-carbon aluminum-killed steel.

SUMMARY AND CONCLUSIONS

The analysis of texture has previously been accomplished either by tedious determination of the orientations of a large number of individual crystallites by Laue or etch-pit methods or by an intuitive process from a number of pole figures or inverse pole figures. The introduction by Roe of a method for obtaining the crystallite orientation distribution in a series of generalized spherical harmonics permits a clear, objective analysis of texture not previously possible.

The augmented Jacobi polynomials required for this expansion have been derived for cubic crystallographic and orthotropic physical symmetry through the sixteenth order and are tabulated in the Appendix. A numerical technique is presented which permits approximate evaluation of the integral equations from a finite array of data points. The primary recrystallization textures of decarburized Fe–3%Si and low-carbon aluminum-killed steels were analyzed by using the respective crystallite orientation distributions.

The strongest single texture component in the Fe–3%Si sample was the {110}⟨001⟩. The sample also contained lesser, nearly uniform amounts of all members of the cube-on-corner group, with slight preference for {111}⟨110⟩.

The strongest texture components observed in the low-carbon aluminum-killed steel sample belong to the cube-on-corner group with a preference for {111}⟨110⟩.

ACKNOWLEDGMENTS

The writers are indebted to R. J. Roe of E. I. duPont de Nemours for correspondence relating to his papers on the subject, to W. G. Granzow for determination of experimental data, to J. W. Flowers for helpful discussion, to J. F. Woodruff and R. H. Heyer for their encouragement, and to the Armco Steel Corporation for permission to publish.

An IBM-360 computer was used for most of the numerical calculations. Figures 2 to 12 were originally plotted with an IBM-1620 computer and a Calcomp-563 digital plotter.

REFERENCES

1. C. G. Dunn, "On the Determination of Preferred Orientations," *J. Appl. Phys.* **30**: 850–857, 1959.
2. Ryong-Joon Roe, "Description of Crystallite Orientation in Polycrystalline Materials. III. General Solution to Pole Figure Inversion," *J. Appl. Phys.* **36**: 2024–2031, 1965.
3. Ryong–Joon Roe, "Inversion of Pole Figures for Materials Having Cubic Crystal Symmetry," *J. Appl. Phys.* **37**: 2069–2072, 1966.
4. Stanley L. Lopata and Eric B. Kula, "A Reflection Method for Pole-Figure Determination," *Trans. AIME* **224**: 865–866, 1962.
5. Eugene S. Meieran, "Use of the Reciprocal Lattice for the Development of a New Pole Figure Technique," *Rev. Sci. Instr.* **33**: 319–322, 1962.
6. L. G. Schulz, "A Direct Method of Determining Preferred Orientation of a Flat Reflection Sample Using a Geiger Counter X-Ray Spectrometer," *J. Appl. Phys.* **20**: 1030–1033, 1949.
7. C. Feng, "Determination of Relative Intensity in X-Ray Reflection Study," *J. Appl. Phys.* **36**: 3432–3435, 1965.
8. A. J. Heckler, J. A. Elias, and A. P. Woods, "Automatic Computer Plotting of Pole Figures and Axis Density Figures," *Trans. AIME* **239**: 1241–1244, 1967.
9. B. F. Decker, E. T. Asp, and D. Harker, "Preferred Orientation Determination Using a Geiger Counter X-Ray Diffraction Goniometer," *J. Appl. Phys.* **19**: 388–392, 1948.
10. Michael Field and M. Eugene Merchant, "Reflection Method of Determining Preferred Orientation on the Geiger-Counter Spectrometer," *J. Appl. Phys.* **20**: 741–745, 1949.
11. C. G. Dunn, *Cold Working of Metals*, ASM, Cleveland, 1949, p. 123.
12. John E. May and David Turnbull, "Secondary Recrystallization in Silicon Iron," *Trans. AIME* **212**: 769–781, 1958.
13. C. G. Dunn and J. L. Walter, *Recrystallization, Grain Growth and Textures*, ASM, Metals Park, Ohio, 1966, p. 509.

APPENDIX

TABLE OF AUGMENTED JACOBI POLYNOMIALS

	L= 0 M= 0 N= 0 K= 1	L= 2 M= 0 N= 0 K= 3	L= 2 M= 2 N= 0 K= 3	L= 4 M= 0 N= 0 K= 7	L= 4 M= 2 N= 0 K= 6	L= 4 M= 2 N= 4 K= 7	L= 4 M= 4 N= 0 K= 7	L= 4 M= 4 N= 4 K= 8	L= 6 M= 0 N= 0 K=10	L= 6 M= 2 N= 0 K=11
D=	2	10	15	18	45	126	315	18	26	2730
S	E(L,M,N,S)									
0	1	1	1	9	3	5	3	35	50	10
1						4	0	56	0	0
2		3	-1	20	4	-4	-4	28	105	17
3						-4	0	8	0	0
4				35	-7	-1	1	1	126	6
6									231	-33

	L= 6 M= 2 N= 4 K=11	L= 6 M= 4 N= 0 K=10	L= 6 M= 4 N= 4 K=11	L= 6 M= 6 N= 0 K=11	L= 6 M= 6 N= 4 K=11	L= 8 M= 0 N= 0 K=15	L= 8 M= 2 N= 0 K=14	L= 8 M= 2 N= 4 K=13	L= 8 M= 2 N= 8 K=14	L= 8 M= 4 N= 0 K=14
D=	195	819	26	6006	429	34	10710	935	17017	11781
S	E(L,M,N,S)									
0	42	10	126	10	42	1225	35	63	45	35
1	16	0	96	0	48	0	0	14	20	0
2	17	5	15	-15	-15	2520	64	64	-64	40
3	72	0	240	0	-40	0	0	90	-36	0
4	-26	-26	338	6	-26	2772	44	-36	20	-36
5	-88	0	176	0	-8	0	0	78	20	0
6	-33	11	33	-1	-1	3432	0	0	0	-104
7								-182	-4	0
8						6435	-143	-91	-1	65

	L= 8 M= 4 N= 4 K=14	L= 8 M= 4 N= 8 K=15	L= 8 M= 6 N= 0 K=14	L= 8 M= 6 N= 4 K=13	L= 8 M= 6 N= 8 K=14	L= 8 M= 8 N= 0 K=15	L= 8 M= 8 N= 8 K=16	L=10 M= 0 N= 0 K=18	L=10 M= 2 N= 0 K=18	L=10 M= 2 N= 4 K=18
D=	34	15470	14586	4641	255	109395	34	42	3465	546
S	E(L,M,N,S)									
0	693	99	35	33	429	35	6435	7938	882	2310
1	308	88	0	22	572	0	11440	0	0	336
2	440	-88	0	0	0	-56	8008	16170	1666	3094
3	1500	-120	0	50	-364	0	4368	0	0	2464
4	324	-36	-84	36	-364	28	1820	17160	1352	-104
5	468	24	0	-42	-196	0	560	0	0	4000
6	2184	24	64	-64	-64	-8	120	19305	741	-1577
7	1820	8	0	-30	-12	0	16	0	0	952
8	455	1	-15	-5	-1	1	1	24310	-442	1122
9										-7752
10								46189	-4199	-4845

	L=10 M= 2	L=10 M= 4	L=10 M= 4	L=10 M= 4	L=10 M= 6	L=10 M= 6	L=10 M= 6	L=10 M= 8	L=10 M= 8	L=10 M=10
	N= 8 K=19	N= 0 K=17	N= 4 K=17	N= 8 K=18	N= 0 K=19	N= 4 K=18	N= 8 K=19	N= 0 K=18	N= 8 K=19	N= 0 K=19
D=	27846	45045	42	2142	90090	21	1071	255255	42	1939938
S					E(L,M,N,S)					
0	770	126	4290	1430	378	12870	4290	126	24310	126
1	224	0	1248	832	0	5616	3744	0	28288	0
2	-238	182	4394	-338	266	6422	-494	-42	1326	-210
3	448	0	7744	1408	0	24288	4416	0	13056	0
4	-1144	-8	8	88	-664	664	7304	-264	49368	120
5	-1600	0	8000	-3200	0	-800	320	0	65280	0
6	589	-249	6889	-2573	-711	19671	-7347	279	49011	-45
7	1232	0	272	352	0	-5848	-7568	0	23232	0
8	118	-374	12342	1298	1054	-34782	-3658	-118	6962	10
9	-304	0	15504	608	0	-23256	-912	0	1216	0
10	-95	323	4845	95	-323	-4845	-95	19	95	-1

	L=10 M=10	L=10 M=10	L=12 M= 0	L=12 M= 2	L=12 M= 2	L=12 M= 2	L=12 M= 2	L=12 M= 4	L=12 M= 4	L=12 M= 4
	N= 4 K=18	N= 8 K=19	N= 0 K=22	N= 0 K=21	N= 4 K=22	N= 8 K=22	N=12 K=22	N= 0 K=23	N= 4 K=24	N= 8 K=23
D=	101745	1995	50	75075	75	113050	6128925	450450	50	169575
S					E(L,M,N,S)					
0	286	4862	106722	1386	90090	2574	546	2310	450450	4290
1	208	7072	0	0	9240	528	168	0	92400	1760
2	-338	1326	216216	2664	137640	888	-888	3720	576600	1240
3	-352	-3264	0	0	72360	2160	-360	0	646416	6432
4	-8	-4488	225225	2325	37045	-4185	465	1195	57121	-2151
5	160	-3264	0	0	144500	-1800	300	0	982600	-4080
6	83	-1581	243100	1700	-49980	-1140	-140	-2380	209916	1596
7	-8	-528	0	0	135660	-5560	-140	0	488376	-6672
8	-22	-118	277134	646	-40698	1346	14	-5814	1098846	-12114
9	-8	-16	0	0	-34884	6696	36	0	23256	-1488
10	-1	-1	352716	-1292	71060	1276	4	-6460	1065900	6380
11					-326876	-2024	-4	0	1961256	4048
12			676039	-7429	-245157	-759	-1	7429	735471	759

	L=12 M= 4	L=12 M= 6	L=12 M= 6	L=12 M= 6	L=12 M= 6	L=12 M= 8	L=12 M= 8	L=12 M= 8	L=12 M=10	L=12 M=10
	N=12 K=24	N= 0 K=22	N= 4 K=23	N= 8 K=22	N=12 K=23	N= 0 K=22	N= 8 K=23	N=12 K=23	N= 0 K=22	N= 4 K=23
D=	36773550	121550	53550	1425	1682450	3464175	50	132825	4408950	532950
S					E(L,M,N,S)					
0	910	2310	7150	24310	2210	462	277134	8398	462	2730
1	560	0	2200	14960	2040	0	227392	10336	0	1400
2	-1240	2520	6200	4760	-2040	168	18088	-2584	-264	-1240
3	-1072	0	12328	43792	-3128	0	434112	-10336	0	1608
4	239	-2205	-1673	22491	-1071	-945	549423	-8721	-825	-1195
5	680	0	8500	-12600	900	0	114912	-2736	0	-6460
6	196	-5380	7532	20444	1076	-380	82308	1444	1100	-2940
7	-168	0	-6612	32248	348	0	618272	2224	0	3612
8	-126	-2166	6498	-25574	-114	1346	905858	1346	-638	3654
9	-8	0	2812	-64232	-148	0	645792	496	0	300
10	20	7980	-20900	-44660	-60	-812	259028	116	188	-940
11	8	0	-19228	-14168	-12	0	56672	16	0	-460
12	1	-3059	-4807	-1771	-1	161	5313	1	-23	-69

	L=12 M=10	L=12 M=10	L=12 M=12	L=12 M=12	L=14 M= 0	L=14 M= 2	L=14 M= 2	L=14 M= 2	L=14 M= 2	L=14 M= 4
	N= 8 K=22	N=12 K=23	N= 0 K=23	N=12 K=24	N= 0 K=25	N= 0 K=26	N= 4 K=27	N= 8 K=27	N=12 K=27	N= 0 K=26
D=	1925	3450	33801950	50	58	39585	16269	18734	42010995	44414370
S					E(L,M,N,S)					
0	25194	58786	462	1352078	736164	56628	180180	180180	4620	1716
1	25840	90440	0	2496144	0	0	13728	27456	1056	0
2	-2584	28424	-792	1961256	1486485	109989	296637	136653	-3333	2937
3	15504	-28424	0	1307504	0	0	111672	147312	792	0
4	43605	-53295	495	735471	1531530	99858	135458	-229086	-5874	1522
5	25992	-47652	0	346104	0	0	247000	39600	-6600	0
6	-21660	-29260	-220	134596	1616615	81719	-40641	-236049	7337	-589
7	-47816	-13244	0	42504	0	0	312816	-327712	8624	0
8	-39034	-4466	66	10626	1763580	52972	-121524	46412	-3116	-2964
9	-18600	-1100	0	2024	0	0	188784	-258336	-5328	0
10	-5452	-188	-12	276	2028117	7429	-24035	25645	155	-4807
11	-920	-20	0	24	0	0	-192280	586960	1672	0
12	-69	-1	1	1	2600150	-74290	170430	170430	274	-4370
13							-681720	-215280	-216	0
14					5014575	-334305	-596505	-94185	-63	6555

	L=14 M= 4 N= 4 K=27	L=14 M= 4 N= 8 K=26	L=14 M= 4 N=12 K=27	L=14 M= 6 N= 0 K=26	L=14 M= 6 N= 4 K=27	L=14 M= 6 N= 8 K=27	L=14 M= 6 N=12 K=27	L=14 M= 8 N= 0 K=25	L=14 M= 8 N= 8 K=26	L=14 M= 8 N=12 K=26
D=	58	18183	163101510	46881835	4959	1914	476905	12785955	58	130065
S					E(L,M,N,S)					
0	3063060	92820	2380	1716	340340	587860	45220	1716	1763580	45220
1	466752	28288	1088	0	77792	268736	31008	0	1074944	41344
2	4443681	62033	-1513	2277	382789	304589	-22287	1353	542963	-13243
3	3499056	139872	752	0	500456	1140304	18392	0	3505728	18848
4	1158242	-59358	-1522	-558	-47182	137826	10602	-2574	1907334	48906
5	6422000	31200	-5200	0	595400	164880	-82440	0	95040	-15840
6	164331	28923	-899	-3337	103447	1037807	-96773	-3421	3191793	-99209
7	5630688	-178752	4704	0	16464	-29792	2352	0	3558016	-93632
8	3814668	-44148	2964	-3732	533676	-352052	70908	1132	320356	-21508
9	1384416	-57408	-1184	0	-319792	755872	46768	0	1492608	30784
10	8724705	-282095	-1705	69	-13915	25645	465	5129	5718835	34565
11	1153680	-106720	-304	0	313720	-1654160	-14136	0	6189760	17632
12	5624190	170430	274	6670	-953810	-1647490	-7946	-4370	3238170	5206
13	14997840	143520	144	0	-1184040	-645840	-1944	0	861120	864
14	6561555	31395	21	-3105	-345345	-94185	-189	1035	94185	63

	L=14 M=10 N= 0 K=26	L=14 M=10 N= 4 K=27	L=14 M=10 N= 8 K=27	L=14 M=10 N=12 K=27	L=14 M=12 N= 0 K=26	L=14 M=12 N=12 K=27	L=14 M=14 N= 0 K=26	L=14 M=14 N= 4 K=27	L=14 M=14 N= 8 K=27	L=14 M=14 N=12 K=27
D=	58815393	2091045	6670	28275	75404350	58	145422675	190285095	5462730	5481
S					E(L,M,N,S)					
0	1716	18564	352716	208012	1716	5200300	1716	3060	19380	742900
1		7072	268736	237728	0	7131840	0	1632	20672	1188640
2	165	1513	13243	-7429	-1287	1448655	-3003	-4539	-13243	482885
3		21432	537168	66424	0	454480	0	-3384	-28272	-227240
4	-3630	-16742	537966	317262	-2574	5624190	2002	1522	-16302	-624910
5		-27560	-83952	321816	0	11840400	0	2600	2640	-657800
6	55	-93	-10263	7337	4147	13830245	-1001	279	10263	-476905
7		-25872	514976	-311696	0	11051040	0	-1008	6688	-263120
8	4460	-34788	252436	-389804	-2964	6476340	364	-468	1132	-113620
9		22128	-575328	-272912	0	2847520	0	144	-1248	-38480
10	-4091	45001	-912293	-126821	1209	936975	-91	165	-1115	-10075
11		10680	-614800	-40280	0	225264	0	24	-464	-1976
12	1550	-12090	-229710	-8494	-274	37538	14	-18	-114	-274
13		-7800	-46800	-1080	0	3888	0	-8	-16	-24
14	-225	-1365	-4095	-63	27	189	-1	-1	-1	-1

	L=16 M= 0	L=16 M= 2	L=16 M= 2	L=16 M= 2	L=16 M= 2	L=16 M= 2	L=16 M= 4	L=16 M= 4	L=16 M= 4	L=16 M= 4
	N= 0 K=31	N= 0 K=29	N= 4 K=28	N= 8 K=29	N=12 K=28	N=16 K=30	N= 0 K=30	N= 4 K=29	N= 8 K=29	N=12 K=29
D=	66	8415	285285	1124838	216315	1944671850	5819814	66	79695	10360350
S					E(L,M,N,S)					
0	41409225	920205	80223	85085	109395	7293	70785	10669659	323323	31977
1			4719	10010	19305	1716	0	1255254	76076	11286
2	83431920	1799512	138424	88088	-12584	-12584	125840	16736720	304304	-3344
3			39325	62062	51051	-4004	0	9831250	443300	28050
4	85357272	1673672	80344	-70840	-200200	8008	80344	6668552	-167992	-36520
5			92323	56210	-115115	4004	0	20126414	350108	-55154
6	88884432	1452360	8680	-135800	54600	-3640	10416	107632	-48112	1488
7			134113	-73402	-69433	-2548	0	22799210	-356524	-25942
8	94645460	1113476	-44436	-48188	121212	1092	-74060	5110140	158332	-30636
9			127995	-190750	274365	1092	0	13567470	-577700	63918
10	104006000	611800	-46920	40600	-78120	-168	-156400	20738640	-512720	75888
11			41745	-34650	-237545	-308	0	1085370	-25740	-13574
12	120349800	-157320	13800	-13320	-9528	-8	-207000	31395000	-865800	-47640
13			-130065	305370	90857	52	0	9104550	-610740	-13978
14	155117520	-1520760	80040	121800	19720	8	-160080	14567280	633360	7888
15			-310155	-134850	-13485	-4	0	56448210	701220	5394
16	300540195	-5892945	-310155	-67425	-4495	-1	310155	28224105	175305	899

	L=16 M= 4	L=16 M= 6	L=16 M= 6	L=16 M= 6	L=16 M= 6	L=16 M= 6	L=16 M= 8	L=16 M= 8	L=16 M= 8	L=16 M= 8
	N=16 K=30	N= 0 K=29	N= 4 K=28	N= 8 K=29	N=12 K=28	N=16 K=30	N= 0 K=30	N= 8 K=30	N=12 K=29	N=16 K=31
D=	931395465	415701	231	22770	740025	266112990	15935205	66	19305	173551950
S					E(L,M,N,S)					
0	10659	135135	2909907	617253	61047	20349	45045	23661365	780045	52003
1	5016	0	513513	217854	32319	14364	0	11134760	550620	48944
2	-16720	200200	3803800	484120	-5320	-26600	48048	13361712	-48944	-48944
3	-11000	0	3592875	1134042	71757	-28140	0	48272952	1018164	-79856
4	7304	19096	226424	-39928	-8680	1736	-42504	10220280	740600	-29624
5	9592	0	5519997	672038	-105869	18412	0	14708040	-772340	26864
6	-496	-191016	-281976	882312	-27288	9096	-97776	51937680	-535440	35696
7	-4760	0	2778055	-304094	-22127	-4060	0	13464024	326564	11984
8	-1380	-312340	3078780	667748	-129204	-5820	-48188	11847364	-764124	-6884
9	1272	0	-2029635	604950	-66933	-1332	0	59405000	-2190900	-8720
10	816	-225400	4269720	-738920	109368	1176	81200	30612400	-1510320	-3248
11	-88	0	-1529385	253890	133889	868	0	1474200	259140	336
12	-200	124200	-2691000	519480	28584	120	119880	57662280	1057608	888
13	-40	0	3251625	-1526850	-34945	-100	0	98939880	754812	432
14	16	560280	-7283640	-2216760	-27608	-56	-146160	66502800	276080	112
15	8	0	-12096045	-1051830	-8091	-12	0	21036600	53940	16
16	1	-310155	-4032015	-175305	-899	-1	40455	2629575	4495	1

	L=16 M=10 N= 0 K=29	L=16 M=10 N= 4 K=28	L=16 M=10 N= 8 K=29	L=16 M=10 N=12 K=28	L=16 M=10 N=16 K=30	L=16 M=12 N= 0 K=30	L=16 M=12 N=12 K=29	L=16 M=12 N=16 K=30	L=16 M=14 N= 0 K=29	L=16 M=14 N= 4 K=28
D=	8580495	148005	150150	10395	3738042	110320650	66	148335	319929885	30045015
S					E(L,M,N,S)					
0	32175	124355	260015	557175	185725	19305	15043725	1002915	6435	10659
1		36575	152950	491625	218500	0	15928650	1415880	0	4389
2	17160	58520	73416	-17480	-87400	-2288	104880	104880	-5720	-8360
3		158125	491970	674475	-264500	0	12562830	-985320	0	1375
4	-58344	-124168	215832	1016600	-203320	-40040	31395000	-1255800	-8008	-7304
5		-41965	-50370	171925	-29900	0	23725650	-825240	0	-20383
6	-45864	-12152	374808	-251160	83720	13104	3229200	-215280	15288	1736
7		-279055	301098	474695	87100	0	4633590	170040	0	24395
8	52780	-93380	-199636	836940	37700	40404	28831140	259740	-12740	9660
9		-8533	25070	-60099	-1196	0	47269170	188136	0	-11607
10	62600	-212840	363080	-1164360	-12520	-52080	43590960	93744	6440	-9384
11		73953	-121014	-1382697	-8964	0	26648230	34552	0	1243
12	-100440	390600	-743256	-886104	-3720	28584	11347848	9528	-2040	3400
13		155295	-718794	-356439	-1020	0	3368698	1928	0	805
14	48024	-112056	-336168	-90712	-184	-7888	670480	272	376	-376
15		-94395	-80910	-13485	-20	0	80910	24	0	-217
16	-8091	-18879	-8091	-899	-1	899	4495	1	-31	-31

	L=16 M=14 N= 8 K=29	L=16 M=14 N=12 K=28	L=16 M=14 N=16 K=30	L=16 M=16 N= 0 K=31	L=16 M=16 N=16 K=32
D=	5598450	4785	2046	9917826435	66
S			E(L,M,N,S)		
0	52003	1002915	9694845	6435	300540195
1	42826	1238895	15967980	0	565722720
2	-24472	52440	7603800	-11440	471435600
3	9982	123165	-1400700	0	347373600
4	29624	1255800	-7283640	8008	225792840
5	-57086	1753635	-8844420	0	129024480
6	-124936	753480	-7283640	-4368	64512240
7	-61418	-871455	-4637100	0	28048800
8	48188	-1818180	-2375100	1820	10518300
9	79570	-1716741	-990756	0	3365856
10	37352	-1078056	-336168	-560	906192
11	-4746	-488047	-91756	0	201376
12	-15096	-161976	-19720	120	35960
13	-8694	-38801	-3220	0	4960
14	-2632	-6392	-376	-16	496
15	-434	-651	-28	0	32
16	-31	-31	-1	1	1

DISCUSSION

R. Baro (Université de Strasbourg, France): In your function w, you only have the orientation features. Am I right?

P. R. Morris: That is correct.

R. Baro: You assume that all the crystallites have the same dimension in this type of study or not?

P. R. Morris: This is not relevant because you are talking about the volume fraction of the material in different orientations.

R. Baro: Would it be possible to use the same method and get information in the p function for the dimension of the crystallites?

P. R. Morris: No.

Chairman W. C. Hagel: I might ask a question about the significance of this work. You have gone through a complicated mathematical treatment so you can detect orientations other than cube on edge. Is that the primary contribution?

P. R. Morris: The method is perfectly general. You get an orientation distribution. We have specified certain so-called idealized textures because people are familiar with interpreting preferred orientation data in terms of the idealized textures. This is not necessary. The mathematics here, incidentally, is due to Roe. We think this is the first application of it. If you would extrapolate, I think that you should be able by a minimal energy consideration to calculate those physical properties of a polycrystal which are describable for single crystals in terms of tensors. I think this is an extension of the method, although we haven't done this yet. Unhappily, many of the interesting commercial properties of steels are not describable in terms of tensors.

Z. W. Wilchinsky (Enjay Polymer Labs.): I should like to make a comment here. This method is completely general, so that, once the coefficients have been determined, it is now possible from these coefficients to generate pole figures for any set of planes, even for planes that you can't measure. Furthermore, you could get a distribution of orientation for the crystallites themselves. We didn't have too much time to go into this. This involves a little more subtlety than orientation of only the crystal planes because it is a more complicated type of orientation.

P. R. Morris: Roe has gone further than this and has considered the case of resolution of planes where peaks overlap, which are not ordinarily derivable in an X-ray diffraction measurement. He has been able to derive these.

R. Baro: How long does it take you to get all this done for one sample?

P. R. Morris: The experimental work requires about an hour per pole figure. We have a number of computer programs; the longest of these is the one which leads to the mapping. To map a 5 by 5 by 5° net takes about 120 min of 360 computer time.

H. Pickett (General Electric Co.): I am interested specifically in getting some feel for the input data. How many pole figures do you evaluate? How many sets of planes?

P. R. Morris: Typically three are used; so we have one redundancy.

H. Pickett: So it's a 110?

P. R. Morris: We have used 110, 220, and I've used 211 and Dr. Heckler has used 222 as the third. The disadvantage of the 222 is the rather weak reflection. If you notice, the errors are greatest in the 222.

H. Pickett: This only applies to a cubic crystal?

P. R. Morris: That is correct. Roe's second paper in 1966 was a particularization of the method for the cubic system. You have to work out the matrix of coefficients for linear dependencies among the $W_{l,m,n}$ for other systems if you want to use them.

TOPOTACTICAL RELATIONSHIPS BETWEEN HEMATITE α-Fe_2O_3 AND THE MAGNETITE Fe_3O_4 WHICH IS FORMED ON IT BY THERMAL DECOMPOSITION UNDER LOW OXYGEN PRESSURE

Raymond Baro, Hervé Moineau, and J. Julien Heizmann

Laboratoire de Metallurgie Structurale
Metz, France

ABSTRACT

The present work was performed on natural monocrystals from Mexico. All crystals presented well-formed natural faces of indices (01.2), (01.4), (11.3), and (00.1). They were reduced to magnetite in a thermobalance by heating at 1200°C under a pressure of 10^{-3} torr. The techniques used in the present work were Laue reflection diagrams and pole figures. The magnetite formed on the hematite showed preferential orientations of three different types. The mutual orientations of the two oxides all present a common zone axis for the lattices formed by the oxygen ions.

INTRODUCTION

The studies concerned with the reduction of iron oxides are quite numerous.[1] Because the number of parameters influencing the kinetics is high, the conclusions of those studies appear very often contradictory.

In order to procure a better understanding of the physical mechanisms of the chemical reduction, we undertook a set of studies with the goal of underlining the importance of the geometrical and structural factors that characterize the samples.[2] The present study was conducted with the aim of showing the influence on the formation of magnetite of the direction of the reduction front relative to the crystalline lattice of hematite.

The reduction step that leads from hematite to magnetite is the most interesting from the structural point of view because the structures of those two oxides are very different (hematite is rhombohedral $R_{\bar{3}c}$ and magnetite is face-centered cubic F_{d3m}). The transformation is not reversible, whereas the transformation of magnetite to wustite is.

Another, more practical reason to undertake the present study was to try to explain why hematite ores are usually more easily and rapidly reduced than magnetite ores in spite of the greater amount of oxygen that has to be removed during the reduction of hematite.[3]

CHARACTERISTICS OF THE SAMPLES USED

The hematite crystals used in the present work were natural monocrystals originated in Mexico. Their faces were well formed and about 5 mm large.

The crystallographic indices of the faces have been determined by goniometry and verified by taking back-reflection Laue diffraction diagrams. The indices were (01.2), (01.4), (11.3), and (00.1).

CONDITIONS OF REDUCTION

The reduction was performed in a thermobalance by heating the sample to 1200°C under a pressure of 10^{-3} torr.[4] It was in this way possible to follow the progression of the reduction process and to obtain the desired reduction rates. The reduction rates varied from 13 to 75%.

EXPERIMENTAL STUDY

X-ray diffraction was used in two different ways: (1) for back-reflection Laue diagrams, and (2) for a texture goniometer.

Laue Diagrams

For the lowest reduction rates, the Laue diagrams (Figures 1 and 2) showed, in addition to the spots corresponding to the hematite crystals, parts of Debye–Scherrer

Figure 1. Back-reflection Laue diagram of hematite before reduction; (01.2) plane perpendicular to the direct beam.

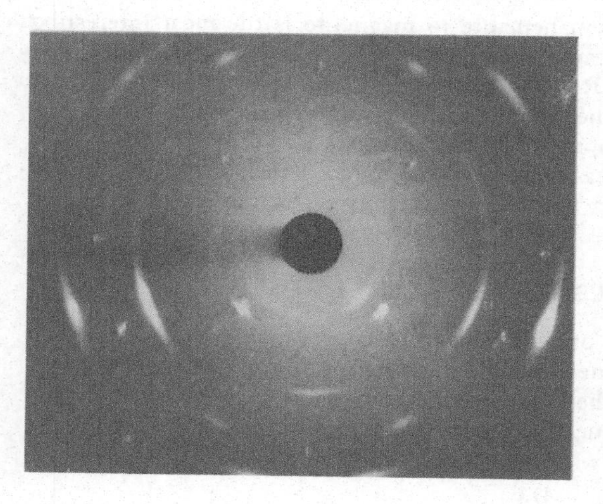

Figure 2. Back-reflection Laue diagram of the hematite crystal of Figure 1 with the same orientation but after a 13% reduction.

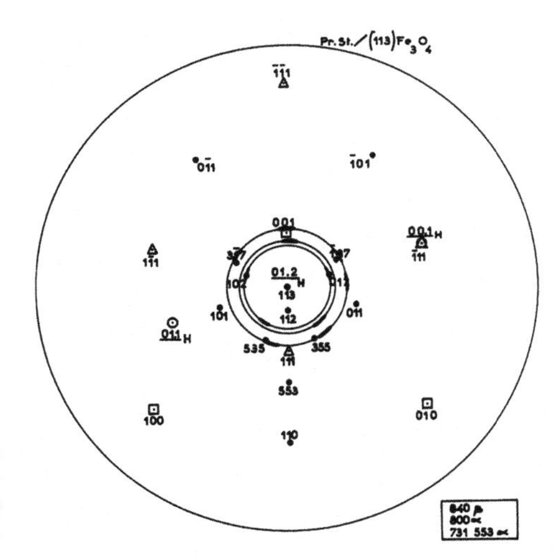

Figure 3. Experimental stereographic projection as deduced from Laue diagram (Figure 2) superposed on the theoretical stereographic projection of hematite with the projection center at $(01.2)_H$.

rings belonging to magnetite. The indices of those rings were $(800)_\alpha$, $(731)_\alpha$, and $(840)_\beta$ for the Fe $K\alpha$ and $K\beta$ radiations.

In order to determine the orientations of the magnetite crystals relative to the underlying monocrystal of hematite, one proceeds as follows:

1. One draws the theoretical stereographic projection of the lattice of hematite obtained by locating the center of projection at the pole of the face concerned in the study.
2. One draws the experimental stereographic projection as it can be deduced from the X-ray diagram.
3. By comparing these two stereographic projections, one has to find out which poles of magnetite are superposed on important poles of hematite (Figure 3).
4. One (or more) superposition having been found, the orientation has to be verified as sufficient to explain the presence of all parts of the Debye–Scherrer diagram.

This very inexpensive method becomes difficult to use as soon as the X-ray diagram is no longer symmetric or the Debye–Scherrer rings are no longer well defined.

This is the reason we used a much more general method utilizing a texture goniometer.

Texture Goniometer

We used a Siemens goniometer that we transformed slightly so that the spiral which is described on the stereographic projection would be tightened. The translation movement was suppressed. The X-rays used were filtered Fe $K\alpha$ radiations.

To study a reduced face, we record the corresponding pole figure, *i.e.*, we determine the positions of the underlying hematite crystal, for which the (113) Bragg reflections of magnetite are received by the Geiger–Müller counter, which is immobile (Figures 4 and 5).

To define the relative orientations of hematite to magnetite, one has to compare this experimental pole figure with the theoretical stereographic projections of the hematite and magnetite structures.

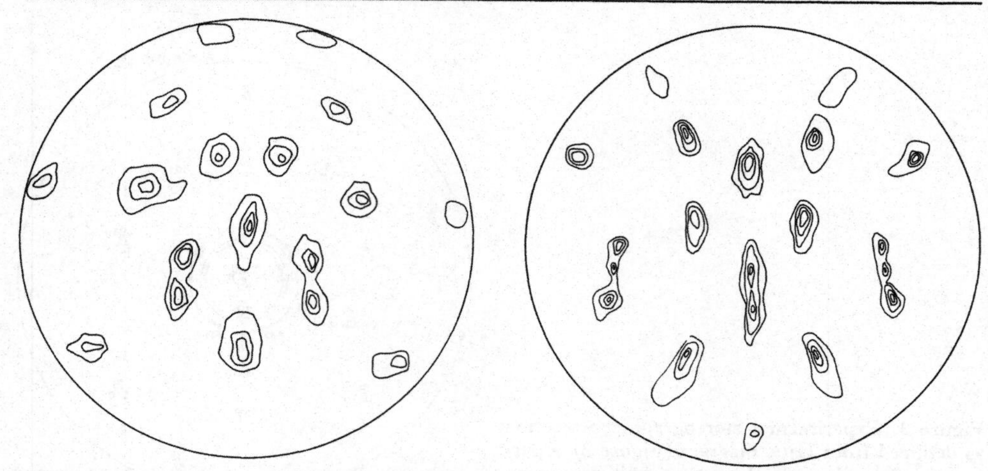

Figure 4. Pole figure of the $(113)_M$ reflections of magnetite obtained by reduction of a $(01.2)_H$ face of hematite.

Figure 5. Pole figure of the $(113)_M$ reflections of magnetite obtained by reduction of a $(11.3)_H$ face of hematite.

RESULTS

Table I summarizes the results for the four faces listed below.

The (01.2) Faces

A $(113)_M$ pole is located close to the $(01.2)_H$ pole. This superposition involves the superposition of the $(111)_M$ and $(00.1)_H$ poles; *i.e.*, the most compact oxygen planes in both structures remain parallel. It is easy to verify that, in addition, the $[10.0]_H$* and $[110]_M$* rows are parallel. We call this *orientation A*:

$$\left.\begin{array}{l}(111)_M \text{ parallel to } (00.1)_H \\ [110]_M^* \text{ parallel to } [10.0]_H^*\end{array}\right\} \text{orientation } A$$

This orientation suffices to explain the whole diffraction diagram.

Table I

Face	Orientation
(01.2)	$(111)_M$ parallel to $(00.1)_H$, $[110]_M$* parallel to $[10.0]_H$* } Orientation *A*
(11.3)	Orientation *A*; $(111)_M$ parallel to $(11.3)_H$, $[1\bar{1}0]_M$ parallel to $[\bar{1}1.0]_H$ } Orientation *B*
(01.4)	Orientation *A*
(00.1)	Orientation *A*; $(112)_M$ parallel to $(00.1)_H$, $[1\bar{1}0]_M$* parallel to $[10.0]_H$* } Orientation *C*

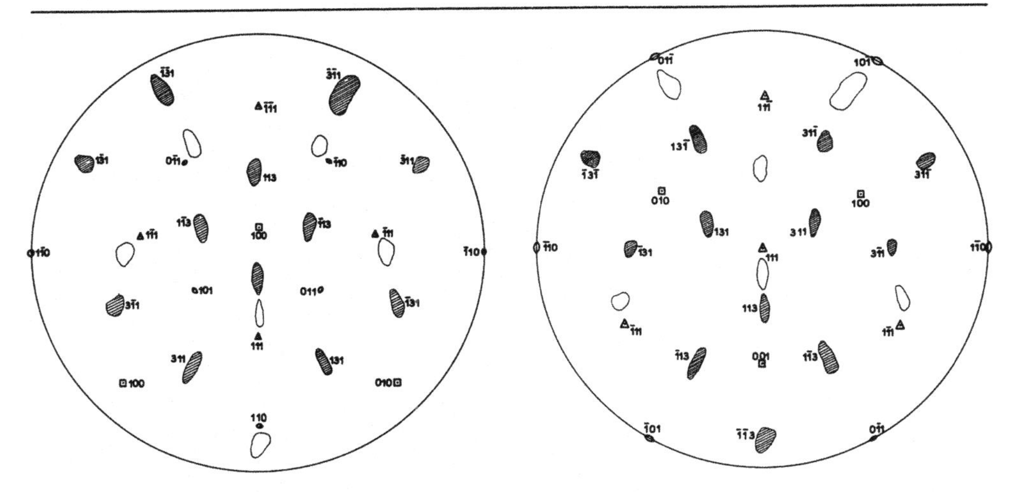

Figure 6. Pole figure of the reduced $(11.3)_H$ face. Hatchings: orientation *A*.

Figure 7. Pole figure of the reduced $(11.3)_H$ face. Hatchings: orientation *B*.

The (11.3) Faces

The orientation *A* is present again here (Figure 6), but it cannot explain all by itself the pole figure. A second orientation is present which we call *orientation B* (Figure 7):

$$\left.\begin{array}{ll}(111)_M & \text{parallel to } (11.3)_H \\ [1\bar{1}0]_M & \text{parallel to } [\bar{1}1.0]_H\end{array}\right\}\text{orientation } B$$

The (01.4) Faces

Only the orientation *A* is found here (Figure 8).

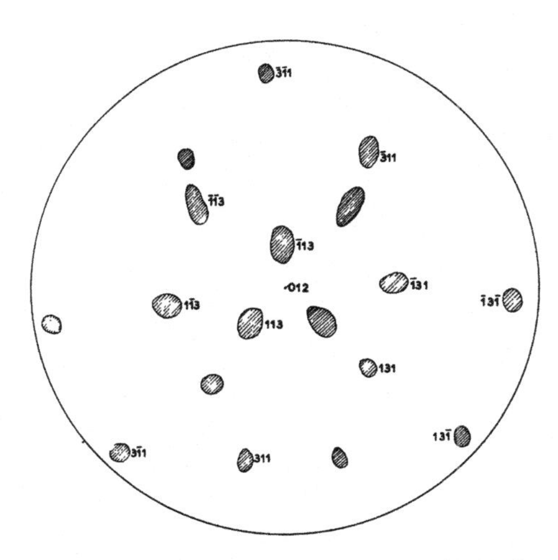

Figure 8. Pole figure of the reduced $(01.4)_H$ face. Hatchings: orientation *A*.

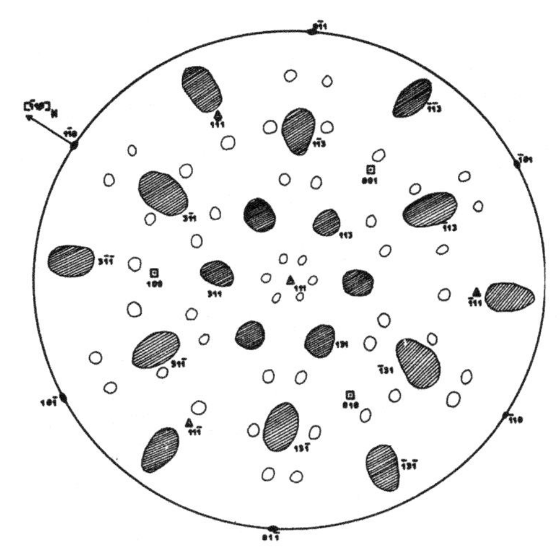

Figure 9. Pole figure of the reduced $(00.1)_H$ face. Hatchings: orientation *A*.

Table II

Parallel planes		Parallel rows		Distances between the rows (Å)		Distances between oxygen ions in rows (Å)		Distance L (Å)	
Fe_3O_4	α-Fe_2O_3	Fe_3O_4	α-Fe_2O_3	$Fe_3O_4 : b$	α-$Fe_2O_3 : a$	b'	a'	$L_1 = \frac{ab}{a - b}$	$L_2 = \frac{a'b'}{a' - b'}$
(113)	(01.2)	$[12\bar{1}]$	[10.0]	2.83	2.71	5.16	5.03	64	200
(012)	(01.4)	$[12\bar{1}]$	[10.0]	3.76	3.60	4.70	5.03	84	72
(111)	(11.3)	$[1\bar{1}0]$	$[\bar{1}1.0]$	2.57	2.62	2.57	2.90	134	13
(111)	(00.1)	$[\bar{1}10]$	$[\bar{1}1.0]$	2.57	2.81	2.57	3.12	30	14
(112)	(00.1)	$[1\bar{1}0]$*	[10.0]*	7.26	2.81	2.96	3.12	4	58

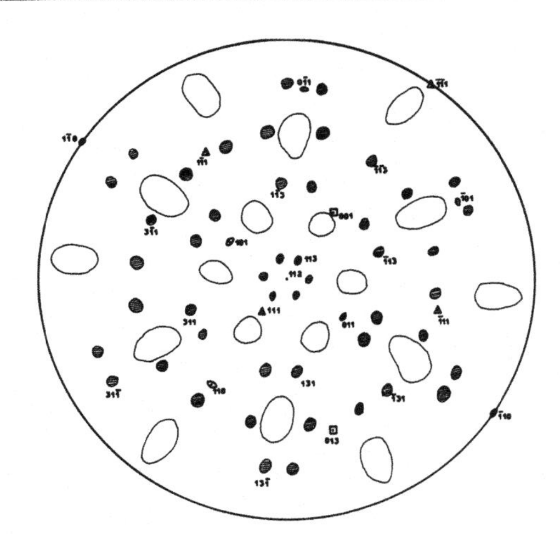

Figure 10. Pole figure of the reduced $(00.1)_H$ face. Hatchings: orientation *C*.

The (00.1) Faces

Here orientation *A* is present (Figure 9) in addition to a new orientation called *orientation C* (Figure 10):

$$\left.\begin{array}{l}(112)_M \text{ parallel to } (00.1)_H \\ [1\bar{1}0]_M^* \text{ parallel to } [10.0]_H^*\end{array}\right\}\text{orientation } C$$

The orientations *A* and *C* had already been found by Keeling and Wick[5] for (00.1) faces of hematite reduced in a CO–CO_2 mixture but they had not been found for the same faces reduced by thermal decomposition under vacuum.

Discussion

The results found[6] are similar to those obtained[7] by a reduction performed in a mixture of CO–CO_2. This means that, at least at the temperatures we used, the type of the reduction process does not significantly determine the orientation of the magnetite phase that is formed relative to the substrate.

The orientations we found all present the common characteristic that compact rows in the two lattices are juxtaposed in the interface in such a manner that the pileup of the oxygen ions is the most compact possible.

The iron ions consequently have at best a secondary influence on the mutual orientation of the two lattices.

REPRESENTATION OF THE OXYGEN IONS IN THE INTERFACE

It is easy to verify that the orientations so found correspond to an optimum matching of the oxygen rows of the two lattices. Nevertheless, one notes that the distances between neighboring atoms are not exactly the same in both lattices. We give in Table II the distances between oxygen ions located in rows that belong to the two lattices but are parallel. The difference in the distance of the ions produces in the interface a strain which can only be eliminated by a breakdown of the lattice of magnetite. This means that, periodically, one oxygen ion of the magnetite lattice will be absent.

Let L be the distance between two such missing ions.

$$L = na = (n + 1)b$$

where a is the distance between oxygen ions in the row of hematite, and b is the corresponding distance in the magnetite row which is parallel to it. It follows that

$$L = \frac{ab}{a - b}$$

We have put in Table II the values of L which correspond to the different orientations we found.

The existence of these ruptures in the periodicity of the magnetite lattice corresponds to a very fine porosity of the magnetite which is formed from the hematite crystal. This magnetite should be in the form of prism-shaped crystals whose sections have a mean diameter of 100 Å.

This porosity should certainly facilitate the flow of the gases to the hematite–magnetite interface and also the reduction of the already formed magnetite. The magnetite which is formed by reducing a hematite crystal is thus equivalent to a very fine powder and is therefore very easy to reduce.

This hypothesis of the existence of a lattice of pores has still to be verified directly. The porosity of the magnetite which is formed depends on the nature of the initial faces of the microcrystals which form the hematite ores. The differences in reductibility of the ores of different origins could thus well be ascribed to the fact that hematite crystals of different origins are often bound by faces of different types.

The crystallography characteristics of the ores seem thus quite important and justify further studies.

REFERENCES

1. J. O. Edstrom, "The Mechanism of Reduction of Iron Oxides," *J. Iron Steel Inst.* (*London*) **175**: 289, 1953.
2. G. Bitsianes and T. L. Joseph, "Topochemical Aspects of Iron Ore Reduction," *Trans. AIME* **203**: 639, 1955.
3. W. M. McKewan, "Kinetics of Iron Oxide Reduction," *Trans. AIME* **218**: 2, 1960.
4. M. Blackmann and G. Kaye, "An Electron Diffraction Study of the Effects of Heat Treatment on Fe_2O_3 (Haematite) Single Crystals," *Proc. Phys. Soc.* (*London*) **75**: 364, 1960.
5. R. O. Keeling and D. A. Wick, "Magnetite: Preferred Orientation on the Basal Plane of Partially Reduced Hematite," *Science* **141**: 1175, 1963.
6. H. Moineau and R. Baro, "Relations topotaxiques entre des cristaux d'hématite α-Fe_2O_3 et la magnétite Fe_3O_4 qui en est issue par dissociation sous vide," *Compt. Rend. Acad. Sci. Paris* **264**: 432, 1967.
7. J. J. Heizmann and R. Baro, "Relations topotaxiques entre des cristaux naturels d'Hématite Fe_2O_3 et la Magnétite Fe_3O_4 qui en est issue par réduction chimique," *Compt. Rend. Acad. Sci. Paris* **263**: 953, 1966.

DISCUSSION

Z. W. Wilchinsky (Enjay Polymer Labs.): Have you run any supplementary experiments to verify the magnitude of L?

R. Baro: Not yet. There are such studies underway. We hope to be able to do this maybe by electron microscopy or by small-angle X-ray scattering. For the time being, we are not able to prepare any samples which are thin enough so that we might look through by electron microscopy. It is a big problem.

Chairman W. C. Hagel: How pure were these single crystals from Mexico? How thick were the magnetite films that you formed; after how long a time?

R. Baro: First, I must confess that those crystals are certainly not pure. I am sorry I have no artificially grown hematite crystals, and I should appreciate very much if anybody could help me get some crystals or get me in touch with people who grow them. So, they were just natural crystals, with well-formed facets. At first, we thought that orientation would change as reduction goes down into the hematite crystal. But this didn't happen. Once the rings are formed, they stabilize almost immediately. From the results, you understand why. The process is always located in the interface, and there is no reason why it should change when the reduction then progresses.

Chairman W. C. Hagel: This is not time-dependent, really, nor thickness-dependent?

R. Baro: No, not at all.

Chairman W. C. Hagel: This is in contrast to the oxidation of magnetite to hematite because there one sees nice Widmanstaetten structures and the formation rate becomes thickness-dependent.

R. Baro: The oxidation of magnetite is certainly not the reverse of the reduction of hematite, certainly not. The mechanisms are completely different.

Chairman W. C. Hagel: A little simpler, in this case.

R. Baro: Maybe.

Chairman W. C. Hagel: You might discuss where you would like to go from here, *i.e.*, the scope of your work.

R. Baro: This is just one part of the work we are doing. Some other people are working on electron microscopy studies. We have devised a method which enables us to determine the orientation of the microcrystals in the ores, and that's quite new because this means that we can now determine the texture of the ores, and not only the texture but also which are the facets which are growing together. And so we hope that we might be able to explain why there are big differences between ores which have about the same composition but are reduced in a completely different manner. We believe that crystallographic and geometrical reasons play an important role.

Chairman W. C. Hagel: Would you carry this through to the wüstite analysis?

R. Baro: Maybe in 10 or 20 years. I think we have much work to do on this.

CRYSTAL STRUCTURE ANALYSIS OF NIOBIUM-DOPED RUTILE SINGLE CRYSTAL

Nobukazu Niizeki

The Electrical Communication Laboratory
Nippon Telegraph and Telephone Public Corporation
Musasino-si, Tokyo, Japan

ABSTRACT

Determination of the atomic sites for the doped niobium atoms by the X-ray diffraction method has been undertaken to obtain the structural basis for the understanding of semiconductive behavior of crystals. Diffraction intensities were measured with spherical samples ground from the undoped and doped (4 mol. %) TiO_2 single-crystal rods grown by the flame-fusion method. Measurements were carried out by an automatic single-crystal diffractometer and observed intensities were corrected for *Lp* and absorption factors. The total number of the independent (*hk*0) reflections was 45 with Mo $K\alpha$ radiation. Structure factors were calculated on the basis of the following atomic arrangement: titanium at equipoint (2*a*) of space group $P4_2/mnm$, O at (4*f*) with $x = 0.306$, and an isoatomic temperature factor $B = 0.70$ were used. Reliability factor *R* for the undoped rutile was 0.037 without further refinement. For the doped crystal, an *R*-value of 0.053 was obtained for the substitutional model, where niobium was assumed to be at $x = 0$, $y = 0$. The *R* value was 0.125 for the interstitial model, where niobium was assumed to be at $x = \frac{1}{2}$, $y = 0$. The difference Fourier map was obtained by subtracting the electron density distribution in a Fourier map of the undoped crystal from that of the doped crystal. Since the intensity measurements were done only on a relative scale, the two Fourier maps were normalized so as to give an identical number of electrons under the oxygen peak. The resulting map revealed a small peak only at the location $x = 0$, $y = 0$. It could be concluded that niobium atoms enter the titanium sites substitutionally, at least in rather heavily doped crystals. Further work on the refinement of the doped structure is in progress.

INTRODUCTION

It is well known that rutile can be made semiconductive (*n* type) by reduction in vacuum or in a reducing atmosphere or by doping cations with valences higher than +4, for example, Nb^{5+}.

Numerous reports have been published on the structural as well as the electronic properties of reduced rutile. The defect structure was discussed by Hurlen,[1] who pointed out the structural basis for interstitial titanium as the cause of oxygen deficiency of the compound. Straumanis *et al.*[2] made precise measurements of lattice constants and densities on samples reduced at the various oxygen pressures. Their results revealed a corresponding number of the oxygen vacancies in these samples. It is now generally agreed that these two kinds of lattice defects may alternatively exist, depending on the degree of the reduction.

In contrast to the availability of data for the reduced rutile, data concerning the niobium-doped rutile are scarce. The electronic properties were discussed by Frederikse[3]

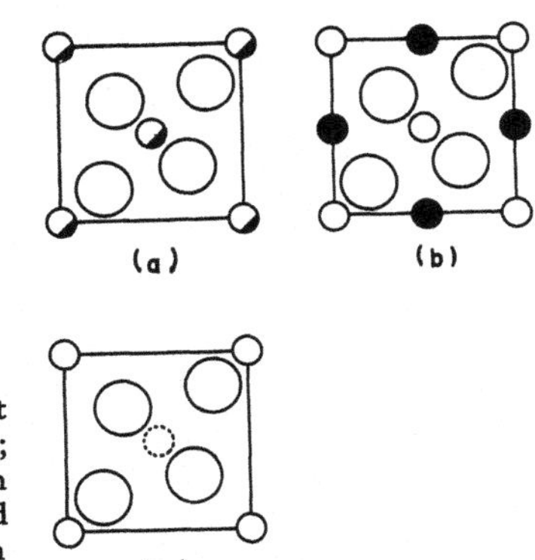

Figure 1. Schematic diagrams of the three possible defect structures of niobium-doped rutile: (a) substitutional model; (b) interstitial model; and (c) substitutional model with cation vacancy. The small open circles denote titanium ions, the solid circles denote niobium ions, the shaded circles denote cation vacancies, and the large open circles denote oxygen.

and by Chester.[4] The Hall effect and thermoelectric power measurements were recently reported by Itakura *et al.*[5] The present paper describes the structural aspects of the samples dealt with in the last paper.

Since the ionic radius of Nb^{5+} is only slightly larger than that of Ti^{4+}, the following three kinds of the lattice defects are conceivable in the doped rutile.

1. Niobium enters into the structure substituting for titanium at the lattice site.
2. Niobium enters at the interstitial site.
3. Four Nb^{5+} and one cation vacancy substitute for five Ti^{4+}.

It is easily evaluated that the number of quasi-free electrons to be found in the above three kinds of niobium incorporation are five, one, and zero per niobium atom, respectively. These three kinds of defect structures are schematically illustrated in Figure 1.

Since the niobium-doped crystal is semiconductive, the third kind of defect associated with no electron may be discarded from the following discussion. As can be seen from Figure 1, the atomic sites for the incorporated niobium are distinguishable in the projection on (001) for the substitutional and interstitial models. Thus, it is expected that the precise measurement of *hk*0 intensities of the diffracted X-ray from the doped crystal may reveal the defect structure.

EXPERIMENTAL

Preparation of Samples

A transparent yellowish rutile single crystal and black niobium-doped single crystal were supplied by the Fuji Crystal Company, Limited. These crystals were grown by the flame-fusion process and subsequently annealed in the air at 800°C for 7 days. The niobium-doped crystal was obtained from the source materials of a mixture of TiO_2 powder (99.9% pure) and 4 mol.% Nb_2O_5 (99.8% pure) powder. However, no chemical analysis was made on the black crystal within the thin sections of which slight color bandings were observed.

Fragments of the crystals were made into spheres by Bond's method.[6] Their crystallographic orientation was determined by precession photography, and two *c*-axis mounted samples were prepared. The radii of the spheres are $112 \pm 3\ \mu$ (transparent pure crystal) and $198 \pm 4\ \mu$ (black doped crystal).

Intensity Measurement

Intensities of the reflections *hk*0 were measured with a tape-controlled automatic single-crystal diffractometer manufactured by the Rigaku Denki Company, Limited.

The radiation used was zirconium-filtered Mo $K\alpha$ radiation, and to avoid the separation of α_1 and α_2 lines, the maximum 2θ angle was set at 90°. All the equivalent reflections in this region were measured and averaged, and, in total, 45 independent *hk*0 intensities were collected. As a detector system, a combination of a scintillation counter and a pulse-height analyzer was used with a set of the absorbing zirconium plates, which was to put all the measured intensities within the linear range of the counter characteristics.

The lattice constants used for the computation of the crystal and counter positionings were determined on the diffractometer. The crystal orientation and the counter position were manually adjusted so as to give the maximum counts from the 10,0,0 reflection. The values obtained are $a_0 = 4.5948$ Å (pure crystal) and $a_0 = 4.5977$ Å (doped crystal).

The measured intensities were corrected for the usual Lorentz-polarization factors and absorption factors. The extinction effect was not corrected, although the effect was observed in the two strongest reflections.

Structure Analysis: Pure Rutile

The observed structure factors of the pure rutile were compared with the calculated values mainly to estimate the accuracy of the present measurement. The calculations were done on the basis of the following well-established atomic arrangement: titanium at equipoint (2*a*) of space group P4/*mnm* and oxygen at (4*f*) with $x = 0.306$. An isoatomic temperature factor, determined by Wilson's method, $B = 0.70$ was used. The observed and calculated values are compared in Table I. The reliability factor was $R = 0.037$. Further refinement of the structure was not tried.

Table I. Comparison of the Observed and Calculated Structure Factors of the Pure Rutile

h k 0	F_o	F_c	*h k* 0	F_o	F_c
0 2 0	22.4	25.1	3 3 0	39.5	41.6
0 4 0	33.5	33.7	3 4 0	7.21	6.42
0 6 0	26.1	27.4	3 5 0	13.7	12.9
0 8 0	8.92	7.83	3 6 0	4.39	−4.06
0 10 0	14.3	15.7	3 7 0	21.6	20.9
1 1 0	46.5	65.4	3 8 0	0	0.80
1 2 0	20.0	19.6	3 9 0	12.3	11.5
1 3 0	30.7	30.9	3 10 0	0	0.50
1 4 0	15.4	−15.2	4 4 0	23.7	23.4
1 5 0	28.6	29.7	4 5 0	0	1.52
1 6 0	8.81	8.61	4 6 0	19.3	18.4
1 7 0	15.9	16.1	4 7 0	5.52	−5.96
1 8 0	3.91	−1.73	4 8 0	12.0	11.6
1 9 0	13.1	12.3	4 9 0	5.82	5.99
1 10 0	2.45	−1.33	5 5 0	26.4	26.8
2 2 0	47.3	56.1	5 6 0	1.34	−1.29
2 3 0	6.79	−6.15	5 7 0	9.18	9.11
2 4 0	26.6	26.5	5 8 0	0.59	−0.93
2 5 0	0	−1.28	5 9 0	9.21	9.46
2 6 0	16.9	16.6	6 6 0	15.4	15.7
2 7 0	3.82	4.41	6 7 0	4.74	4.40
2 8 0	19.9	19.9	6 8 0	7.71	7.32
2 9 0	4.11	−4.26	7 7 0	11.3	12.8
2 10 0	5.79	5.71			

Defect Structure

As can be seen from Figure 1, niobium ions incorporated in the rutile structure are located at equipoint (2*a*) in the substitutional model and in equipoints (4*d*) or (4*e*) in the interstitial model. In the projection on (001), the last two sites coincide at $x = 0, y = \frac{1}{2}$. The observed structure factors are compared in Table II with the two sets of the calculated values. The reliability factors were 0.125 for the interstitial model and 0.053 for the substitutional model.

Results and Discussions

The accuracy of the present method of intensity measurement is estimated to be less than about 5% in intensity from the analysis of the pure rutile. Thus, the difference obtained between two values of reliability factor for the two kinds of defects is significant, and it is concluded that the substitutional kind is the defect structure of the doped rutile, at least for the rather heavily doped one.

The location of the doped niobium ions may be directly observed in the differences of the Fourier map between those representing the pure and doped structures. A difficulty in this approach is that the two sets of intensities were collected on relative and independent scales. The method used in the present study to normalize the two Fourier maps is as follows: the relative number of the electron densities under the oxygen peaks, *i.e.*, within circles of radius 1.40 Å, were summed up in two Fourier maps, and they were made equal by a proper scaling factor. A Fourier map obtained from the maps thus normalized is shown in Figure 2.

Table II. Comparison of the Observed and Calculated Structure Factors of the Niobium-Doped Rutile, F_c^s and F_c^i Representing the Substitutional and Interstitial Models, Respectively

h	k	0	F_o	F_c^s	F_c^i	h	k	0	F_o	F_c^s	F_c^i
0	2	0	23.4	25.5	28.1	3	3	0	38.1	41.0	35.2
0	4	0	34.0	33.8	37.2	3	4	0	7.66	5.89	6.48
0	6	0	28.7	29.3	32.2	3	5	0	16.3	15.6	9.39
0	8	0	11.8	11.5	12.6	3	6	0	5.28	−3.91	−4.30
0	10	0	20.9	20.9	23.0	3	7	0	25.1	24.2	20.8
1	1	0	38.4	57.7	49.9	3	8	0	1.04	1.18	1.30
1	2	0	20.1	17.2	19.0	3	9	0	14.6	15.9	13.2
1	3	0	30.6	30.8	22.1	4	4	0	25.5	25.3	27.9
1	4	0	16.0	−13.7	−15.1	4	5	0	0.80	1.78	1.96
1	5	0	31.2	30.8	25.2	4	6	0	22.6	21.5	23.6
1	6	0	9.98	8.15	8.97	4	7	0	6.49	−6.01	−6.61
1	7	0	19.3	19.2	14.8	4	8	0	15.2	15.7	17.2
1	8	0	4.29	−2.36	−2.59	4	9	0	7.82	6.51	7.17
1	9	0	16.8	16.5	13.5	5	5	0	29.9	29.6	26.2
1	10	0	3.54	−2.26	−2.48	5	6	0	1.78	−1.29	−1.42
2	2	0	44.9	53.0	36.9	5	7	0	11.6	13.0	9.31
2	3	0	7.46	−5.50	−6.06	5	8	0	0	0.66	0.58
2	4	0	28.6	27.5	30.2	5	9	0	13.6	14.2	11.8
2	5	0	0.58	−1.37	−1.50	6	6	0	19.4	19.7	21.7
2	6	0	20.0	19.3	21.3	6	7	0	5.79	5.64	5.11
2	7	0	4.55	4.33	4.76	6	8	0	11.4	11.7	12.8
2	8	0	24.4	23.8	26.2	7	7	0	17.6	17.6	15.3
2	9	0	5.10	−4.50	−4.95						
2	10	0	8.48	9.96	11.0						

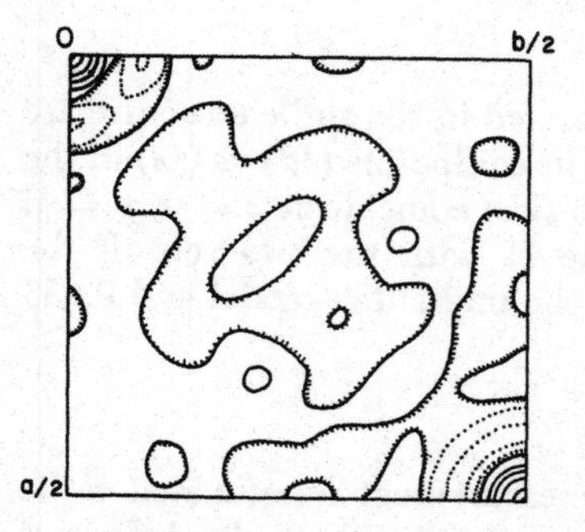

Figure 2. Difference Fourier map representing ρ (doped rutile)$-\rho$ (pure rutile). The contours are drawn in arbitrary scales.

An assumption made in this treatment is that there is no difference in the concentration of the oxygen vacancies in both kinds of crystal. It is concluded that x atoms of niobium, entering the structure as pentavalent ions, reduce x Ti^{4+} ions into trivalent state, Ti^{3+}, and that the compensated rutile may be represented by a formula $(Ti^{4+}_{1-2x}Ti^{3+}_{x}Nb^{5+}_{x})O_2 = (Ti_{1-x}Nb_x)O_2$.

The difference in the defect structure may be detected, in principle, in the following properties of the crystal: lattice constants, density, and number of free carriers. These properties were measured on the ceramic samples with various concentrations of niobium. These ceramics were obtained by calcination at 1200°C for 2 hr, followed by sintering at 1400°C for 3 hr in a nitrogen flow of 1 liter/min. The results of the measurements are shown in Figures 3 and 4.

The linear increase of the lattice constants shown in Figure 3 may be interpreted as the result of the increasing average ionic radius of cation. Although the substitutional model may explain this increase in a straightforward way, the changes in lattice constants due to the interstitial niobium are difficult to evaluate.

The three calculated lines in Figure 3, showing the changes in density as a function

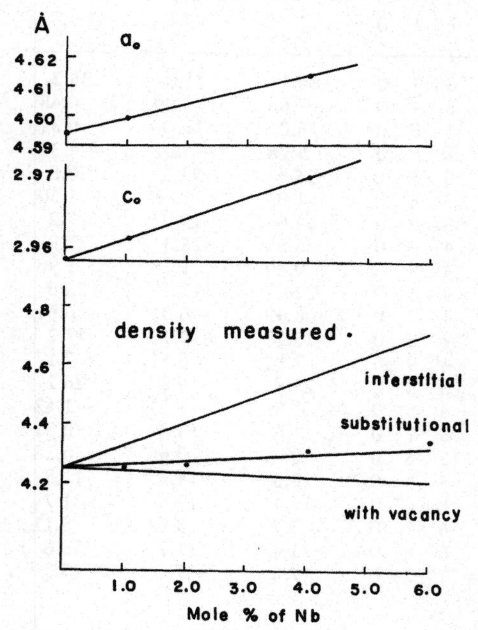

Figure 3. Lattice constants and densities of the ceramic rutile with different niobium concentrations.

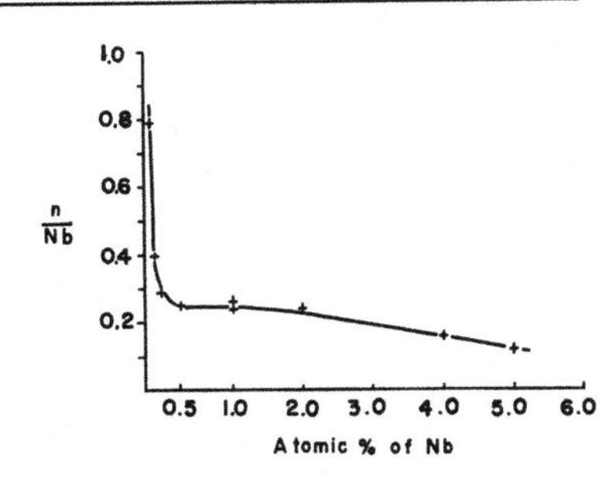

Figure 4. The concentration ratio of free carrier *n* to doped niobium atom at room temperature.

of the niobium concentration, behave quite differently. The observed values are in good agreement with the one representing the substantial model.

In Figure 4 is shown the concentration ratio of the free carrier *n* to the doped niobium ions at room temperature.[5] It is evident that a small fraction of the quasi-free electrons are activated as free carriers. The determination of the defect structure from this kind of data is impossible without the detailed knowledge concerning the conduction and scattering of the carriers.

The exact composition of the doped rutile sphere used in the structure analysis has not been determined, although the value of 4 mol.% was assumed for calculating the structure factors. This problem may be clarified either by the least-squares treatment of the intensities, with this concentration as an additional parameter to be refined, or by X-ray microanalysis of the sample.

ACKNOWLEDGMENT

The author is grateful to Mr. T. Kanno for his help in the measurement of diffraction intensities.

REFERENCES

1. T. Hurlen, "On the Defect Structure of Rutile," *Acta Chem. Scand.* **13**: 365, 1959.
2. M. E. Straumanis, T. Ejima, and W. J. James, "The TiO_2 Phase Explored by the Lattice Constants and Density Method," *Acta Cryst.* **14**: 493, 1961.
3. H. P. R. Frederikse, "Recent Studies of Rutile," *J. Appl. Phys. Suppl.* **32**: 2211, 1961.
4. P. F. Chester, "Electron Spin Resonance in Semiconductive Rutile," *J. Appl. Phys. Suppl.* **32**: 2233, 1961.
5. M. Itakura, N. Niizeki, H. Toyoda, and H. Iwasaki, "Hall Effect and Thermoelectric Power in Semiconductive TiO_2," *Japan. J. Appl. Phys.* **6**: 311, 1967.
6. W. L. Bond, "Making Small Spheres," *Rev. Sci. Instr.* **22**: 344, 1951.

AUTHOR INDEX

Bold numbers refer to papers in this volume

SUBJECT INDEX

Printed by Printforce, the Netherlands